Cell Signaling in Model Plants

Cell Signaling in Model Plants

Editors

Jen-Tsung Chen
Parviz Heidari

MDPI • Basel • Beijing • Wuhan • Barcelona • Belgrade • Manchester • Tokyo • Cluj • Tianjin

Editors
Jen-Tsung Chen
National University of Kaohsiung
Taiwan

Parviz Heidari
Shahrood University of Technology
Iran

Editorial Office
MDPI
St. Alban-Anlage 66
4052 Basel, Switzerland

This is a reprint of articles from the Special Issue published online in the open access journal *International Journal of Molecular Sciences* (ISSN 1422-0067) (available at: https://www.mdpi.com/journal/ijms/special_issues/CSMP).

For citation purposes, cite each article independently as indicated on the article page online and as indicated below:

LastName, A.A.; LastName, B.B.; LastName, C.C. Article Title. *Journal Name* **Year**, *Volume Number*, Page Range.

ISBN 978-3-0365-1954-8 (Hbk)
ISBN 978-3-0365-1955-5 (PDF)

Contents

International Journal of
Molecular Sciences

Editorial

Cell Signaling in Model Plants

Jen-Tsung Chen [1],* and Parviz Heidari [2]

[1]   Department of Life Sciences, National University of Kaohsiung, Kaohsiung 811, Taiwan
[2]   Department of Agronomy and Plant Breeding, Faculty of Agriculture, Shahrood University of Technology, Shahrood 3619995161, Iran; heidarip@shahroodut.ac.ir
*   Correspondence: jentsung@nuk.edu.tw

Received: 20 August 2020; Accepted: 20 August 2020; Published: 23 August 2020

Plants as sessile organisms are not able to move and must cope with adverse environmental conditions and stresses such as extreme temperatures, drought, high soil salinity, oxidative stress, pathogen attack, and so on. To respond in an appropriate manner to a specific environmental stimulus, plants possess signal transduction pathways, which are complex networks of interactions involving signal elements transmitting through the plant cell. In fact, cell signaling affects virtually every aspect of plant cell structure and function. For building elegant, complicated, and interconnected regulating networks, a huge number of components are involved, including receptors, secondary messengers, protein kinases, transcription factors, reactive oxygen species (ROS), and plant hormones that regulate or stimulate other components. Therefore, to unveil a global picture of plant cell signaling networks and underlying master regulators and machinery remains a challenge for researchers. For enriching our understanding of plant cell signaling with the assistance of modern molecular tools, this Special Issue "Cell Signaling in Model Plants" collects recent innovative original research and reviews, with an emphasis on the studies using model plants and crops. A total of 14 publications were published, which can be divided into five subtopics, as described below.

1. Arabidopsis Thaliana Transmembrane Receptor-Like Kinases (RLKs)

Transmembrane receptor-like kinases (RLKs) are conserved protein kinases that play critical roles in transducing the signal from the outside to the inside of plant cells, particularly to the nucleus. Jose et al. refined the recent advances of transmembrane RLKs in *A. thaliana* to unveil how stress responses as well as plant development are regulated by signaling pathways related to different groups of RLKs [1].

2. Functional Analysis of Signaling Components

2.1. Flowering Locus C (FLC) Homologs

To reveal the underlying mechanisms controlling the annual flowering of fruit trees, Kagaya et al. analyzed the function of *FLOWERING LOCUS C* (*FLC*) homologs in apple [2]. They declared that homologs of *FLC* might be involved in flowering and associated with juvenility.

2.2. Della/Gai Functions

Gibberellin (GA) signaling plays a vital role in regulating plant growth and development. Wang et al. investigated the function of GAI-1, one of the proteins with the DELLA amino acid motif (aspartic acid–glutamic acid–leucine–leucine–alanine), as a negative regulator involved in GA signaling, using the *gai-1* mutant line of *A. thaliana* [3]. They indicated that the *gai-1* mutant line is more tolerant of drought than the wild type. They also found a strong interaction between GAI proteins and ABA-responsive element (ABRE)-binding transcription factors. Jung et al. characterized *SLR1*, encoding the DELLA protein in rice, using CRISPR/Cas-9 genome editing [4]. They stated that the expression of $GA_{20}OX_2$ (gibberellin oxidase) and GA_3OX_2, as GA-related genes, was upregulated in the edited mutant lines. Finally, they indicated that in the created mutant lines, *slr1-d7* and *slr1-d8*, cell elongation is limited.

2.3. Dwarf Gene Mini Plant 1 (MNP1)

Dwarf phenotypes are widely used in crop breeding to increase resistance and yield. Guo et al. defined the function of *MNP1* in *Medicago truncatula* as a model legume plant [5]. They concluded that in the *mnp1* mutant line, the cell number of internodes and cell length are reduced. They also found that MNP1 is committed in GA biosynthesis.

2.4. MiPEP165a

MicroRNAs (miRNAs) play a significant role in the regulation of gene expression. Ormancey et al. investigated the role of a miRNA-encoded peptide, miPEP165a, in *A. thaliana* [6]. The authors found that passive diffusion followed by an endocytosis process are two functions of entry of miPEP165a.

3. Cell Signaling in Response to Biotic Stresses

Ethylene is a gaseous phytohormone that is involved in response to biotic and abiotic stresses. Zhang et al. studied the effect of ethylene signaling on resistance to some aphids in *Medicago truncatula* using an ethylene-insensitive mutant called sickle mutant [7]. Their results revealed that the sickle mutant can cause a moderate resistance to some aphids from the independent pathway of R-genes such as *AKR* (*Acyrthosiphon kondoi* resistance), *APR* (*Acyrthosiphon pisum* resistance), and *TTR* (*Therioaphis trifolii* resistance).

Botrytis cinerea is a necrotrophic fungus causing grey mold disease in many plant species. Maqsood et al. examined the effect of iprodione, as a fungicide, on the molecular mechanisms of *B. cinerea* resistant mutant [8]. By analyzing the whole transcriptome sequencing, they pronounced that genes involved in metabolism, production of detoxification enzymes, mitogen-activated protein kinases (MAPK) signaling, transporter function, catalytic activity, and drug efflux are linked with resistance to iprodione.

4. Cell Signaling Related to Plant Acclimation

4.1. Brassinosteroid Signaling

Brassinosteroids (BRs) as steroid hormones play critical roles in regulating the plant growth and development stages. Mao and Li reviewed the regulatory mechanisms of brassinosteroid-insensitive 1 (BRI1), BRI1-associated receptor kinase (BAK1), and brassinosteroid-insensitive 2 (BIN2), as three key kinases, involved in the BR signaling cascade [9]. They stated that BIN2, as an interface kinase, plays an important role in BR signal transduction from receptor kinase, BRI1/BAK1, to nucleus through two transcription factors, BRI1-EMS-supressor 1 (BES1), and brassinazole-resistant 1 protein (BZR1).

BRs have many interactions with other phytohormones that regulate the downstream pathways related to plant growth or response to stresses. Bulgakov and Avramenko reviewed the links between BR and abscisic acid (ABA) signaling that affect the stress-acclimation processes [10]. They proposed three interconnected mechanisms that in the first mechanism, BIN2, as a kinase of BR signaling, is responsible for interaction with ABA signaling.

4.2. Cyclic AMP Signaling

The cyclic nucleotide, cAMP (3′,5′-cyclic adenosine monophosphate) as a signaling molecule, is involved in molecular processes linked to response to environmental stresses. Blanco et al. reviewed the current knowledge of cAMP signaling [11]. They indicated that the main signaling mechanism of this cyclic nucleotide is exchange of cAMP into Ca^{2+} signals through cyclic nucleotides-gated channels.

4.3. Hydrogen Sulfide

Hydrogen Sulfide (H_2S) is a toxic gaseous molecule. Recent studies revealed that H_2S plays a positive role in regulating plant growth. Xuan et al. considered the H_2S roles in cellular processes and also indicated all possible crosstalk between H_2S and phytohormones [12]. Interestingly, the authors suggested that H_2S may affect the protein activities and subcellular localization by contributing to post-translational modification.

5. Effects of Selenium and Sedimentary Calcite-Processed Particles on Cell Signaling

Selenium (Se) is identified as a beneficial element that can be included in improving the plant's resistance facing adverse environmental conditions. Kamran et al. reviewed the protectant roles of Se in response to soil salinity [13]. They indicated that Se can improve salinity tolerance by decreasing Na^+ ion accumulation through the expression induction of the Na^+/H^+ antiport and increasing the antioxidants.

Tran et al. investigated the early cellular responses of bright yellow2 tobacco cultured cells under the application of calcite processed particles (CaPPs) [14]. Their results revealed that CaPPs act such as nanoparticles and induce various signaling pathways. CaPPs firstly induced ROS and then, increased the cytosolic Ca^{2+} and activation of anion channels.

6. Conclusions and Perspectives

This Special Issue provides new and in-depth insights into molecular aspects of plant cell signaling in response to biotic, such as aphid- and grey mold disease-resistance, and abiotic stresses, such as soil salinity and drought stress, and additionally, functional analysis on signaling components involved in flowering, juvenility, GA signaling, and biosynthesis, and miRNA-regulated gene expression. Furthermore, plant acclimation was reported, with emphasis on mechanistic insights into the roles of brassinosteroids, cyclic AMP, and hydrogen sulfide, and the recent advances of transmembrane receptor-like kinases were refined. Clearly, plant cell signaling is an intensive topic and whether it is now or in the future, the emerging technology in functional analysis such as genome editing technologies, high-throughput technologies, integrative multiple-omics as well as bioinformatics can assist researchers to reveal novel aspects of the regulatory mechanisms of plant growth and development, and acclimation to environmental and biotic stresses. The achievement of such research will be useful in improving crop stress tolerances to increase agricultural productivity and sustainability for the food supply of the world.

Funding: This research received no external funding.

Conflicts of Interest: The author declares no conflict of interest.

References

1. Jose, J.; Ghantasala, S.; Roy Choudhury, S. Arabidopsis Transmembrane Receptor-Like Kinases (RLKs): A Bridge between Extracellular Signal and Intracellular Regulatory Machinery. *Int. J. Mol. Sci.* **2020**, *21*, 4000. [CrossRef] [PubMed]
2. Kagaya, H.; Ito, N.; Shibuya, T.; Komori, S.; Kato, K.; Kanayama, Y. Characterization of FLOWERING LOCUS C Homologs in Apple as a Model for Fruit Trees. *Int. J. Mol. Sci.* **2020**, *21*, 4562. [CrossRef] [PubMed]
3. Wang, Z.; Liu, L.; Cheng, C.; Ren, Z.; Xu, S.; Li, X. GAI functions in the plant response to dehydration stress in A. thaliana thaliana. *Int. J. Mol. Sci.* **2020**, *21*, 819. [CrossRef] [PubMed]
4. Jung, Y.J.; Kim, J.H.; Lee, H.J.; Kim, D.H.; Yu, J.; Bae, S.; Cho, Y.-G.; Kang, K.K. Generation and Transcriptome Profiling of Slr1-d7 and Slr1-d8 Mutant Lines with a New Semi-Dominant Dwarf Allele of SLR1 Using the CRISPR/Cas9 System in Rice. *Int. J. Mol. Sci.* **2020**, *21*, 5492. [CrossRef] [PubMed]
5. Guo, S.; Zhang, X.; Bai, Q.; Zhao, W.; Fang, Y.; Zhou, S.; Zhao, B.; He, L.; Chen, J. Cloning and Functional Analysis of Dwarf Gene Mini Plant 1 (MNP1) in Medicago truncatula. *Int. J. Mol. Sci.* **2020**, *21*, 4968. [CrossRef] [PubMed]

6. Ormancey, M.; Le Ru, A.; Duboé, C.; Jin, H.; Thuleau, P.; Plaza, S.; Combier, J.-P. Internalization of miPEP165a into A. thaliana Roots Depends on both Passive Diffusion and Endocytosis-Associated Processes. *Int. J. Mol. Sci.* **2020**, *21*, 2266. [CrossRef] [PubMed]

7. Zhang, L.; Kamphuis, L.G.; Guo, Y.; Jacques, S.; Singh, K.B.; Gao, L.-L. Ethylene Is Not Essential for R-Gene Mediated Resistance but Negatively Regulates Moderate Resistance to Some Aphids in Medicago truncatula. *Int. J. Mol. Sci.* **2020**, *21*, 4657. [CrossRef] [PubMed]

8. Maqsood, A.; Wu, C.; Ahmar, S.; Wu, H. Cytological and Gene Profile Expression Analysis Reveals Modification in Metabolic Pathways and Catalytic Activities Induce Resistance in Botrytis cinerea Against Iprodione Isolated From Tomato. *Int. J. Mol. Sci.* **2020**, *21*, 4865. [CrossRef] [PubMed]

9. Mao, J.; Li, J. Regulation of Three Key Kinases of Brassinosteroid Signaling Pathway. *Int. J. Mol. Sci.* **2020**, *21*, 4340. [CrossRef] [PubMed]

10. Bulgakov, V.P.; Avramenko, T.V. Linking Brassinosteroid and ABA Signaling in the Context of Stress Acclimation. *Int. J. Mol. Sci.* **2020**, *21*, 5108. [CrossRef] [PubMed]

11. Blanco, E.; Fortunato, S.; Viggiano, L.; de Pinto, M.C. Cyclic AMP: A Polyhedral Signalling Molecule in Plants. *Int. J. Mol. Sci.* **2020**, *21*, 4862. [CrossRef] [PubMed]

12. Xuan, L.; Li, J.; Wang, X.; Wang, C. Crosstalk between Hydrogen Sulfide and Other Signal Molecules Regulates Plant Growth and Development. *Int. J. Mol. Sci.* **2020**, *21*, 4593. [CrossRef] [PubMed]

13. Kamran, M.; Parveen, A.; Ahmar, S.; Malik, Z.; Hussain, S.; Chattha, M.S.; Saleem, M.H.; Adil, M.; Heidari, P.; Chen, J.-T. An Overview of Hazardous Impacts of Soil Salinity in Crops, Tolerance Mechanisms, and Amelioration through Selenium Supplementation. *Int. J. Mol. Sci.* **2020**, *21*, 148. [CrossRef] [PubMed]

14. Tran, D.; Zhao, T.; Arbelet-Bonnin, D.; Kadono, T.; Meimoun, P.; Cangémi, S.; Noûs, C.; Kawano, T.; Errakhi, R.; Bouteau, F. Early Cellular Responses Induced by Sedimentary Calcite-Processed Particles in Bright Yellow 2 Tobacco Cultured Cells. *Int. J. Mol. Sci.* **2020**, *21*, 4279. [CrossRef] [PubMed]

 International Journal of
Molecular Sciences

Article

Generation and Transcriptome Profiling of Slr1-d7 and Slr1-d8 Mutant Lines with a New Semi-Dominant Dwarf Allele of *SLR1* Using the CRISPR/Cas9 System in Rice

Yu Jin Jung [1,2], Jong Hee Kim [1], Hyo Ju Lee [1], Dong Hyun Kim [1], Jihyeon Yu [3], Sangsu Bae [3], Yong-Gu Cho [4] and Kwon Kyoo Kang [1,2,*]

[1]   Division of Horticultural Biotechnology, Hankyong National University, Anseong 17579, Korea; yuyu1216@hknu.ac.kr (Y.J.J.); gllmon@naver.com (J.H.K.); ju950114@naver.com (H.J.L.); skullmask@naver.com (D.H.K.)
[2]   Institute of Genetic Engineering, Hankyong National University, Anseong 17579, Korea
[3]   Department of Chemistry, Hanyang University, Seoul 04763, Korea; muner8146@gmail.com (J.Y.); sangsubae@hanyang.ac.kr (S.B.)
[4]   Department of Crop Science, Chungbuk National University, Cheongju 28644, Korea; ygcho@chungbuk.ac.kr
*   Correspondence: kykang@hknu.ac.kr; Tel.: +82-31-670-5104

Received: 30 June 2020; Accepted: 29 July 2020; Published: 31 July 2020

Abstract: The rice *SLR1* gene encodes the DELLA protein (protein with DELLA amino acid motif), and a loss-of-function mutation is dwarfed by inhibiting plant growth. We generate slr1-d mutants with a semi-dominant dwarf phenotype to target mutations of the DELLA/TVHYNP domain using CRISPR/Cas9 genome editing in rice. Sixteen genetic edited lines out of 31 transgenic plants were generated. Deep sequencing results showed that the mutants had six different mutation types at the target site of the TVHYNP domain of the *SLR1* gene. The homo-edited plants selected individuals without DNA (T-DNA) transcribed by segregation in the T1 generation. The slr1-d7 and slr1-d8 plants caused a gibberellin (GA)-insensitive dwarf phenotype with shrunken leaves and shortened internodes. A genome-wide gene expression analysis by RNA-seq indicated that the expression levels of two GA-related genes, $GA_{20}OX_2$ (Gibberellin oxidase) and GA_3OX_2, were increased in the edited mutant plants, suggesting that $GA_{20}OX_2$ acts as a convert of GA_{12} signaling. These mutant plants are required by altering GA responses, at least partially by a defect in the phytohormone signaling system process and prevented cell elongation. The new mutants, namely, the slr1-d7 and slr1-d8 lines, are valuable semi-dominant dwarf alleles with potential application value for molecule breeding using the CRISPR/Cas9 system in rice.

Keywords: CRISPR/Cas9; GA; DELLA/TVHYNP; Dwarf; $GA_{20}OX_2$; GA signaling

---

1. Introduction

Rice, being one of the major food crops consumed by nearly half of the world's population, is grown annually at about 4.5 million hectares. Rice consumption per capita is particularly high in Asia, where it provides 60–70% calories per day (Food and Agriculture Organization (FAO), 2004). Therefore, among the breeding program priorities of rice breeders is to improve its tolerance to abiotic stress, such as tolerance to lodging. To date, the properties of many dwarf mutants found in plants have been associated with genes on the biosynthesis and signaling pathways of gibberellin (GA), indole-3-acetic acid (IAA), brassinolide (BR), and other hormones [1]. GA is one of the important plant hormones acting as a group of diterpenoid compounds that regulate during various growth and development processes in the higher plants, including stem elongation, germination, dormancy, flowering, flower development, leaf, and fruit aging [1–3]. The phenotypes of mutants deficient in GA biosynthesis or signaling usually exhibit dark green and rough leaves in rice [4]. So far, several genes related to defective mutants on the GA biosynthetic pathway, namely, *d18*, *d35*, *sd1*, and *eui*, have been isolated and characterized in rice [5–8]. Molecular genetic studies of GA-sensitive rice and Arabidopsis mutants have identified important factors for GA signaling, which seems to be well conserved among flowering plants [9,10]. The most important regulator of the GA signaling pathway is the DELLA protein, which is known as the repressor of GA action [11,12]. The DELLA proteins belong to the GRAS family as a transcription factor and are known to contain the N-terminal DELLA/TVHYNP amino acid motif and the C-terminal GRAS domain [13,14]. In addition, the genes encoding the GA receptor GA-INSENSITIVE DWARF1 (GID1), the F-box protein GA-INSENSITIVE DWARF2 (GID2) and the DELLA protein have been cloned, and an integrated GA signal transduction pathway has emerged [15–17]. Furthermore, it has been reported that DELLA family proteins interact with growth-related transcription factors such as PIF (phytochrome interaction factor) to control plant cell and organ size [18]. In general, GID1–GA–DELLA complexes in plant cells recognize GA by receptors. However, in the case of rice, the F-box protein of GID2 additionally interacts with the DELLA protein, which is polyubiquitinated by E3 ubiquitin-ligase (GID2) and then degraded through the 26S proteasome [19]. It is known that internode elongation is facilitated by GA signaling through GID1 and the DELLA protein in rice [20]. To date, accumulating evidence highlighted the N-terminus of DELLA as necessary for the inhibition of GA action. It has also recently been shown that DELLA N-terminus is required to interact with the GA receptor GID1 and consequent degradation [21]. In rice, the slr1-d1, -d2, -d3, -d4, -d5 and-d6 mutants in the GA signal transduction inhibitor DELLA protein N-terminal region consequently result in a dominant, semi-dwarf phenotype [22]. These mutants are known to have an amino acid modified by one bp substitution in the conserved DELLA/TVHYNP domain of the DELLA protein [22].

The CRISPR/Cas9 system, a recently developed genome modification tool, has been widely used for genome editing of several major crops due to its high accuracy and efficiency [23,24]. Furthermore, CRISPR/Cas9 has not only been used to knock out target genes in cells but also to introduce fragments of a certain size into the gene [25,26].

In this study, the CRISPR/Cas9 system was employed for targeting the TVHYNP domain of the *OsSLR1* gene, known as the DELLA protein. A total of six homozygous edited plants with new different allelic variants, namely, slr1-d7, slr1-d8, slr1-d9, slr1-d10, slr1-d11, and slr1-d12, showed dwarfism. In addition, mutants slr1-d7 and slr1-d8 were further investigated at transcriptome levels using RNA-sequencing.

2. Results

2.1. Editing of the TVHYNP Domain Encoding the OsSLR1 Gene and the CRISPR/Cas9 System

According to the structure of the *OsSLR1* gene, DELLA and TVHYNP domains are well conserved at the N-terminus (Figure 1A, Supplementary Figure S1 and S2). Sixteen mutants were identified by single guide RNA (sgRNA) region which targeted the *OsSLR1* gene in the positive transgenic T_0 plants (Supplementary Table S1, Figure S3 and S4). Deep sequence analyses detected 6 homozygous mutations, 2 heterozygous mutations, and 8 bi-allelic mutations (Supplementary Table S2). All the T_0 mutants were dwarf, producing many tillers. Six homozygous mutants were identified, among which four were characterized by few bp deletions and two by a few bp insertions. Specifically, the following were observed: a 3-bp deletion and mutant named slr1-d7, a 1-bp deletion that was designated as slr1-d8, a 5-bp deletion named slr1d9 and a 14-bp deletion called slr1-d10. The insertion mutants were a T-insertion named slr1d11 and a C-insertion named slr1-d12 (Figure 1B). Theoretically, the slr-d7 mutant encodes a protein without serine (Ser, S) and the slr1-d8~slr1-d12 are knockout mutants with a stop codon that cannot encode the protein (Figure 1C). A single-base deletion and insertion are predicted to cause a frameshift, resulting in the knockout of the *OsSLR1* gene. However, all the mutations did not affect the core sequence TVHYNP domain of the *OsSLR1* gene. The sgRNA was also investigated using Cas-OFFinder (http://www.rgenome.net/cas-offinder/) [27], and two potential off-target sites were chosen. Interestingly, no mutations were detected in these loci. These mutations are either untranslated or modified SLR1 proteins and have different mutant sites compared to the previously reported slr1-d1~slr1-d6 allele [28–30].

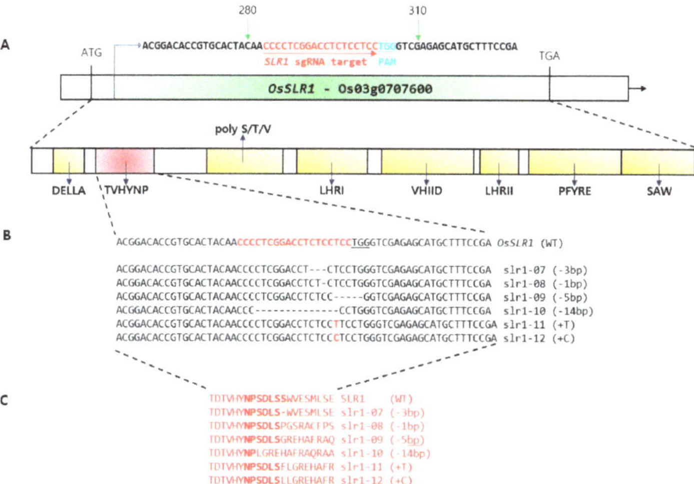

Figure 1. Genome editing in the rice *OsSLR1* gene. (**A**) Design of single guide RNA (sgRNA) sites in the TVHYNP motif; the nuclease cleavage site is represented by the red arrow and the Protospacer Adjacent Motif (PAM) (NGG) appears in blue. DELLA protein organization representing the conserved domains. (**B**) Nucleotide sequence alignment by deep sequence analysis of the sgRNA target region in six mutant lines of transformed rice plants. Deletion and insertion indicated by dash and red letters, respectively. (**C**) Amino acid sequences of the target region in six mutant lines.

2.2. New Allelic Slr-d7~Slr-d8 Mutant Plants Showed Dwarfing

The new dwarf mutants (slr1-d7~slr1-d12) showed several deficiencies in addition to reducing plant size (Supplementary Figure S5). Compared to the wild type (WT), these mutants had a slow growth rate, showed dwarfed and shriveled leaves (Figure 2A). The stomata are the key channels that regulate gas exchange and water evaporation in the leaves. As a result of observing stomata sizes by SEM (scanning electron microscopy) images, the slr1-d7 and slr1-d8 lines were smaller than that of WT (Figure 2B). To observe cytological differences in the stem internodes of these mutants, paraffin sections of the stem internodes were investigated from two mutants (slr1-d7 and slr1-d8) and WT. These dwarf mutants showed that the cell size was significantly reduced, and the internode thickened as the cell layer was increased (Figure 2C). In addition, the length of all internodes of the slr1-d7 and slr1-d8 lines were reduced compared to WT (Figure 2D). These results are similar to the characteristics of *dn*-type rice dwarf mutants previously reported by Takeda [31]. Thus, slr1-d7 and slr1-d8 lines were semi-dominant dwarf mutants, indicating that a decrease in cell length may be a direct cause of shortened culm length in dwarf mutant plants. Furthermore, to know the cause of dwarfism, the length of the leaf sheath was measured according to GA$_3$ concentration treatment in slr-d7, slr-d8, and WT. The results showed that the slr-d7 and slr-d8 variants produced a more extended leaf sheath following GA$_3$ treatment, but a reduced length extension compared to WT (Figure 2E).

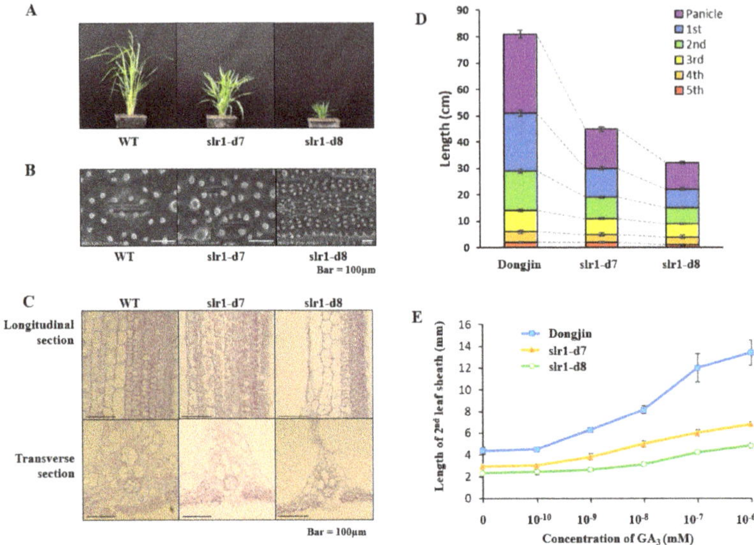

Figure 2. Phenotypic analysis of wild-type (WT) and slr1 mutant plants. (**A**) Phenotype of mature WT and mutant plant lines. (**B**) SEM (scanning electron microscopy) images of rice stomata in slr1-d7, slr1-d8 and WT. (**C**) Longitudinal tissue sections of the main stem at the mature stage in WT and slr1 mutant using paraffin section. Bar: 100 μm. (**D**) Length of internodes in slr1-d7, slr1-d8 and WT. (**E**) Elongation of the second leaf sheath of slr1-d7 and slr1-d8 in response to exogenous treatment with different concentrations of GA$_3$. Error bars are SD from the mean ($n = 3$).

2.3. Altered Transcriptome Profiling in Slr-d7 and Slr-8 Mutants

To understand the impact of dwarfism on gene expression at the whole-genome level, RNA-Seq was conducted to detect transcription profiling changes in WT, slr1-d7 and slr1-d8 lines. RNA-seq results showed that gene expression was altered significantly between WT and the slr1-d7 and slr1-d8 lines (Figure 3A). There are 214 genes upregulated and 154 genes downregulated in the slr1-d7 mutant compared with WT plants. By comparison, 334 genes were upregulated and 104 genes were downregulated in the slr1-d8 line (Figure 3B). Venn diagram analysis revealed 806 genes expressed in both WT and slr1-d7 or slr1-d8 mutants, which may explain the effects of knocking out *SLR1* on plants (Figure 3B,C). Gene ontology (GO) enrichment analysis of the 806 annotated up- and down-regulated genes identified 193 significantly (false discovery rate (FDR < 0.05)) enriched GO terms for the biological process, cellular component, and molecular function categories (Figure 4). Within the biological process category, the enriched differentially expressed genes (DEGs) were mainly associated with the response to the oligopeptide transport (GO:0006857), the intracellular protein transport (GO:0006886), karrikin (GO:0080167), and salt stress (GO:0009651). Within the cellular component category, the enriched DEGs were mainly associated with the plasma membrane (GO:0005886), the membrane (GO:0016020), cytosol (GO:0005829), and the integral component of the membrane (GO:0016021). Within the molecular function category, the DEGs were associated with protein serine/threonine kinase activity (GO:0004674), ATP binding (GO:0005524), and protein binding (GO:0005515) (Supplementary Table S3). To confirm the results from the RNA-seq analysis, 38 DEGs in the enriched GO terms were selected in the slr1-d7 and slr1-d8 lines, and their expression levels were confirmed by qRT-PCR analysis. The qRT-PCR results showed that the transcription levels of these genes were consistent with the RNA-seq results (Figure 5). These results indicated that dwarfism of the slr1-d7 and slr1-d8 lines mediates gene expression levels involved in regulating the plant hormone (GA, salicylic acid (SA), jasmonate (JA), IAA, cytokinin (CT) and ethylene (ET)) metabolism, signal transduction and transport (Supplementary Table S4).

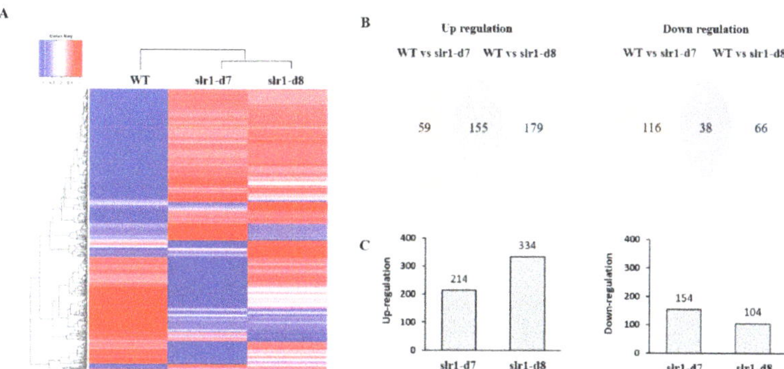

Figure 3. Global gene expression changes in knockout *OsSLR1* in rice. (**A**) Heat map of gene expression between WT vs. slr1-d7 and WT vs. slr1-d8 lines. Red denotes samples with relatively high expression of a given gene and blue denotes samples with relatively low expression. (**B**) Comparison of the number of differentially expressed genes (DEGs) in WT vs. slr1-d7 and WT vs. slr1-d8. (**C**) The number of DEGs up- and down-regulated between WT vs. mutant lines.

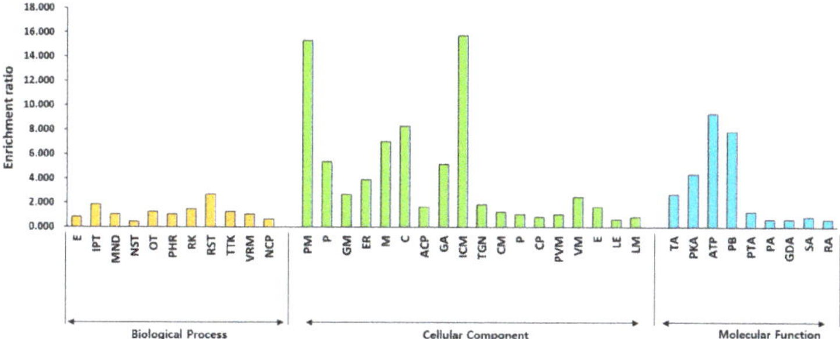

Figure 4. Gene ontology (GO) classification of DEGs shared by WT vs. slr1-d7 and WT vs. slr1-d8 lines. The *x*-axis shows user-selected GO terms, and the *y*-axis shows the enrichment ratio. Biological process: E, exocytosis; IPT, intracellular protein transport; MND, mitotic nuclear division; NST, nitric oxide mediated signal transduction; OT, oligopeptide transport; PHR, plant-type hypersensitive response; RK, response to karrikin; RST, response to salt stress; TTK, transmembrane receptor protein tyrosine kinase signaling pathway; VRM, vegetative to reproductive phase transition of meristem; NCP, nuclear-transcribed mRNA catabolic process. Cellular component: PM, plasma membrane; P, plasmodesma; GM, Golgi membrane; ER, endoplasmic reticulum; M, membrane; C, cytosol; ACP, anchored component of plasma membrane; GA, Golgi apparatus; ICM, integral component of membrane; TGN, trans-Golgi network; CM, chloroplast membrane; P, phragmoplast; CP, cytoplasmic mRNA processing body; PVM, plant-type vacuole membrane; VM, vacuolar membrane; E, endosome; LE, late endosome; EM, endosome membrane. Molecular function: TA, transporter activity; PKA, protein serine/threonine kinase activity; ATP, ATP binding; PB, protein binding; PTA, protein transporter activity; PA, potassium: proton antiporter activity; GDA, glucan endo-1,3-beta-D-glucosidase activity; SA, symporter activity; RA, ribonuclease activity.

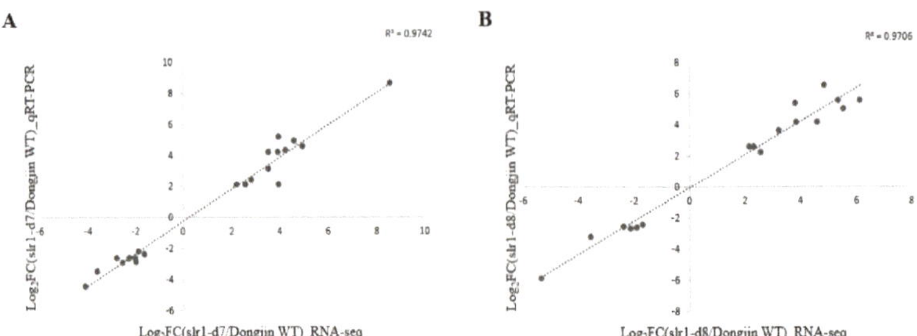

Figure 5. Correlation analysis of gene expression pattern by RNA-Seq and qRT-PCR. (**A**) WT vs. slr1-d7 line, (**B**) WT vs. slr1-d8 line.

2.4. Key DEGs Related to Biosynthesis and Signaling Pathway of Plant Hormone

In the RNA-seq analysis, the key DEGs related to plant hormone biosynthesis and signal transduction pathways between slr1-d7 vs. WT and slr1-d8 vs. WT were investigated. DEGs between slr1-d7 vs. WT were down-regulated and included the following: gibberellin-regulated protein 2, gibberellin 2-beta-dioxygenase 8 (*GA₂OX₈*), E3 ubiquitin-protein ligase (*XERICO*), gibberellin 2-beta-dioxygenase 1 (*GA₂OX₁*) in GA biosynthesis, *ERF03*, *ERF110*, *BBM1* in ethylene biosynthesis, and *ILR1* in auxin biosynthesis. Additionally, DEGs between slr1-d7 vs. WT were up-regulated and included the following: gibberellin 2-beta dioxygenase 8, *PIF1*, *PIF4*, *GA₂₀OX₂*, *GAMYB*, *GA₃OX₂* in GA biosynthesis, *ERF109*, *ERF39* in ethylene biosynthesis, *LOGL1* in cytokinin biosynthesis, *IAA7* in the auxin biosynthesis pathway, and *JAR1* in jasmonic acid (JA) biosynthesis (Supplementary Table S5). Among the DEGs associated with cell plate and leaf morphogenesis, Os05g0432200 and Os04g0407800 seemed to be important for the difference between WT and edited lines. The expression levels of two GA-related genes, especially *GA₂₀OX₂* (gibberellin oxidase) and *GA₃OX₂*, were increased in the edited mutant plants compare to WT (Figure 5). The edited mutant lines are required by altering GA responses, at least partially by a defect in the phytohormone signaling process and prevented cell elongation.

3. Discussion

CRISPR technology, as a powerful and highly efficient genome editing tool for breeding programs, has been utilized to enable the modification of gene(s) of interest. We report here the CRISPR/Cas9 mutation that potentially could confer a desirable dwarf trait using CRISPR technology. For a long time, plant breeders have used the DELLA gene mutant to reduce plant height [32]. The DELLA protein is one of the main components of the GA signaling pathway and acts as an inhibitor of the GA response. To date, information on dwarfism has shown results in a dominant, semi-dwarf phenotype, such as the observation that the GA signaling repressor DELLA protein or deletion in the N-terminal region suppresses GA signaling [33]. In rice, a total of six slr1-d mutants are known, and these mutants have dark green leaves, reduced internode elongation, and reduced response to GA treatment [28–30,32]. Most slr1-d alleles had 1 bp substitutions, resulting in amino acid substitutions in the conserved TVHYNP motif of *SLR1* (Supplementary Figure S6A). In this study, we generated and characterized six new alleles, namely, slr1-d7~slr1-d12, of the dwarf mutants in rice (Supplementary Figure S6B). These mutants showed the same phenotype for leaf color, GA response, and internode elongation with the previously reported slr1-d1~slr1-d6 mutants [28]. Among these mutants, the slr1-d7 gene had a deletion of three nucleotides, resulting in a serine deletion following core sequencing of the TVHYNP motif. Furthermore, the slr1-d8 mutant showed a 1 bp substitution (+T/+T). These two mutants displayed the most obvious and significant mutant phenotypes. The knockout of the slr1-d8 mutant showed a difference in plant height compared to the deletion of the serine residue of the slr1-d7 mutant (Figure 1, Supplementary Figure S7). This suggests the importance of the TVHYNP domain sequence, as the absence of these amino acids affects the normal metabolism of the SLR1 protein, but the degree of reduction is weaker than previously reported for the slr1-d6 mutant series [28]. We also performed paraffin sections to investigate cell length and cell layer using slr1-d7, slr1-d8 and WT. As a result, it was found that slr1-d7 and slr1-d8 not only showed significantly reduced cell length, but also node thickening, as cell layers increased as compared to WT (Figure 2). In addition, these dwarf mutants showed a decrease in the whole internode length containing a panicle when compared to WT. This result is similar to the characteristics of *dn*-type rice dwarf mutants reported by Takeda [31] and, as a semi-dominant, a decrease in cell length may be a direct cause of shortening clum length in dwarf mutant plants. In shoot elongation tests of dwarf mutant reaction by exogenous GA_3, the length of the secondary leaf sheath was elongated by GA treatment in WT, but not observed in slr1-d7 and slr1-d8 lines. However, there was a difference between the mutations. These results were similar to those reported in barley *Sln1D* and corn *dwarf 8*, suggesting that a single amino acid deletion or exchange mutation showed an intermediate phenotype depending on plant growth and GAI protein stabilization [13,17,34]. In RNA-seq analysis, the key DEGs related to plant hormone biosynthesis and the signal transduction pathways between slr1-d7 vs. WT and slr1-d8 vs. WT were investigated. DEGs between slr1-d7 vs. WT were down-regulated and included the following: gibberellin-regulated protein 2, gibberellin 2-beta-dioxygenase 8 (GA_2OX_8), E3 ubiquitin-protein ligase (*XERICO*), gibberellin 2-beta-dioxygenase 1 (GA_2OX_1) in GA biosynthesis [35], *ERF03*, *ERF110*, *BBM1* in ethylene biosynthesis [36], and *ILR1* in auxin biosynthesis [37]. In RT-PCR and RNA-seq analysis, the expression levels of two GA-related genes, $GA_{20}OX_2$ and GA_3OX_2, increased in the edited mutant line compared to WT, suggesting that these genes convert in the GA_{12} signaling system (Figure 5). The phenomenon of inhibiting cell elongation by altering the GA response due to defects in the signal transduction process of plant hormones was consistent with the results of the *Arabidopsis* mutants [35]. Furthermore, DEGs between slr1-d7 vs. WT were up-regulated and included the following: gibberellin 2-beta dioxygenase 8, *PIF1*, *PIF4*, $GA_{20}OX_2$, *GAMYB*, GA_3OX_2 in GA biosynthesis [38–43], *ERF109*, *ERF39* in ethylene biosynthesis [36], *LOGL1* in cytokinin biosynthesis [44], *IAA7* in the auxin biosynthesis pathway [45], and *JAR1* in jasmonic acid (JA) biosynthesis [46]. In summary, our results showed that slr1-d7 and slr1-d8 caused a defect in the phytohormone signaling system process and prevented cell elongation. Furthermore, we suggested that the new slr1-d7~slr1-d12 allelic variants are valuable semi-dominant dwarf alleles with potential application value for molecule breeding using the CRISPR/Cas9 system in rice.

4. Materials and Methods

4.1. Plant Materials

Rice variety Dongjin (*Oryza sativar* L., ssp. Japonica) was used for transformation experiments. Plants were grown in GMO greenhouse facilities and rice fields at Hankyong National University in Korea. Harvested seeds were dried to ~14% moisture content and kept in dry conditions at 4 °C.

4.2. CRISPR/Cas9 Vector Construction and Rice Transformation

SgRNAs were designed as described in Park et al. [47] to target the TVHYNP motif. The TVHYNPSD amino acid sequence of the *SLR1* gene is encoded by ACCGTGCACTACAACCCCTCGGAC, and the target sequence ACCCCTCGGACCTCTCCCTCCTGG with TGG as the PAM was selected. The 20nt sgRNA scaffold sequence was synthesized by Bioneer co., LTD (Dajeon, Korea). The slr-sgRNA templates were annealed using two primers, 5′-ggcagACCCCTCGGACCTCTCCTCC-3′ and 5′-aaacGGAGGAGAGGTCCGAGGGGTc-3′, and cloned into an *AarI*-digested *OsU3*:pBOsC binary vector. The Ti-plasmid vector for sgRNA expression, OsU3:*slr1*-sgRNA/pBOsC, and its flanking sequences were confirmed by the Sanger sequencing method and mobilized into *Agrobacterium-tumefaciens* strain EHA105. Transgenic plants were regenerated following a previously described protocol [48]. To confirm the transgene, the independent and transformed lines were analyzed by PCR. Plants derived from tissue culture were rooted and potted into 7 cm pots placed in the glasshouse and gradually acclimatized to the glasshouse conditions.

4.3. Targeted Deep Sequencing and Mutation Analysis

Total DNA extraction from plant tissues was performed using the DNA Quick Plant Kit (Inclone, Korea). Targeted deep sequencing analysis was performed following the method described by Jung et al. [46]. All primers used for targeted deep sequencing are listed in Supplementary Table S1. Paired-end read sequencing by PCR amplicons was produced with MiniSeq (Illumina, San Diego, CA, USA). All data derived from MiniSeq were analyzed by Cas-Analyzer (http://www.rgenome.net/cas-analyzer), as previously reported by Park et al. [49].

4.4. RNA-Seq and Data Analysis

To investigate the transcriptome of edited lines obtained by the *OsSLR1* gene via the CRISPR/Cas9 system, WT, slr1-d7(T/T), and slr1-d8 (−3/−3) plants were used for RNA-seq analysis. Four-week-old leaf tissues were used for RNA extraction, as previously reported by Wang et al. [50]. RNA concentration (A260/A280 and A260/A230) was measured with spectrophotometry (Nanodrop 2000, Thermo Scientific, Hudson, NH, USA). A Bioanalyzer (Agilent Technologies, Santa Clara, CA, USA) was used to evaluate the RNA qualities. Leaf samples of 100 mg were collected from three plants (WT, slr1-d7, and slr1-d8) for RNA-seq analysis. RNA-sequencing was carried out by Macrogen (Seoul, Korea, https://dna.macrogen.com/). Clean reads were produced by removing low-quality reads and mapped to the reference genome (https://plants.ensembl.org/) using TopHat2 (https://ccb.jhu.edu/software/tophat) [51]. Based on location information of the mapped reads, gene expression levels were normalized to reads per kilobase per million mapped reads (RPKM). DEG analyses between the edited plant RNA (slr1-d7, slr1-d8 and WT) were performed using the standard fold change (FC) ≥2 and FDR <0.05. GO analysis was performed as previously reported by Chow et al. [52].

4.5. Validation Test of Selected DEGs

To validate the accuracy of the RNA-sequencing data, qRT-PCR was conducted on twenty-one selected genes. The slr1-d7 and slr1-d8 lines were assessed according to WT, and relative gene expression levels were normalized by the *Actin* gene (XM_015761709). All assays for each gene were performed in triplicate with the same conditions and the RNA-seq data were deposited into the NCBI database.

4.6. GA$_3$ Treatment

The slr1-d7, slr1-d8, and WT seedlings grown in pots for 4 weeks were sprayed with 50 μM GA$_3$ (Sigma-Aldrich, Seoul, Korea). The stock solution of GA$_3$ was dissolved in ethanol and added to autoclaved water after cooling to approximately 60 °C to make the final 50 μM solution. The WT plants were treated with water containing equal amounts of ethanol.

4.7. Light Microscopy

For the paraffin section, stems and leaves were harvested from slr1-d7, slr1-d8, and WT plants. First, stem tissues were treated with 15% hydrofluoric acid, followed by dehydration with 70% ethanol, removal of it, infiltration, and embedding. For imaging, a 10 μm microtome section was placed on glass slides and floated in a 37 °C water bath containing deionized water. The sections were floated onto clean glass slides and microwaved at 65 °C for 15 min. Following this, the tissues were bound to the glass and each slide was used in chemical staining immediately.

Supplementary Materials: Supplementary Materials can be found at http://www.mdpi.com/1422-0067/21/15/5492/s1. **Figure S1.** Nucleotide and amino acid sequences of the *SLR1* gene in rice. **Figure S2.** Amino acid sequence alignment of the coding region of the *SLR1* gene from *Oryza sativa*, *Sorghum bicolor*, *Zea mays*, *Panicum miliaceum* and *Triticum aestivum*. **Figure S3.** Confirmation of the efficiency of sgRNA using T7-endonuclease I enzyme. **Figure S4.** CRISPR/Cas9 binary vector construction and rice transformation. **Figure S5.** Confirmation of the mutant phenotype and the conserved domain region of the DELLA protein. **Figure S6.** CRISPR/Cas9-induced mutations in the *OsSLR1* gene and the phenotype of the edited plants. **Figure S7.** Appearance of the panicle, leaves and grains after harvesting slr1-d7 and slr1-d8 lines compared to WT. **Table S1.** Design of sgRNAs for CRISPR genome editing on the *OsSRL1* gene in rice. **Table S2.** Mutation rate and edited plant types for the *OsSLR1* gene using the CRISPR/Cas9 system. **Table S3.** GO enrichment analysis of DEGs by RNA-seq analysis. **Table S4.** Key DEGs of up- and down-regulated genes related to phytohormone biosynthesis and signaling transduction pathway by RNA seqs in WT vs slr1-d7 and WT vs slr1-d8 lines. **Table S5.** List of primers and gene sequences in DEGs, correlation analysis of gene expression pattern by RNA-Seq and qRT-PCR.

Author Contributions: Formal analysis, Y.J.J., J.H.K., H.J.L., D.H.K. and J.Y.; investigation, Y.-G.C.; resources, S.B. and K.K.K.; software, H.J.L., J.Y. and S.B.; supervision, K.K.K.; writing-original draft, Y.J.J.; writing-review and editing, K.K.K. All authors have read and agreed to the published version of the manuscript.

Funding: This research was funded by "Cooperative Research Program for Agriculture Science & Technology Development (Project No. PJ01477203)" Rural Development Administration, Korea.

Conflicts of Interest: The authors declare no conflict of interest. The funders had no role in the design of the study; in the collection, analyses, or interpretation of data; in the writing of the manuscript, or in the decision to publish the results.

References

1. Wang, Y.; Li, J. The plant architecture of rice (*Oryza sativa*). *Plant Mol. Biol.* **2005**, *59*, 75–84. [CrossRef]
2. Yamaguchi, S. Gibberellin metabolism and its regulation. *Annu. Rev. Plant Biol.* **2008**, *59*, 225–251. [CrossRef]
3. Sun, T.P. The molecular mechanism and evolution of the GA–GID1–DELLA signaling module in plants. *Curr. Biol.* **2011**, *21*, R338–R345. [CrossRef] [PubMed]
4. Sakamoto, T.; Miura, K.; Itoh, H.; Tatsumi, T.; Ueguchi-Tanaka, M.; Ishiyama, K.; Miyao, A. An overview of gibberellin metabolism enzyme genes and their related mutants in rice. *Plant Physiol.* **2004**, *134*, 1642–1653. [CrossRef] [PubMed]
5. Itoh, H.; Ueguchi-Tanaka, M.; Sentoku, N.; Kitano, H.; Matsuoka, M.; Kobayashi, M. Cloning and functional analysis of two gibberellin 3β-hydroxylase genes that are differently expressed during the growth of rice. *Proc. Natl. Acad. Sci. USA* **2001**, *98*, 8909–8914. [CrossRef]
6. Itoh, H.; Tatsumi, T.; Sakamoto, T.; Otomo, K.; Toyomasu, T.; Kitano, H.; Matsuoka, M. A rice semi-dwarf gene, *Tan-Ginbozu (D35)*, encodes the gibberellin biosynthesis enzyme, *ent*-kaurene oxidase. *Plant Mol. Biol.* **2004**, *54*, 533–547. [CrossRef]
7. Monna, L.; Kitazawa, N.; Yoshino, R.; Suzuki, J.; Masuda, H.; Maehara, Y.; Minobe, Y. Positional cloning of rice semidwarfing gene, *sd-1*: Rice "green revolution gene" encodes a mutant enzyme involved in gibberellin synthesis. *DNA Res.* **2002**, *9*, 11–17. [CrossRef]
8. Zhu, Y.; Nomura, T.; Xu, Y.; Zhang, Y.; Peng, Y.; Mao, B.; Zhu, X. *ELONGATED UPPERMOST INTERNODE* encodes a cytochrome P450 monooxygenase that epoxidizes gibberellins in a novel deactivation reaction in rice. *Plant Cell* **2006**, *18*, 442–456. [CrossRef]
9. Hirano, K.; Nakajima, M.; Asano, K.; Nishiyama, T.; Sakakibara, H.; Kojima, M.; Banks, J.A. The GID1-mediated gibberellin perception mechanism is conserved in the *lycophyte Selaginella moellendorffii* but not in the bryophyte *Physcomitrella patens*. *Plant Cell* **2007**, *19*, 3058–3079. [CrossRef]
10. Yasumura, Y.; Crumpton-Taylor, M.; Fuentes, S.; Harberd, N.P. Step-by-step acquisition of the gibberellin-DELLA growth-regulatory mechanism during land-plant evolution. *Curr. Biol.* **2007**, *17*, 1225–1230. [CrossRef]
11. Peng, J.; Carol, P.; Richards, D.E.; King, K.E.; Cowling, R.J.; Murphy, G.P.; Harberd, N.P. The *Arabidopsis GAI* gene defines a signaling pathway that negatively regulates gibberellin responses. *Genes Dev.* **1997**, *11*, 3194–3205. [CrossRef] [PubMed]
12. Ikeda, A.; Ueguchi-Tanaka, M.; Sonoda, Y.; Kitano, H.; Koshioka, M.; Futsuhara, Y.; Yamaguchi, J. *slender* rice, a constitutive gibberellin response mutant, is caused by a null mutation of the *SLR1* gene, an ortholog of the height-regulating gene *GAI/RGA/RHT/D8*. *Plant Cell* **2001**, *13*, 999–1010. [CrossRef] [PubMed]
13. Chandler, P.M.; Marion-Poll, A.; Ellis, M.; Gubler, F. Mutants at the *Slender1* locus of barley cv Himalaya. Molecular and physiological characterization. *Plant Physiol.* **2002**, *129*, 181–190. [CrossRef]
14. Hirano, K.; Ueguchi-Tanaka, M.; Matsuoka, M. GID1-mediated gibberellin signaling in plants. *Trends Plant Sci.* **2008**, *13*, 192–199. [CrossRef]

15. Ueguchi-Tanaka, M.; Ashikari, M.; Nakajima, M.; Itoh, H.; Katoh, E.; Kobayashi, M.; Matsuoka, M. *GIBBERELLIN INSENSITIVE DWARF1* encodes a soluble receptor for gibberellin. *Nature* **2005**, *437*, 693–698. [CrossRef]

16. Sun, T.P. Gibberellin-GID1-DELLA: A pivotal regulatory module for plant growth and development. *Plant Physiol.* **2010**, *154*, 567–570. [CrossRef]

17. Willige, B.C.; Ghosh, S.; Nill, C.; Zourelidou, M.; Dohmann, E.M.N.; Maier, A.; Schwechheimer, C. The DELLA domain of GA INSENSITIVE mediates the interaction with the GA INSENSITIVE DEARF1A gibberellin receptor of *Arabidopsis*. *Plant Cell* **2007**, *19*, 1209–1220. [CrossRef]

18. Epstein, E.; Bloom, A. *Mineral Nutrition of Plants: Principles and Perspectives*, 2nd ed.; Sinauer Associates: Sunderland, MA, USA, 2005; p. 400.

19. Liang, Y.; Sun, W.; Zhu, Y.G.; Christie, P. Mechanisms of silicon-mediated alleviation of abiotic stresses in higher plants: A review. *Environ. Pollut.* **2007**, *147*, 422–428. [CrossRef]

20. Harberd, N.P.; Belfield, E.; Yasumura, Y. The angiosperm gibberellin-GID1-DELLA growth regulatory mechanism: How an "inhibitor of an inhibitor" enables flexible response to fluctuating environments. *Plant Cell* **2009**, *21*, 1328–1339. [CrossRef]

21. Guntzer, F.; Keller, C.; Meunier, J.D. Benefits of plant silicon for crops: A review. *Agron. Sustain. Dev.* **2012**, *32*, 201–213. [CrossRef]

22. Rodrigues, F.A.; Datnoff, L.E. Silicon and rice disease management. *Fitopatol. Bras.* **2005**, *30*, 457–469. [CrossRef]

23. Alagoz, Y.; Gurkok, T.; Zhang, B.; Unver, T. Manipulating the biosynthesis of bioactive compound alkaloids for next-generation metabolic engineering in opium poppy using CRISPR-Cas 9 genome editing technology. *Sci. Rep.* **2016**, *6*, 30910. [CrossRef]

24. Liu, X.; Wu, S.; Xu, J.; Sui, C.; Wei, J. Application of CRISPR/Cas9 in plant biology. *Acta Phys. Sin.* **2017**, *7*, 292–302. [CrossRef]

25. Hayut, S.F.; Bessudo, C.M.; Levy, A.A. Targeted recombination between homologous chromosomes for precise breeding in tomato. *Nat. Commun.* **2017**, *8*, 1–9.

26. Čermák, T.; Curtin, S.J.; Gil-Humanes, J.; Čegan, R.; Kono, T.J.; Konečná, E.; Voytas, D.F. A multipurpose toolkit to enable advanced genome engineering in plants. *Plant Cell* **2017**, *29*, 1196–1217. [CrossRef]

27. Bae, S.; Park, J.; Kim, J.S. Cas-OFFinder: A fast and versatile algorithm that searches for potential off-target sites of Cas9 RNA-guided endonucleases. *Bioinformatics* **2014**, *30*, 1473–1475. [CrossRef]

28. Asano, K.; Hirano, K.; Ueguchi-Tanaka, M.; Angeles-Shim, R.B.; Komura, T.; Satoh, H.; Ashikari, M. Isolation and characterization of dominant dwarf mutants, Slr1-d, in rice. *Mol. Genet. Genom.* **2009**, *281*, 223–231. [CrossRef]

29. Hirano, K.; Asano, K.; Tsuji, H.; Kawamura, M.; Mori, H.; Kitano, H.; Matsuoka, M. Characterization of the molecular mechanism underlying gibberellin perception complex formation in rice. *Plant Cell* **2010**, *22*, 2680–2696. [CrossRef]

30. Zhang, Y.H.; Bian, X.F.; Zhang, S.B.; Ling, J.; Wang, Y.J.; Wei, X.Y.; Fang, X.W. Identification of a novel gain-of-function mutant allele, slr1-d5, of rice DELLA protein. *J. Integr. Agric.* **2016**, *15*, 1441–1448. [CrossRef]

31. Takeda, K. Internode elongation and dwarfism in some graminaeous plants. *Gamma Field Symp.* **1977**, *16*, 1–18.

32. Wu, Z.; Tang, D.; Liu, K.; Miao, C.; Zhuo, X.; Li, Y.; Cheng, Z. Characterization of a new semi-dominant dwarf allele of *SLR1* and its potential application in hybrid rice breeding. *J. Exp. Bot.* **2018**, *69*, 4703–4713. [CrossRef] [PubMed]

33. Peng, J.; Richards, D.E.; Hartley, N.M.; Murphy, G.P.; Devos, K.M.; Flintham, J.E.; Sudhakar, D. 'Green revolution'genes encode mutant gibberellin response modulators. *Nature* **1999**, *400*, 256–261. [CrossRef] [PubMed]

34. Gubler, F.; Chandler, P.M.; White, R.G.; Llewellyn, D.J.; Jacobsen, J.V. Gibberellin signaling in barley aleurone cells. Control of *SLN1* and *GAMYB* expression. *Plant Physiol.* **2002**, *129*, 191–200. [CrossRef]

35. Ko, J.H.; Yang, S.H.; Han, K.H. Upregulation of an *Arabidopsis RING-H2* gene, *XERICO*, confers drought tolerance through increased abscisic acid biosynthesis. *Plant J.* **2006**, *47*, 343–355. [CrossRef]

36. Boutilier, K.; Offringa, R.; Sharma, V.K.; Kieft, H.; Ouellet, T.; Zhang, L.; Custers, J.B. Ectopic expression of BABY BOOM triggers a conversion from vegetative to embryonic growth. *Plant Cell* **2002**, *14*, 1737–1749. [CrossRef]

37. Widemann, E.; Miesch, L.; Lugan, R.; Holder, E.; Heinrich, C.; Aubert, Y.; Heitz, T. The amidohydrolases *IAR3* and *ILL6* contribute to jasmonoyl-isoleucine hormone turnover and generate 12-hydroxyjasmonic acid upon wounding in *Arabidopsis* leaves. *J. Biol. Chem.* **2013**, *288*, 31701–31714. [CrossRef]

38. Schomburg, F.M.; Bizzell, C.M.; Lee, D.J.; Zeevaart, J.A.; Amasino, R.M. Overexpression of a novel class of gibberellin 2-oxidases decreases gibberellin levels and creates dwarf plants. *Plant Cell* **2003**, *15*, 151–163. [CrossRef]

39. Huq, E.; Al-Sady, B.; Hudson, M.; Kim, C.; Apel, K.; Quail, P.H. Phytochrome-interacting factor 1 is a critical bHLH regulator of chlorophyll biosynthesis. *Science* **2004**, *305*, 1937–1941. [CrossRef]

40. Pedmale, U.V.; Huang, S.S.C.; Zander, M.; Cole, B.J.; Hetzel, J.; Ljung, K.; Ecker, J.R. Cryptochromes interact directly with PIFs to control plant growth in limiting blue light. *Cell* **2016**, *164*, 233–245. [CrossRef] [PubMed]

41. Rieu, I.; Ruiz-Rivero, O.; Fernandez-Garcia, N.; Griffiths, J.; Powers, S.J.; Gong, F.; Phillips, A.L. The gibberellin biosynthetic genes $AtGA_{20}ox_1$ and $AtGA_{20}ox_2$ act, partially redundantly, to promote growth and development throughout the *Arabidopsis* life cycle. *Plant J.* **2008**, *53*, 488–504. [CrossRef] [PubMed]

42. Kaneko, M.; Inukai, Y.; Ueguchi-Tanaka, M.; Itoh, H.; Izawa, T.; Kobayashi, Y.; Matsuoka, M. Loss-of-function mutations of the rice *GAMYB* gene impair α-amylase expression in aleurone and flower development. *Plant Cell* **2004**, *16*, 33–44. [CrossRef] [PubMed]

43. Yamaguchi, S.; Smith, M.W.; Brown, R.G.; Kamiya, Y.; Sun, T.P. Phytochrome regulation and differential expression of gibberellin 3β-hydroxylase genes in germinating *Arabidopsis* seeds. *Plant Cell* **1998**, *10*, 2115–2126. [CrossRef] [PubMed]

44. Kuroha, T.; Tokunaga, H.; Kojima, M.; Ueda, N.; Ishida, T.; Nagawa, S.; Sakakibara, H. Functional analyses of LONELY GUY cytokinin-activating enzymes reveal the importance of the direct activation pathway in *Arabidopsis*. *Plant Cell* **2009**, *21*, 3152–3169. [CrossRef]

45. Liscum, E.; Reed, J.W. Genetics of Aux/IAA and ARF action in plant growth and development. *Plant Mol. Biol.* **2002**, *49*, 387–400. [CrossRef]

46. Pena-Cortés, H.; Albrecht, T.; Prat, S.; Weiler, E.W.; Willmitzer, L. Aspirin prevents wound-induced gene expression in tomato leaves by blocking jasmonic acid biosynthesis. *Planta* **1993**, *191*, 123–128. [CrossRef]

47. Park, J.; Bae, S.; Kim, J.S. Cas-Designer: A web-based tool for choice of CRISPR-Cas9 target sites. *Bioinformatics* **2015**, *31*, 4014–4016. [CrossRef]

48. Jung, Y.J.; Lee, H.J.; Kim, J.H.; Kim, D.H.; Kim, H.K.; Cho, Y.G.; Kang, K.K. CRISPR/Cas9-targeted mutagenesis of *F3' H, DFR* and *LDOX*, genes related to anthocyanin biosynthesis in black rice (*Oryza sativa* L.). *Plant Biotechnol. Rep.* **2019**, *13*, 521–531. [CrossRef]

49. Park, J.; Lim, K.; Kim, J.S.; Bae, S. Cas-analyzer: An online tool for assessing genome editing results using NGS data. *Bioinformatics* **2017**, *33*, 286–288. [CrossRef]

50. Wang, L.; Feng, Z.; Wang, X.; Wang, X.; Zhang, X. DEGseq: An R package for identifying differentially expressed genes from RNA-seq data. *Bioinformatics* **2010**, *26*, 136–138. [CrossRef]

51. Kim, K.H.; Kang, Y.J.; Shim, S.; Seo, M.J.; Baek, S.B.; Lee, J.H.; Park, S.K.; Jun, T.H.; Moon, J.K.; Lee, S.H.; et al. Genome-wide RNA-seq analysis of differentially expressed transcription factor genes against bacterial leaf pustule in soybean. *Plant Breed. Biotech.* **2015**, *3*, 197–207. [CrossRef]

52. Chow, K.S.; Wan, K.L.; Isa, M.N.M.; Bahari, A.; Tan, S.H.; Harikrishna, K.; Yeang, H.Y. Insights into rubber biosynthesis from transcriptome analysis of *Hevea brasiliensis* latex. *J. Exp. Bot.* **2007**, *58*, 2429–2440. [CrossRef] [PubMed]

International Journal of
Molecular Sciences

Review

Linking Brassinosteroid and ABA Signaling in the Context of Stress Acclimation

Victor P. Bulgakov [1,*] and **Tatiana V. Avramenko** [1,2]

[1] Federal Scientific Center of the East Asia Terrestrial Biodiversity (Institute of Biology and Soil Science), Far Eastern Branch of the Russian Academy of Sciences, 159 Stoletija Str., Vladivostok 690022, Russia; avramenko.dvo@gmail.com

[2] Far Eastern Federal University, Sukhanova Str. 8, Vladivostok 690950, Russia

* Correspondence: bulgakov@ibss.dvo.ru; Tel.: +7-423-2375279

Received: 2 July 2020; Accepted: 17 July 2020; Published: 20 July 2020

Abstract: The important regulatory role of brassinosteroids (BRs) in the mechanisms of tolerance to multiple stresses is well known. Growing data indicate that the phenomenon of BR-mediated drought stress tolerance can be explained by the generation of stress memory (the process known as 'priming' or 'acclimation'). In this review, we summarize the data on BR and abscisic acid (ABA) signaling to show the interconnection between the pathways in the stress memory acquisition. Starting from brassinosteroid receptors brassinosteroid insensitive 1 (BRI1) and receptor-like protein kinase BRI1-like 3 (BRL3) and propagating through BR-signaling kinases 1 and 3 (BSK1/3) → BRI1 suppressor 1 (BSU1) —‖ brassinosteroid insensitive 2 (BIN2) pathway, BR and ABA signaling are linked through BIN2 kinase. Bioinformatics data suggest possible modules by which BRs can affect the memory to drought or cold stresses. These are the BIN2 → SNF1-related protein kinases (SnRK2s) → abscisic acid responsive elements-binding factor 2 (ABF2) module; BRI1-EMS-supressor 1 (BES1) or brassinazole-resistant 1 protein (BZR1)–TOPLESS (TPL)–histone deacetylase 19 (HDA19) repressor complexes, and the BZR1/BES1 → flowering locus C (FLC)/flowering time control protein FCA (FCA) pathway. Acclimation processes can be also regulated by BR signaling associated with stress reactions caused by an accumulation of misfolded proteins in the endoplasmic reticulum.

Keywords: ABA signaling; brassinosteroid signaling cascade; drought tolerance; priming; stress adaptation; stress memory

---

1. Introduction

In the last several years, there has been increased interest in the signaling system of brassinosteroids (BRs), and data has appeared on plant resistance to a lack of water upon activation of individual BR components [1,2]. The current model of BR signaling is that heterodimerization of protein brassinosteroid insensitive 1 (BRI1) and BRI1-associated receptor kinase (BAK1) initiates a signaling cascade that controls BR-responsive genes mainly through two homologous transcription factors, BRI1-EMS-supressor 1 (BES1) and brassinazole-resistant 1 protein (BZR1) [3]. The signal from the receptor is transmitted via brassinosteroid insensitive 2 (BIN2), a GSK3-like kinase taking the central place in BR signaling [4,5]. In the absence of BR, BIN2 is active and phosphorylates BZR1 and BES1, leading to loss of their DNA binding activity, exclusion from the nucleus by the 14-3-3 proteins, and degradation by the proteasome [3]. BR binding to the extracellular domain of BRI1 induces association and inter-activation between BRI1 and BAK1. Activated BRI1 then phosphorylates BSK1, which in turn dissociates from the receptor complex and interacts with BRI1 suppressor 1 (BSU1). BSU1 inactivates BIN2 by dephosphorylating its pTyr200, allowing the accumulation of unphosphorylated BZR1 and BES1. Dephosphorylated BZR1 and BES1 translocate to the nucleus and bind to their target genes to induce the BR response. Both BZR1 and BES1 bind to the BRRE (CGTGT/CG) and E-box (CANNTG) promoter elements through the conserved N-terminal DNA-binding domain and target a series of common genes to regulate BR-related responses [3,6–16].

However, this classic pathway leads to a decrease in growth, due to the relocation of resources in favor of protective reactions [2]. Fàbregas et al. [2] published an intriguing investigation to show that the *Arabidopsis thaliana* vascular brassinosteroid receptor BRL3 (receptor-like protein kinase BRI1-like 3) confers drought tolerance without decreasing growth. Authors observed that BRL3-overexpressing plants (BRL3ox) contained high levels of proline, sugars, and other osmoprotectants in non-stressed conditions and thus were better prepared for water deficiency due to a phenomenon known as priming [2]. The authors showed that drought resistance is under the control of cell-type specific BR signaling and that *BRL3* overexpression activates an alternative pathway of BR signaling. Analysis of BR signaling failed to provide a linear picture of the involvement of BRs in adaptation to drought stress [2]. As noted by the authors, overexpression of the canonical BRI1 pathway and its downregulation can both confer abiotic stress resistance. The phenotype of *Arabidopsis* BRL3ox plants demonstrates an active mechanism of drought tolerance driven by expression of the BRL3 receptor, but not the phenomenon known as drought avoidance (changes in stomatal conductance, leaf area, and leaf orientation).

BRL3 forms stable hetero-oligomers with BAK1, but not with BRI1, although BRL3 can complement BRI1 in different cell types and under different conditions [17]. The formation of distinct BR receptor complexes is interesting in itself, but it apparently does not explain the BRL3ox priming phenomenon. Analysis of integration with other signaling systems may be useful to unravel the mechanism of drought tolerance. In particular, *BRL3* overexpression caused an altered gene response of the ABA pathway.

ABA is a key phytohormone that regulates physiological and molecular responses to drought stress, including the accumulation of osmoprotectants [18]. Previous investigations of the BR signaling pathway showed a connection with ABA signaling (discussed below), with participation of other hormonal and light signaling systems [19,20]. Indeed, ABA signaling is closely related to abiotic stress resistance, and it can be assumed that ABA signaling interacts with the BRI1/BRL3 pathway. In brief, stress induces ABA accumulation and binding to its receptors of the PYL family to inhibit protein phosphatases 2C (PP2Cs). PP2C inactivation activates class 3 sucrose nonfermenting-1-related protein kinases (SnRK2s) that phosphorylate ABA-responsive element binding factors (ABFs). Activated ABFs initiate expression of responsive genes by binding to the *cis*-acting ABA response element (ABRE) [21].

Zhang et al. [22] noted: "Whether BR and ABA interaction is through modification or intersection of their signaling components or by independent or parallel pathways … remains a big mystery". They found that ABA regulation of BR signaling depends on ABA signaling proteins, ABI1 and ABI2 [22]. The authors hypothesized that an activated BRI1 complex inhibits BIN2 kinase through an unknown mechanism and that ABA signaling is involved in BR signaling by regulating the GSK3-like kinase BIN2 or related proteins. Recently, Ren and colleagues showed how this happens (see section "Linking BRI1/BRL3 to the ABA signaling pathway") [10]. In this review, we summarize data about links between BRI1 and BRL3 and the ABA signaling pathway at the level of protein–protein interactions. We propose new research trends in the study of the BR signaling pathway in relation to stress adaptation.

2. Linking BRI1/BRL3 to the ABA Signaling Pathway

There is still little data on the difference between BRI1 and BRL3 at the level of protein–protein interactions in the signaling cascade that regulates downstream reactions (Figure 1). Both BRI1 and BRL3 open the brassinosteroid signaling cascade by binding brassinolide [23] and might be linked to the ABA signaling system via the following pathway: BRI1/BRL3 → BR-signaling kinases 1 and 3 (BSK1/3) → BIN2 (Figure 1). However, BIN2 phosphorylates BSK1/3 [10,24,25], but not vice versa, and therefore we consider that the BRI1/BRL3 → BSK1/3 signaling module could not be related to the ABA signaling system. Instead, this module enters the branching signaling pathway related to plant immunity and development via somatic embryogenesis receptor kinases (SERKs) and the LRR receptor-like serine/threonine-protein kinase FLS2 (Figure 1; see also interactions in [26]).

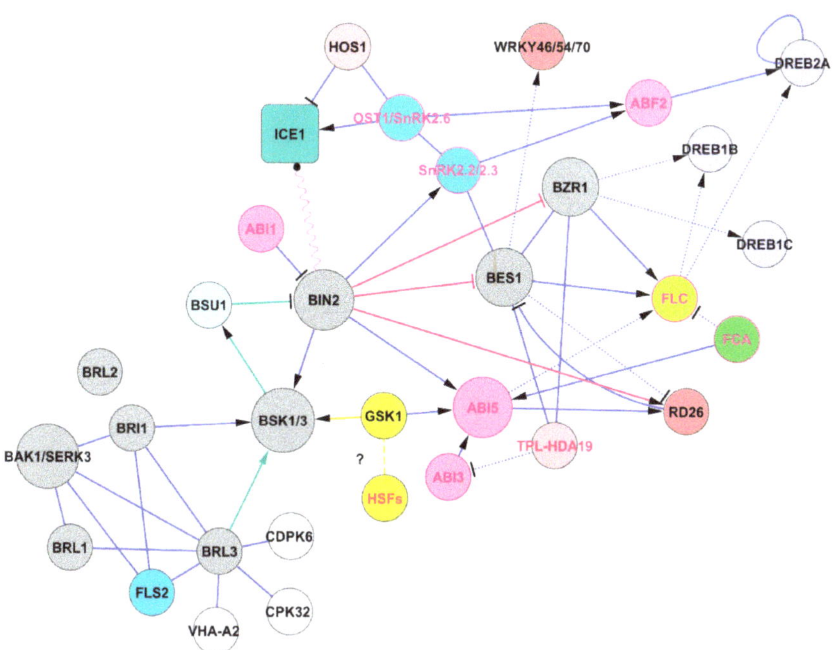

Figure 1. The pathway of BRL3 signaling. Brassinosteroid receptors BRL1, BRL3, and BRI1/BAK1 trigger the BR signaling pathway (proteins of the BR signaling system are shown in gray). BRL2 is not connected to this system. Solid lines represent protein–protein interactions presented in PAIR, IntAct, and BioGRID, and dashed lines represent possible interactions taken from STRING. Dotted lines represent transcriptional regulation. Green lines indicate signaling in the module BRL3 → BSK1/3 → BSU1 —‖ BIN2. BIN2 regulates expression of BZR1 and BES1. BIN2 regulates drought tolerance directly by activating RD26, indirectly via BZR1-DREB1B and SnRK2.2/2.3-ABF2-DREB2A pathways. BIN2 also interacts with ICE1, implementing time-dependent regulation of the SnRK2.6/OST1-HOS1-ICE1 cold signaling module. Finally, BIN2 activates ABI5, an important concentrator of ABA signals. Red protein labels indicate that these proteins are involved in stress memory generation. These interactions were visualized using the program Cytoscape as described previously [27]. The data loaded into the program were obtained from PAIR version 3.3 [http://www.cls.zju.edu.cn/pair/]. The protein–protein interactions presented in PAIR were supplemented with data from BioGRID [http://thebiogrid.org/], UniProtKB [https://www.uniprot.org/], TAIR [https://www.arabidopsis.org/], IntAct [https://www.ebi.ac.uk/intact/interactors/], and STRING [https://string-db.org/] databases. Abbreviations: ABI1/3/5, ABA insensitive 1/3/5; ABF2, abscisic acid responsive elements-binding factor 2; BAK1, BRI1-associated receptor serine/threonine kinase; BES1, brassinazole-resistant 2; BIN2, brassinosteroid insensitive 2; BRI1, brassinosteroid insensitive 1; BRL1/2/3, serine/threonine-protein kinase BRI1-like 1/2/3; BSK1/3, BR-signaling kinases 1 and 3; BSU1, BRI1 suppressor 1; BZR1, brassinazole-resistant 1; CDPK6 and CPK32, calcium-dependent protein kinases; DREB1B,1C,2A, dehydration-responsive element-binding proteins; FCA, flowering time control protein; FLC, flowering locus C; FLS2, LRR receptor-like serine/threonine-protein kinase; GSK1, shaggy-related protein kinase iota; HDA19, histone deacetylase 19; HOS1, E3 ubiquitin-protein ligase HOS1; HSFs, heat shock factors; ICE1, inducer of *CBP* expression 1; RD26, NAC transcription factor; TPL, TOPLESS; SnRK2.2/2.3, SNF1-related protein kinases 2.2 and 2.3; VHA-A2, vacuolar proton ATPase.

Next, we focused on the study of BRL3-interacting partners and considered the possibility that proteins interacting with BRL3 perform a protective function. We manually checked all 45 BRL3-interacting proteins using BioGrid and TAIR annotations and found that almost all of them are signaling components related to plant immunity and development, with the exception of several proteins (https://thebiogrid.org/5872/summary/arabidopsis-thaliana/brl3.html). These include two calcium-dependent protein kinases (CDPK6 and CPK32), the vacuolar proton ATPase VHA-A2, and plasma membrane H^+-ATPase 2 (AHA2). CDPK6 and CPK32 are involved in the ABA signaling pathway (BioGrid annotation), but their functionality in regards to BR signaling is unknown. Both VHA-A2 and AHA2 are important ATPases in establishing plant ion homeostasis under saline-alkali environmental conditions and act through the Salt-Overly-Sensitive signaling pathway and CBL-dependent calcium signaling [28–30]. Forty-sixth BRL3-interacting protein (not included in the BioGRID annotation) is a regulator of G-protein signaling 1 (RGS1) [31]. BRL3 phosphorylates RGS1 and thus functions in glucose sensing [31].

It is interesting to note that AHA2 also physically interacts with the serine/threonine-protein kinase BRI1-like 2 (BRL2). Paradoxically, the hub-type protein BRL2 (93 known interaction, BioGrid) is almost totally unrelated to BRL3, except for one common interaction, namely BRL3-AHA2. BRL2 interacts with numerous responsive proteins, including peroxidases, catalase CAT2, dehydrin ERD10, caffeic acid/5-hydroxyferulic acid O-methyltransferase (OMT1) and others, while BRL3 does not. These data are in accordance with the observation that BRL2, in contrast to BRI1 and BRL3, does not encode a functional BR receptor [23]. Summarizing the above information, we presume that the interaction of BRL3 with nearby proteins poorly explains BRL3-mediated drought tolerance. Thus, the unique effect of BRL3 on drought tolerance should be sought in long-distance signaling pathways.

A recent report by Ren et al. [10] indicates that BSK3 upregulates the serine/threonine-protein phosphatase BSU1 transcript and protein levels. Because BSU1 dephosphorylates and inactivates BIN2 [3,11], a signaling shunt, BSK1/3 → BSU1 —‖ BIN2, may be established. The signaling module joining two subsystems is as follow: BRI1/BRL3 → BSK1/3 → BSU1 —‖ BIN2 → BSK1/3 (Figure 1).

BSK3 physically interacts with BIN2 at the plasma membrane. In this interaction, BSK3 is a substrate of BIN2 kinase [10]. BSK3 phosphorylation by BIN2 allows the formation of BSK3/BSK1 heterodimer, BSK3/BSK3 homodimer, BSK3/BRI1 interaction, and BSK3/BSU1 interaction. If BIN2 is inhibited in this cascade, there will be consequences, since BIN2 blocks the activity of BZR1 and BES1 [3] and activates important components of the ABA signaling pathway (see below, section BIN2-based module). The BRL3 → BSK1/3 → BSU1 —‖ BIN2 module can work independently of BRI1/BAK1 because BSK3 can activate BR signaling without a functional BRI1 receptor [10].

Previously, BSK3 had been described as a partially redundant regulator of brassinosteroid signaling [25] and now it is considered a scaffold protein to regulate overall BR signaling [10]. It is of interest as a participant in a "systemic foraging strategy" that increases the soil volume explored by the root system for the adaptation of plants to low nitrogen concentrations [32]. Therefore, BSKs could be central factors mediating the effects of the BRL3 receptor. BSKs join BRL3 to ABA signaling by modulating BIN2 activity because BIN2 interacts with central components of the ABA signaling pathway, such as the bZIP transcription factor ABI5 [33], protein phosphatase 2C ABI1 [34], with transcription factor ICE1 [35], and phosphorylates SNF1-related protein kinases SnRK2.2 and SnRK2.3 [36]. The interaction of BR signaling components with ABA signaling components can result in the generation of stress memory, i.e., the phenomenon described by Fàbregas et al. [2] as "acclimation", in which ABA signaling components such as protein phosphatases 2C, ABI5, and SnRK2 kinases are involved in stress memory generation [27].

3. BIN2-Based Module

In the BR signaling pathway, BIN2 phosphorylates BES1 and BZR1 transcription factors to inhibit BR signaling through degradation of BES1 and BZR1 and by inhibiting their binding to DNA [12,13]. According to the conventional model of BR signaling, BRs act via BES1, which cooperates with WRKY46, WRKY54, and WRKY70, as well as other transcription factors, to activate plant growth-related genes and repress drought-responsive genes [14–16]. Under normal growth conditions, WRKY46/54/70 and BES1 positively regulate growth-related genes and negatively regulate the expression of drought-responsive genes. Under drought stress, WRKY46/54/70 and BES1 are destabilized which causes the repression of growth-related genes and activation of drought-related genes, which results in enhanced drought tolerance [15]. BES1 and the stress-responsive NAC transcription factor RD26 bind to a common promoter element, thus mutually inhibiting each other's transcriptional activity ([14], see also Figure 1). The antagonistic interaction between BES1 and RD26 means that plant growth is reduced when plants are under water deficit, which induces RD26 to inhibit BR-induced growth, thus allowing the reallocation of resources to resist drought stress [14].

BIN2 positively regulates drought tolerance by upregulation of RD26 [34]. BIN2 directly interacts and phosphorylates RD26. In this way, we can see the involvement of ABA signaling components because protein phosphatase 2C ABI1 from the ABA pathway inhibits BIN2 kinase activity by dephosphorylation. The water deficit eliminates the ABI1-induced inhibition of BIN2 and further triggers drought tolerance by RD26. It should be noted that the expression of RD26 is also activated in BRL3-ox roots under a water deficit [2].

BIN2 negatively regulates the freezing tolerance, whereas BZR1 positively modulates the freezing tolerance [14,37]. BIN2 phosphorylates SnRK2.2 and SnRK2.3 (but not SnRK2.6/OST1), acting as a positive regulator of the ABA signaling pathway [36]. BZR1 acts via the CBF-dependent cold signaling pathway, directly activating *CBF1/DREB1B* and *CBF2/DREB1C* expression and by regulation of other cold-responsive (COR) genes [37]. Moreover, the freezing tolerance is regulated by the well-known SnRK2.6/OST1-HOS1-ICE1 signaling module that controls freezing tolerance via the CBF-dependent cold signaling pathway ([38], see also Figure 1). BIN2 interacts with SnRK2.6/OST1 but cannot phosphorylate it, suggesting that BIN2 acts through a non-conventional transphosphorylation site of SnRK2.6 [36]. BIN2 also interacts with ICE1, providing the attenuation of *CBF* expression (by time-dependent downregulation of ICE1 abundance) during the later stages of the cold stress response [14]. The silencing of BIN2 increases the resistance of plants to cold, while BIN2 overexpression results in hypersensitivity to freezing stress [37]. This effect was observed not only for acclimated but also for non-acclimated conditions [37].

The above information indicates that the signaling pathways passing through BIN2 lead to the regulation of both drought and cold protective reactions. This is in agreement with data from Fabregas et al. [2] regarding the enhanced expression of genes in BRL3ox compared to WT plants, in Gene Ontology (GO) categories, such as Response To Water Deprivation, Response to Temperature Stimulus, and Response To Cold or Cold Acclimation. As shown in Figure 1, BIN2 attenuation occurs in two ways, by the BR signaling component (BSU1) and the ABA signaling component (ABI1), which leads to both the weakening and strengthening of protective reactions to balance growth under stress conditions. The regulatory logic of this balance is not yet fully understood. To date, there is no data allowing discriminate functions of BRL3 and BRI1 in relation to the signaling chain BRL3 (or BRI1) → BSK1/3 → BSU1 —‖ BIN2. Both receptors, BRI1 and BRL3 act through BIN2. We searched for other links between the BRI1 or BRL3 and BES1 or BZR in different databases and reports, and found no results, except for one mention in STRING, namely in the category "Co-Mentioned in PubMed Abstracts" (https://string-db.org/network/3702.AT4G39400.1). In earlier work, Kim et al. [3] suggested the existence of missing components in brassinosteroid signaling. It could be assumed that these missing components are numerous kinases with unknown functions that interact with BSK1/3. There are putative LRR receptor-like serine/threonine-protein kinases, AT1G51800 and AT5G10290, with an unknown function, as well as brassinosteroid-signaling kinases 5 and 8 and others, identified by Sreeramulu et al. [25] as BSK1/3-interacting proteins.

4. The Priming Phenomenon

Fàbregas et al. [2] hypothesized that the priming phenomenon might be a reason for drought tolerance and normal growth of BRL3-overexpressing plants. They based this assumption on the fact that the roots of BRL3ox plants are pre-loaded with osmoprotectant metabolites under normal conditions, and therefore they are better prepared for stress. Indeed, from a theoretical point of view, a plant can achieve this state (physiological equilibrium between growth and protection) by optimizing biochemical processes using memory generation processes. In higher plants, the stress memory phenomenon, known as 'priming' or 'acclimation', is achieved by chromatin modifications [39–41]. Describing the role of chromatin in water stress responses of plants, Han and Wagner [42] mentioned the role of histone modifications, histone (de)acetylases, histone lysine methyltransferases, histone arginine methyltransferases, histone variants, DNA methylation, and ATP-dependent chromatin remodeling complexes in memory generation. Most of these processes involve ABA signaling components [42].

Three types of stress-memory genes were described by Forestan et al. [43]: "transcriptional memory" genes, which have stable transcriptional changes persisting after recovery; "epigenetic memory candidate" genes, where stress-induced chromatin changes persist longer than the stimulus; and "delayed memory" genes, which are not immediately affected by the stress, but their expression patterns are perceived, stored, and later retrieved via chromatin remodeling for a delayed response.

The growing body of information indicates the involvement of BR signaling components in memory generation to stress. Shigeta et al. [44] suggested chromatin remodeling as a mechanism for the functioning of the BR pathway based on proteomic experiments. The authors proposed two mechanisms, specifically through the involvement of ATP-dependent chromatin remodeling complexes (CRC) or chromatin-modifying enzymes, such as histone deacetylases. Further, it was confirmed that histone modifying enzymes mediate the transcriptional activation of genes by components of the BR pathway [45,46]. Recently, Li et al. [47] showed that components of the BR pathway antagonize Polycomb silencing, thus introducing an epigenetic aspect in BR signaling. Possible mechanisms for generating memory in the BR signaling pathway are presented in Figure 2. The proteins involved in stress memory are marked in red.

Currently, there is no data to distinguish the specificity of the action of different BR receptors (BRI1, BRL1, and BRL3) at the level of protein–protein interactions. Therefore, the circuit shown in Figure 2 is applicable for the common BR pathway. The activated BR pathway leads to a state where BSU1 phosphatase inactivates BIN2, thus allowing activation of BZR and BES1 [10]. The regulator in ABA pathway, ABI1 phosphatase, can also dephosphorylate and destabilize BIN2 to inhibit BIN2 kinase activity [34]. Therefore, BIN2 functions as an important node in ABA-modulated BR signaling [22,34]. Activated BZR and BES1 in this pathway can in turn interact with stress-memory generating factors, such as TPL-HDA19, FLC/FCA and histone H3K27 demethylase (Figure 2). As components of the ABA pathway affect the BR pathway, the BR components also affect the ABA pathway. Specifically, ABI5 is regulated by BIN2 and GSK1, BIN2 regulates the function of SnRK2 kinases, and BZR1-TPL-HDA19 complex regulates the expression of *ABI3*. The DNA templates that carry response elements for binding factors are very different (ABRE, DRE, BRRE, and others), and they are not indicated in Figure 2. Note that the *cis*- and *trans*-regulatory logic of transcription factors involved is not considered, since it is not fully understood.

Since BRL3ox plants are more resistant to drought, they demonstrate a more stable rate of photosynthesis and transpiration during drought conditions, and have a larger preconditioned osmoprotective pool than WT plants [2]. It can be proposed that BRI1/BRL3 acts through memory factors that alter chromatin structure. It is not yet clear how the memory signal is passed, however it may be through the BRI1/BRL3 → BSK1/3 → BSU1 —|| BIN2 signaling pathway or others yet unknown. The BZR1/BES1 → TPL-HDA19 and BZR1/BES1 → FLC/FCA modules may be involved in such interactions. Below we consider possible options for these interactions.

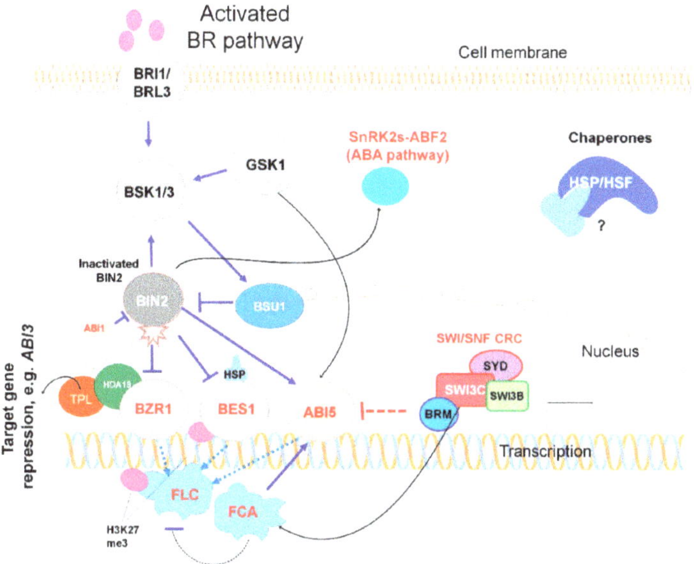

Figure 2. A model of stress memory generation by BR signaling. Solid lines represent protein–protein interactions and dotted lines represent transcriptional regulation. Proteins, involved in stress memory generation, are FLC and FCA (which substantially reduce plant water use and are important for heat and cold adaptation), TPL/HDA19 complex (ensures the epigenetic link between BR and ABA signaling through BZR1/BES1-ABI3-ABI5 interactions), and key components of the ABA signaling system such as SnRK2.2/2.3 and OST1/SnRK2.6, ABA-responsive element binding factors ABI3, ABI5 and ABF2 (involved in abiotic stress defense and stress memory). BZR1 recognizes and binds to a BRRE *cis* element in *FLC* and recruits H3K27 demethylase to dynamically modulate plant response to BR signals and environmental cues. SWI/SNF CRC is also a possible memory generator in this scheme. HSF function in BR signaling is possible, but has not been studied. Abbreviations: ABI5, ABA insensitive 5; BES1, brassinazole-resistant 2; BIN2, brassinosteroid insensitive 2; BRI1, brassinosteroid insensitive 1; BRL3, serine/threonine-protein kinase BRI1-like 3; BSK1/3, BR-signaling kinases 1 and 3; BRM, ATP-dependent helicase BRAHMA; BSU, BRI1 suppressor 1; BZR1, brassinazole-resistant 1; SWI/SNF CRC, (Switch/Sucrose non-fermenting, ATP-dependent chromatin remodeling complex); FCA, flowering time control protein; FLC, flowering locus C; GSK1, shaggy-related protein kinase iota; HDA19, histone deacetylase 19; HSP, heat shock protein; HSF, heat shock factor; SWI3B/3C, chromatin remodeling complex subunits; SYD, SWI2/SNF2-type ATPase; TPL, TOPLESS; SnRK2s, SNF1-related protein kinases 2.

BR-mediated repression of gene expression requires that histone deacetylases interact with TOPLESS (TPL) and that BZR1 associates with TPL and histone deacetylase HDA19 in vivo [45]. BZR1 recruits the TPL-HDA19 complex to BR-repressed promoters and mediates transcriptional repression via chromatin modification. The important role of BES1 is to create the BR-activated BES1-TPL-HDA19 repressor complex that controls epigenetic silencing of *ABI3* and *ABI5* [46]. This complex allows the suppression of ABA signaling during seedling development. Formation of a protein complex between BES1 or BZR1 and HDA19 is essential for regulation of drought stress tolerance [48].

The interaction of BR signaling components with FLC (MADS-box transcription factor encoded by flowering locus C) provides a new mechanism for drought resistance, where FLC substantially reduces plant water use [49]. FLC is also involved in long-term cold adaptation mediated by the epigenetic memory mechanism [50]. In the presence of BR, BZR1 and BES1-interacting MYC-like proteins (BIMs) bind to a BR-responsive element in the first intron of *FLC* and further recruits a histone 3 lysine 27 (H3K27) demethylase to suppress levels of the H3K27 trimethylation mark and thus antagonize Polycomb silencing at *FLC* [47]. FLC binds to numerous target genes to regulate their expression, including those involved in response to water deprivation, such as *CBF1/DREB1B* and *CBF3/DREB1A* [51]. The functioning of the *FLC* is regulated not only by BZR1 and BES1, but also by ABI5 [52], which establishes an additional connection between the ABA and BR pathways (Figure 2).

Another player in the BR-induced stress memory is the RNA-binding protein FCA, a component of flowering pathways in *Arabidopsis* and a regulator of *FLC* [53]. It has been shown previously that FCA interacts with SWI3A and SWI3B, components of the Switch/Sucrose non-fermenting, ATP-dependent chromatin remodeling complex (SWI/SNF CRC) [54]. FCA interacts with ABI5 and is essential for proper expression of ABI5-regulated genes involved in antioxidant defense and thermotolerance [55]. FCA not only regulates the function of many genes involved in adaptation to stress-induced ROS, heat, cold, and drought conditions via FLC and ABI5, but also adjusts the function of protective genes by itself, through chromatin modification and RNA metabolism [55]. Histone acetylation is important in the FCA-mediated thermal adaptation of developing seedlings, chlorophyll biosynthesis, and seedling photosynthetic fitness [56]. The FLC/FCA module functions not only in hot conditions, but also in cold, providing adaptation to winter conditions through an *FLC* antisense transcript *COOLAIR* [53,57]. This may explain why BRL3ox plants demonstrated high gene expression not only in the category "Response to Water Deprivation", but also in the categories "Response to Temperature Stimulus" and "Cold Acclimation" [2]. It is interesting to note that these categories were also supplemented with GO category "Secondary Metabolic Process" [2]. Upregulation of genes related to secondary metabolism in BRL3ox plants can be explained by formation of the BR-activated BES1-TPL-HDA19 repressor complex, which acts via TPL on the jasmonate signaling system [58]. Additionally, activation might be mediated by FCA, which upregulates a number of secondary metabolism-specific biosynthetic genes and related transcription factors [55].

It is possible that the BR and ABA signaling pathways work simultaneously to ensure the priming effect on the drought tolerance of BRL3ox plants. In BRL3ox plants, ABA-dependent genes are upregulated, such as those that encode galactinol synthase 2 (GOLS2), dehydrin Xero 2 (LTI30), Em-like protein GEA1 (EM1), NAC transcription factor RD26, and cold and ABA inducible protein KIN1 [2]. Most of them are involved in the Response To Water Deprivation, Response To Cold or Cold Acclimation, and Response To Osmotic Stress (GO and BioGrid annotations). Of the ABA core signaling genes, protein phosphatases PP2C (ABI1, ABI2, and HAB1), transcription factors ABI3 and ABI5, SnRK2.2/2.3 kinases, and AREB/ABF transcription factors (such as ABF1, ABF2, ABF3, and ABF4) were shown to be involved in stress memory through interaction with SWI/SNF CRC [42,59]. The mechanism for memory generation through these interactions (Figure 2) may be realized via the ABA-chaperone pathway, where ABA-responsive elements (ABREs) recruit the SWI/SNF CRC to the chromatin template via ABFs and through the heat-shock transcription factors' (HSFs) interaction with SWI/SNF CRC, histone-modifying enzymes, and other cofactors [27]. Indeed, the CRISPR/Cas9-mediated activation of AREB1/ABF2 through histone acetylation was shown to be useful for improving drought stress tolerance [60]. We must also consider that SWI/SNF chromatin remodelers interact with many players of the BR-ABA network, such as PP2Cs, SnRK2s, ABFs, BIM1, and others [27]. Han et al. [40] discovered that plants with mutated ATP-dependent helicase BRAHMA (BRM, a component of SWI/SNF CRC) acquired ABA hypersensitivity and increased drought resistance. BRM represses *ABI5* expression ([40], Figure 2). The authors suggested that the physiological role of BRM is to help plants avoid stress responses in the absence of stress. BRM is considered an important element in determining the allocation of resources between drought tolerance and growth [40].

5. Stabilization of Endoplasmic Proteins

Little is known about the connection between BR signaling and chaperones, which are necessary for stress adaptation. An important interaction occurs through BSK1/3 and GSK1, the shaggy-related protein kinase iota (synonyms: BIN2-LIKE 2, BIL2; Figure 1). We noted that the STRING database provides data indicating numerous interactions between *Arabidopsis* GSK1 and proteins known as heat-shock transcription factors (HSFs). The STRING data were accessed from the interaction of GSK and HSF homologs in non-plant organisms, such as *Saccharomyces cerevisiae*, *Caenorhabditis elegans*, and *Homo sapiens* (https://string-db.org/network/3702.AT1G06390.1). In many cases, these interactions were associated with stress reactions caused by an accumulation of misfolded proteins in the endoplasmic reticulum (ER).

These data prompted us to examine in more detail the literature about relationships between BR signaling and ER stress. In plants, there are no known interactions between GSKs and HSFs, but the association of BR signaling with ER stress signaling is well documented. A connection between ER stress signaling and BR-mediated growth and stress acclimation was shown by Che et al. [61]. They reported that *Arabidopsis* bZIP17 and bZIP28 transcription factors activate ER chaperone genes and BR signaling, which was required for stress acclimation and growth. Furthermore, Cui et al. [62] showed that UBC32, a stress-induced ubiquitin conjugation enzyme, connects the ER-associated protein degradation (ERAD) process, BR-mediated growth promotion, and salt stress tolerance [62]. BRI1 was also shown to be involved in this process.

The *Arabidopsis ethyl methanesulfonate-mutagenized brassinosteroid-insensitive 1 suppressor 7* (*EBS7*) gene, which encodes an ER membrane-localized ERAD component, is connected to the function of BRI1 and to stress tolerance via Hrd1a (ERAD-associated E3 ubiquitin-protein ligase Hrd1a), one of the central components of the *Arabidopsis* ERAD machinery [63]. Unlike in yeast and animal model systems, *Arabidopsis* ERAD components are just beginning to be studied, however recent investigations have revealed new important players and there is support for a connection between ER stress signaling and stress tolerance [64,65]. Our search in the databases showed that the interactome of eukaryotic organisms is enriched with numerous protein–protein interactions involving Hrd proteins, while there are no such interactions in the *Arabidopsis* interactome. In other words, *Arabidopsis* is underexplored in this regard. Summing up these data, we can hypothesize that BR signaling can increase stress resistance by stabilizing endoplasmic proteins. In the case of BRL3, it may be by the BRL3 (GSK1) → BSK1/3 pathway.

The chaperone signaling system comprises predominantly HSFs and heat-shock proteins (HSP), and peptidyl-prolyl cis-trans isomerases (PPIase), also called immunophilins [27]. Since the chaperone-type immunophilin FKBP42/TWD1 positively regulates the BRI1/BAK1 function and acts together with HSP90, it is possible that FKBP42/TWD1 and HSP90 assist the folding of membrane proteins [66]. Mutation in the *HvBRI1* gene causes a decreased HSP level and decreased *HSP* gene expression [67]. Although it is known that HSPs interact with the BR core components [68,69], there is no evidence of such interaction with HSFs. If it is established that BR signaling components interact with HSFs, then studies of BRs in terms of the implementation of stress memory will receive a new direction, since HSFs are the main sculptors of the epigenetic landscape [70].

6. Conclusions

In this review, we examined a key finding, recently reported by Fàbregas and colleagues [2], in which an increased expression of the BRL3 receptor provided resistance to a lack of water but did not impair plant development. Such cases, in relation to any stress, are quite rare, since resistance to any stressful condition is usually accompanied by growth retardation. Drought tolerance of plants is controlled by numerous signaling modules forming a branched network of protein–protein interactions. In this study, we examined all known interactions of the BR and ABA signaling pathways, but left out the signaling pathways of gibberellins and the light signaling system, which undoubtedly affect stress tolerance, but would greatly complicate the understanding of the described phenomenon.

In such cases as the one that was described by Fàbregas et al. [2], we should look for adaptation processes caused by memory generation. Ding et al. [39] postulated that under natural conditions, stress memory is activated by the previous dehydration stress, continues during the recovery period, and prepares the plant's response to the next dehydration stress. This is surprising, but so far the

BR pathway has not been extensively studied with respect to epigenetic changes and the generation of stress memory, and only the first steps have been taken on this path [45–47,68,71]. For the closest animal relatives of BRs, glucocorticoids, we observe an extensive field of research related specifically to changes in chromatin structure [72]. At present, it is still unclear whether BRs have a lesser effect on adaptive chromatin rearrangements or if they are simply underexplored in this regard.

We hypothesized that BRs may be involved in stress acclimation by three interconnected mechanisms. The first mechanism is that the signal passes through the module BRL3 (or BRI1) → BSK1/3 → BSU1 —‖ BIN2 → BSK1/3, in which BIN2 is responsible for communication with ABA signaling and the BSK proteins serve as signal concentrators. The second mechanism is priming through chromatin modifications, in which BRL3 and other BR receptors could act collectively to ensure stress memory via BIN2 → SnRK2s → ABF2, BES1 or BZR1–TPL –HDA19 repressor complexes and BZR1/BES1 → FLC/FCA pathway. The third mechanism is stress acclimation by the BR-mediated stabilization of endoplasmic proteins in the ERAD process. New research prospects involving the BR signaling pathway in relation to stress adaptation are very intriguing and include the study of BR and ABA interaction pathways, chromatin modifications, and the ERAD process.

Author Contributions: Conception, data analysis, manuscript writing, V.P.B.; analysis and interpretation of data, revising for important intellectual content, final approval, T.V.A. All authors have read and agreed to the published version of the manuscript.

Funding: This research was funded by the Russian Science Foundation, grant number 18-44-08001 (VPB).

Conflicts of Interest: The authors declare no conflict of interest.

Abbreviations

ABA	Abscisic acid
ABREs	ABA-responsive elements
BRs	Brassinosteroids
ER	Endoplasmic reticulum
ERAD	ER-associated protein degradation
GO	Gene Ontology
HSFs	Heat-shock factors
HSPs	Heat-shock proteins
PPIases	Peptidyl-prolyl cis-trans isomerases
SERK	Somatic embryogenesis receptor kinase
SWI/SNF CRC	Switch/Sucrose non-fermenting, ATP-dependent chromatin remodeling complex

References

1. Nolan, T.; Chen, J.; Yin, Y. Cross-talk of brassinosteroid signaling in controlling growth and stress responses. *Biochem. J.* **2017**, *474*, 2641–2661. [CrossRef] [PubMed]
2. Fàbregas, N.; Lozano-Elena, F.; Blasco-Escámez, D.; Tohge, T.; Martínez-Andújar, C.; Albacete, A.; Osorio, S.; Bustamante, M.; Riechmann, J.L.; Nomura, T.; et al. Overexpression of the vascular brassinosteroid receptor BRL3 confers drought resistance without penalizing plant growth. *Nat. Commun.* **2018**, *9*, 4680. [CrossRef] [PubMed]
3. Kim, T.W.; Guan, S.; Sun, Y.; Deng, Z.; Tang, W.; Shang, J.X.; Sun, Y.; Burlingame, A.L.; Wang, Z.Y. Brassinosteroid signal transduction from cell-surface receptor kinases to nuclear transcription factors. *Nat. Cell Biol.* **2009**, *11*, 1254–1260. [CrossRef] [PubMed]
4. Ryu, H.; Cho, H.; Kim, K.; Hwang, I. Phosphorylation dependent nucleocytoplasmic shuttling of BES1 is a key regulatory event in brassinosteroid signaling. *Mol. Cells* **2010**, *29*, 283–290. [CrossRef] [PubMed]
5. Kim, T.W.; Wang, Z.Y. Brassinosteroid signal transduction from receptor kinases to transcription factors. *Annu. Rev. Plant Biol.* **2010**, *61*, 681–704. [CrossRef] [PubMed]
6. Wang, Z.Y.; Nakano, T.; Gendron, J.; He, J.; Chen, M.; Vafeados, D.; Yang, Y.; Fujioka, S.; Yoshida, S.; Asami, T. Nuclear-localized BZR1 mediates brassinosteroid-induced growth and feedback suppression of brassinosteroid biosynthesis. *Dev. Cell* **2002**, *2*, 505–513. [CrossRef]

7. He, J.X.; Gendron, J.M.; Sun, Y.; Gampala, S.S.; Gendron, N.; Sun, C.Q.; Wang, Z.Y. BZR1 is a transcriptional repressor with dual roles in brassinosteroid homeostasis and growth responses. *Science* **2005**, *307*, 1634–1638. [CrossRef]

8. Sun, Y.; Fan, X.Y.; Cao, D.M.; Tang, W.; He, K.; Zhu, J.Y.; He, J.X.; Bai, M.Y.; Zhu, S.; Oh, E.; et al. Integration of brassinosteroid signal transduction with the transcription network for plant growth regulation in *Arabidopsis*. *Dev. Cell* **2010**, *19*, 765–777. [CrossRef]

9. Yu, X.; Li, L.; Zola, J.; Aluru, M.; Ye, H.; Foudree, A.; Guo, H.; Anderson, S.; Aluru, S.; Liu, P.; et al. A brassinosteroid transcriptional network revealed by genome-wide identification of BESI target genes in *Arabidopsis thaliana. Plant J.* **2011**, *65*, 634–646. [CrossRef]

10. Ren, H.; Willige, B.C.; Jaillais, Y.; Geng, S.; Park, M.Y.; Gray, W.M.; Chory, J. BRASSINOSTEROID-SIGNALING KINASE 3, a plasma membrane-associated scaffold protein involved in early brassinosteroid signaling. *PLoS Genet.* **2019**, *15*, e1007904. [CrossRef]

11. Kim, T.W.; Guan, S.; Burlingame, A.L.; Wang, Z.Y. The CDG1 kinase mediates brassinosteroid signal transduction from BRI1 receptor kinase to BSU1 phosphatase and GSK3-like kinase BIN2. *Mol. Cell* **2011**, *43*, 561–571. [CrossRef] [PubMed]

12. He, J.X.; Gendron, J.M.; Yang, Y.; Li, J.; Wang, Z.Y. The GSK3-like kinase BIN2 phosphorylates and destabilizes BZR1, a positive regulator of the brassinosteroid signaling pathway in *Arabidopsis. Proc. Natl. Acad. Sci. USA* **2002**, *99*, 10185–10190. [CrossRef] [PubMed]

13. Vert, G.; Chory, J. Downstream nuclear events in brassinosteroid signalling. *Nature* **2006**, *441*, 96–100. [CrossRef] [PubMed]

14. Ye, H.; Liu, S.; Tang, B.; Chen, J.; Xie, Z.; Nolan, T.M.; Jiang, H.; Guo, H.; Lin, H.Y.; Li, L.; et al. RD26 mediates crosstalk between drought and brassinosteroid signalling pathways. *Nat. Commun.* **2017**, *8*, 14573. [CrossRef] [PubMed]

15. Chen, J.; Nolan, T.M.; Ye, H.; Zhang, M.; Tong, H.; Xin, P.; Chu, J.; Chu, C.; Li, Z.; Yin, Y. Arabidopsis WRKY46, WRKY54, and WRKY70 transcription factors are involved in brassinosteroid-regulated plant growth and drought responses. *Plant Cell* **2017**, *29*, 1425–1439. [CrossRef]

16. Xie, Z.; Nolan, T.; Jiang, H.; Tang, B.; Zhang, M.; Li, Z.; Yin, Y. The AP2/ERF transcription factor TINY modulates brassinosteroid-regulated plant growth and drought responses in *Arabidopsis. Plant Cell* **2019**, *31*, 1788–1806. [CrossRef]

17. Fàbregas, N.; Li, N.; Boeren, S.; Nash, T.E.; Goshe, M.B.; Clouse, S.D.; de Vries, S.; Caño-Delgado, A.I. The brassinosteroid insensitive1-like3 signalosome complex regulates *Arabidopsis* root development. *Plant Cell* **2013**, *25*, 3377–3388. [CrossRef]

18. Takahashi, F.; Kuromori, T.; Sato, H.; Shinozaki, K. Regulatory gene networks in drought stress responses and resistance in plants. *Adv. Exp. Med. Biol.* **2018**, *1081*, 189–214. [CrossRef]

19. Zhou, J.; Wang, J.; Li, X.; Xia, X.J.; Zhou, Y.H.; Shi, K.; Chen, Z.; Yu, J.Q. H_2O_2 mediates the crosstalk of brassinosteroid and abscisic acid in tomato responses to heat and oxidative stresses. *J. Exp. Bot.* **2014**, *65*, 4371–4383. [CrossRef]

20. Saini, S.; Sharma, I.; Pati, P.K. Versatile roles of brassinosteroid in plants in the context of its homoeostasis, signaling and crosstalks. *Front. Plant Sci.* **2015**, *6*, 950. [CrossRef]

21. Fujii, H.; Chinnusamy, V.; Rodrigues, A.; Rubio, S.; Antoni, R.; Park, S.Y.; Cutler, S.R.; Sheen, J.; Rodriguez, P.L.; Zhu, J.K. In vitro reconstitution of an abscisic acid signalling pathway. *Nature* **2009**, *462*, 660–664. [CrossRef]

22. Zhang, S.; Cai, Z.; Wang, X. The primary signaling outputs of brassinosteroids are regulated by abscisic acid signaling. *Proc. Natl. Acad. Sci. USA* **2009**, *106*, 4543–4548. [CrossRef] [PubMed]

23. Caño-Delgado, A.; Yin, Y.; Yu, C.; Vafeados, D.; Mora-García, S.; Cheng, J.C.; Nam, K.H.; Li, J.; Chory, J. BRL1 and BRL3 are novel brassinosteroid receptors that function in vascular differentiation in *Arabidopsis. Development* **2004**, *131*, 5341–5351. [CrossRef] [PubMed]

24. Tang, W.; Kim, T.W.; Oses-Prieto, J.A.; Sun, Y.; Deng, Z.; Zhu, S.; Wang, R.; Burlingame, A.L.; Wang, Z.Y. BSKs mediate signal transduction from the receptor kinase BRI1 in *Arabidopsis. Science* **2008**, *321*, 557–560. [CrossRef]

25. Sreeramulu, S.; Mostizky, Y.; Sunitha, S.; Shani, E.; Nahum, H.; Salomon, D.; Hayun, L.B.; Gruetter, C.; Rauh, D.; Ori, N.; et al. BSKs are partially redundant positive regulators of brassinosteroid signaling in *Arabidopsis. Plant J.* **2013**, *74*, 905–919. [CrossRef]

26. Smakowska-Luzan, E.; Mott, G.A.; Parys, K.; Stegmann, M.; Howton, T.C.; Layeghifard, M.; Neuhold, J.; Lehner, A.; Kong, J.; Grünwald, K.; et al. An extracellular network of *Arabidopsis* leucine-rich repeat receptor kinases. *Nature* **2018**, *553*, 342–346. [CrossRef] [PubMed]

27. Bulgakov, V.P.; Wu, H.C.; Jinn, T.L. Coordination of ABA and chaperone signaling in plant stress responses. *Trends Plant Sci.* **2019**, *24*, 636–651. [CrossRef] [PubMed]

28. Tang, R.J.; Liu, H.; Yang, Y.; Yang, L.; Gao, X.S.; Garcia, V.J.; Luan, S.; Zhang, H.X. Tonoplast calcium sensors CBL2 and CBL3 control plant growth and ion homeostasis through regulating V-ATPase activity in *Arabidopsis*. *Cell Res.* **2012**, *22*, 1650–1665. [CrossRef]

29. Ji, H.; Pardo, J.M.; Batelli, G.; Van Oosten, M.J.; Bressan, R.A.; Li, X. The Salt Overly Sensitive (SOS) pathway: Established and emerging roles. *Mol. Plant* **2013**, *6*, 275–286. [CrossRef]

30. Yang, Y.; Wu, Y.; Ma, L.; Yang, Z.; Dong, Q.; Li, Q.; Ni, X.; Kudla, J.; Song, C.; Guo, Y. The Ca2+ sensor SCaBP3/CBL7 modulates plasma membrane H+-ATPase activity and promotes alkali tolerance in *Arabidopsis*. *Plant Cell* **2019**, *31*, 1367–1384. [CrossRef]

31. Tunc-Ozdemir, M.; Jones, A.M. BRL3 and AtRGS1 cooperate to fine tune growth inhibition and ROS activation. *PLoS ONE* **2017**, *12*, e0177400. [CrossRef] [PubMed]

32. Jia, Z.; Giehl, R.; Meyer, R.C.; Altmann, T.; von Wirén, N. Natural variation of BSK3 tunes brassinosteroid signaling to regulate root foraging under low nitrogen. *Nat. Commun.* **2019**, *10*, 2378. [CrossRef] [PubMed]

33. Hu, Y.; Yu, D. BRASSINOSTEROID INSENSITIVE2 interacts with ABSCISIC ACID INSENSITIVE5 to mediate the antagonism of brassinosteroids to abscisic acid during seed germination in *Arabidopsis*. *Plant Cell* **2014**, *26*, 4394–4408. [CrossRef] [PubMed]

34. Jiang, H.; Tang, B.; Xie, Z.; Nolan, T.; Ye, H.; Song, G.Y.; Walley, J.; Yin, Y. GSK3-like kinase BIN2 phosphorylates RD26 to potentiate drought signaling in *Arabidopsis*. *Plant J.* **2019**, *100*, 923–937. [CrossRef]

35. Ye, K.; Li, H.; Ding, Y.; Shi, Y.; Song, C.; Gong, Z.; Yang, S. BRASSINOSTEROID-INSENSITIVE2 negatively regulates the stability of transcription factor ICE1 in response to cold stress in *Arabidopsis*. *Plant Cell* **2019**, *31*, 2682–2696. [CrossRef]

36. Cai, Z.; Liu, J.; Wang, H.; Yang, C.; Chen, Y.; Li, Y.; Pan, S.; Dong, R.; Tang, G.; Barajas-Lopez, J.; et al. GSK3-like kinases positively modulate abscisic acid signaling through phosphorylating subgroup III SnRK2s in *Arabidopsis*. *Proc. Natl. Acad. Sci. USA* **2014**, *111*, 9651–9656. [CrossRef]

37. Li, H.; Ye, K.; Shi, Y.; Cheng, J.; Zhang, X.; Yang, S. BZR1 Positively Regulates Freezing Tolerance via CBF-Dependent and CBF-Independent Pathways in *Arabidopsis*. *Mol. Plant* **2017**, *10*, 545–559. [CrossRef]

38. Ding, Y.; Li, H.; Zhang, X.; Xie, Q.; Gong, Z.; Yang, S. OST1 kinase modulates freezing tolerance by enhancing ICE1 stability in *Arabidopsis*. *Dev. Cell* **2015**, *32*, 278–289. [CrossRef]

39. Ding, Y.; Fromm, M.; Avramova, Z. Multiple exposures to drought 'train' transcriptional responses in *Arabidopsis*. *Nat. Commun.* **2012**, *3*, 740. [CrossRef]

40. Han, S.K.; Sang, Y.; Rodrigues, A.; Wu, M.F.; Rodriguez, P.L.; Wagner, D. The SWI2/SNF2 chromatin remodeling ATPase BRAHMA represses abscisic acid responses in the absence of the stress stimulus in *Arabidopsis*. *Plant Cell* **2012**, *24*, 4892–4906. [CrossRef]

41. Avramova, Z. Transcriptional 'memory' of a stress: Transient chromatin and memory (epigenetic) marks at stress-response genes. *Plant J.* **2015**, *83*, 149–159. [CrossRef] [PubMed]

42. Han, S.K.; Wagner, D. Role of chromatin in water stress responses in plants. *J. Exp. Bot.* **2014**, *65*, 2785–2799. [CrossRef] [PubMed]

43. Forestan, C.; Farinati, S.; Zambelli, F.; Pavesi, G.; Rossi, V.; Varotto, S. Epigenetic signatures of stress adaptation and flowering regulation in response to extended drought and recovery in *Zea mays*. *Plant Cell Environ.* **2019**, *43*, 55–75. [CrossRef] [PubMed]

44. Shigeta, T.; Yoshimitsu, Y.; Nakamura, Y.; Okamoto, S.; Matsuo, T. Does brassinosteroid function require chromatin remodeling? *Plant Signal. Behav.* **2011**, *6*, 1824–1827. [CrossRef] [PubMed]

45. Oh, E.; Zhu, J.Y.; Ryu, H.; Hwang, I.; Wang, Z.Y. TOPLESS mediates brassinosteroid-induced transcriptional repression through interaction with BZR1. *Nat. Commun.* **2014**, *5*, 4140. [CrossRef]

46. Ryu, H.; Cho, H.; Bae, W.; Hwang, I. Control of early seedling development by BES1/TPL/HDA19-mediated epigenetic regulation of ABI3. *Nat. Commun.* **2014**, *5*, 4138. [CrossRef]

47. Li, Z.; Ou, Y.; Zhang, Z.; Li, J.; He, Y. Brassinosteroid signaling recruits Histone 3 Lysine-27 demethylation activity to FLOWERING LOCUS C chromatin to inhibit the floral transition in *Arabidopsis*. *Mol. Plant* **2018**, *11*, 1135–1146. [CrossRef]

48. Kim, H.; Shim, D.; Moon, S.; Lee, J.; Bae, W.; Choi, H.; Kim, K.; Ryu, H. Transcriptional network regulation of the brassinosteroid signaling pathway by the BES1-TPL-HDA19 co-repressor complex. *Planta* **2019**, *250*, 1371–1377. [CrossRef]

49. Ferguson, J.N.; Meyer, R.C.; Edwards, K.D.; Humphry, M.; Brendel, O.; Bechtold, U. Accelerated flowering time reduces lifetime water use without penalizing reproductive performance in *Arabidopsis*. *Plant Cell Environ.* **2019**, *42*, 1847–1867. [CrossRef]

50. Qüesta, J.I.; Antoniou-Kourounioti, R.L.; Rosa, S.; Li, P.; Duncan, S.; Whittaker, C.; Howard, M.; Dean, C. Noncoding SNPs influence a distinct phase of Polycomb silencing to destabilize long-term epigenetic memory at *Arabidopsis* FLC. *Genes Dev.* **2020**, *34*, 446–461. [CrossRef]

51. Deng, W.; Ying, H.; Helliwell, C.A.; Taylor, J.M.; Peacock, W.J.; Dennis, E.S. FLOWERING LOCUS C (FLC) regulates development pathways throughout the life cycle of *Arabidopsis*. *Proc. Natl. Acad. Sci. USA* **2011**, *108*, 6680–6685. [CrossRef] [PubMed]

52. Xiong, F.; Ren, J.J.; Yu, Q.; Wang, Y.Y.; Lu, C.C.; Kong, L.J.; Otegui, M.S.; Wang, X.L. AtU2AF65b functions in abscisic acid mediated flowering via regulating the precursor messenger RNA splicing of *ABI5* and *FLC* in *Arabidopsis*. *New Phytol.* **2019**, *223*, 277–292. [CrossRef] [PubMed]

53. Tian, Y.; Zheng, H.; Zhang, F.; Wang, S.; Ji, X.; Xu, C.; He, Y.; Ding, Y. PRC2 recruitment and H3K27me3 deposition at FLC require FCA binding of *COOLAIR*. *Sci. Adv.* **2019**, *5*, eaau7246. [CrossRef] [PubMed]

54. Sarnowski, T.J.; Swiezewski, S.; Pawlikowska, K.; Kaczanowski, S.; Jerzmanowski, A. AtSWI3B, an *Arabidopsis* homolog of SWI3, a core subunit of yeast Swi/Snf chromatin remodeling complex, interacts with FCA, a regulator of flowering time. *Nucleic Acids Res.* **2002**, *30*, 3412–3421. [CrossRef] [PubMed]

55. Lee, S.; Lee, H.J.; Jung, J.H.; Park, C.M. The Arabidopsis thaliana RNA-binding protein FCA regulates thermotolerance by modulating the detoxification of reactive oxygen species. *New Phytol.* **2015**, *205*, 555–569. [CrossRef]

56. Ha, J.H.; Lee, H.J.; Jung, J.H.; Park, C.M. Thermo-induced maintenance of photo-oxidoreductases underlies plant autotrophic development. *Dev. Cell* **2017**, *41*, 170–179. [CrossRef]

57. Jiao, F.; Pahwa, K.; Manning, M.; Dochy, N.; Geuten, K. Cold induced antisense transcription of *FLOWERING LOCUS C* in distant grasses. *Front. Plant Sci.* **2019**, *10*, 72. [CrossRef]

58. Bulgakov, V.P.; Avramenko, T.V.; Tsitsiashvili, G.S. Critical analysis of protein signaling networks involved in the regulation of plant secondary metabolism: Focus on anthocyanins. *Crit. Rev. Biotechnol.* **2017**, *37*, 685–700. [CrossRef]

59. Peirats-Llobet, M.; Han, S.K.; Gonzalez-Guzman, M.; Jeong, C.W.; Rodriguez, L.; Belda-Palazon, B.; Wagner, D.; Rodriguez, P.L. A direct link between abscisic acid sensing and the chromatin-remodeling ATPase BRAHMA via core ABA signaling pathway components. *Mol. Plant* **2016**, *9*, 136–147. [CrossRef]

60. Roca Paixão, J.F.; Gillet, F.X.; Ribeiro, T.P.; Bournaud, C.; Lourenço-Tessutti, I.T.; Noriega, D.D.; Melo, B.P.; de Almeida-Engler, J.; Grossi-de-Sa, M.F. Improved drought stress tolerance in *Arabidopsis* by CRISPR/dCas9 fusion with a Histone AcetylTransferase. *Sci. Rep.* **2019**, *9*, 8080. [CrossRef]

61. Che, P.; Bussell, J.D.; Zhou, W.; Estavillo, G.M.; Pogson, B.J.; Smith, S.M. Signaling from the endoplasmic reticulum activates brassinosteroid signaling and promotes acclimation to stress in *Arabidopsis*. *Sci. Signal.* **2010**, *3*, ra69. [CrossRef] [PubMed]

62. Cui, F.; Liu, L.; Zhao, Q.; Zhang, Z.; Li, Q.; Lin, B.; Wu, Y.; Tang, S.; Xie, Q. Arabidopsis ubiquitin conjugase UBC32 is an ERAD component that functions in brassinosteroid-mediated salt stress tolerance. *Plant Cell* **2012**, *24*, 233–244. [CrossRef] [PubMed]

63. Liu, Y.; Zhang, C.; Wang, D.; Su, W.; Liu, L.; Wang, M.; Li, J. EBS7 is a plant-specific component of a highly conserved endoplasmic reticulum-associated degradation system in *Arabidopsis*. *Proc. Natl. Acad. Sci. USA* **2015**, *112*, 12205–12210. [CrossRef]

64. Park, J.H.; Kang, C.H.; Nawkar, G.M.; Lee, E.S.; Paeng, S.K.; Chae, H.B.; Chi, Y.H.; Kim, W.Y.; Yun, D.J.; Lee, S.Y. EMR, a cytosolic-abundant ring finger E3 ligase, mediates ER-associated protein degradation in *Arabidopsis*. *New Phytol.* **2018**, *220*, 163–177. [CrossRef]

65. Lin, L.; Zhang, C.; Chen, Y.; Wang, Y.; Wang, D.; Liu, X.; Wang, M.; Mao, J.; Zhang, J.; Xing, W.; et al. PAWH1 and PAWH2 are plant-specific components of an *Arabidopsis* endoplasmic reticulum-associated degradation complex. *Nat. Commun.* **2019**, *10*, 3492. [CrossRef] [PubMed]

66. Chaiwanon, J.; Garcia, V.J.; Cartwright, H.; Sun, Y.; Wang, Z.Y. Immunophilin-like FKBP42/TWISTED DWARF1 interacts with the receptor kinase BRI1 to regulate brassinosteroid signaling in *Arabidopsis*. *Mol. Plant* **2016**, *9*, 593–600. [CrossRef]

67. Sadura, I.; Libik-Konieczny, M.; Jurczyk, B.; Gruszka, D.; Janeczko, A. HSP transcript and protein accumulation in brassinosteroid barley mutants acclimated to low and high temperatures. *Int. J. Mol. Sci.* **2020**, *21*, 1889. [CrossRef]

68. Shigeta, T.; Zaizen, Y.; Sugimoto, Y.; Nakamura, Y.; Matsuo, T.; Okamoto, S. Heat shock protein 90 acts in brassinosteroid signaling through interaction with BES1/BZR1 transcription factor. *J. Plant Physiol.* **2015**, *178*, 69–73. [CrossRef]

69. Samakovli, D.; Roka, L.; Plitsi, P.K.; Kaltsa, I.; Daras, G.; Milioni, D.; Hatzopoulos, P. Active BR signaling adjusts the subcellular localization of BES1/HSP90 complex formation. *Plant Biol.* **2020**, *22*, 129–133. [CrossRef]

70. Miozzo, F.; Sabéran-Djoneidi, D.; Mezger, V. HSFs, Stress sensors and sculptors of transcription compartments and epigenetic landscapes. *J. Mol. Biol.* **2015**, *427*, 3793–3816. [CrossRef]

71. Guo, H.; Li, L.; Aluru, M.; Aluru, S.; Yin, Y. Mechanisms and networks for brassinosteroid regulated gene expression. *Curr. Opin. Plant Biol.* **2013**, *16*, 545–553. [CrossRef] [PubMed]

72. Jing, D.; Huang, Y.; Liu, X.; Sia, K.; Zhang, J.C.; Tai, X.; Wang, M.; Toscan, C.E.; McCalmont, H.; Evans, K.; et al. Lymphocyte-specific chromatin accessibility pre-determines glucocorticoid resistance in acute lymphoblastic leukemia. *Cancer Cell* **2018**, *34*, 906–921. [CrossRef] [PubMed]

International Journal of
Molecular Sciences

Article

Cloning and Functional Analysis of Dwarf Gene *Mini Plant 1* (*MNP1*) in *Medicago truncatula*

Shiqi Guo [1,2,†], Xiaojia Zhang [1,2,†], Quanzi Bai [1,2], Weiyue Zhao [1,2], Yuegenwang Fang [1,2],
Shaoli Zhou [1,2], Baolin Zhao [1], Liangliang He [1,2,*] and Jianghua Chen [1,*]

[1] CAS Key Laboratory of Tropical Plant Resources and Sustainable Use, Xishuangbanna Tropical Botanical Garden, Chinese Academy of Sciences, 88 Xuefu Road, Kunming 650223, China; guoshiqi17@mails.ucas.ac.cn (S.G.); zhangxiaojia0512@163.com (X.Z.); baiquanzi@xtbg.ac.cn (Q.B.); zhaoweiyue@xtbg.ac.cn (W.Z.); fangyuegw@163.com (Y.F.); zhoushaoli@xtbg.ac.cn (S.Z.); zhaobaolin@xtbg.ac.cn (B.Z.)

[2] College of Life Science, University of Chinese Academy of Sciences, Beijing 100049, China

[*] Correspondence: heliangliang@xtbg.ac.cn (L.H.); jhchen@xtbg.ac.cn (J.C.)

[†] These authors contributed equally to this work.

Received: 31 May 2020; Accepted: 3 July 2020; Published: 14 July 2020

Abstract: Plant height is a vital agronomic trait that greatly determines crop yields because of the close relationship between plant height and lodging resistance. Legumes play a unique role in the worldwide agriculture; however, little attention has been given to the molecular basis of their height. Here, we characterized the first dwarf mutant *mini plant 1* (*mnp1*) of the model legume plant *Medicago truncatula*. Our study found that both cell length and the cell number of internodes were reduced in a *mnp1* mutant. Using the forward genetic screening and subsequent whole-genome resequencing approach, we cloned the *MNP1* gene and found that it encodes a putative copalyl diphosphate synthase (CPS) implicated in the first step of gibberellin (GA) biosynthesis. MNP1 was highly homologous to *Pisum sativum* LS. The subcellular localization showed that MNP1 was located in the chloroplast. Further analysis indicated that GA$_3$ could significantly restore the plant height of *mnp1-1*, and expression of *MNP1* in a *cps1* mutant of *Arabidopsis* partially rescued its mini-plant phenotype, indicating the conservation function of MNP1 in GA biosynthesis. Our results provide valuable information for understanding the genetic regulation of plant height in *M. truncatula*.

Keywords: dwarfism; gene cloning; *MNP1*; CPS; *Medicago truncatula*

---

1. Introduction

Dwarf phenotypes have been widely used to improve lodging resistance and enhance harvest index in crops. For this reason, the proper modulation of plant height has always been a priority for breeders. Although many factors regulate plant height, gibberellin (GA) plays a leading role and is also known as the "green revolution phytohormone" because of its great contribution to the cultivation of high yields and lodging resistant crop varieties. The "green revolution" gene *semi-dwarf 1* (*sd1*) encoding GA biosynthesis enzyme GA 20-oxidase (GA20ox) is always important in rice breeding from the 1960s [1]. The *reduced height 1* (*rht-B1b* and *rht-D1b*) mutants showed a semi-dwarfing phenotype due to insensitivity to GA and were also used to breed for lodging resistance and yield increase in wheat [2].

GA is involved in various processes of plant growth and development, including leaf expansion, seed germination, induction of flowering and stem elongation [3–7]. With the extensive characterization of dwarf mutants related to GA, numerous genes encoding GA biosynthetic enzymes have been identified [8,9]. Bioactive GA biosynthesis is divided into three stages. In the first stage, geranylgeranyl diphosphate (GGDP), the precursor of GA, is catalyzed by the copalyl diphosphate synthase (CPS) and *ent*-kaurene synthase (KS) to form *ent*-kaurene, and this process takes place in the plastid [10,11].

Then, in the second stage, *ent*-kaurene is converted to GA_{12} by *ent*-kaurene oxidase (KO) and *ent*-kaurenoic acid oxidase (KAO), both of which are cytochrome P450 enzymes [11–13]. In the final stage, GA_{12} is catalyzed by GA20ox and converted to GA_9 via GA_{15} and GA_{24}, and then GA_9 is converted to GA_4 by GA 3β-hydroxylase (GA3ox) [14–17]. GA_{12} is also converted to GA_1 through the 13-hydroxylation pathway [9]. The biosynthesis of bioactive GA_1 and GA_4 occurs in the cytoplasm. Because the early-step genes of GA biosynthesis, *CPS1/GA1*, *KS* and *KO*, are single copy in *Arabidopsis*, mutations of these genes usually induce severely dwarf phenotype with greatly impaired fertility [12,18]. In contrast, the loss-of-function mutants of *GA20ox* and *GA3ox* (the late-step genes of GA biosynthesis) show a semi-dwarf phenotype due to the functional redundancy of multiple copies of genes [19,20]. In addition, the mutants altered in GA degradation and signal transduction pathway also show various degrees of dwarf phenotype and are valuable in molecular breeding [21–25].

Previous studies have suggested that, in addition to GAs, other plant hormones, such as brassinosteroids (BRs) [26,27] and strigolactones (SLs) [28,29], also play important roles in plant height development. The dwarf mutants related to these hormones can be divided into two types, hormone-sensitive and hormone-insensitive. The plant height of those hormone-sensitive mutants could be restored by exogenous hormones because their hormone content is reduced due to the disorder of hormone metabolic pathways [26,29–32]. The hormone-insensitive mutants are not sensitive to the hormone due to the abnormal signaling pathway [27,33–36]. So far, the molecular genetic pathways underlying the plant height regulation are well-characterized in *Arabidopsis* and rice, but only a few studies have been conducted on other species.

Legumes are the second most important economic crops after cereals and provide the major sources of plant proteins and oils for humans and animals [37]. Investigations on dwarf mutants in peas and soybeans have strongly suggested that the GA pathway plays a conserved role in determining the plant height of legumes [38–40]. *Medicago truncatula*, a diploid model legume plant, has been sequenced [41,42], but little attention was given to the basis of its height and the involved regulatory mechanisms of the GA pathway.

In this study, we characterized the severely dwarf mutant *mnp1* with two alleles isolated from the *Tnt1* retrotransposon-tagged mutant population of *M. truncatula*. Through forward genetic screening and the subsequent whole-genome resequencing approach, we cloned the *MNP1/Medtr7g011663* gene and found that it was well-clustered with the homologous genes encoding *Pisum sativum* LS, *Solanum lycopersicum* GIB-1, *Arabidopsis thaliana* CPS1/GA1, *Oryza sativa* OsCPS1 and *Zea mays* An1, all of which are the enzymes involved in the first step of GA biosynthesis. Because the dwarf phenotype of *mnp1* was significantly restored by exogenous application of GA_3, and the mini-plant phenotype of the *Arabidopsis cps1* mutant was partially rescued by the expression of *MNP1*, we proposed a conserved function of MNP1 in GA biosynthesis. Given the evidence that both the *mnp1* and the pea *ls* mutants are fertile and there are multiple possible copies of *MNP1/LS* in *M. truncatula*, peas and soybeans, it is reasonable to hypothesize that the duplication of *CPS* genes and the subsequent functional divergence may have occurred in legumes during evolution [43]. The result has significant implications for the legume breeding programs and provides a good model to further study the regulatory mechanism of height regulation in *M. truncatula*.

2. Results

2.1. Mini Plant 1 Mutants Were Severely Dwarfed Due to Shorter and Fewer Cells

Legumes are the third largest family of angiosperms, including many important crops, such as soybeans and peanuts [44]. To gain a better understanding of the molecular basis of plant height regulation in legumes, we screened the *Tnt1* retrotransposon insertion mutant collection of the model plant *M. truncatula* [45] to isolate mutants with significant changes in plant height. Two allelic mutants with similar severely dwarf phenotypes were identified and designated as *mini plant 1-1* (*mnp1-1*) and *mini plant 1-2* (*mnp1-2*), respectively, because all F_1 progenies derived from a cross between *mnp1-1* and *mnp1-2* were dwarf plants. Compared with the wild type, the mutants are severely dwarf, with increased branches and dark green leaves (Figure 1A–C and Figure 2C).

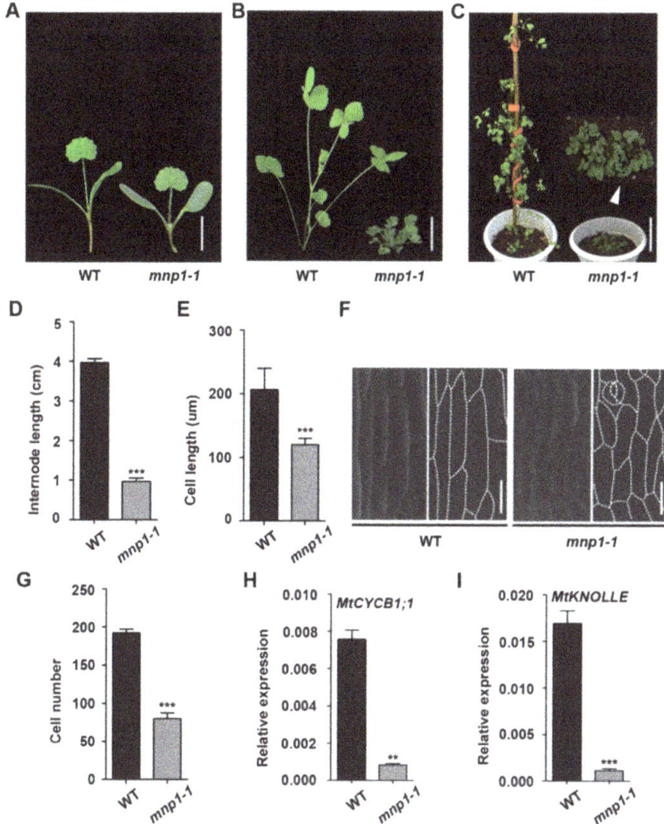

Figure 1. Phenotypic characterization of *mnp1-1* mutant. (**A–C**) Morphologies of wild type (WT) and *mnp1-1* mutant at different developmental stages. (**A**) Ten days after sowing. Scale bar = 0.75 cm. (**B**) Six weeks after sowing. Scale bar = 2 cm. (**C**) The reproductive stage of plants. Scale bar = 6.5 cm. (**D**) The length of the third internode beneath the shoot apex. Values are means ± *SD* (*n* = 20 internodes). Two-sample *t*-test, *** $p < 0.001$. (**E**) The length of epidermal cells of the third internode beneath the shoot apex. Values are means ± *SD* (*n* = 20 cells from three biological replicates). Two-sample *t*-test, *** $p < 0.001$. (**F**) Scanning electron microscope images and cell outlines of a representative third internode beneath the shoot apex. Scale bar = 50 um. (**G**) Number of epidermal cells in the third internode beneath the shoot apex. The cell number was calculated from the ratio of the average internode length (**D**) to the average cell length (**E**). Error bars represent the standard deviation of the cell number of 20 independent internodes. Two-sample *t*-test, *** $p < 0.001$. (**H**) Expression analysis of cell division marker gene *MtCYCB1;1*. Values are means ± *SD*. Two-sample *t*-test, ** $p < 0.01$. (**I**) Expression analysis of cell division marker gene *MtKNOLLE*. Values are means ± *SD*. Two-sample *t*-test, *** $p < 0.001$.

During the growth and development from seedlings to adult plants, the height gap between the wild type and *mnp1-1* mutant was becoming bigger (Figure 1A–C). By measuring the length of the third internode beneath the shoot apex, we confirmed that the *mnp1-1* mutants have reduced internode length compared with the wild type (Figure 1D). Then, scanning electron microscopy (SEM) analysis was used to determine the reasons for the shorter internode of *mnp1-1* mutants. The epidermal cells of the *mnp1-1* internode were considerably shorter than those of the wild type (Figure 1E,F). In addition, the number of internode cells was also greatly reduced in *mnp1-1* mutants (Figure 1G), indicating that cell division was significantly suppressed. This speculation would be in agreement with the quantitative analysis of the reduced cell cycle activity of the *mnp1-1* internode. The expression of the G2/M phase cell cycle marker *MtCYCB1;1* and the cytokinesis marker *MtKNOLLE* [46] were both dramatically lower in *mnp1-1* than that of wild type (Figure 1H,I). Therefore, both decreased length and number of internode cells contributed to the shortened stem of *mnp1-1*. In addition to the decrease of stem length, the petiole of *mnp1-1* was shortened as well (Figure S1). In conclusion, these results demonstrated that *MNP1* plays an important role in the length determination of stem and petiole in *M. truncatula*.

2.2. Molecular Cloning of MNP1 Gene

Analysis of the F_2 generation resulting from a cross between *mnp1-2* and wild type showed a segregation ratio of 3:1 between wild-type-like and dwarf phenotypes (36:13, $\chi^2 = 0.0068 < \chi^2_{0.05} = 3.84$) (Figure S2A), indicating that the *mnp1-2* phenotype was controlled by a single recessive gene. To clone the target gene corresponding for the mutant phenotype, *mnp1-1* and *mnp1-2* were backcrossed with the wild type, respectively, and mutant plants were isolated from both F_2 populations, followed by whole-genome resequencing at 20× coverage. Then, the resequencing data were analyzed using the bioinformatics tool Identification of Transposon Insertion Sites (ITIS) as previously described (Table S1) [47]. ITIS identified nine and seventy-one *Tnt1* insertions in the genomes of the *mnp1-1* and *mnp1-2* mutants, respectively. There were two *Tnt1* insertion sites on chromosome 7 that appeared to be nearby from the genomic sequence data of *mnp1-1* and *mnp1-2*; one was inserted into an intergenic region, and another was inserted into a genic region corresponding to the *Medtr7g011663* gene (annotated in A17 genome v4.0) (Figure 2A; Table S1). Then, PCR-based genotyping and sequencing analysis confirmed that the *mnp1-1* and *mnp1-2* mutants harbored *Tnt1* insertions in the fourth exon and the seventh exon of the candidate gene/*Medtr7g011663*, respectively (Figure 2B–D and Figure S2B). To determine whether the mutation of *Medtr7g011663* is responsible for the *mnp1* mutants' phenotype, an additional mutant line with a predicted *Tnt1* insertion in *Medtr7g011663* locus was identified via BLAST searching of the public mutant database [45], and thus was designated as *mnp1-3*. The *mnp1-3* plants displayed a severely dwarfed phenotype similar to *mnp1* alleles when growing in the greenhouse (Figure 2C). PCR-based sequencing confirmed that there is indeed a *Tnt1* insertion in the sixth exon of *Medtr7g011663* in *mnp1-3* (Figure 2D and Figure S2B). Thus, we considered *Medtr7g011663* as the putative *MNP1* gene.

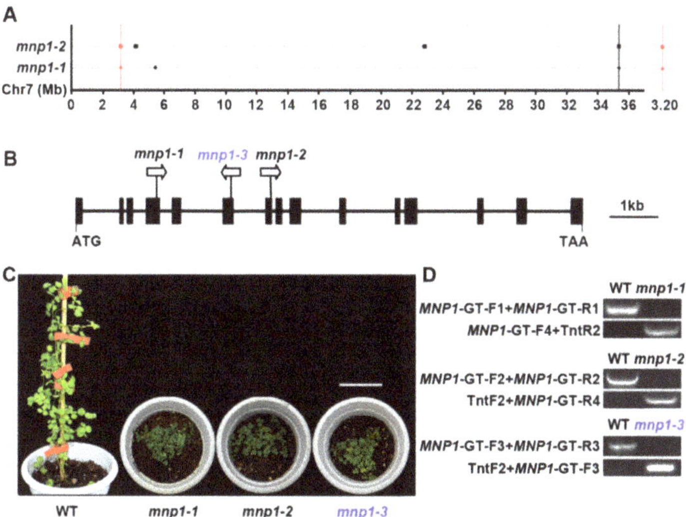

Figure 2. Molecular cloning of the *MNP1* gene. (**A**) Adjacent *Tnt1* insertion sites were found on chromosome 7 of *mnp1-1* and *mnp1-2*. The *x*-axis represents chromosome 7. Rhombus and squares represent *Tnt1* insertions in *mnp1-1* and *mnp1-2*, respectively. The rhombus and square on a black line show nearby *Tnt1* insertions in an intergenic region. The rhombus and square on a red line show nearby *Tnt1* insertions in *Medtr7g011663* and the right image is an enlarged view in the same region. (**B**) Schematic illustration of *MNP1* gene structure and *Tnt1* insertion sites in *mnp1* alleles. The *mnp1-3* (blue color) mutants were screened from the *Tnt1* population using a reverse genetics approach. Filled black boxes represent exons and lines between them denote introns. Arrows indicate *Tnt1* orientation. (**C**) The phenotype of *mnp1* alleles. Scale bar = 6.5 cm. (**D**) Genotyping of *mnp1* alleles. The primers (*MNP1*-GT-F/R) were designed for detecting *MNP1* genomic fragments, and the primer pair TntF2/R2 were *Tnt1*-specific primers.

2.3. MNP1 Encodes a Putative CPS Protein in M. truncatula

To figure out the type of protein encoded by *MNP1*, phylogenetic analysis of MNP1 and its homologous proteins from *M. truncatula* and related legume plants (pea and soybean), dicotyledonous model plants (*Arabidopsis* and tomato) and grasses (rice and maize) was performed. MNP1 protein was closely grouped with numerous homologs from legumes, and each selected legume species has at least two homologous copies. When compared to the reported homologous proteins, MNP1 showed the most homology to the pea LS and significant homology to the GIB-1 in tomatoes, CPS1/GA1 in *Arabidopsis*, OsCPS1 in rice and An1 in maize (Figure 3A), all of which are in the CPS family belonging to type-B cyclase and take part in the first step of GA biosynthesis [48–52]. The loss-of-function mutants of *ls*, *gib-1*, *cps1/ga1*, *Oscps1* and *an1* all show dwarfed phenotypes. In addition, the alignment of multiple amino acid sequences shows that MNP1 exhibits a high degree of amino acid sequence identities with these CPS proteins (Figure S3). Furthermore, there is an aspartate-rich motif DXDD near the N-terminal region of MNP1 (Figure 3B), which is conserved among type-B cyclase and important for the catalysis of the type-B cyclization reactions [52,53]. Taken together, we believe that MNP1 would be a conserved CPS protein involved in the GA biosynthesis pathway in *M. truncatula*.

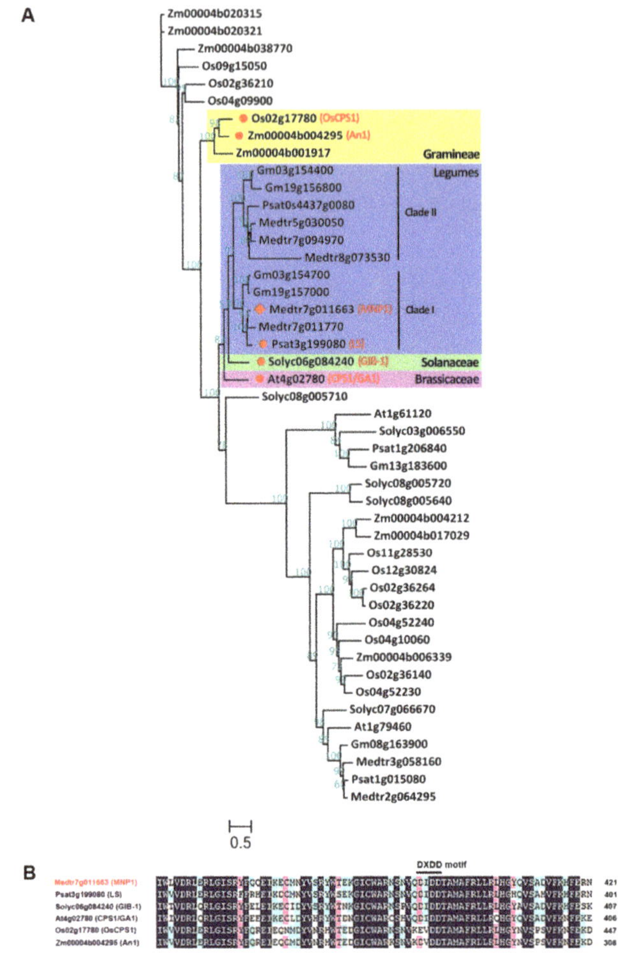

Figure 3. Phylogenetic analysis and sequences alignment of MNP1 and its closely related homologs. (**A**) Phylogenetic analysis of MNP1 and its homologs. Proteins from the species *Medicago truncatula* (Medtr), *Pisum sativum* (Psat), *Glycine max* (Gm), *Solanum lycopersicum* (Solyc), *Arabidopsis thaliana* (At), *Oryza sativa* (Os) and *Zea mays* (Zm). Bootstrap values are indicated upon the branches. Red rhombus indicates MNP1 protein and red circles indicate the reported CPS proteins. (**B**) The sequences alignment of MNP1 and the reported CPS proteins. The amino acid color indicates the homology of sequences between these species: black = 100%, pink ≥ 75% and blue ≥ 50%. The DXDD motifs in the sequences are indicated by the black line.

2.4. Subcellular Localization of MNP1

The CPS1/GA1 has been reported to be localized on plastids in *Arabidopsis* with a chloroplast transit peptide (cTP) at its *N*-terminus [50]. Then, we carried out cTP prediction using the ChloroP program (http://www.cbs.dtu.dk/services/ChloroP) and found that the MNP1 is also highly predicted to have a cTP at its *N*-terminus, with a score of 0.591 (strong).

Based on the ChloroP prediction results, the sequence encoding the N-terminal truncation of 1–100 amino acids of MNP1 (TPMNP1) was used to generate *p35S::TPMNP1-GFP* constructs, which was then transiently expressed in epidermal cells of tobacco (*Nicotiana benthamia*). The green fluorescence signal of the fusion protein was observed in a chloroplast (Figure 4). This result was further confirmed by the subcellular localization analysis of the GFP fusion protein with full-length MNP1 (Figure S4). Thus, these data suggest that MNP1 may play the same role as *Arabidopsis* CPS1 in chloroplasts and participate in GA biosynthesis.

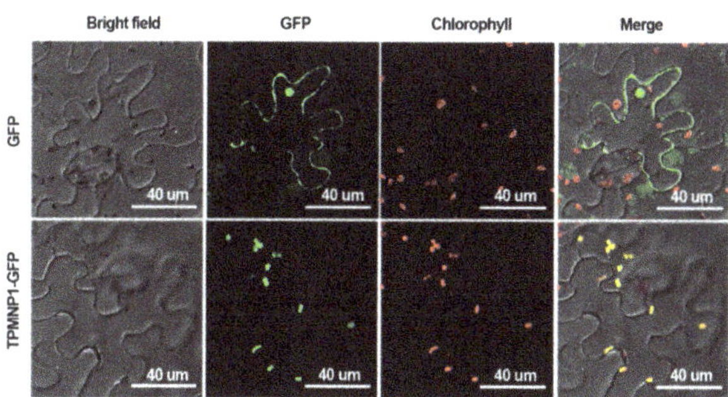

Figure 4. Subcellular localization of MNP1. According to ChloroP prediction, there is a chloroplast transit peptide (cTP) at the *N*-terminus of MNP1 protein, so a sequence encoding 100 amino acids containing cTP was used to generate *p35S::TPMNP1-GFP* constructs. Then, the constructs were transformed into tobacco (*Nicotiana benthamia*) leaf epidermal cells by *Agrobacterium*-mediated transformation. *p35S::GFP* was used as a positive control. Images were taken 36 h after transformation with dual GFP (green) and chlorophyll (red) channels. Scale bar = 40 um.

2.5. Genes of GA Biosynthesis Pathway Are Significantly Up-Regulated in mnp1-1

Based on the above evidence, we conclude that *MNP1* is the putative gene encoding a CPS protein that participates in GA biosynthesis in *M. truncatula*. Therefore, exogenous GA_3 was used to investigate whether *mnp1* is a GA-sensitive mutant. As expected, the plant height of *mnp1-1* sprayed with GA_3 was significantly higher than that of the control group without GA treatment (Figure 5A). Besides, the blade size and petiole length of *mnp1-1* mutants were also significantly restored after GA treatment (Figure S5). Thus, it could be stated that the lack of GA leads to the dwarf phenotype of *mnp1*. The GA biosynthesis pathway involves many genes besides *CPS* (Figure 5B). According to the reference [54], we tested the expression level of the putative genes (Table S2) in the GA biosynthesis pathway of *M. truncatula* stem tissue. The results showed that most GA biosynthesis genes were highly upregulated in the *mnp1-1* mutant, while a small proportion of the genes (*MtKS*, *MtKAO1*, *MtCYP714_A1*, *MtCYP714_C2*) showed a low level of upregulation, with no significant difference. Among these upregulated genes, *MtGA20ox7* was most significant and was thousands of times higher than that of the wild type (Figure 5C), which is coincident with the earlier statement that GA20ox has an important role in GA homeostasis regulation in plants [20]. The upregulation of the genes that lie downstream of the GA biosynthesis pathway in *mnp1-1* implied a negative feedback response to the low GA content, and *MtGA20ox7* may play a key role in GA feedback regulation in *mnp1-1*.

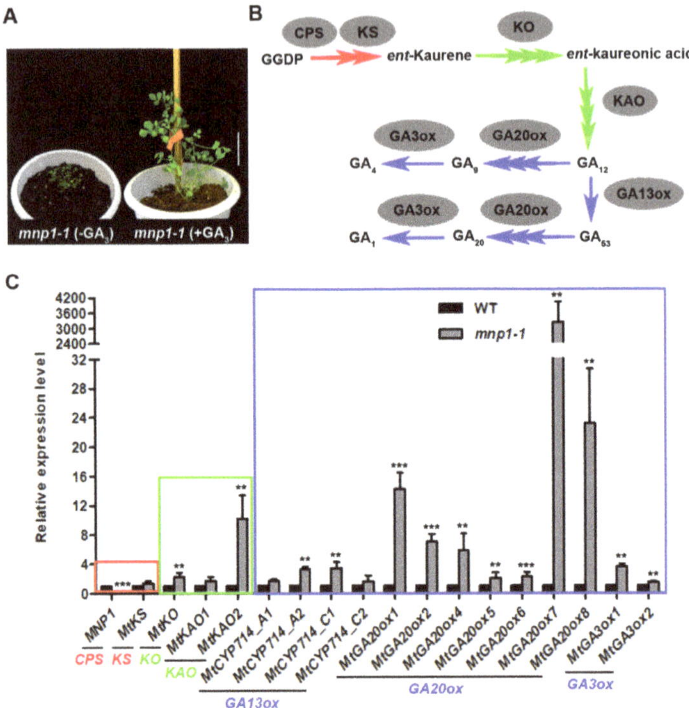

Figure 5. Expression analysis of GA biosynthesis genes in GA-sensitive mutant *mnp1-1*. (**A**) From left to right are *mnp1-1* without GA_3 treatment and *mnp1-1* with 70 uM GA_3 treatment. Scale bar = 4 cm. (**B**) GA biosynthesis pathway schematic diagram. The red, green and blue arrows represent the three stages of GA biosynthesis pathway. Gray ovals represent enzymes. (**C**) Relative expression levels of GA biosynthesis genes in the stem of WT and *mnp1-1*. The red, green and blue boxes represent the three stages of GA biosynthesis pathway as in (**B**). The significant difference was determined by unpaired two-sample *t*-test (** $p < 0.01$, *** $p < 0.001$).

2.6. MNP1 Could Partially Rescue the Phenotype of Arabidopsis cps1 Mutant

The *cps1/ga1* mutant of *Arabidopsis* shows a severely dwarfed and sterile phenotype due to the loss of CPS function [55]. To examine the extent of the functional conservation between *M. truncatula* and *A. thaliana* CPS proteins, we obtained a homozygous *T-DNA* insertion mutant of *At4g02780* (SALK_109115) from the *Arabidopsis* Biological Resource Center (ABRC), namely the *cps1* mutant. The *cps1* mutant showed extremely dwarf as expected, and was able to produce inflorescences, but no fertile seeds (Figure 6A). Next, we introduced *p35S::MNP1-GFP* constructs into *cps1* heterozygotes by the floral dip method. Through resistance screening and PCR genotyping, we isolated the *p35S::MNP1-GFP* transgenic plants in the *cps1* homozygous background, and found that the size of transgenic plants was partially restored (Figure 6B,C). RT-PCR analysis confirmed the expression of *MNP1* gene in the transgenic plants (Figure 6D). These results indicated that *MNP1* could partially recover the mini-plant phenotype of the *cps1* mutant, suggesting that CPS has functional conservation between *M. truncatula* and *A. thaliana*.

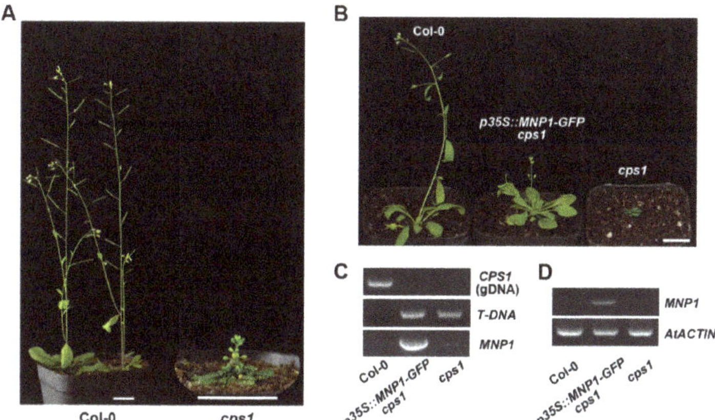

Figure 6. *MNP1* partially rescued the mini-plant phenotype of *Arabidopsis cps1* mutant. (**A**) The phenotype of *Arabidopsis* wild type (Col-0) and *cps1* mutant. The Col-0 and *cps1* mutant were 7 and 12 weeks old, respectively. Scale bar = 2 cm. (**B**) The *p35S::MNP1-GFP* transgenic plant of the *cps1* homozygous background partially restored the mini-plant phenotype of *cps1* mutant. The plants were 6 weeks old. Scale bar = 2 cm. (**C**) Genotyping of the transgenic plant. The homozygous *T-DNA* insertion in *CPS1/At4g02780* locus and *MNP1* coding sequence were detected in the transgenic plant. (**D**) RT-PCR amplification of *MNP1* from Col-0, the transgenic plant and *cps1* mutant. *AtACTIN* was used as an internal control.

3. Discussion

Although the genes encoding CPS have been identified in many species [48–52], the pea LS is the only CPS protein characterized from legumes in general before the present study, and *mnp1* appears to be the first dwarf mutant related to GA biosynthesis in *M. truncatula*. We found that the *mnp1* dwarf phenotype is caused by the decrease of the cell elongation and cell division in the stem. This result is consistent with the previously reported function of GA in promoting cell elongation and division. Cell elongation is regulated by cell wall-loosening protein expansin (EXP) and xyloglucan endo-transglycosylases (XET) which play a role in cell wall reconstruction. Some *XET* and *EXP* genes have been shown to be specifically upregulated by GA, which is believed to cause cell elongation in *Arabidopsis* and rice [56–59]. GA also promotes plant growth via upregulating the transcription levels of cell division-related genes including cell cycle genes *CYCA1;1* and *CDC2Os-3* in deepwater rice [60]. However, the underlying mechanism by which GA regulates the expression of these genes remains to be studied. The identification of *mnp1* provides a very good model to further study this mechanism in *M. truncatula*.

Focusing on the phylogenetic analysis of MNP1 and its homologous proteins, we found that CPS proteins belonging to legumes were grouped into two clades (clade I and II), and each clade was identified in all selected legumes, suggesting that a lineage-specific duplication of CPS genes may have occurred in legumes during the evolution process. There are just single copies from *Arabidopsis* and the tomato outgroup of the legume CPS proteins, while the CPS proteins of grasses gather together and are significantly separated from those of eudicots (Figure 3A). Consistent with this result, the conserved DXDD motif of CPS shows some degree of sequence divergence between monocots and eudicots (Figure 3B). In *M. truncatula*, MNP2/Medtr7g011770 appears to be a very close paralogue of MNP1/Medtr7g011663, because MNP1 and MNP2 are tightly clustered on chromosome 7 and shared high sequence identity. MNP1 and MNP2 belong to the legume CPS clade I, while in the legume CPS clade II, two members were found in the *M. truncatula* genome, namely MNP3/Medtr7g094970 and MNP4/Medtr5g030050. Since a highly conserved DXDD motif existed in all these four CPS proteins of

M. truncatula (Figure S6), it will be interesting to explore the possibilities of functional redundancy and diversification between MNP1 and the rest members.

GA homeostasis is important for the regulation of many developmental processes and has been found to be maintained by feedback regulation of GA metabolism genes in a variety of plant species [61]. *GA20ox* and *GA3ox* are the main participants in the negative feedback regulation of GA. The expression of these two kinds of genes was upregulated in the GA biosynthesis deletion mutant [20]. In our study, we found that the expression of *MtGA20ox7* was significantly upregulated, up to thousands of times in *mnp1-1* compared with the wild type. Therefore, it seems that MtGA20ox7 may be a key member in the regulation of GA homeostasis in *M. truncatula*.

The *Arabidopsis cps1* mutants are male sterility caused by defective pollen development [55]. In *M. truncatula*, the expression of *MNP1* gene was also detected in stamens (Figure S7), suggesting that *MNP1* may play a potentially important role in stamen development. However, *mnp1* is fully fertile in *M. truncatula*, with its flower organ, pods and seeds being relatively smaller when compared with the wild type (Figure S8A–E). Pollen viability, tested by Alexander's staining, indicated no significant difference between the wild type and *mnp1-1* as well (Figure S8C). It is a common phenomenon that GA deficiency leads to dwarfing and male sterility in various species, such as *Arabidopsis*, maize and tomato, but this scenario does not appear to be the case in legumes. Fertile pollens can be produced in all the reported dwarf mutants with GA deficiency in peas [43,51], and *mnp1* is similar to pea *ls* mutants, unlike *cps* mutants of other species. Given that a legume species usually contain multiple *CPS* genes, it can be argued that the mechanism of GA biosynthesis for plant height and pollen development in legumes may be conserved and distinct from that of other species. In terms of the dwarfed but fertile phenotypes of *mnp1* and *ls* mutants, the identification of *MNP1/LS* and other key genes involved in GA metabolism would be of great potential utility in legume breeding.

4. Materials and Methods

4.1. Plant Materials and Growth Conditions

M. truncatula ecotype R108 and *A. thaliana* ecotype Col-0 were used for this study. The *mnp1-1* (NF0500), *mnp1-2* (NF13564) and *mnp1-3* (NF10616) mutants (all in ecotype R108 background) were isolated from the *Tnt1* retrotransposon-tagged mutant collection of *M. truncatula* as previously reported [45]. Among them, the *mnp1-3* mutants were screened from the *Tnt1* population by a reverse genetic approach. Seeds of *Arabidopsis cps1* mutant (SALK_109115) were purchased from the *Arabidopsis* Biological Resource Center (ABRC). The *p35S::MNP1-GFP* transgenic plants were generated in *cps1* background.

Arabidopsis GA-deficient mutant *cps1* cannot germinate in the soil. For this reason, *Arabidopsis* plants need to be grown on solid 1/2 MS medium for approximately 2 weeks and then be transplanted into the soil. All plants (*M. truncatula* and *A. thaliana*) were grown under the following greenhouse conditions: 16 h day/8 h night cycle, 150 uE/m^2/s light intensity, 22 °C day/18 °C night temperature and 70% humidity.

4.2. Statistical Analysis of Cell Length and Number

For the measurement of the internode length, twenty individual plants of both the wild type and *mnp1-1* genotypes were grown simultaneously in the same greenhouse, and the third internode beneath the shoot apex of each plant (2-month-old) was collected and considered as an independent biological sample. Thus, a total of twenty internodes were processed to calculate the average length. Then, three internodes of each genotype were randomly selected from the above twenty samples, and were submerged in fixative solution (5% formaldehyde, 5% acetic acid and 50% ethanol) for over 12 h at room temperature. Subsequently, the samples were dehydrated in a graded ethanol series (50%, 70%, 90%, 95%, 100%), critical-point dried in liquid CO_2 and sputter-coated with gold. The three dried internodes for each genotype were individually examined using scanning electron microscopy

(SEM) by an EVO LS10 (Zeiss, Oberkochen, Germany) at an accelerating voltage of 5 kV. Therefore, three SEM images of the third internode were obtained for each genotype (wild type and *mnp1-1*).

For the measurement of epidermal cell length of the internode, 20 cells were randomly selected from the SEM images (6–7 cells per image) for both the wild type and *mnp1-1* genotypes, and the lengths were measured by ImageJ.

The cell number was calculated from the ratio of the average internode length (that was evaluated from a total of 20 internodes) to the average cell length (that was evaluated from 20 cells of three biological replicates).

All above experiments were repeated twice independently with similar results.

4.3. Molecular Cloning of MNP1

The molecular cloning of the *MNP1* gene referred to the method reported previously [62]. We screened the *Tnt1* retrotransposon insertion mutant collection of *M. truncatula* (ecotype R108) and isolated two *mnp1* alleles (*mnp1-1*, NF0500 and *mnp1-2*, NF13564) with severely dwarfed phenotypes. Then, these two *mnp1* alleles were backcrossed with the wild type to purify the genetic background for reducing incoherent *Tnt1* insertions, and *mnp1-1* and *mnp1-2* F$_2$ segregation populations were generated, respectively. Equal amounts of leaf material were harvested from 12 independent mutant individuals of each population to make two mixed samples. The genomic DNA of the mixed samples was extracted using the Plant Genomic DNA Kit (Tiangen, Beijing, China). Whole-genome resequencing was carried out at 20× coverage. Then, the data of whole-genome resequencing were analyzed by a novel bioinformatics tool, Identification of Transposon Insertion Sites (ITIS) to identify all *Tnt1* insertion sites in the genome [47]. The common *Tnt1* insertion sites in *mnp1-1* and *mnp1-2* genomes were found in *Medtr7g011663* locus (annotated in A17 genome v4.0). Subsequently, PCR experiments using *mnp1-1* and *mnp1-2* genomic DNA as templates were performed to verify the insertion of *Tnt1* in *Medtr7g011663*. An additional allele *mnp1-3* (NF10616) was screened from the *Tnt1* population by a reverse genetics approach, which also displayed a mini-plant phenotype. Genomic PCR analysis confirmed that the *mnp1-3* mutant does carry a *Tnt1* insertion in *Medtr7g011663*. Thus, *Medtr7g011663* was regarded as the putative *MNP1* gene. The analysis data of ITIS and the primers used for PCR are shown in Tables S1 and S3, respectively.

4.4. Phylogenetic Analysis and Sequences Alignment

The sequences of MNP1 homologs were identified through BLAST from Phytozome (https://phytozome.jgi.doe.gov/pz/portal.html), URGI (https://urgi.versailles.inra.fr/Species/Pisum) and maizeGDB (https://maizegdb.org/) in the protein databases of *Medicago truncatula*, *Glycine max*, *Solanum lycopersicum*, *Arabidopsis thaliana*, *Oryza sativa*, *Pisum sativum* and *Zea mays*. Multiple amino acid sequences were aligned by ClustalX2 (v2.1) at default parameters and beautified by DNAMAN V6. The phylogenetic tree was performed by the maximum likelihood method with IQTREE v1.6.10 as previously reported [63]. The JTT + F + G4 model was selected as suggested by the IQTREE model test tool (BIC criterion) with 1000 ultrafast bootstrap replicates and 5000 iterations.

4.5. Exogenous GA$_3$ Application Method

Bioactive GA$_3$ (Genview, Lot: 5209010140) was dissolved in ethanol (0.1 M) and diluted with water before being applied [64]. About 600 mL of 70 uM bioactive GA$_3$ working solution was sprayed to a total of twelve *mnp1-1* mutant plant one time. The first spray was applied at 10-day-old seedlings after sowing, and the later sprays performed once a week for two months in total. An equivalent group (n = 12) of *mnp1-1* mutant plants was treated similarly with a solution without GA$_3$ at each same time. All *mnp1-1* mutants with the treatments (+GA$_3$ and −GA$_3$) were grown simultaneously in the same greenhouse. Experiments were repeated twice independently with similar results.

4.6. RNA Extraction, RT-PCR and Quantitative RT-PCR (qRT-PCR)

Medicago stem tissues and *Arabidopsis* rosette leaves for RNA extraction were harvested from 7-week-old and 6-week-old plants, respectively. Total RNA was isolated using TransZol (TransGen, Beijing, China) according to the manufacturer's protocol and then was reverse transcribed into cDNA by HiScript® II 1st Strand cDNA Synthesis Kit (Vazyme, Nanjing, China). The resulting cDNAs were used as templates for RT-PCR and qRT-PCR. *AtACTIN* (*Actin2/At3g18780*) was used as an internal control for *Arabidopsis* RT-PCR. qRT-PCR was performed using 2 × T5 Fast qPCR mix (SYBR Green I) (TsingKe, Beijing, China) on the Roche Light Cycler 480II real-time PCR machine (95 °C, 1 min; 95 °C, 10 s, 60 °C, 10 s, 72 °C, 15 s, 40 cycles). *MtACTIN* (*Medtr3g095530*) was used as an internal control for *Medicago* qPCR. Three independent biological replicates were used for RNA extraction and subsequent cDNA synthesis. All samples were selected randomly under the same greenhouse conditions. Three technical replicates for each biological replicate were used in qRT-PCR analysis. The genes involved in this study and the primers used for qPCR are listed in Tables S2 and S3, respectively.

4.7. Plasmid Construction

Coding sequences of target genes were isolated by RT-PCR from wild type root tissue of seedlings (3 weeks old). For subcellular localization experiments, the coding sequences of the *N*-terminus of MNP1 (100-amino acid, TPMNP1) and the full-length coding sequence of MNP1 were inserted into the *pCAMBIA3301MP* vector between NcoI and AvrII site via the ClonExpress II One Step Cloning Kit (Vazyme) to generate *p35S::TPMNP1-GFP* and *p35S::MNP1-GFP* constructs, respectively. The *p35S:: MNP1-GFP* constructs were also used for plant transformation. The primers used for plasmid construction are listed in Table S3.

4.8. Subcellular Localization

The constructs, *p35S::TPMNP1-GFP* and *p35S::GFP*, were introduced into *Agrobacterium tumefaciens* EHA105 strain, and then they were transiently expressed in tobacco (*Nicotiana benthamia*) leaves by *Agrobacterium*-mediated transformation [65]. The *p35S::GFP* constructs were served as a positive control. The TPMNP1-GFP fusion protein was examined using a confocal laser scanning microscope (FV1000; Olympus, Japan). This experiment was repeated three times independently with similar results.

4.9. Plant Transformation

The *p35S::MNP1-GFP* constructs were introduced into *Agrobacterium tumefaciens* EHA105 strain, which was subsequently used to transform *cps1* heterozygotes by *Agrobacterium*-mediated transformation using the floral dip method [66]. Through resistance screening with 20 mg/L Basta (BBI Life Sciences, Lot: C707BA0017) and the subsequent PCR genotyping, the *p35S::MNP1-GFP* transgenic lines in the *cps1* homozygous background were isolated. The primers used for PCR are shown in Table S3.

4.10. Alexander's Staining

Mature pollens were stained with Alexander's staining solution as previously described [67]. Mature anthers of WT and *mnp1-1* from the same developmental stage were immersed directly in a drop of staining solution, covered with a coverslip respectively, and then kept them in an oven at 50 °C for 1 h. Next, a microscopic examination was conducted via a fluorescence microscope (Olympus BX63) using the bright field channel. The fertile pollen would be stained red to deep red, while aborted pollen would be green. Experiments were repeated more than three times independently.

Supplementary Materials: Supplementary materials can be found at http://www.mdpi.com/1422-0067/21/14/4968/s1.

Author Contributions: S.G., X.Z. and J.C. designed the research. Q.B. and S.Z. supported constructive comments on the research. S.G. and X.Z. performed most of the experiments. W.Z. and Y.F. assisted the experiments. S.G., X.Z., L.H. and B.Z. analyzed the data. S.G. wrote the original manuscript. J.C. and L.H. revised the manuscript. All authors have read and agreed to the published version of the manuscript.

Funding: This research was funded by the National Natural Science Foundation of China, grant number 31700244, U1702234 and 31471171. L.H. and J.C. were also supported by the STS of CAS (KFJ-STS-ZDTP-076).

Acknowledgments: We are grateful to Zhijia Gu and Yanxia Jia (Kunming Institute of Botany, Chinese Academy of Sciences) for their technical help in SEM and laser confocal experiments.

Conflicts of Interest: The authors declare no conflict of interest.

References

1. Spielmeyer, W.; Ellis, M.H.; Chandler, P.M. Semidwarf (sd-1), "green revolution" rice, contains a defective gibberellin 20-oxidase gene. *Proc. Natl. Acad. Sci. USA* **2002**, *99*, 9043–9048. [CrossRef] [PubMed]
2. Peng, J.; Richards, D.E.; Hartley, N.M.; Murphy, G.P.; Devos, K.M.; Flintham, J.E.; Beales, J.; Fish, L.J.; Worland, A.J.; Pelica, F.; et al. 'Green revolution' genes encode mutant gibberellin response modulators. *Nature* **1999**, *400*, 256–261. [CrossRef] [PubMed]
3. Nelissen, H.; Rymen, B.; Jikumaru, Y.; Demuynck, K.; Van Lijsebettens, M.; Kamiya, Y.; Inze, D.; Beemster, G.T. A local maximum in gibberellin levels regulates maize leaf growth by spatial control of cell division. *Curr. Biol.* **2012**, *22*, 1183–1187. [CrossRef]
4. Cao, D.; Cheng, H.; Wu, W.; Soo, H.M.; Peng, J. Gibberellin mobilizes distinct DELLA-dependent transcriptomes to regulate seed germination and floral development in Arabidopsis. *Plant Physiol.* **2006**, *142*, 509–525. [CrossRef] [PubMed]
5. King, R.W.; Moritz, T.; Evans, L.T.; Junttila, O.; Herlt, A.J. Long-day induction of flowering in Lolium temulentum involves sequential increases in specific gibberellins at the shoot apex. *Plant Physiol.* **2001**, *127*, 624–632. [CrossRef] [PubMed]
6. Lo, S.F.; Yang, S.Y.; Chen, K.T.; Hsing, Y.I.; Zeevaart, J.A.; Chen, L.J.; Yu, S.M. A novel class of gibberellin 2-oxidases control semidwarfism, tillering, and root development in rice. *Plant Cell* **2008**, *20*, 2603–2618. [CrossRef]
7. Davidson, S.E.; Elliott, R.C.; Helliwell, C.A.; Poole, A.T.; Reid, J.B. The pea gene NA encodes ent-kaurenoic acid oxidase. *Plant Physiol.* **2003**, *131*, 335–344. [CrossRef]
8. Yamaguchi, S. Gibberellin metabolism and its regulation. *Annu. Rev. Plant Biol.* **2008**, *59*, 225–251. [CrossRef]
9. Magome, H.; Nomura, T.; Hanada, A.; Takeda-Kamiya, N.; Ohnishi, T.; Shinma, Y.; Katsumata, T.; Kawaide, H.; Kamiya, Y.; Yamaguchi, S. CYP714B1 and CYP714B2 encode gibberellin 13-oxidases that reduce gibberellin activity in rice. *Proc. Natl. Acad. Sci. USA* **2013**, *110*, 1947–1952. [CrossRef]
10. Duncan, J.D.; West, C.A. Properties of Kaurene Synthetase from Marah macrocarpus Endosperm: Evidence for the Participation of Separate but Interacting Enzymes. *Plant Physiol.* **1981**, *68*, 1128–1134. [CrossRef]
11. Helliwell, C.A.; Sullivan, J.A.; Mould, R.M.; Gray, J.C.; Peacock, W.J.; Dennis, E.S. A plastid envelope location of Arabidopsis ent-kaurene oxidase links the plastid and endoplasmic reticulum steps of the gibberellin biosynthesis pathway. *Plant J.* **2001**, *28*, 201–208. [CrossRef]
12. Helliwell, C.A.; Sheldon, C.C.; Olive, M.R.; Walker, A.R.; Zeevaart, J.A.; Peacock, W.J.; Dennis, E.S. Cloning of the Arabidopsis ent-kaurene oxidase gene GA3. *Proc. Natl. Acad. Sci. USA* **1998**, *95*, 9019–9024. [CrossRef] [PubMed]
13. Helliwell, C.A.; Chandler, P.M.; Poole, A.; Dennis, E.S.; Peacock, W.J. The CYP88A cytochrome P450, ent-kaurenoic acid oxidase, catalyzes three steps of the gibberellin biosynthesis pathway. *Proc. Natl. Acad. Sci. USA* **2001**, *98*, 2065–2070. [CrossRef] [PubMed]
14. Xu, Y.L.; Li, L.; Wu, K.; Peeters, A.J.; Gage, D.A.; Zeevaart, J.A. The GA5 locus of Arabidopsis thaliana encodes a multifunctional gibberellin 20-oxidase: Molecular cloning and functional expression. *Proc. Natl. Acad. Sci. USA* **1995**, *92*, 6640–6644. [CrossRef] [PubMed]
15. Lange, T.; Hedden, P.; Graebe, J.E. Expression cloning of a gibberellin 20-oxidase, a multifunctional enzyme involved in gibberellin biosynthesis. *Proc. Natl. Acad. Sci. USA* **1994**, *91*, 8552–8556. [CrossRef] [PubMed]

16. Chiang, H.H.; Hwang, I.; Goodman, H.M. Isolation of the Arabidopsis GA4 locus. *Plant Cell* **1995**, *7*, 195–201. [CrossRef] [PubMed]

17. Israelsson, M.; Mellerowicz, E.; Chono, M.; Gullberg, J.; Moritz, T. Cloning and overproduction of gibberellin 3-oxidase in hybrid aspen trees. Effects on gibberellin homeostasis and development. *Plant Physiol.* **2004**, *135*, 221–230. [CrossRef]

18. Regnault, T.; Daviere, J.M.; Heintz, D.; Lange, T.; Achard, P. The gibberellin biosynthetic genes AtKAO1 and AtKAO2 have overlapping roles throughout Arabidopsis development. *Plant J.* **2014**, *80*, 462–474. [CrossRef]

19. Rieu, I.; Ruiz-Rivero, O.; Fernandez-Garcia, N.; Griffiths, J.; Powers, S.J.; Gong, F.; Linhartova, T.; Eriksson, S.; Nilsson, O.; Thomas, S.G.; et al. The gibberellin biosynthetic genes AtGA20ox1 and AtGA20ox2 act, partially redundantly, to promote growth and development throughout the Arabidopsis life cycle. *Plant J.* **2008**, *53*, 488–504. [CrossRef]

20. Hedden, P.; Phillips, A.L. Gibberellin metabolism: New insights revealed by the genes. *Trends Plant Sci.* **2000**, *5*, 523–530. [CrossRef]

21. Griffiths, J.; Murase, K.; Rieu, I.; Zentella, R.; Zhang, Z.L.; Powers, S.J.; Gong, F.; Phillips, A.L.; Hedden, P.; Sun, T.P.; et al. Genetic characterization and functional analysis of the GID1 gibberellin receptors in Arabidopsis. *Plant Cell* **2006**, *18*, 3399–3414. [CrossRef] [PubMed]

22. Willige, B.C.; Ghosh, S.; Nill, C.; Zourelidou, M.; Dohmann, E.M.; Maier, A.; Schwechheimer, C. The DELLA domain of GA INSENSITIVE mediates the interaction with the GA INSENSITIVE DWARF1A gibberellin receptor of Arabidopsis. *Plant Cell* **2007**, *19*, 1209–1220. [CrossRef] [PubMed]

23. Varbanova, M.; Yamaguchi, S.; Yang, Y.; McKelvey, K.; Hanada, A.; Borochov, R.; Yu, F.; Jikumaru, Y.; Ross, J.; Cortes, D.; et al. Methylation of gibberellins by Arabidopsis GAMT1 and GAMT2. *Plant Cell* **2007**, *19*, 32–45. [CrossRef] [PubMed]

24. Zhu, Y.; Nomura, T.; Xu, Y.; Zhang, Y.; Peng, Y.; Mao, B.; Hanada, A.; Zhou, H.; Wang, R.; Li, P.; et al. ELONGATED UPPERMOST INTERNODE encodes a cytochrome P450 monooxygenase that epoxidizes gibberellins in a novel deactivation reaction in rice. *Plant Cell* **2006**, *18*, 442–456. [CrossRef] [PubMed]

25. Wuddineh, W.A.; Mazarei, M.; Zhang, J.; Poovaiah, C.R.; Mann, D.G.; Ziebell, A.; Sykes, R.W.; Davis, M.F.; Udvardi, M.K.; Stewart, C.N., Jr. Identification and overexpression of gibberellin 2-oxidase (GA2ox) in switchgrass (Panicum virgatum L.) for improved plant architecture and reduced biomass recalcitrance. *Plant Biotechnol. J.* **2015**, *13*, 636–647. [CrossRef] [PubMed]

26. Szekeres, M.; Nemeth, K.; Koncz-Kalman, Z.; Mathur, J.; Kauschmann, A.; Altmann, T.; Redei, G.P.; Nagy, F.; Schell, J.; Koncz, C. Brassinosteroids rescue the deficiency of CYP90, a cytochrome P450, controlling cell elongation and de-etiolation in Arabidopsis. *Cell* **1996**, *85*, 171–182. [CrossRef]

27. Li, J.; Wen, J.; Lease, K.A.; Doke, J.T.; Tax, F.E.; Walker, J.C. BAK1, an Arabidopsis LRR receptor-like protein kinase, interacts with BRI1 and modulates brassinosteroid signaling. *Cell* **2002**, *110*, 213–222. [CrossRef]

28. Zhou, F.; Lin, Q.; Zhu, L.; Ren, Y.; Zhou, K.; Shabek, N.; Wu, F.; Mao, H.; Dong, W.; Gan, L.; et al. D14-SCF(D3)-dependent degradation of D53 regulates strigolactone signalling. *Nature* **2013**, *504*, 406–410. [CrossRef]

29. Lin, H.; Wang, R.; Qian, Q.; Yan, M.; Meng, X.; Fu, Z.; Yan, C.; Jiang, B.; Su, Z.; Li, J.; et al. DWARF27, an iron-containing protein required for the biosynthesis of strigolactones, regulates rice tiller bud outgrowth. *Plant Cell* **2009**, *21*, 1512–1525. [CrossRef]

30. Yokota, T. The structure, biosynthesis and function of brassinosteroids. *Trend Plant Sci.* **1997**, *2*, 137–143. [CrossRef]

31. Choe, S.; Dilkes, B.P.; Gregory, B.D.; Ross, A.S.; Yuan, H.; Noguchi, T.; Fujioka, S.; Takatsuto, S.; Tanaka, A.; Yoshida, S.; et al. The Arabidopsis dwarf1 mutant is defective in the conversion of 24-methylenecholesterol to campesterol in brassinosteroid biosynthesis. *Plant Physiol.* **1999**, *119*, 897–907. [CrossRef] [PubMed]

32. Arite, T.; Iwata, H.; Ohshima, K.; Maekawa, M.; Nakajima, M.; Kojima, M.; Sakakibara, H.; Kyozuka, J. DWARF10, an RMS1/MAX4/DAD1 ortholog, controls lateral bud outgrowth in rice. *Plant J.* **2007**, *51*, 1019–1029. [CrossRef]

33. Clouse, S.D. Brassinosteroid signal transduction: From receptor kinase activation to transcriptional networks regulating plant development. *Plant Cell* **2011**, *23*, 1219–1230. [CrossRef]

34. Li, J.; Nam, K.H.; Vafeados, D.; Chory, J. BIN2, a new brassinosteroid-insensitive locus in Arabidopsis. *Plant Physiol.* **2001**, *127*, 14–22. [CrossRef] [PubMed]

35. Jiang, L.; Liu, X.; Xiong, G.; Liu, H.; Chen, F.; Wang, L.; Meng, X.; Liu, G.; Yu, H.; Yuan, Y.; et al. DWARF 53 acts as a repressor of strigolactone signalling in rice. *Nature* **2013**, *504*, 401–405. [CrossRef] [PubMed]

36. Ishikawa, S.; Maekawa, M.; Arite, T.; Onishi, K.; Takamure, I.; Kyozuka, J. Suppression of tiller bud activity in tillering dwarf mutants of rice. *Plant Cell Physiol.* **2005**, *46*, 79–86. [CrossRef] [PubMed]

37. Maphosa, Y.; Jideani, V.A. The Role of Legumes in Human Nutrition. In *Functional Food*; Chavarri, M., Ed.; IntechOpen: London, UK, 2017; pp. 103–121. [CrossRef]

38. Li, Z.F.; Guo, Y.; Ou, L.; Hong, H.; Wang, J.; Liu, Z.X.; Guo, B.; Zhang, L.; Qiu, L. Identification of the dwarf gene GmDW1 in soybean (Glycine max L.) by combining mapping-by-sequencing and linkage analysis. *TAG Theor. Appl. Genet.* **2018**, *131*, 1001–1016. [CrossRef]

39. Reid, J.B.; Ross, J.J. A Mutant-Based Approach, Using Pisum sativum, to Understanding Plant Growth. *Int. J. Plant Sci.* **1993**, *154*, 22–34. [CrossRef]

40. Yaxley, J.R.; Ross, J.J.; Sherriff, L.J.; Reid, J.B. Gibberellin biosynthesis mutations and root development in pea. *Plant Physiol.* **2001**, *125*, 627–633. [CrossRef]

41. Tang, H.; Krishnakumar, V.; Bidwell, S.; Rosen, B.; Chan, A.; Zhou, S.; Gentzbittel, L.; Childs, K.L.; Yandell, M.; Gundlach, H.; et al. An improved genome release (version Mt4.0) for the model legume Medicago truncatula. *BMC Genom.* **2014**, *15*, 312. [CrossRef]

42. Pecrix, Y.; Staton, S.E.; Sallet, E.; Lelandais-Brière, C.; Moreau, S.; Carrère, S.; Blein, T.; Jardinaud, M.F.; Latrasse, D.; Zouine, M.; et al. Whole-genome landscape of Medicago truncatula symbiotic genes. *Nat. Plants* **2018**, *4*, 1017–1025. [CrossRef] [PubMed]

43. Swain, S.M.; Ross, J.J.; Reid, J.B.; Kamiya, Y. Gibberellins and pea seed development: Expression of the lhˆi, ls and leˆ5839 mutations. *Planta* **1995**, *195*, 426–433. [CrossRef]

44. Burks, D.; Azad, R.; Wen, J.; Dickstein, R. The Medicago truncatula Genome: Genomic Data Availability. *Methods Mol. Biol.* **2018**, *1822*, 39–59. [CrossRef] [PubMed]

45. Tadege, M.; Wen, J.; He, J.; Tu, H.; Kwak, Y.; Eschstruth, A.; Cayrel, A.; Endre, G.; Zhao, P.X.; Chabaud, M.; et al. Large-scale insertional mutagenesis using the Tnt1 retrotransposon in the model legume Medicago truncatula. *Plant J.* **2008**, *54*, 335–347. [CrossRef] [PubMed]

46. Hacham, Y.; Holland, N.; Butterfield, C.; Ubeda-Tomas, S.; Bennett, M.J.; Chory, J.; Savaldi-Goldstein, S. Brassinosteroid perception in the epidermis controls root meristem size. *Development* **2011**, *138*, 839–848. [CrossRef] [PubMed]

47. Jiang, C.; Chen, C.; Huang, Z.; Liu, R.; Verdier, J. ITIS, a bioinformatics tool for accurate identification of transposon insertion sites using next-generation sequencing data. *BMC Bioinform.* **2015**, *16*, 72. [CrossRef] [PubMed]

48. Bensen, R.J.; Johal, G.S.; Crane, V.C.; Tossberg, J.T.; Schnable, P.S.; Meeley, R.B.; Briggs, S.P. Cloning and characterization of the maize An1 gene. *Plant Cell* **1995**, *7*, 75–84. [CrossRef]

49. Jacobsen, S.E.; Olszewski, N.E. Characterization of the Arrest in Anther Development Associated with Gibberellin Deficiency of the gib-1 Mutant of Tomato. *Plant Physiol.* **1991**, *97*, 409–414. [CrossRef]

50. Sun, T.P.; Kamiya, Y. The Arabidopsis GA1 locus encodes the cyclase ent-kaurene synthetase A of gibberellin biosynthesis. *Plant Cell* **1994**, *6*, 1509–1518. [CrossRef]

51. Ait-Ali, T.; Swain, S.M.; Reid, J.B.; Sun, T.; Kamiya, Y. The LS locus of pea encodes the gibberellin biosynthesis enzyme ent-kaurene synthase A. *Plant J.* **1997**, *11*, 443–454. [CrossRef]

52. Otomo, K.; Kenmoku, H.; Oikawa, H.; Konig, W.A.; Toshima, H.; Mitsuhashi, W.; Yamane, H.; Sassa, T.; Toyomasu, T. Biological functions of ent- and syn-copalyl diphosphate synthases in rice: Key enzymes for the branch point of gibberellin and phytoalexin biosynthesis. *Plant J.* **2004**, *39*, 886–893. [CrossRef] [PubMed]

53. Prisic, S.; Xu, J.; Coates, R.M.; Peters, R.J. Probing the Role of the DXDD Motif in Class II Diterpene Cyclases. *ChemBioChem* **2007**, *8*, 869–874. [CrossRef] [PubMed]

54. Igielski, R.; Kepczynska, E. Gene expression and metabolite profiling of gibberellin biosynthesis during induction of somatic embryogenesis in Medicago truncatula Gaertn. *PLoS ONE* **2017**, *12*, e0182055. [CrossRef]

55. Cheng, H.; Qin, L.; Lee, S.; Fu, X.; Richards, D.E.; Cao, D.; Luo, D.; Harberd, N.P.; Peng, J. Gibberellin regulates Arabidopsis floral development via suppression of DELLA protein function. *Development* **2004**, *131*, 1055–1064. [CrossRef] [PubMed]

56. Xu, W.; Purugganan, M.M.; Polisensky, D.H.; Antosiewicz, D.M.; Fry, S.C.; Braam, J. Arabidopsis TCH4, regulated by hormones and the environment, encodes a xyloglucan endotransglycosylase. *Plant Cell* **1995**, *7*, 1555–1567. [CrossRef] [PubMed]

57. Uozu, S.; Tanaka-Ueguchi, M.; Kitano, H.; Hattori, K.; Matsuoka, M. Characterization of XET-related genes of rice. *Plant Physiol.* **2000**, *122*, 853–859. [CrossRef]
58. Lee, Y.; Kende, H. Expression of beta-expansins is correlated with internodal elongation in deepwater rice. *Plant Physiol.* **2001**, *127*, 645–654. [CrossRef]
59. Lee, Y.; Kende, H. Expression of alpha-expansin and expansin-like genes in deepwater rice. *Plant Physiol.* **2002**, *130*, 1396–1405. [CrossRef]
60. Fabian, T.; Lorbiecke, R.; Umeda, M.; Sauter, M. The cell cycle genes cycA1;1 and cdc2Os-3 are coordinately regulated by gibberellin in planta. *Planta* **2000**, *211*, 376–383. [CrossRef]
61. Gallego-Giraldo, L.; Ubeda-Tomas, S.; Gisbert, C.; Garcia-Martinez, J.L.; Moritz, T.; Lopez-Diaz, I. Gibberellin homeostasis in tobacco is regulated by gibberellin metabolism genes with different gibberellin sensitivity. *Plant Cell Physiol.* **2008**, *49*, 679–690. [CrossRef]
62. Zhao, B.; He, L.; Jiang, C.; Liu, Y.; He, H.; Bai, Q.; Zhou, S.; Zheng, X.; Wen, J.; Mysore, K.S.; et al. Lateral Leaflet Suppression 1 (LLS1), encoding the MtYUCCA1 protein, regulates lateral leaflet development in Medicago truncatula. *New Phytol.* **2020**, *227*, 613–628. [CrossRef]
63. He, L.; Liu, Y.; He, H.; Liu, Y.; Qi, J.; Zhang, X.; Li, Y.; Mao, Y.; Zhou, S.; Zheng, X.; et al. A molecular framework underlying the compound leaf pattern of Medicago truncatula. *Nat. Plants* **2020**, *6*, 511–521. [CrossRef] [PubMed]
64. Dalmadi, A.; Kalo, P.; Jakab, J.; Saskoi, A.; Petrovics, T.; Deak, G.; Kiss, G.B. Dwarf plants of diploid Medicago sativa carry a mutation in the gibberellin 3-beta-hydroxylase gene. *Plant Cell Rep.* **2008**, *27*, 1271–1279. [CrossRef] [PubMed]
65. Kapila, J.; Rycke, R.D.; Montagu, M.V.; Angenon, G. An Agrobacterium-mediated transient gene expression system for intact leaves. *Plant Sci.* **1997**, *122*, 101–108. [CrossRef]
66. Clough, S.J.; Bent, A.F. Floral dip: A simplified method for Agrobacterium-mediated transformation of Arabidopsis thaliana. *Plant J.* **1998**, *16*, 735–743. [CrossRef]
67. Alexander, M.P. Differential staining of aborted and nonaborted pollen. *Stain Technol.* **1969**, *44*, 117–122. [CrossRef]

International Journal of
Molecular Sciences

Review

Cyclic AMP: A Polyhedral Signalling Molecule in Plants

Emanuela Blanco [1],*, Stefania Fortunato [2], Luigi Viggiano [2] and Maria Concetta de Pinto [2],*

[1] Institute of Biosciences and Bioresources, National Research Council, Via G. Amendola 165/A, 70126 Bari, Italy

[2] Department of Biology, University of Bari Aldo Moro, Via E. Orabona 4, 70125 Bari, Italy; stefania.fortunato@uniba.it (S.F.); luigi.viggiano@uniba.it (L.V.)

* Correspondence: emanuela.blanco@ibbr.cnr.it (E.B.); mariaconcetta.depinto@uniba.it (M.C.d.P.); Tel.: +39-(080)-5583400 (E.B.); +39-(080)-5442156 (M.C.d.P.)

Received: 14 June 2020; Accepted: 7 July 2020; Published: 9 July 2020

Abstract: The cyclic nucleotide cAMP (3′,5′-cyclic adenosine monophosphate) is nowadays recognised as an important signalling molecule in plants, involved in many molecular processes, including sensing and response to biotic and abiotic environmental stresses. The validation of a functional cAMP-dependent signalling system in higher plants has spurred a great scientific interest on the polyhedral role of cAMP, as it actively participates in plant adaptation to external stimuli, in addition to the regulation of physiological processes. The complex architecture of cAMP-dependent pathways is far from being fully understood, because the actors of these pathways and their downstream target proteins remain largely unidentified. Recently, a genetic strategy was effectively used to lower cAMP cytosolic levels and hence shed light on the consequences of cAMP deficiency in plant cells. This review aims to provide an integrated overview of the current state of knowledge on cAMP's role in plant growth and response to environmental stress. Current knowledge of the molecular components and the mechanisms of cAMP signalling events is summarised.

Keywords: abiotic stress; cAMP; cyclic nucleotides-gated channels; plant innate immunity

---

1. Introduction

The role of 3′,5′-cyclic adenosine monophosphate (cAMP) as second messenger in a wide variety of physiologic responses has long been unravelled in animals, bacteria, fungi and algae. By contrast, comprehensive knowledge of cAMP signal transduction in higher plants is still lacking. However, over the last twenty years, several pieces of evidence about cAMP biological functions in plants have been reported. Recent advances in plant biology research, supported by biochemical, genetic and omic studies, have led to the characterisation of cAMP as a polyhedral molecule, critically involved in the signalling pathways of both plant development and environmental stress response.

The recognition of cAMP existence in mammals was the first step towards the identification of its role in living organisms [1]. Thereafter, the molecular structure and conformation of cAMP, which are key factors defining cAMP chemical properties, the specificity of target recognition sites and hence its biological activity, have been determined [2,3]. cAMP responses are extremely complex: different stimuli able to change cAMP levels might lead to different physiological outcomes [4]. The high level of cell compartmentalisation of cAMP signalling pathways is the physiological basis of such numerous and diversified responses to cAMP, as signal response elements are differentially localised and temporally regulated [5]. In animals, several interconnected signalling pathways encompass cAMP and cyclic nucleotides activity in the regulation of cellular events, such as cell proliferation, differentiation, death and migration, as well as complex functions, e.g., memory [6,7].

Even though earlier comparative studies put forward a similar role for cAMP in plants, compared to mammalian organisms, cAMP presence and activity in plants has been a matter of controversy for decades. In fact, in plant cells, cAMP is present in nanomolar concentrations, which are one order of magnitude lower than in mammalian cells. In early studies, cAMP was hardly detectable because cellular levels were below the detection limits of available analytical methods [8,9]. The conclusive proof of cAMP existence and activity in plant extracts could be achieved later, through the advances in high performance liquid chromatography and electrospray mass spectrometry, with a lower detection limit of 25 femtomoles for cyclic nucleotide quantification [10,11].

More recently, studies in plant cells focused on the biosynthetic molecular components of cAMP production and breakdown, which are able to switch on and off the signal encoded by cAMP. The lifecycle of the cAMP molecule includes a source, several regulatory factors with specific cAMP binding domains to transduce the signal and breakdown enzymes to avoid the accumulation of cAMP and terminate the signal [4]. The identification of plant biosynthetic enzymes was not straightforward and took a lot of efforts because of the low homology with previously characterised animal systems were not straightforward. In animals, cAMP is produced in the cytoplasm from adenosine triphosphate by plasma-membrane associated or soluble adenylate cyclases (ACs). Once generated inside the cell, cAMP transduces signals acting through a few cellular effectors, which are responsible for the divergence of cAMP signalling. Changes in intracellular cAMP levels affect cAMP-dependent protein kinases activity (PKA) [12]. Furthermore, cAMP binds to cyclic nucleotide binding proteins, as cyclic nucleotides-gated channels (CNGCs) and hyperpolarisation-activated cyclic nucleotides-modulated channels [13], or to specific transporters and transcription factors in the nucleus. The cAMP levels are regulated, in terms of both lifetime and cell sub-localisation, by cytoplasmic phosphodiesterases (PDEs), which hydrolyse it into AMP, switching off the signal.

After the discovery of the first plant AC in *Agapanthus umbellatus* [14], it took considerable efforts before other plant ACs were identified and characterised [15–18]. Similar to plant guanylate cyclases, also indicated as "moonlighting" proteins [19], which are multifunctional enzymes and hold diverse domain structures, plant ACs also harbour multiple catalytically active AC centres, which co-function with other functional domains [18,20–22].

Although physiological and biochemical studies provided evidence for enzyme activation by cyclic nucleotides [23], the lack of genetic information on their molecular identity has hitherto prevented the characterisation of PDEs and PKAs orthologs in plants. However, both PDEs and PKAs are postulated to form complexes with other enzymes [9,23]. There is only one molecularly confirmed PDE in liverwort *Marchantia polymorpha*, which exhibits both AC and PDE activities, but no homologues were found in other plant species [24]. Moreover, many studies point to light PKA activity in many plant species, but these observations still await molecular confirmation [23].

Cyclic nucleotides have a direct effect on cation fluxes (K^+, Na^+ and Ca^{2+}), and CNGCs are key components of cAMP signal transduction pathways [25]. These ion channels take part to plant reproductive processes, leaf senescence and plant responses to abiotic and biotic stresses [26–30]. They are sensitive to intracellular alterations of the cAMP level and can turn cAMP variations into changes in membrane potential and ion concentrations. CNGCs have different cellular localisation, thus defining the spatial regulation of intracellular cAMP levels.

These findings on plant cAMP biosynthesis and regulation shed light on cAMP role and cAMP-dependent signal transduction mechanisms in plants. However, the understanding of cellular function requires an integrated analysis of context-specific, spatiotemporal data from diverse sources. In this context, the availability of more reliable methods to monitor and/or alter intracellular cAMP levels, without interfering with cell physiological processes, is of utmost importance. Indeed, the roles of cAMP in plants have been mainly established by studies that utilise pharmacological approaches. A recently developed non-invasive method to alter cellular cAMP levels overcame the concern about the effects exerted by the high non-physiological concentrations of exogenously applied cAMPs analogues in both animals and plant systems [31–33].

Ion homeostasis [34–36], cell division [37,38], pollen tube growth and reorientation [14] and stomatal opening [39,40] are all plant processes involving cAMP level alterations. Proteomic analyses on Arabidopsis plants highlighted the involvement of cAMP in the regulation of photosynthesis and photorespiration, as well as in the energy-transducing pathways and ATP generation [41–43]. These studies, while unravelling cAMP role in plant cell development and growth, also pointed out the unavoidable influence of environment in plant life, emphasising cAMP involvement in perception of abiotic and biotic stimuli and in boosting plant stress responses.

In this review, we offer a comprehensive portrayal of molecular mechanisms behind cAMP-dependent signalling events in plant growth and in plant response to abiotic and biotic stress, taking advantage of advanced analytical tools and the newly developed methods successfully applied in plants.

2. The cAMP-Sponge, a New Genetic Tool to Unravel cAMP Functions in Plants

Since the assessment of cAMP presence in plants, pharmacological approaches were used to elicit cAMP level alteration inside the cell and to observe associated metabolism changes. Various cell-permeable cAMP analogues and known mammalian activators or inhibitors of ACs or PDEs were initially used to explore cAMP role first in animals and then in plants.

At first, tissues or whole plants were incubated with cAMP for hours or even days, to discover the effects of cAMP on developmental processes, but there was no monitoring of the effective cAMP intake and/or consequent cAMP degradation during the long time of incubation. In plants, the pharmacological approach has been initially used to study cAMP involvement in different physiological processes, spanning from the synthesis of phytoalexins to the control of cell cycle progression [37,38,44]. Alongside, whole-cell patch-clamp assays were performed, where regulators could be directly introduced into the cell or added to the solution, as done for the first time in *Vicia faba*, to investigate cAMP influence on K^+ channel activities [45]. In addition to these early studies, many other works, even recently, rely on pharmacological methodologies with exogenously applied compounds to alter endogenous cAMP, emphasising cAMP involvement in different plant processes, response to environmental stimuli and in signalling events [41–43].

A few concerns still exist about the reliability of these pharmacological approaches, especially in deciphering cAMP-dependent signalling mechanisms. Indeed, these methodologies do not consider the importance of cAMP physiological concentrations, which are well below the exogenously applied cAMP at micromolar levels. Hence, secondary effects cannot be excluded [16]. Moreover, the pharmacological approach cannot fully dissect cAMP-activated signalling pathways, as it does allow taking into account endogenous cAMP fluctuations.

More recently, a sophisticated molecular approach was successfully applied in animal systems to investigate simultaneous multiple signalling pathways: the engineering of a buffering molecule able to selectively bind one specific component of the investigated system, directly inside the cell [31,46]. It is the case of the first genetically encoded buffer for cAMP, called "cAMP-sponge", based on the high-affinity cAMP binding portions of the regulatory subunits of human protein kinase A (PKA-RIβ) [31]. The PKA-RIβ C-terminus binds cAMP with high affinity, but it is unable to generate dimers or to bind the PKA catalytic inhibitory domain located at N-terminus [31,47]. The choice of the human PKA domains to buffer cAMP was well considered: the affinity had to be high enough to compete with endogenous effectors of the cAMP signal (Epac, PKA and CNGCs), but lower than the resting free levels of cAMP. Furthermore, the fragment was tagged with the fluorescent protein mCherry, which is spectrally compatible with FRET-based sensors for cAMP, allowing the simultaneous detection of both expressed buffer and cAMP, at a single cell level. The recombinant probe can also be specifically targeted to a specific subcellular compartment. Lefkimmiatis and co-workers generated both a non-targeted construct and a cytosolic cAMP-sponge construct, the latter bearing the N-terminal nuclear exclusion signal, highlighting the possibility of restricting the cAMP sponge expression to cell compartments by the addition of targeting motifs [31].

The cAMP-sponge was shown to bind specifically cAMP in vitro with sub micromolar affinity and it was insensitive to cyclic cGMP (3′,5′-cyclic guanosine monophosphate). It was validated at the single cell level, using a FRET-based imaging approach. The cAMP sponge was able to buffer agonist-induced cAMP signals and to block the downstream activation of PKA [31]. This molecular approach offered the opportunity to give a glance on cAMP functioning in living cells, providing information about endogenous cAMP changes.

The cAMP-sponge has been recently used as non-invasive tool in two plant model organisms, *Nicotiana tabacum* Bright Yellow-2 (BY-2) cells and *Arabidopsis thaliana* plants, to obtain a new portrayal of cAMP role in plants, through the in vivo depletion of cAMP in plant cells [32,33].

In both model systems, the cAMP-sponge was successfully expressed under the control of the 35S Cauliflower Mosaic Virus (CaMV) constitutive promoter. Three and two stable and independent transformed lines (cAS lines) of transgenic tobacco BY-2 cells and Arabidopsis plants, respectively, were obtained (Figure 1). The integration of the transgene in the nuclear genome and the in vivo presence of the cAMP-sponge protein were confirmed by the detection of the mCherry fluorescence. The transgenic lines showed the same total cAMP content of the wild type (WT) ones, likely because the cAMP bound to the cAMP-sponge was released during the procedure of total cAMP extraction. By contrast, the measurement of free cAMP content showed significant differences, with transgenic cAS lines displaying about half the free cAMP compared with WT lines (Figure 1).

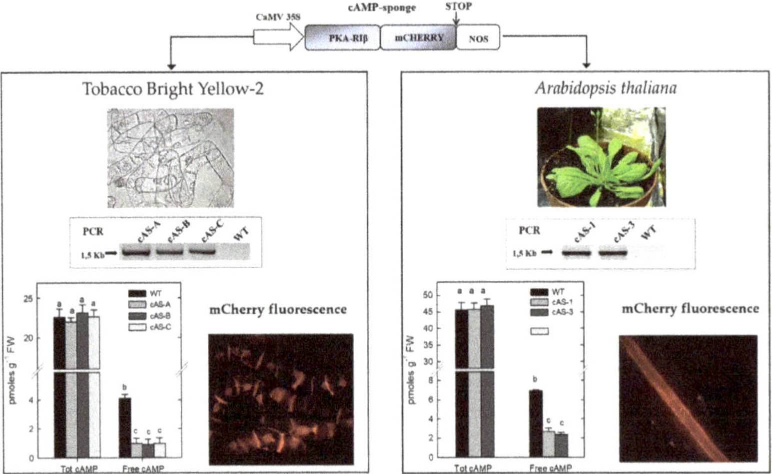

Figure 1. Schematic representation of the "cAMP sponge" overexpression in *Nicotiana tabacum* Bright yellow-2 (BY-2) cells and *Arabidopsis thaliana* plants. The cAMP-sponge construct used for tobacco BY-2 and Arabidopsis genetic transformations is reported on the top of the figure. The two following panels illustrate the characterisation of different transgenic lines (cAS lines) overexpressing the cAMP sponge in tobacco BY-2 cells (left) and in Arabidopsis plants (right). RT-PCR products show the integration of the transgene in the transformed cAS lines. Total and free cAMP content in wild type (WT) and cAS lines are reported in the histogram graphs. The presence of cAMP sponge protein is visualised by mCherry fluorescence. (Adapted from Sabetta et al. (2016) and Sabetta et al. (2019) [32,33]).

The characterisation of the transgenic cAS lines of tobacco BY-2 cells showed that cAMP dampening inhibited cell growth and this was due to mitosis inhibition, rather than a decrease in cell viability. Moreover, transgenic cells showed enhanced antioxidant levels indicating that these cells sense cAMP deficiency as a stress condition [32]. On the other hand, cAS Arabidopsis transgenic lines did not exhibit any phenotype in physiological conditions, showing the same germination time, number, colour and size of rosette leaves, as well as time and height of inflorescence as WT lines. These observations supported a non-pleiotropic effect of cAMP-sponge and the specificity of this genetic approach [33]. A comprehensive proteomic analysis conducted on the transformed tobacco BY-2 cells in the exponential phase of growth highlighted that 29 and 65 proteins were over- and under-accumulated, respectively, compared to WT cells [32]. By contrast, the proteomic analysis on Arabidopsis leaves from six-week-old plants indicated that only four proteins were differentially accumulated in cAS plants compared with WT, and among these phospholipase C was heavily downregulated [33]. Despite the absence of phenotype at resting conditions, cAS Arabidopsis plants showed reduced resistance to the avirulent pathogen *Pseudomonas syringae* pv. tomato DC3000 carrying the avirulence gene AvrB (PstAvrB), confirming that cAMP is required for the correct immune response activation [33].

These findings demonstrate the potential of the cAMP-sponge tool to unravel cAMP roles and signalling mechanisms in plants.

3. cAMP in Plant Physiological Processes

Although in plants the key actors of cAMP signal transduction are still not well defined, increasing evidence demonstrates that cAMP could affect several physiological processes (Figure 2).

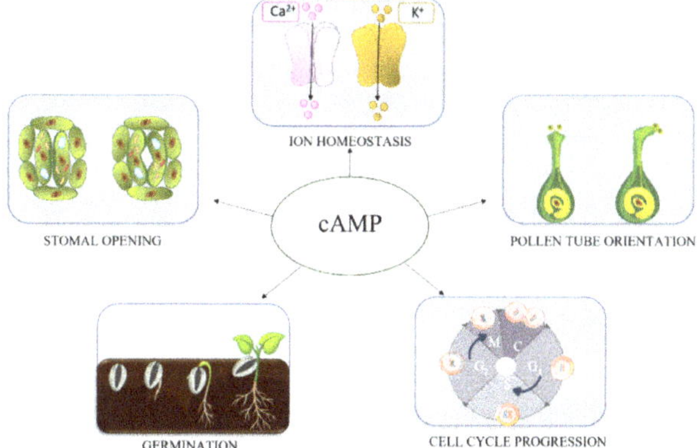

Figure 2. cAMP involvement in plant physiological processes. Literature data indicate a role for cAMP in ion homeostasis, mainly through the regulation of membrane-localised ion channels [29,34–36,40,48–54] and in stomatal opening, through Ca^{2+} and K^+ flux regulation [39,40,52,53], cAMP was also shown to influence pollen tube orientation and growth, by the regulation of Ca^{2+} channels and choline acetyltransferase activity [14,54–58]. Seed germination [59–63] and cell cycle progression [32,33,37,38,64] are also regulated by cAMP. More details are provided in the text.

Many papers highlight that cAMP is a potential regulator of ion homeostasis. The first substantial evidence for this cAMP function was obtained through the observation of whole-cell patch-clamp current in *Vicia faba* mesophyll after the application of 1 mM or higher concentrations of cAMP. Since this concentration is higher than the physiological cAMP concentrations, Li and co-workers supposed that plant cells could have high levels of PDE activity that hydrolyses the exogenous cAMP. They demonstrated that the application of micromolar cAMP concentration alongside the PDE inhibitor isobutyl-1-methylxanthine (IBMX) modulated an outward K^+-current. Moreover, using inhibitors of animal PKA and the catalytic subunit of PKA, the authors obtained indirect evidence that the modulation of K^+ channel activity could be mediated by a cAMP-regulated protein kinase [45]. Moreover, cAMP decreased cytosolic calcium in guard cell protoplasts, enhancing stomatal aperture in both light and darkness, in a protein kinase-dependent manner [39]. A successive work showed that cAMP participates in stomatal opening, antagonising the effect of abscisic acid (ABA) and Ca^{2+} on the inhibition of the inward K^+-current [40].

A clear demonstration of changes in Ca^{2+} homeostasis and subsequent protoplast swelling in response to cAMP was shown in *Nicotiana plumbaginofolia*. It should be noted that the same effects were obtained with cGMP. Both cyclic mononucleotides induced a raise in cytosolic Ca^{2+}, suggesting that the release of intracellular and extracellular Ca^{2+} stores acted as a signal at the crossroad of transduction pathways of these second messengers [48]. A direct effect of cAMP in regulating calcium conductance in leaf guard and mesophyll cells was shown in Arabidopsis [49]. By means of excised outside-out patches, Lemtiri-Chlieh and colleagues showed that the addition of permeable cAMP analogues stimulated a channel with fast gating kinetics. The results indicate that the increase of cytosolic Ca^{2+} was due to a plasma membrane-localised Ca^{2+} channel, suggesting the existence of functional CNGCs in these cells [49].

More recently, it was recognised that cAMP-dependent alteration of ion homeostasis could occur through the binding of CNGCs, which represent key sites where cyclic nucleotide interacts with ion signalling pathways [29,50]. For instance, the cAMP-activated inward of Ca^{2+} current through the plasma membrane is impaired in leaves of the CNGC2 loss of function mutant, *dnd1* [51]. CNGC2 and influx of apoplastic Ca^{2+} was also shown to be implicated in jasmonic acid-dependent rise in cytosolic cAMP, involved in the signalling of Arabidopsis guard cells [52,53].

A transient rise in cytosolic Ca^{2+} concentration after cAMP addition also occurred in pollen tube of *Agapanthus umbellatus* [54]. In the same species, cAMP was proposed as a signalling molecule involved in pollen tube reorientation. Growing pollen tubes showed an uniform cAMP concentration of 100–150 nM and changes in tube growth direction resulted from transient elevation in the apical region. The cAMP changes are due to PSiP, a putative AC cloned from *A. umbellatus* pollen. The antisense assays, achieved with oligos against this AC, caused the loss of pollen tube growth, suggesting that cAMP synthesis was a requirement for this event [14]. In *Pyrus pyrifolia*, through patch-clamp studies, it was shown that cAMP activated Ca^{2+} channel with a consequent increase in cytosolic Ca^{2+} of pollen tube protoplast. This event was specific for cAMP since cGMP failed to provoke the same effect. The cAMP-dependent opening of plasma membrane Ca^{2+} channels and the cytosolic Ca^{2+} increase affected pollen tube growth [55]. A different mechanism for cAMP control of pollen tube elongation was proposed in *Lilium longiflorum*. Application of exogenous cAMP at physiological concentration, as well as AC activators and PDE inhibitors, promoted the elongation of pollen tubes after self-incompatible pollination [56,57]. In addition, the content of endogenous cAMP in pistils after self-pollination was lower than that observed with cross-pollination and this difference reflected the different activities of AC and PDE [57]. Successively, it was demonstrated that cAMP stimulated the activity of choline acetyltransferase, which controls the synthesis of acetylcholine, a molecule that, together with other choline derivatives, promotes the elongation of *Lilium longiflorum* pollen tube. Thus, the low levels of cAMP and the subsequent low activity of choline acetyltransferase caused the self-incompatibility in *Lilium longiflorum* [58].

A pivotal role for cAMP in the control of cell cycle progression and cell division was also reported. In synchronised tobacco BY-2 cells, peaks of cAMP level were observed in S and G1 phases of cell cycle. The treatment with indomethacin, which is an inhibitor of prostaglandin-dependent adenylyl cyclase in animal cells, inhibited cAMP accumulation and mitosis [37]. The data on the expression of histone H4 and cyclin A, together with flow cytometric analyses, showed that indomethacin inhibits G1/S transition. [38]. However, the addition of exogenous cAMP failed to rescue indomethacin blocked cells, suggesting that indomethacin might affect other prostaglandin regulated activities [37]. The need to inhibit cAMP accumulation with methods independent of prostaglandin metabolism was overcome by using BY-2 cells overexpressing the cAMP sponge [32]. In vivo cAMP dampening in BY-2 cells caused a reduction in cell growth, mainly due to the mitosis inhibition, which occurred in parallel with a reduction in the cytoskeletal proteins, alpha- and beta-tubulin and actin depolymerisation factor, which are critical for cell division [65]. In parallel with cAMP deficiency, the expression of cell cycle genes was downregulated, suggesting that mitotic inhibition was due to a delay in cell cycle progression, which can occur at the G1/S and G2/M checkpoints. The delay of cell cycle progression was also supported by proteomic analysis [32]. The need of cAMP for a correct cell cycle progression and mitosis was also shown with pharmacological approaches in two-day-old seedling roots of *Raphanus sativus*. Domanska and colleagues suggested that different concentrations of cAMP are required for the start of DNA replication and mitosis and that cAMP can be involved in cell cycle transition during both replication and mitosis phases [64]. cAMP-dependent regulation of cell proliferation and differentiation was proposed for the formation of leguminous roots nodules. Plants with symbiotic nodules contained high levels of cAMP in the root nodules and cAMP contents increased during nodule development and decreased with nodule senescence [66,67].

Another possible role of cAMP in higher plants is the promotion of seed germination, suggested by the relationship observed between this second messenger and gibberellins (GA). The first evidence of this relationship was noticed in barley aleurone layers, where cAMP was shown to be able to substitute GA in the induction of α-amylase [63]. Early studies also showed that both cAMP and GA promoted germination of light-sensitive lettuce seeds and mannitol-treated weed seeds [60,61]. More recently, it was shown that cAMP acts downstream GA in the germination of the root parasitic plant *Orobanche minor* [62]. The *O. minor* seeds, prior to exposure to stimulants released from roots of host plants, need conditioning, which is a preincubation in a warm moist environment. Endogenous cAMP accumulated in the conditioned seeds. Moreover, exposure to light or supra-optimal temperature, throughout the conditioning period, led to cAMP decrease and low germination rates, which could be restored by GA treatments [62]. Similar results were also revealed during the seed germination of non-parasitic plant *Phacelia tanacetifolia* [63]. Under optimal light and temperature conditions, the seeds showed a transient cAMP accumulation before germination, which could be blocked by an inhibitor of GA biosynthesis. When the seeds were exposed to non-optimal conditions, inhibition of cAMP accumulation and germination occurred. Thus, cAMP could play a key role in favouring or blocking germination in response to environmental signals [63].

4. cAMP Involvement in Plant Response to Abiotic Stress

The establishing of plant responses to environmental stimuli requires the activation of multiple reactions at gene, transcript and protein level, interconnected by the action of signalling messengers [68]. In environmentally stressed plants, cellular metabolism faces a remarkable rearrangement allowing stress acclimation. Early alarm stages of plant abiotic stress response include the onset of oxidative stress and the induction of stress-responsive signalling pathways. Following the acclimation phase, with the biosynthesis of stress-protective compounds, cells encounter new recovering homeostasis, at the expense of cellular energy [68–70]. In this scenario, cAMP may act as stress sensors and/or modulator of cellular metabolism, mainly, but not only, through its influence on ion channels and the resulting regulation of ion fluxes [16] (Table 1).

Table 1. Proposed role of cAMP in the acquisition of stress tolerance.

Stress	Mechanisms	Molecular Players	References
Salinity	Limitation of Na$^+$ influx	VICs; CNGCs	[71,72]
Aluminium	K$^+$ current permitting malate outflow	Cation channels	[73]
K$^+$ deficiency	K$^+$ homeostasis regulation	AtKUP5; AtKUP7; CNGCs.	[20,21]
Heat	Ca^{2+} influx and HSPs expression	CNGCs; HSPs.	[74]
Drought	Synthesis of protective polypeptides	ABA signalling	[75]
Wounding	Regulations of the phenylpropanoid pathway	PAL; 4CL; CHS.	[76]
ROS	Reduction of Ca^{2+} influx and K$^+$ efflux	CNGCs	[72]

Abbreviations: ABA, abscisic acid; CHS, chalcone synthase; 4CL, 4-coumarate:coenzyme A ligase; CNGCs, cyclic nucleotide gated channels; HSPs, heat shock proteins; PAL, phenylalanine ammonia lyase; VICs, voltage-independent non-selective channels.

In Arabidopsis, the improvement of plant salinity tolerance involves cAMP, which causes the deactivation of voltage-independent non-selective channels, limiting Na$^+$ influx [71]. In wheat, tolerance to aluminium requires cAMP-dependent outward-rectifying K^+ current, which permits malate outflow that chelates this toxic metal [73].

The important link between K^+ flux and cAMP production was further defined in *Arabidopsis thaliana* by the isolation and characterisation of two K^+-uptake permeases, AtKUP5 and AtKUP7. Both the K^+-uptake permeases have a dual function, harbouring also a functional AC catalytic domain [20]. AtKUP7 is a K^+ transporter in roots, functionally active under K^+-limited conditions [77,78]. In addition, AtKUP7 was defined as a proton-coupled carrier with AC function, but it is still unclear if cAMP production is dependent on K^+ fluxes and/or if cAMP can modulate K^+ fluxes [20]. AtKUP5 causes a K^+ flux-dependent cAMP accumulation in the cytosol, which can in turn activate downstream components essential for K^+ homeostasis, including CNGCs [21].

cAMP involvement in abiotic stress response often goes through the regulation of CNGCs [29]. Remarkably, these ion channels, having overlapped binding domains for cyclic nucleotides and calmodulin, favour the crosstalk between the signalling of these second messengers [79,80]. Functional characterisation of Arabidopsis CNGC2 shows that cAMP activation of AtCNGC2 currents could be reversed by calmodulin, suggesting that the physical interaction of Ca^{2+} and calmodulin with CNGCs stops cyclic nucleotide activation of the channels. Therefore, the cytosolic cAMP, Ca^{2+} and calmodulin can operate in an integrated way to gate currents through CNGCs. [81].

CNGCs allow the influx of K^+, Na^+ and Ca^{2+} into the cell, with different selectivities; hence, they work downstream the environmental stimuli perception to mediate plant tolerance to drought, salinity and extreme temperature, which affect ionic and osmotic cellular homeostasis [29]. AtCNGC2 was shown to partially complement the yeast mutant at low K^+ concentration only in the presence of membrane-permeable cAMP [82]. AtCNGC10, AtCNGC19 and AtCNGC20 were shown to be involved in plant tolerance to salt stress [83,84]. The antisense lines of AtCNGC10 showed altered K^+ and Na^+ levels in shoots and were less tolerant to salt stress [83]. AtCNGC19 and AtCNGC20, participating in the re-allocation of Na^+ in the plants, might permit their survival to high salt levels [84].

Arabidopsis CNGC16 was shown to confer thermotolerance to germinating pollen, linking cyclic nucleotide signalling to heat stress response. In the *cngc16* mutants, the reduced transmission of pollen at high temperature was linked to a weakened expression of crucial stress-responsive genes. [85]. The role of CNGCs in plant thermotolerance was also validated in the vegetative tissue of plants. Mutants in CNGC2 showed hypersensitive heat-responsive Ca^{2+} influx, which conferred acquired thermotolerance at milder heat stress than in wild-type plants [86]. Mutation in Arabidopsis CNGC6 led to impaired heat stress response, which suggests its involvement in the acquisition of thermotolerance [74]. In addition, in Arabidopsis, it was shown that a heat shock caused an increase in intracellular cAMP levels, which, in turn, stimulating CNGC6, triggered a cytosolic Ca^{2+} influx. Furthermore, the treatment with an exogenous cAMP analogue induced the expression of some heat shock proteins, indicating the contribution of this second messenger in plant heat stress response [74].

Proteomic studies also supported a role of cAMP in controlling plant response to temperature, as well as to light. Thomas, Alqurashi, and their colleagues, suggested that cAMP participates as signalling molecule to the photosynthetic process of acclimation. [41,42]. The analyses revealed that, after cAMP treatment, the most enriched proteins belonged to the GO categories "Response to stress", "Response to abiotic stimulus", "Response to salt" and "Response to cold". Moreover, there was an enrichment of the category "Photosynthesis and light reaction processes" in both up- and downregulated cAMP responsive genes [41]. cAMP involvement in photosynthetic pathways was also described by Donaldoson and colleagues [43], who reported the interaction between cAMP and enzymes involved in Calvin cycle and photorespiration pathway. This is of interest since in *Nicotiana tabacum*, through a quantitative method based on mass spectrometric analysis, AC activity was observed in chloroplasts [87]. Moreover, in oat seedlings, it was shown that light influenced cAMP accumulation, pointing out that cAMP could take part in the phytochrome signalling pathway [88].

A role for cAMP in plant response to drought was also proposed in wheat. Indeed, the exogenous application of both cAMP and ABA promoted the synthesis of polypeptides whose accumulation is stimulated by dehydration, suggesting that cAMP signalling is possibly involved in the effect of ABA on protein synthesis during drought [75].

cAMP was shown to be involved in response to wounding in *Hippeastrum x hybridum*. In this plant, the transcriptional activity of the HpAC1 gene, which encodes a functional AC, as well as the level of cAMP, showed two peaks in response to mechanical damage. The authors proposed that the first rapid induction of HpAC1, and the concomitant transient changes in cAMP, might function as an "alarm" that alerts plant cells against the damage. The later increase in HpAC1 expression and cAMP accumulation might be linked to the induction of systemic responses and, in particular, to the induction of phenylalanine ammonia lyase (PAL) involved in the production of phytoalexins, which protect damaged tissue against potential pathogen attacks [76]. Together with PAL induction, cAMP was shown to be involved in the stimulation of the expression of 4-coumarate:coenzyme A ligase and chalcone synthase, enzymes of the phenylpropanoid pathway, which participates to plant response to a multiplicity of environmental stimuli, including nutrient depletion, UV irradiation, extreme temperatures and heavy metal toxicity [89].

Oxidative stress is a common feature associated with various abiotic stress factors, and reactive oxygen species (ROS) have an important biological role in sensing and activating acclimation mechanisms [68,90,91]. The superoxide-generating NADPH oxidase integrates Ca^{2+} and ROS signalling, which in turn may be connected to cyclic nucleotides through CNGCs [92]. Each messenger mutually enhances the induction of the other during abiotic stress conditions, resulting in the propagation of ROS and Ca^{2+} waves across the plasma membrane to establish the proper acclimation response, to which cAMP may directly or indirectly participate [93,94].

A correlation among cAMP, ROS and ion homeostasis was demonstrated in plant response to salt stress [72]. Several studies indicated that, in roots under salt stress, ROS accumulation could be due to the disturbance of mitochondrial function, as well as to activation of NADPH oxidases [95,96]. Furthermore, under salt stress, Na^+-influx into the cell causes a significant loss of cytosolic K^+, which can be responsible for important metabolic alterations [97,98]. The treatment of Arabidopsis roots with H_2O_2 induced a rapid Ca^{2+}-influx and K^+-efflux, which were reduced by pre-treatment with cAMP. Moreover, coherently with the accumulation of H_2O_2 level in salt-stressed roots [95,96], pre-treatment with cAMP decreased salt-dependent K^+-efflux [94]. Ordonez and colleagues proposed that CNGCs, proved to be involved in plant responses to salt stress [84], could be in part responsible of the H_2O_2-dependent K^+- efflux, which was reduced by cyclic nucleotides [72].

5. Role of cAMP in Plant Innate Immunity

Plants are continuously exposed to a variety of invading microorganisms, including viruses, bacteria and fungi. Although plants are lacking mobile sentinel cells, distinctive of the animal immune systems, they can perceive and keep away pathogens, through a two-layer innate immune system [99]. In the first layer of defence, called pattern-triggered immunity (PTI), membrane pattern recognition receptors (PRRs) recognise pathogen/microbe-associated molecular patterns (PAMPs/MAMPs) or endogenous damage-associated molecular patterns (DAMPs) [100,101]. This recognition initiates a series of defence responses, including ROS production, Ca^{2+} influx and activation of kinases as Ca^{2+}-dependent protein kinases and mitogen-activated protein kinase, leading to the upregulation of defence genes [101,102]. However, pathogens can secrete into plant cells effectors, namely virulence factors encoded by avirulence (avr) genes, which can suppress PTI. The effector recognition by intracellular receptors encoded by resistance genes activates the second layer of defence, the effector-triggered immunity (ETI). Defence responses of ETI are typically stronger than PTI and often culminate with the hypersensitive response (HR), a form of programmed cell death, occurring at the infection site with the aim to narrow pathogen infection [99,103]. An increase in the antimicrobial phytoalexins, as well as in salicylic acid (SA) and pathogenesis-related (PR) proteins, occurs locally in the site of infection, and systemically in uninfected tissues [104].

Several studies indicated the involvement of cAMP in plant immune response [33,105–110]. Considering all the literature data until now reported, possible cAMP-mediated mechanisms activated during plant-immunity are discussed (Figure 3).

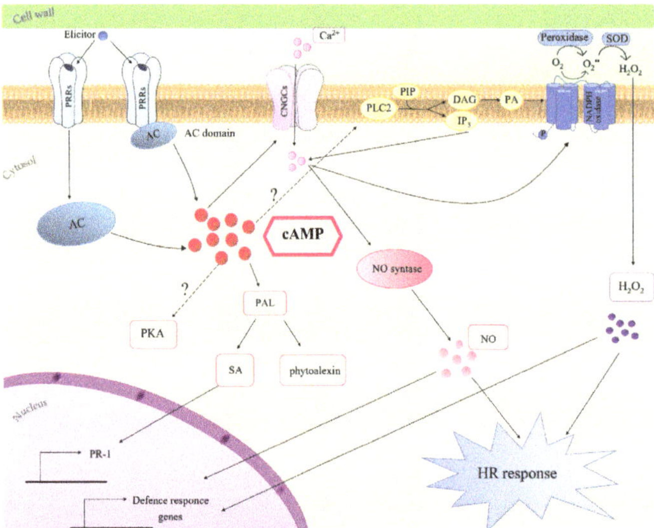

Figure 3. Molecular mechanisms of cAMP involvement in plant innate immunity. Elicitor recognition elevates cytosolic cAMP, which can activate CNGCs or PLC2, inducing Ca^{2+} accumulation and oxidative burst, through the activation of NADPH oxidase. cAMP-dependent oxidative burst can also be due to apoplastic peroxidases. Ca^{2+} stimulates NO production, which, together with ROS, induces defence response and HR. cAMP accumulation also activates PAL expression and production of SA and phytoalexins. More details are provided in the text. Question marks indicate pathways not completely characterised. Abbreviations: AC, adenylate cyclase; cAMP, 3′,5′-cyclic adenosine monophosphate; CNGCs, cyclic nucleotides-gated channels; DAG, diacylglycerol; HR, hypersensitive response; IP_3, inositol triphosphate; NO, nitric oxide; PA, phosphatidic acid; PAL, phenylalanine ammonia lyase; PIP, monophosphatidylinosotol; PKA, protein kinase A, PLC_2, phospholipase C_2; PR-1, pathogenesis-related genes; PRRs, pattern recognition receptors; SA, salicylic acid; SOD, superoxide dismutase.

Initially, a role for cAMP in the biosynthesis of phytoalexins was proposed. In carrot cell culture, the addition of the permeable dibutyryl cAMP, or forskolin and cholera toxin, activators of adenylate cyclase and G proteins, respectively, induced the biosynthesis of the antifungal phytoalexin 6-methoxymellein. Interestingly, the cAMP-dependent production of this phytoalexin was inhibited by Ca^{2+} channel blockers, as well as by inhibitors of calmodulin-dependent processes, suggesting that the increase in cAMP content in carrot cells induces Ca^{2+} influx across the plasma membrane [44,105]. In *Cupressus lusitanica* cell cultures, cAMP is involved in elicitor-induced production of the phytoalexin, β-thujaplicin. The authors suggested that cAMP-dependent β-thujaplicin accumulation involves Ca^{2+} and K^+ fluxes since it was inhibited by K^+ and Ca^{2+} channel blockers. This study also indicated a contribution of protein kinase cascades in cAMP signalling processes leading to β-thujaplicin accumulation [107]. The cAMP-dependent production of phytoalexins was also shown in *Medicago sativa*. In this case, the treatment with an elicitor of the phytopathogenic fungus, *Verticillium alboatrum*, caused a dose-dependent increase in the activity of AC and in intracellular cAMP content. Moreover, the treatment of Medicago cells with cAMP enhanced PAL activity and the synthesis of the phytoalexin medicarpin [106]. Consistently, in Arabidopsis, the treatment of seedlings with the permeable cAMP analogue 8-Br-cAMP increased, up to 40-fold and 2-fold, respectively, the expression of PAL2 and PAL1 [87]. PAL, the expression of which increased in response to diverse pathogens and elicitors, also plays a key role in SA synthesis [111–113] (Figure 3). Remarkably, cAMP elevation in Arabidopsis increased the endogenous SA level in response against Verticillium secreted toxins. The treatment of Arabidopsis with an AC inhibitor strongly reduced SA accumulation and PR-1 expression caused by Verticillium toxins. Both 8-Br-cAMP and SA enhanced resistance of Arabidopsis to the toxins, but cAMP acts upstream SA, since it was not able to potentiate the resistance of Arabidopsis plants deficient in SA [108]. In line with a role for cAMP in SA-dependent defence responses, the upregulation of PR-1 gene expression, occurring in response to an avirulent strain of *Pseudomonas syringae*, was decreased in cAS plants with low cAMP levels [33].

During plant immune responses, an oxidative burst arises in two phases, the first occurring within few minutes after pathogen perception and the second occurring later and with a higher amplitude [114]. ROS play several roles in response to pathogens, such as the reinforcement of cell wall, the activation MAP kinase pathways, the induction of HR and the triggering of systemic responses [115,116]. Two main mechanisms including NADPH oxidases and peroxidases have been proposed for ROS generation in response to pathogens [116,117]. Many literature data suggest an involvement of cAMP in pathogen/elicitor induced oxidative burst (Figure 3). In French bean cell culture, cAMP level increased upon the addition of an elicitor of the fungus *Colletotrichum lindemuthianum* and cAMP itself induced ROS accumulation. The cAMP-mediated apoplastic oxidative burst was increased by cholera toxin and inhibited by Ca^{2+} channel blockers. Bindschedler and co-workers suggested that G proteins and cAMP are involved in extracellular alkalisation and Ca^{2+} influx, essential for the pH-dependent apoplastic peroxidases, which mediate the oxidative burst [118]. Likewise, the treatment of *Arabidopsis thaliana* cells with forskolin enhanced the oxidative burst occurring in response to an elicitor from *Fusarium oxysporum* [119]. ROS generation induced by the PAMP lipopolysaccharide in Arabidopsis was prevented by the addition of an AC inhibitor [109]. Similarly, cAMP dampening in Arabidopsis cAS plants caused a delay in H_2O_2 increase at the early stage of response to an avirulent strain of *Pseudomonas syringae* [33].

Genetic evidence supports a role for CNGCs in pathogen-induced HR and disease resistance (Figure 3). In Arabidopsis, the mutation in DND1 (defence-no-death), which encodes AtCNGC2, failed to induce HR in response to an avirulent strain of *P. syringae*. Moreover, *dnd1* mutants showed constitutive systemic resistance and elevated levels of SA [120]. HLM1, encoding AtCNGC4, which works as a K^+- and Na^+-permeable channel activated by cGMP or cAMP, was upregulated in response to pathogen infection. *hlm1* mutant plants showed a lesion-mimic phenotype and an altered HR in response to avirulent *P. syringae* pv tomato (Pst) strains harbouring the avrRps4 or avrRpm1 genes [121]. In Arabidopsis, cAMP-activated AtCNGC11 and AtCNGC12 are positive mediators of resistance against the avirulent *Hyaloperonospora parasitica*. In the *cpr22* (constitutive expresser of PR genes22) mutant, a 3-kb deletion that fuses AtCNGC11 and AtCNGC12, generates the chimeric gene ATCNGC11/12, which confers the constitutive activation of defence responses [122].

An increase in cytosolic Ca^{2+}, due to influx across the plasma membrane or to efflux from intracellular stores, represents a primary event in plant immune signalling [123–126]. Interestingly, in *dnd1* mutant cells, the deficiency of cAMP-activated inward Ca^{2+} influx is associated with reduced production of nitric oxide (NO) [51], which was defined as the concertmaster in the HR and defence-gene activation [127,128]. *dnd1* mutants showed a weakened HR, and the addition of exogenous NO complements this phenotype [51]. Application of pathogens or PAMPS elevated cytosolic cAMP and the addition of exogenous cAMP led to Ca^{2+} elevation, NO generation and defence response in the absence of the non-self pathogen signal. Inoculation of *dnd1* plants with Pst containing the avrRpm1 or avrRpt2 genes led to a reduction in Ca^{2+} influx and to an impairment in immune response [51,109]. The weakening of pathogen-associated cytosolic Ca^{2+} influx also occurred by blocking cAMP synthesis in plants exposed to the pathogen, with a corresponding impairment in HR. On the contrary, co-infiltration with IBMX along with avirulent pathogens enhanced plant immune response, increasing HR. Thus, it was suggested that elevation of cytosolic cAMP, acting upstream from Ca^{2+}, is a key signal in the transduction of pathogen perception and in the downstream signalling cascade of defence responses [109]. Furthermore, the cAMP dampening, occurring in Arabidopsis cAS plants, delayed cytosolic Ca^{2+} elevation and reduced HR in response to PstAvrB. Sabetta and co-workers suggested that the delay in Ca^{2+} elevation could be due to a failure in the activation of CNGCs, but also to the down-accumulation of phospholipase C2 (PLC2) occurring in cAS plants [33] (Figure 3). Consistently, it is known that cytosolic Ca^{2+} accumulation in response to numerous elicitors of plant defence involves phosphatidylinositol-specific PLCs [125]. Moreover, since it was reported that PLCs significantly contribute to pathogen/elicitor induced oxidative burst [129–131], the low level of PLC2 in cAS plants could also contribute to the delayed H_2O_2 increase in the first phase of PstAvrB infection [33]. The low availability of cAMP, and the subsequent delay in Ca^{2+} influx, could be responsible for an incorrect temporal modulation of the AtSR1 [33], a Ca^{2+}-dependent calmodulin binding transcription factor, repressing the expression of target genes [131–133]. Consequently, some defence proteins, such as HSP90, CRK14 and DJ1E [134–137], were not accumulated in cAS cells after pathogen infection, weakening defence response [33].

The involvement of cAMP in plant immunity was supported by the isolation of ACs involved in plant response to pathogens. The silencing of NbAC, a gene encoding an AC in *Nicotiana benthamiana*, suppresses the necrotic lesions induced by tabtoxinine-β-lactam, a non-specific bacterial toxin, produced by *P. syringae* pv. Tabaci [138]. The expression of HpAC1, a gene encoding an AC from *Hippeastrum x hybridum*, and the levels of cAMP, increased in response to *Phoma narcissi* infection [76]. Recently, a leucine-rich repeat protein, AtLRRAC1, harbouring multiple catalytically active AC centres, was identified in Arabidopsis. AtLRRAC1 was able to complement AC-deficient *Escherichia coli* and to generate cAMP in vitro [18,22]. Interestingly, *atlrrac1* mutants showed compromised immune responses to biotrophic fungi and hemibiotrophic bacteria. The expression of early-induced immune-related genes after elicitation with the PAMP flg22 was strongly inhibited in *atlrrac1* plants, suggesting an involvement of AtLRRAC1 in PTI [22].

6. Conclusions

cAMP is the object of intense scientific interest, both in animal systems, where much more progress was achieved in defining its role, and in plants, becoming lately the centre of a bustling research. cAMP is nowadays recognised as a relevant signalling molecule in plant development as well as in responses to environmental stimuli, of both biotic and abiotic nature. As cAMP-signalling networks and their spatial and temporal regulation are extremely complex, future research must deal with the nature of cAMP signals in terms of strength, duration and frequency, considering also the crosstalk between this second messenger and other intracellular regulators [139]. Since the existence of cAMP-regulated processes in plants and the first evidence of compartmentalised cAMP signals in animals, the need for reliable cAMP detection methods able to reveal cAMP waves in living systems arose. Recent advances in modern biotechnologies and synthetic biology, alongside newly developed detection methods and instrumentations, offer a wide range of possibilities to unravel cAMP role in living cells.

The cAMP-sponge represents a cutting-edge genetically encoded tool, used to exploit cAMP fluctuations for the first time in living plant organisms and specific cell compartments. It overcomes major concerns on biochemical assays and pharmacological studies performed so far in plants [31–33]. Other developed genetically-encoded tools employed in bacteria and isolated plant cells are the promoter reporter systems, based on the plant protein Oligopeptide TransporterX promoter, which measure alterations in downstream gene expression following changes in intracellular levels of cyclic nucleotides. Unfortunately, this system cannot discriminate between cGMP and cAMP [140].

Taking advantage of the progress reached in animal systems, many other strategies and their combination may help in elucidating cAMP signalling in plant systems. Indeed, optogenetic approaches and genetically encoded fluorescent biosensors are effectively used to monitor and modulate cAMP levels [141,142]. Photoactivated ACs and light-regulated PDEs, or even their association, are successfully used in animal cells [143,144]. The generation of stable plant lines, expressing the combination of optimised sensors for cAMP and concomitant or downstream messengers, may provide a comprehensive view of the signalling event investigated.

Another important requirement is a clear identification and functional characterisation of cAMP-binding proteins involved in the signalling of this second messenger. Nowadays, many lines of evidence indicate that, in plants, the conversion of cAMP into Ca^{2+} signals via CNGCs is the main signalling mechanism of this cyclic nucleotide. However, although indications for bona fide PKA are lacking, its presence in plants cannot be excluded. New bioinformatics algorithms and molecular tools may provide opportunities to extend the presently scarce knowledge of cAMP-dependent protein kinases [16,23]. Moreover, studies on cAMP-dependent changes in transcriptomes, proteomes and phosphoproteomes, as well as metabolomes, will improve the understanding of cAMP involvement in plant physiological processes, along with acclimation to adverse environmental conditions.

Author Contributions: Conceptualisation, M.C.d.P. and E.B.; writing—original draft preparation, E.B., S.F., L.V. and M.C.d.P.; writing—review and editing, E.B., S.F., L.V. and M.C.d.P.; visualisation, S.F.; supervision, M.C.d.P. funding acquisition, E.B. and M.C.d.P. All authors have read and agreed to the published version of the manuscript.

Funding: This research was funded by CNR, grant number FOE-2019 DBA.AD003.139, and by University of Bari Aldo Moro, grant number H95E10000710005.

Acknowledgments: The authors thank M. Hofer (Boston Healthcare System, Brigham and Women's Hospital and Harvard Medical School of Massachusetts, USA) for kindly providing "cAMP-sponge".

Conflicts of Interest: The authors declare no conflict of interest.

References

1. Rall, T.W.; Sutherland, E.W.; Berthet, J. The relation of epinephrine and glucagon to liver phosphorylase. IV. Effect of epinephrine and glucagon on the reactivation of phosphorylase in liver homogenates. *J. Biol. Chem.* **1957**, *224*, 1987–1995.

2. Shabb, J.B.; Corbin, J.D. Cyclic nucleotide-binding domains in proteins having diverse functions. *J. Biol. Chem.* **1992**, *267*, 5723–57236. [PubMed]

3. Newton, R.P.; Smith, C.J. Cyclic nucleotides. *Phytochemistry* **2004**, *65*, 2423–2437. [CrossRef]

4. Gancedo, J.M. Biological roles of cAMP: Variations on a theme in the different kingdoms of life. *Biol. Rev. Camb. Philos. Soc.* **2013**, *88*, 645–668. [CrossRef] [PubMed]

5. Arora, K.; Sinha, C.; Zhang, W.; Ren, A.; Moon, C.S.; Yarlagadda, S.; Naren, A.P. Compartmentalization of cyclic nucleotide signaling: A question of when, where, and why? *Pflugers Arch.* **2013**, *465*, 1397–1407. [CrossRef] [PubMed]

6. Cheepala, S.; Hulot, J.S.; Morgan, J.A.; Sassi, Y.; Zhang, W.; Naren, A.P.; Schuetz, J.D. Cyclic nucleotide compartmentalization: Contributions of phosphodiesterases and ATP-binding cassette transporters. *Annu. Rev. Pharmacol. Toxicol.* **2013**, *53*, 231–235. [CrossRef]

7. Stork, P.J.S.; Schmitt, J.M. Crosstalk between cAMP and MAP kinase signaling in the regulation of cell proliferation. *Trends Cell Biol.* **2002**, *12*, 258–266. [CrossRef]

8. Amrhein, N. The current status of cyclic AMP in higher plants. *Ann. Rev. Plant Physiol.* **1977**, *28*, 123–132. [CrossRef]

9. Brown, E.G.; Newton, R.P. Cyclic AMP and higher plants. *Phytochemistry* **1981**, *20*, 2453–2463. [CrossRef]

10. Witters, E.; Roef, L.; Newton, R.P.; VanDongen, W.; Esmans, E.L.; VanOnckelen, H.A. Quantitation of cyclic nucleotides in biological samples by negative electrospray tandem mass spectrometry coupled to ion suppression liquid chromatography. *Rapid Commun. Mass Spectrom.* **1996**, *10*, 225–231. [CrossRef]

11. Witters, E.; Vanhoutte, K.; Dewitte, W.; Machackova, I.; Benkova, E.; Van Dongen, W.; Esmans, E.L.; Van Onckelen, H.A. Analysis of cyclic nucleotides and cytokinins in minute plant samples using phase-system switching capillary electrospray-liquid chromatography-tandem mass spectrometry. *Phytochem. Anal.* **1999**, *10*, 143–151. [CrossRef]

12. Beavo, J.A.; Brunton, L.L. Cyclic nucleotide research—Still expanding after half a century. *Nat. Rev. Mol. Cell Biol.* **2002**, *3*, 710–718. [CrossRef] [PubMed]

13. Craven, K.B.; Zagotta, W.N. CNG and HCN channels: Two peas, one pod. *Annu. Rev. Physiol.* **2006**, *68*, 375–401. [CrossRef] [PubMed]

14. Moutinho, A.; Hussey, P.J.; Trewavas, A.J.; Malho, R. cAMP acts as a second messenger in pollen tube growth and reorientation. *Proc. Natl. Acad. Sci. USA* **2001**, *98*, 10481–10486. [CrossRef]

15. Gehring, C. Adenyl cyclases and cAMP in plant signaling—Past and present. *Cell Commun. Signal.* **2010**, *8*, 15. [CrossRef] [PubMed]

16. Gehring, C.; Turek, I.S. Cyclic Nucleotide monophosphates and their cyclases in plant signaling. *Front. Plant Sci.* **2017**, *8*, 1704. [CrossRef]

17. Wong, A.; Tian, X.C.; Gehring, C.; Marondedze, C. Discovery of novel functional centers with rationally designed amino acid motifs. *Comput. Struct. Biotechnol. J.* **2018**, *16*, 70–76. [CrossRef] [PubMed]

18. Ruzvidzo, O.; Gehring, C.; Wong, A. New perspectives on plant adenylyl cyclases. *Front. Mol. Biosci.* **2019**, *6*, 136. [CrossRef]

19. Irving, H.R.; Cahill, D.M.; Gehring, C. Moonlighting proteins and their role in the control of signaling microenvironments, as exemplified by cGMP and Phytosulfokine Receptor 1 (PSKR1). *Front. Plant Sci.* **2018**, *9*, 415. [CrossRef]

20. Al-Younis, I.; Wong, A.; Gehring, C. The *Arabidopsis thaliana* K$^+$-uptake permease 7 (AtKUP7) contains a functional cytosolic adenylate cyclase catalytic centre. *FEBS Lett.* **2015**, *589*, 3848–3852. [CrossRef]

21. Al-Younis, I.; Wong, A.; Lemtiri-Chlieh, F.; Schrnockel, S.; Tester, M.; Gehring, C.; Donaldson, L. The *Arabidopsis thaliana* K$^+$-uptake permease 5 (AtKUP5) contains a functional cytosolic adenylate cyclase essential for K$^+$ transport. *Front. Plant Sci.* **2018**, *9*, 1645. [CrossRef] [PubMed]

22. Bianchet, C.; Wong, A.; Quaglia, M.; Alqurashi, M.; Gehring, C.; Ntoukakis, V.; Pasqualini, S. An *Arabidopsis thaliana* leucine-rich repeat protein harbors an adenylyl cyclase catalytic center and affects responses to pathogens. *J. Plant Physiol.* **2019**, *232*, 12–22. [CrossRef] [PubMed]

23. Swiezawska, B.; Duszyn, M.; Jaworski, K.; Szmidt-Jaworska, A. Downstream targets of cyclic nucleotides in plants. *Front. Plant Sci.* **2018**, *9*, 1428. [CrossRef] [PubMed]

24. Kasahara, M.; Suetsugu, N.; Urano, Y.; Yamamoto, C.; Ohmori, M.; Takada, Y.; Okuda, S.; Nishiyama, T.; Sakayama, H.; Kohchi, T.; et al. An adenylyl cyclase with a phosphodiesterase domain in basal plants with a motile sperm system. *Sci. Rep.* **2016**, *6*, 39232. [CrossRef] [PubMed]

25. Ma, Y.; He, K.; Berkowitz, G.A. Editorial: From structure to signalsomes: New perspectives about membrane receptors and channels. *Front. Plant Sci.* **2019**, *10*, 682. [CrossRef]

26. Martinez-Atienza, J.; Van Ingelgem, C.; Roef, L.; Maathuis, F.J. Plant cyclic nucleotide signalling: Facts and fiction. *Plant Signal. Behav.* **2007**, *2*, 540–543. [CrossRef]

27. Ma, W.; Smigel, A.; Walker, R.K.; Moeder, W.; Yoshioka, K.; Berkowitz, G.A. Leaf senescence signaling: The Ca2+-conducting Arabidopsis cyclic nucleotide gated channel2 acts through nitric oxide to repress senescence programming. *Plant Physiol.* **2010**, *154*, 733–743. [CrossRef]

28. Ma, W.; Berkowitz, G.A. Ca^{2+} conduction by plant cyclic nucleotide gated channels and associated signaling components in pathogen defense signal transduction cascades. *New Phytol.* **2011**, *190*, 566–572. [CrossRef]

29. Jha, S.K.; Sharma, M.; Pandey, G.K. Role of Cyclic nucleotide gated channels in stress management in plants. *Curr. Genomics* **2016**, *17*, 315–329. [CrossRef]

30. Duszyn, M.; Swiezawska, B.; Szmidt-Jaworska, A.; Jaworski, K. Cyclic nucleotide gated channels (CNGCs) in plant signalling-Current knowledge and perspectives. *J. Plant Physiol.* **2019**, *241*, 153035. [CrossRef]

31. Lefkimmiatis, K.; Moyer, M.P.; Curci, S.; Hofer, A.M. "cAMP Sponge": A buffer for cyclic adenosine 3′,5′-monophosphate. *PLoS ONE* **2009**, *4*, e7649. [CrossRef] [PubMed]

32. Sabetta, W.; Vannini, C.; Sgobba, A.; Marsoni, M.; Paradiso, A.; Ortolani, F.; Bracale, M.; Viggiano, L.; Blanco, E.; de Pinto, M.C. Cyclic AMP deficiency negatively affects cell growth and enhances stress-related responses in tobacco Bright Yellow-2 cells. *Plant Mol. Biol.* **2016**, *90*, 467–483. [CrossRef] [PubMed]

33. Sabetta, W.; Vandelle, E.; Locato, V.; Costa, A.; Cimini, S.; Bittencourt Moura, A.; Luoni, L.; Graf, A.; Viggiano, L.; De Gara, L.; et al. Genetic buffering of cyclic AMP in *Arabidopsis thaliana* compromises the plant immune response triggered by an avirulent strain of *Pseudomonas syringae* pv. tomato. *Plant J.* **2019**, *98*, 590–606. [CrossRef] [PubMed]

34. Anderson, J.A.; Huprikar, S.S.; Kochian, L.V.; Lucas, W.J.; Gaber, R.F. Functional expression of a probable *Arabidopsis-thaliana* potassium channel in *Saccharomyces-cerevisiae*. *Proc. Natl. Acad. Sci. USA* **1992**, *89*, 3736–3740. [CrossRef]

35. Bolwell, G.P. Cyclic-AMP, the reluctant messenger in plants. *Trends Biochem. Sci.* **1995**, *20*, 492–495. [CrossRef]
36. Trewavas, A.J. Plant cyclic AMP comes in from the cold. *Nature* **1997**, *390*, 657–658. [CrossRef]
37. Ehsan, H.; Reichheld, J.P.; Roef, L.; Witters, E.; Lardon, F.; Van Bockstaele, D.; Van Montagu, M.; Inze, D.; Van Onckelen, H. Effect of indomethacin on cell cycle dependent cyclic AMP fluxes in tobacco BY-2 cells. *FEBS Lett.* **1998**, *422*, 165–169. [CrossRef]
38. Ehsan, H.; Roef, L.; Witters, E.; Reichheld, J.P.; Van Bockstaele, D.; Inze, D.; Van Onckelen, H. Indomethacin-induced G1/S phase arrest of the plant cell cycle. *FEBS Lett.* **1999**, *458*, 349–353. [CrossRef]
39. Curvetto, N. Effect of two cAMP analogs on stomatal opening in *Vicia faba*: Possible relationship with cytosolic calcium concentration. *Plant Physiol. Biochem.* **1994**, *32*, 365–372.
40. Jin, X.C.; Wu, W.H. Involvement of cyclic AMP in ABA- and Ca^{2+}-mediated signal transduction of stomatal regulation in *Vicia faba*. *Plant Cell Physiol.* **1999**, *40*, 1127–1133. [CrossRef]
41. Thomas, L.; Marondedze, C.; Ederli, L.; Pasqualini, S.; Gehring, C. Proteomic signatures implicate cAMP in light and temperature responses in *Arabidopsis thaliana*. *J. Proteom.* **2013**, *83*, 47–59. [CrossRef]
42. Alqurashi, M.; Gehring, C.; Marondedze, C. Changes in the *Arabidopsis thaliana* proteome implicate cAMP in biotic and abiotic stress responses and changes in energy metabolism. *Int. J. Mol. Sci.* **2016**, *17*, 852. [CrossRef] [PubMed]
43. Donaldson, L.; Meier, S.; Gehring, C. The arabidopsis cyclic nucleotide interactome. *Cell Commun. Signal.* **2016**, *14*, 10. [CrossRef] [PubMed]
44. Kurosaki, F.; Tsurusawa, Y.; Nishi, A. The elicitation of phytoalexins by Ca^{2+} and cyclic AMP in carrot cells. *Phytochemistry* **1987**, *26*, 1919–1923. [CrossRef]
45. Li, W.W.; Luan, S.; Schreiber, S.L.; Assmann, S.M. Cyclic-AMP stimulates K^+ channel activity in mesophyll-cells of *Vicia-faba*-K. *Plant Physiol.* **1994**, *106*, 957–961. [CrossRef] [PubMed]
46. Uchiyama, T.; Yoshikawa, F.; Hishida, A.; Furuichi, T.; Mikoshiba, K. A novel recombinant hyperaffinity inositol 1,4,5-trisphosphate (IP3) absorbent traps IP3, resulting in specific inhibition of IP3-mediated calcium signaling. *J. Biol. Chem.* **2002**, *277*, 8106–8113. [CrossRef]
47. Saraswat, L.D.; Ringheim, G.E.; Bubis, J.; Taylor, S.S. Deletion mutants as probes for localizing regions of subunit interaction in cAMP-dependent protein-kinase. *J. Biol. Chem.* **1988**, *263*, 18241–18246.
48. Volotovski, I.D.; Sokolovsky, S.G.; Molchan, O.V.; Knight, M.R. Second messengers mediate increases in cytosolic calcium in tobacco protoplasts. *Plant Physiol.* **1998**, *117*, 1023–1030. [CrossRef]
49. Lemtiri-Chlieh, F.; Berkowitz, G.A. Cyclic adenosine monophosphate regulates calcium channels in the plasma membrane of Arabidopsis leaf guard and mesophyll cells. *J. Biol. Chem.* **2004**, *279*, 35306–35312. [CrossRef]
50. Talke, I.N.; Blaudez, D.; Maathuis, F.J.M.; Sanders, D. CNGCs: Prime targets of plant cyclic nucleotide signalling? *Trends Plant Sci.* **2003**, *8*, 286–293. [CrossRef]
51. Ali, R.; Ma, W.; Lemtiri-Chlieh, F.; Tsaltas, D.; Leng, Q.; von Bodman, S.; Berkowitz, G.A. Death don't have no mercy and neither does calcium: Arabidopsis CYCLIC NUCLEOTIDE GATED CHANNEL2 and innate immunity. *Plant Cell* **2007**, *19*, 1081–1095. [CrossRef] [PubMed]
52. Munemasa, S.; Hossain, M.A.; Nakamura, Y.; Mori, I.C.; Murata, Y. The Arabidopsis calcium-dependent protein kinase, CPK6, functions as a positive regulator of methyl jasmonate signaling in guard cells. *Plant Physiol.* **2011**, *155*, 553–561. [CrossRef]
53. Lu, M.; Zhang, Y.Y.; Tang, S.K.; Pan, J.B.; Yu, Y.K.; Han, J.; Li, Y.Y.; Du, X.H.; Nan, Z.J.; Sun, Q.P. AtCNGC2 is involved in jasmonic acid-induced calcium mobilization. *J. Exp. Bot.* **2016**, *67*, 809–819. [CrossRef] [PubMed]
54. Malho, R.; Camacho, L.; Moutinho, A. Signalling pathways in pollen tube growth and reorientation. *Ann. Bot.* **2000**, *85*, 59–68. [CrossRef]
55. Wu, J.Y.; Qu, H.Y.; Jin, C.; Shang, Z.L.; Wu, J.; Xu, G.H.; Gao, Y.B.; Zhang, S.L. cAMP activates hyperpolarization-activated Ca^{2+} channels in the pollen of *Pyrus pyrifolia*. *Plant Cell Rep.* **2011**, *30*, 1193–1200. [CrossRef]
56. Tezuka, T.; Hiratsuka, S.; Takahashi, S.Y. Promotion of the growth of self-incompatible pollen tubes in lily by cAMP. *Plant Cell Physiol.* **1993**, *34*, 955–958. [CrossRef]
57. Tsuruhara, A.; Tezuka, T. Relationship between the self-incompatibility and cAMP level in *Lilium longiflorum*. *Plant Cell Physiol.* **2001**, *42*, 1234–1238. [CrossRef]
58. Tezuka, T.; Akita, I.; Yoshino, N. Self-incompatibility involved in the level of acetylcholine and cAMP. *Plant Signal. Behav.* **2007**, *2*, 475–476. [CrossRef]

59. Duffus, C.M.; Duffus, J.H. A possible role for cyclic AMP in gibberellic acid triggered release of alpha-amylase in barley endosperm slices. *Experientia* **1969**, *25*, 581. [CrossRef]

60. Hall, K.A.; Galsky, A.G. The action of cyclic-AMP on GA3 controlled responses IV. Characteristics of the promotion of seed germination in *Lactuca sative* variety 'Spartan Lake' by gibberellic acid and cyclic 3,5′-adenosine monophosphate. *Plant Cell Physiol.* **1973**, *14*, 565–571. [CrossRef]

61. Holm, R.; Miller, M. Hormonal control of weed seed germination. *Weed Sci.* **1972**, *20*, 209–212. [CrossRef]

62. Uematsu, K.; Nakajima, M.; Yamaguchi, I.; Yoneyama, K.; Fukui, Y. Role of cAMP in gibberellin promotion of seed germination in *Orobanche minor* smith. *J. Plant Growth Regul.* **2007**, *26*, 245–254. [CrossRef]

63. Uematsu, K.; Fukui, Y. Role and regulation of cAMP in seed germination of *Phacelia tanacetifolia*. *Plant Physiol. Biochem.* **2008**, *46*, 768–774. [CrossRef] [PubMed]

64. Domanska, A.; Godlewski, M.; Aniol, P. Db-cAMP, forskolin or 2′-d3′-AMP influence on proliferation of *Raphanus sativus* root meristem cells. *Caryologia* **2009**, *62*, 267–275. [CrossRef]

65. Moulding, D.A.; Blundell, M.P.; Spiller, D.G.; White, M.R.; Cory, G.O.; Calle, Y.; Kempski, H.; Sinclair, J.; Ancliff, P.J.; Kinnon, C.; et al. Unregulated actin polymerization by WASp causes defects of mitosis and cytokinesis in X-linked neutropenia. *J. Exp. Med.* **2007**, *204*, 2213–2224. [CrossRef]

66. Terakado, J.; Okamura, M.; Fujihara, S.; Ohmori, M.; Yoneyama, T. Cyclic AMP in rhizobia and symbiotic nodules. *Ann. Bot-Lond.* **1997**, *80*, 499–503. [CrossRef]

67. Terakado, J.; Fujihara, S.; Yoneyama, T. Changes in cyclic nucleotides during nodule formation. *Soil Sci. Plant Nutr.* **2003**, *49*, 459–462. [CrossRef]

68. He, M.; He, C.Q.; Ding, N.Z. Abiotic stresses: General defenses of land plants and chances for engineering multistress tolerance. *Front. Plant Sci.* **2018**, *9*, 1771. [CrossRef]

69. Kosova, K.; Vitamvas, P.; Prasil, I.T.; Renaut, J. Plant proteome changes under abiotic stress-contribution of proteomics studies to understanding plant stress response. *J. Proteomics* **2011**, *74*, 1301–1322. [CrossRef]

70. Lamers, J.; van der Meer, T.; Testerink, C. How plants sense and respond to stressful environments. *Plant Physiol.* **2020**, *182*, 1624–1635. [CrossRef]

71. Maathuis, F.J.; Sanders, D. Sodium uptake in Arabidopsis roots is regulated by cyclic nucleotides. *Plant Physiol.* **2001**, *127*, 1617–1625. [CrossRef] [PubMed]

72. Ordonez, N.M.; Marondedze, C.; Thomas, L.; Pasqualini, S.; Shabala, L.; Shabala, S.; Gehring, C. Cyclic mononucleotides modulate potassium and calcium flux responses to H_2O_2 in Arabidopsis roots. *FEBS Lett.* **2014**, *588*, 1008–1015. [CrossRef] [PubMed]

73. Zhang, W.H.; Ryan, P.R.; Tyerman, S.D. Malate-permeable channels and cation channels activated by aluminum in the apical cells of wheat roots. *Plant Physiol.* **2001**, *125*, 1459–1472. [CrossRef] [PubMed]

74. Gao, F.; Han, X.W.; Wu, J.H.; Zheng, S.Z.; Shang, Z.L.; Sun, D.Y.; Zhou, R.G.; Li, B. A heat-activated calcium-permeable channel—Arabidopsis cyclic nucleotide-gated ion channel 6—Is Involved in heat shock responses. *Plant J.* **2012**, *70*, 1056–1069. [CrossRef]

75. Maksyutova, N.N.; Viktorova, L.V. A comparative study of the effect of abscisic acid and cAMP on protein synthesis in wheat caryopses under drought conditions. *Biochemistry (Mosc)* **2003**, *68*, 424–428. [CrossRef]

76. Swieiawska, B.; Jaworski, K.; Pawelek, A.; Grzegorzewska, W.; Szewczuk, P.; Szmidt-Jaworska, A. Molecular cloning and characterization of a novel adenylyl cyclase gene, HpAC1, involved in stress signaling in *Hippeastrum x hybridum*. *Plant Physiol. Biochem.* **2014**, *80*, 41–52. [CrossRef]

77. Ahn, S.J.; Shin, R.; Schachtman, D.P. Expression of KT/KUP genes in Arabidopsis and the role of root hairs in K^+ uptake. *Plant Physiol.* **2004**, *134*, 1135–1145. [CrossRef]

78. Han, M.; Wu, W.; Wu, W.H.; Wang, Y. Potassium transporter KUP7 is involved in K(+) acquisition and translocation in Arabidopsis root under K(+)-limited conditions. *Mol. Plant* **2016**, *9*, 437–446. [CrossRef]

79. Kohler, C.; Merkle, T.; Neuhaus, G. Characterisation of a novel gene family of putative cyclic nucleotide- and calmodulin-regulated ion channels in Arabidopsis thaliana. *Plant J.* **1999**, *18*, 97–104. [CrossRef]

80. Arazi, T.; Kaplan, B.; Fromm, H. A high-affinity calmodulin-binding site in a tobacco plasma-membrane channel protein coincides with a characteristic element of cyclic nucleotide-binding domains. *Plant Mol. Biol.* **2000**, *42*, 591–601. [CrossRef]

81. Hua, B.G.; Mercier, R.W.; Zielinski, R.E.; Berkowitz, G.A. Functional interaction of calmodulin with a plant cyclic nucleotide gated cation channel. *Plant Physiol. Biochem.* **2003**, *41*, 945–954. [CrossRef]

82. Leng, Q.; Mercier, R.W.; Yao, W.; Berkowitz, G.A. Cloning and first functional characterization of a plant cyclic nucleotide-gated cation channel. *Plant Physiol.* **1999**, *121*, 753–761. [CrossRef] [PubMed]

83. Kugler, A.; Kohler, B.; Palme, K.; Wolff, P.; Dietrich, P. Salt-dependent regulation of a CNG channel subfamily in Arabidopsis. *BMC Plant Biol.* **2009**, *9*, 140. [CrossRef] [PubMed]

84. Guo, K.M.; Babourina, O.; Christopher, D.A.; Borsics, T.; Rengel, Z. The cyclic nucleotide-gated channel, AtCNGC10, influences salt tolerance in Arabidopsis. *Physiol. Plant* **2008**, *134*, 499–507. [CrossRef]

85. Tunc-Ozdemir, M.; Tang, C.; Ishka, M.R.; Brown, E.; Groves, N.R.; Myers, C.T.; Rato, C.; Poulsen, L.R.; McDowell, S.; Miller, G.; et al. A cyclic nucleotide-gated channel (CNGC16) in pollen is critical for stress tolerance in pollen reproductive development. *Plant Physiol.* **2013**, *161*, 1010–1020. [CrossRef] [PubMed]

86. Finka, A.; Cuendet, A.F.; Maathuis, F.J.; Saidi, Y.; Goloubinoff, P. Plasma membrane cyclic nucleotide gated calcium channels control land plant thermal sensing and acquired thermotolerance. *Plant Cell* **2012**, *24*, 3333–3348. [CrossRef]

87. Witters, E.; Valcke, R.; van Onckelen, H. Cytoenzymological analysis of adenylyl cyclase activity and 3′:5′-cAMP immunolocalization in chloroplasts of Nicotiana tabacum. *New Phytol.* **2005**, *168*, 99–107. [CrossRef]

88. Molchan, O.V.; Sokolovsky, S.G.; Volotovsky, I.D. The phytochrome control of the cAMP endogenous level in oat seedlings. *Russ. J. Plant Physl.* **2000**, *47*, 463–467.

89. Pietrowska-Borek, M.; Nuc, K. Both cyclic-AMP and cyclic-GMP can act as regulators of the phenylpropanoid pathway in *Arabidopsis thaliana* seedlings. *Plant Physiol. Biochem.* **2013**, *70*, 142–149. [CrossRef]

90. Miller, G.; Suzuki, N.; Rizhsky, L.; Hegie, A.; Koussevitzky, S.; Mittler, R. Double mutants deficient in cytosolic and thylakoid ascorbate peroxidase reveal a complex mode of interaction between reactive oxygen species, plant development, and response to abiotic stresses. *Plant Physiol.* **2007**, *144*, 1777–1785. [CrossRef]

91. de Pinto, M.C.; Locato, V.; Paradiso, A.; De Gara, L. Role of redox homeostasis in thermo-tolerance under a climate change scenario. *Ann. Bot.* **2015**, *116*, 487–496. [CrossRef] [PubMed]

92. Miller, G.; Schlauch, K.; Tam, R.; Cortes, D.; Torres, M.A.; Shulaev, V.; Dangl, J.L.; Mittler, R. The plant NADPH oxidase RBOHD mediates rapid systemic signaling in response to diverse stimuli. *Sci. Signal.* **2009**, *2*, ra45. [CrossRef] [PubMed]

93. Gilroy, S.; Suzuki, N.; Miller, G.; Choi, W.G.; Toyota, M.; Devireddy, A.R.; Mittler, R. A tidal wave of signals: Calcium and ROS at the forefront of rapid systemic signaling. *Trends Plant Sci.* **2014**, *19*, 623–630. [CrossRef]

94. Steinhorst, L.; Kudla, J. Calcium and reactive oxygen species rule the waves of signaling. *Plant Physiol.* **2013**, *163*, 471–485. [CrossRef] [PubMed]

95. Bose, J.; Rodrigo-Moreno, A.; Shabala, S. ROS homeostasis in halophytes in the context of salinity stress tolerance. *J. Exp. Bot.* **2014**, *65*, 1241–1257. [CrossRef]

96. Pottosin, I.; Velarde-Buendia, A.M.; Bose, J.; Zepeda-Jazo, I.; Shabala, S.; Dobrovinskaya, O. Cross-talk between reactive oxygen species and polyamines in regulation of ion transport across the plasma membrane: Implications for plant adaptive responses. *J. Exp. Bot.* **2014**, *65*, 1271–1283. [CrossRef]

97. Zhu, J.K. Regulation of ion homeostasis under salt stress. *Curr. Opin. Plant Biol.* **2003**, *6*, 441–445. [CrossRef]

98. Shabala, S.; Cuin, T.A. Potassium transport and plant salt tolerance. *Physiol. Plant* **2008**, *133*, 651–669. [CrossRef]

99. Jones, J.D.G.; Dangl, J.L. The plant immune system. *Nature* **2006**, *444*, 323–329. [CrossRef]

100. Wu, Y.; Zhou, J.M. Receptor-like kinases in plant innate immunity. *J. Integr. Plant Biol.* **2013**, *55*, 1271–1286. [CrossRef]

101. Tang, D.Z.; Wang, G.X.; Zhou, J.M. Receptor kinases in plant-pathogen interactions: More than pattern recognition. *Plant Cell* **2017**, *29*, 618–637. [CrossRef]

102. Couto, D.; Zipfel, C. Regulation of pattern recognition receptor signalling in plants. *Nat. Rev. Immunol.* **2016**, *16*, 537–552. [CrossRef]

103. Cui, H.T.; Tsuda, K.; Parker, J.E. Effector-Triggered Immunity: From pathogen perception to robust defense. *Annu. Rev. Plant Biol.* **2015**, *66*, 487–511. [CrossRef] [PubMed]

104. Wang, W.; Feng, B.M.; Zhou, J.M.; Tang, D.Z. Plant immune signaling: Advancing on two frontiers. *J. Integr. Plant Biol.* **2020**, *62*, 2–24. [CrossRef]

105. Kurosaki, F.; Nishi, A. Stimulation of calcium influx and calcium cascade by cyclic AMP in cultured carrot cells. *Arch. Biochem. Biophys.* **1993**, *302*, 144–151. [CrossRef] [PubMed]

106. Cooke, C.J.; Smith, C.J.; Walton, T.J.; Newton, R.P. Evidence that cyclic-AMP is involved in the hypersensitive response of *Medicago-sativa* to a fungal elicitor. *Phytochemistry* **1994**, *35*, 889–895. [CrossRef]

107. Zhao, J.; Guo, Y.Q.; Fujita, K.; Sakai, K. Involvement of cAMP signaling in elicitor-induced phytoalexin accumulation in *Cupressus lusitanica* cell cultures. *New Phytol.* **2004**, *161*, 723–733. [CrossRef]

108. Jiang, J.; Fan, L.W.; Wu, W.H. Evidences for involvement of endogenous cAMP in Arabidopsis defense responses to Verticillium toxins. *Cell Res.* **2005**, *15*, 585–592. [CrossRef] [PubMed]

109. Ma, W.; Qi, Z.; Smigel, A.; Walker, R.K.; Verma, R.; Berkowitz, G.A. Ca2+, cAMP, and transduction of non-self perception during plant immune responses. *Proc. Natl. Acad. Sci. USA* **2009**, *106*, 20995–21000. [CrossRef]

110. Lomovatskaya, L.A.; Kuzakova, O.V.; Romanenko, A.S.; Goncharova, A.M. Activities of adenylate cyclase and changes in cAMP concentration in root cells of pea seedlings infected with mutualists and phytopathogens. *Russ. J. Plant Physiol.* **2018**, *65*, 588–597. [CrossRef]

111. Mauch-Mani, B.; Slusarenko, A.J. Production of salicylic acid precursors is a major function of phenylalanine ammonia-lyase in the resistance of Arabidopsis to *Peronospora parasitica*. *Plant Cell* **1996**, *8*, 203–212. [CrossRef] [PubMed]

112. Huang, J.; Gu, M.; Lai, Z.; Fan, B.; Shi, K.; Zhou, Y.H.; Yu, J.Q.; Chen, Z. Functional analysis of the Arabidopsis PAL gene family in plant growth, development, and response to environmental stress. *Plant Physiol.* **2010**, *153*, 1526–1538. [CrossRef] [PubMed]

113. Shine, M.B.; Yang, J.W.; El-Habbak, M.; Nagyabhyru, P.; Fu, D.Q.; Navarre, D.; Ghabrial, S.; Kachroo, P.; Kachroo, A. Cooperative functioning between phenylalanine ammonia lyase and isochorismate synthase activities contributes to salicylic acid biosynthesis in soybean. *New Phytol.* **2016**, *212*, 627–636. [CrossRef]

114. Torres, M.A.; Jones, J.D.G.; Dangl, J.L. Reactive oxygen species signaling in response to pathogens. *Plant Physiol.* **2006**, *141*, 373–378. [CrossRef] [PubMed]

115. Torres, M.A. ROS in biotic interactions. *Physiol. Plant* **2010**, *138*, 414–429. [CrossRef]

116. Qi, J.S.; Wang, J.L.; Gong, Z.Z.; Zhou, J.M. Apoplastic ROS signaling in plant immunity. *Curr. Opin. Plant Biol.* **2017**, *38*, 92–100. [CrossRef]

117. Bolwell, G.P. Role of active oxygen species and NO in plant defence responses. *Curr. Opin. Plant Biol.* **1999**, *2*, 287–294. [CrossRef]

118. Bindschedler, L.V.; Minibayeva, F.; Gardner, S.L.; Gerrish, C.; Davies, D.R.; Bolwell, G.P. Early signalling events in the apoplastic oxidative burst in suspension cultured French bean cells involve cAMP and Ca2+. *New Phytol.* **2001**, *151*, 185–194. [CrossRef]

119. Davies, D.R.; Bindschedler, L.V.; Strickland, T.S.; Bolwell, G.P. Production of reactive oxygen species in *Arabidopsis thaliana* cell suspension cultures in response to an elicitor from *Fusarium oxysporum*: Implications for basal resistance. *J. Exp. Bot.* **2006**, *57*, 1817–1827. [CrossRef]

120. Clough, S.J.; Fengler, K.A.; Yu, I.C.; Lippok, B.; Smith, R.K.; Bent, A.F. The Arabidopsis dnd1 "defense, no death" gene encodes a mutated cyclic nucleotide-gated ion channel. *Proc. Natl. Acad. Sci. USA* **2000**, *97*, 9323–9328. [CrossRef]

121. Balague, C.; Lin, B.Q.; Alcon, C.; Flottes, G.; Malmstrom, S.; Kohler, C.; Neuhaus, G.; Pelletier, G.; Gaymard, F.; Roby, D. HLM1, an essential signaling component in the hypersensitive response, is a member of the cyclic nucleotide-gated channel ion channel family. *Plant Cell* **2003**, *15*, 365–379. [CrossRef]

122. Yoshioka, K.; Moeder, W.; Kang, H.G.; Kachroo, P.; Masmoudi, K.; Berkowitz, G.; Klessig, D.F. The chimeric Arabidopsis CYCLIC NUCLEOTIDE-GATED ION CHANNEL11/12 activates multiple pathogen resistance responses. *Plant Cell* **2006**, *18*, 747–763. [CrossRef] [PubMed]

123. Dangl, J.L.; Dietrich, R.A.; Richberg, M.H. Death don't have no mercy: Cell death programs in plant-microbe interactions. *Plant Cell* **1996**, *8*, 1793–1807. [CrossRef]

124. Grant, M.; Brown, I.; Adams, S.; Knight, M.; Ainslie, A.; Mansfield, J. The RPM1 plant disease resistance gene facilitates a rapid and sustained increase in cytosolic calcium that is necessary for the oxidative burst and hypersensitive cell death. *Plant J.* **2000**, *23*, 441–450. [CrossRef]

125. Lecourieux, D.; Raneva, R.; Pugin, A. Calcium in plant defence-signalling pathways. *New Phytol.* **2006**, *171*, 249–269. [CrossRef] [PubMed]

126. Lamotte, O.; Courtois, C.; Dobrowolska, G.; Besson, A.; Pugin, A.; Wendehenne, D. Mechanisms of nitric-oxide-induced increase of free cytosolic Ca2+ concentration in *Nicotiana plumbaginifolia* cells. *Free Radic. Biol. Med.* **2006**, *40*, 1369–1376. [CrossRef]

127. Delledonne, M.; Xia, Y.J.; Dixon, R.A.; Lamb, C. Nitric oxide functions as a signal in plant disease resistance. *Nature* **1998**, *394*, 585–588. [CrossRef]

128. Durner, J.; Wendehenne, D.; Klessig, D.F. Defense gene induction in tobacco by nitric oxide, cyclic GMP, and cyclic ADP-ribose. *Proc. Natl. Acad. Sci. USA* **1998**, *95*, 10328–10333. [CrossRef]

129. de Jong, C.F.; Laxalt, A.M.; Bargmann, B.O.R.; de Wit, P.J.G.M.; Joosten, M.H.A.J.; Munnik, T. Phosphatidic acid accumulation is an early response in the Cf-4/Avr4 interaction. *Plant J.* **2004**, *39*, 1–12. [CrossRef]

130. Andersson, M.X.; Kourtchenko, O.; Dangl, J.L.; Mackey, D.; Ellerstrom, M. Phospholipase-dependent signalling during the AvrRpm1- and AvrRpt2-induced disease resistance responses in *Arabidopsis thaliana*. *Plant J.* **2006**, *47*, 947–959. [CrossRef]

131. D'Ambrosio, J.M.; Couto, D.; Fabro, G.; Scuffi, D.; Lamattina, L.; Munnik, T.; Andersson, M.X.; Alvarez, M.E.; Zipfel, C.; Laxalt, A.M. Phospholipase C2 affects MAMP-triggered immunity by modulating ROS production. *Plant Physiol.* **2017**, *175*, 970–981. [CrossRef]

132. Yang, T.B.; Poovaiah, B.W. A calmodulin-binding/CGCG box DNA-binding protein family involved in multiple signaling pathways in plants. *J. Biol. Chem.* **2002**, *277*, 45049–45058. [CrossRef] [PubMed]

133. Du, L.Q.; Ali, G.S.; Simons, K.A.; Hou, J.G.; Yang, T.B.; Reddy, A.S.N.; Poovaiah, B.W. Ca^{2+}/calmodulin regulates salicylic-acid-mediated plant immunity. *Nature* **2009**, *457*, 1154–1158. [CrossRef]

134. Zhou, W.B.; Freed, C.R. DJ-1 up-regulates glutathione synthesis during oxidative stress and inhibits A53T alpha-synuclein toxicity. *J. Biol. Chem.* **2005**, *280*, 43150–43158. [CrossRef]

135. Wei, H.R.; Persson, S.; Mehta, T.; Srinivasasainagendra, V.; Chen, L.; Page, G.P.; Somerville, C.; Loraine, A. Transcriptional coordination of the metabolic network in Arabidopsis. *Plant Physiol.* **2006**, *142*, 762–774. [CrossRef]

136. Shirasu, K. The HSP90-SGT1 Chaperone complex for NLR immune sensors. *Annu. Rev. Plant Biol.* **2009**, *60*, 139–164. [CrossRef]

137. Yadeta, K.A.; Elmore, J.M.; Creer, A.Y.; Feng, B.M.; Franco, J.Y.; Rufian, J.S.; He, P.; Phinney, B.; Coaker, G. A Cysteine-rich protein kinase associates with a membrane immune complex and the cysteine residues are required for cell death. *Plant Physiol.* **2017**, *173*, 771–787. [CrossRef]

138. Ito, M.; Takahashi, H.; Sawasaki, T.; Ohnishi, K.; Hikichi, Y.; Kiba, A. Novel type of adenylyl cyclase participates in tabtoxinine-beta-lactam-induced cell death and occurrence of wildfire disease in *Nicotiana benthamiana*. *Plant Signal. Behav.* **2014**, *9*, 27420. [CrossRef]

139. Rich, T.C.; Webb, K.J.; Leavesley, S.J. Can we decipher the information content contained within cyclic nucleotide signals? *J. Gen. Physiol.* **2014**, *143*, 17–27. [CrossRef]

140. Wheeler, J.I.; Freihat, L.; Irving, H.R. A cyclic nucleotide sensitive promoter reporter system suitable for bacteria and plant cells. *BMC Biotechnol.* **2013**, *13*, 97. [CrossRef]

141. Jiang, J.Y.; Falcone, J.L.; Curci, S.; Hofer, A.M. Interrogating cyclic AMP signaling using optical approaches. *Cell Calcium* **2017**, *64*, 47–56. [CrossRef] [PubMed]

142. Klausen, C.; Kaiser, F.; Stuven, B.; Hansen, J.N.; Wachten, D. Elucidating cyclic AMP signaling in subcellular domains with optogenetic tools and fluorescent biosensors. *Biochem. Soc. Trans.* **2019**, *47*, 1733–1747. [CrossRef] [PubMed]

143. Rahman, N.; Buck, J.; Levin, L.R. pH sensing via bicarbonate-regulated "soluble" adenylyl cyclase (sAC). *Front. Physiol.* **2013**, *4*, 343. [CrossRef]

144. Paramonov, V.M.; Mamaeva, V.; Sahlgren, C.; Rivero-Muller, A. Genetically-encoded tools for cAMP probing and modulation in living systems. *Front. Pharmacol.* **2015**, *6*, 196. [CrossRef] [PubMed]

International Journal of
Molecular Sciences

Article

Cytological and Gene Profile Expression Analysis Reveals Modification in Metabolic Pathways and Catalytic Activities Induce Resistance in *Botrytis cinerea* Against Iprodione Isolated From Tomato

Ambreen Maqsood [1,2], Chaorong Wu [1], Sunny Ahmar [3] and Haiyan Wu [1,*]

[1] Guangxi Key Laboratory of Agric-Environment and Agric-Products Safety, Agricultural College of Guangxi University, Nanning 530004, Guangxi, China; ambreenagrarian@gmail.com (A.M.); nongyekejian@163.com (C.W.)

[2] Department of Plant Pathology, University College of Agriculture and Environmental Sciences, The Islamia University of Bahawalpur, Bahawalpur 63100, Pakistan

[3] National Key Laboratory of Crop Improvement Genetics, College of Plant Sciences and Technology, Huazhong Agricultural University, Wuhan 430070, Hubei, China; sunny.ahmar@yahoo.com

* Correspondence: whyzxb@gmail.com; Tel.: +86-771-3235-612

Received: 11 May 2020; Accepted: 6 July 2020; Published: 9 July 2020

Abstract: Grey mold is one of the most serious and catastrophic diseases, causing significant yield losses in fruits and vegetables worldwide. Iprodione is a broad spectrum agrochemical used as a foliar application as well as a seed protectant against many fungal and nematode diseases of fruits and vegetables from the last thirty years. The extensive use of agrochemicals produces resistance in plant pathogens and is the most devastating issue in food and agriculture. However, the molecular mechanism (whole transcriptomic analysis) of a resistant mutant of *B. cinerea* against iprodione is still unknown. In the present study, mycelial growth, sporulation, virulence, osmotic potential, cell membrane permeability, enzymatic activity, and whole transcriptomic analysis of UV (ultraviolet) mutagenic mutant and its wild type were performed to compare the fitness. The EC_{50} (half maximal effective concentration that inhibits the growth of mycelium) value of iprodione for 112 isolates of *B. cinerea* ranged from 0.07 to 0.87 µg/mL with an average (0.47 µg/mL) collected from tomato field of Guangxi Province China. Results also revealed that, among iprodione sensitive strains, only B67 strain induced two mutants, M0 and M1 after UV application. The EC_{50} of these induced mutants were 1025.74 µg/mL and 674.48 µg/mL, respectively, as compared to its wild type 1.12 µg/mL. Furthermore, mutant M0 showed higher mycelial growth sclerotia formation, virulence, and enzymatic activity than wild type W0 and M1 on potato dextrose agar (PDA) medium. The *bctubA* gene in the mutant M0 replaced TTC and GAT codon at position 593 and 599 by TTA and GAA, resulting in replacement of phenyl alanine into leucine (transversion C/A) and aspartic acid into glutamic acid (transversion T/C) respectively. In contrast, in *bctubB* gene, GAT codon at position 646 is replaced by AAT and aspartic acid converted into asparagine (transition G/A). RNA sequencing of the mutant and its wild type was performed without (M0, W0) and with iprodione treatment (M-ipro, W-ipro). The differential gene expression (DEG) identified 720 unigenes in mutant M-ipro than W-ipro after iprodione treatment (FDR $\leq$ 0.05 and log2FC $\geq$ 1). Seven DEGs were randomly selected for quantitative real time polymerase chain reaction to validate the RNA sequencing genes expression (log fold 2 value). The gene ontology (GO) enrichment and Kyoto encyclopedia genes and genomes (KEGG) pathway functional analyses indicated that DEG's mainly associated with lysophopholipase, carbohydrate metabolism, amino acid metabolism, catalytic activity, multifunctional genes (MFO), glutathione-S transferase (GST), drug sensitivity, and cytochrome P450 related genes are upregulated in mutant type (M0, M-ipro) as compared to its wild type (W0, W-ipro), may be related to induce resistant in mutants of *B. cinerea* against iprodione.

Keywords: *Botrytis cinerea*; tomato; iprodione; mutant; transcriptome analysis; metabolism; catalytic activity

---

1. Introduction

Grey mold disease caused by *Botrytis cinerea* is one of the most destructive disease on more than 200 plant species including various economically important crops like tomato, grapevines, pepper, cucumber and strawberry [1,2]. The annual economic losses of *B. cinerea* is more than $10 billion worldwide including fresh fruits and vegetables [3]. *B. cinerea* is a necrotrophic fungal plant pathogen of pre and post-harvest diseases with broad host range reproduces sexually and asexually [4]. *B. cinerea* produced micro and macro conidia on the surface of host plant cells [5,6]. These one-celled spores borne on multiple branches expressed its obvious symptoms as greyish to light brown mold leaves, stem, flowers and fruits of host plant [7]. Grey mold is more protruding during persistent rainy, heavy dew, and foggy weather around temperature 18–24 °C [8].

B. cinerea has a unique ability to survive in different environmental conditions in the form of conidia and seclerotia that make this fungus very destructive and hazardous of resistance development [1,9,10]. The tremendous results against *B. cinerea* have been attained by applying various agrochemicals on tomato, cucumber, vine grapes etc. Every year more than half billion-dollar cost of fungicides are sprayed against different pests [2]. Chemical control of grey mold currently approached by seven classes of fungicides including anilinopyrimidines (APs), dicarboximide (Dc's), methyl benzimidazole, carbamates (MBCs), hydroxyanilides (HAs), quinone outside inhibitors (QoIs), succinate dehydrogenase inhibitors (SDHIs), and phenylpyrroles (PPs) [11,12]. Extensive use of agrochemicals induces resistance in plant pathogens in field. Nowadays we selected those site specific fungicides that have high efficacy, low toxicity and little human health risk [13]. However, these characteristics somewhat offset by their susceptibility to resistance development [14]. Iprodione, a dicarboximide (Dc) fungicide, has been used commercially for more than 30 years to control a wide variety of fungal pathogens. Among dicarboximides (Dcs), iprodione is a broad spectrum contact fungicide used as a foliar application and seed protectant for many fruits and vegetable crops. It has both preventive and curative action [15]. It was first manufactured in 1990. The mode of action of iprodione is to obstruct the synthesis of RNA and DNA during the germination of many fungal spores as well as lower the activity of enzyme NADH cytochrome c reductase, so the production of lipids and membrane restricted, ultimately mycelial growth inhibited. It can be used as a wettable powder, granules dispersed in water, flow able for all crops at the rate of 450–750 g/L [16].

Dc-resistant strains (iprodione) have been reported in many plant pathogenic fungal species [17]. Previously Hamada et al. [18] obtained iprodione resistant mutants of *Rhizoctonia cerealis* collected from wheat in China, *Botryosphaeria dothidea* from pistachio in California [19], *Alternaria* isolates collected from pistachio in California [20], and *B. cinerea* from strawberry and tomato in Hubei province, China [21].

The point mutation in the two components of histidine kinase genes (Bcos1) responsible for resistance to iprodione has been identified in *B. cinerea* isolated from strawberry fruit [21]. Substitution at codon I365 was dominant to cause resistance against iprodione among various strains of *B. cinerea* [22–24]. BcNoxA and BcNoxB catalytic subunits are responsible for pathogenicity and the formation of spores, which provide a favorable environment to fungal sclerotic to survive under adverse environmental conditions [25]. Various isolates of *B. cinerea* may possess low, moderate, or high levels of resistance to iprodione [26]. Mostly field isolates possess moderate to low level resistance, whereas laboratory mutants have high level resistance [27].

B. cinerea is a famous plant pathogen for its aptitude to become resistant against different fungicides. Although resistance against a different group of fungicides induced by different mechanisms like structural alteration in binding sites of pathogens reduces the affinity of fungicides, up regulation of fungicide target genes, decomposition of active ingredients reduces the efflux of fungicide concentration [28]. The study of genomics, transcriptome, bioinformatics and proteomic provide a new array to explore resistance attributes. Rendering RNA sequencing the most effective and direct way to explore resistance genes in mutant species of *B. cinerea*.

Guangxi province of China is one of the major tomato producing area [29] and facing several challenges of fungal diseases including *B. cinerea*. Iprodione, fludioxonil, and tebuconazole fungicides have been widely used against fungal diseases in Guangxi province for many years. Thus it is necessary to study the sensitivity of *B. cinerea* against these fungicides and asses the resistance in isolates for these chemicals before they are widely used to control *B. cinerea*. In this study, *B. cinerea* isolates collected from tomato plants in Guangxi Province, China. The key objectives of the present study were to: (i) determine the prevalence and frequency of *B.cinerea*. (ii) determine the iprodione sensitivity of *B. cinerea* isolates; (iii) preliminarily evaluate the risk of *B. cinerea* resistance to iprodione and to characterize the iprodione induced mutants; (iv) asses the fitness stability and pathogenicity of iprodione resistant mutants; and (v) investigate the molecular mechanism responsible for the development of resistance in *B. cinerea* against iprodione.

2. Material and Methods

2.1. Collection of Samples and Chemicals

There were one hundred and twelve different isolates of *B. cinerea* collected from different locations of Tian Dong, Tian Yang County, Baise City, Guangxi province, a southern region of China during 2016–2018 (Supplementary Table S1). Single spore isolation was accomplished from diseased leaves and fruits, as described by Fernández et al. [30]. All isolates were stored at 20 °C on dried filter paper [31]. Iprodione (96.7%) active ingredient (a.i) fludioxonil (99.2% a.i.) tebuconazole (97% a.i.); original drug, (Shandong Weifang Runfeng Chemical Co., Ltd., Jinan, Shandong, China) used in this experiment. Iprodione (0.103 g) was dissolved in acetone (10 mL) for the preparation of the stock solution and stored at 4 °C at dark for further use.

2.2. Sensitivity of B. cinerea to Iprodione

To evaluate the drug sensitivity, all strains preliminarily tested at 0.1, 1, 10, 100 μg/mL different concentrations of iprodione based upon previous findings of Grabke et al. [17]. The particular above mentioned concentrations in acetone were amended into PDA medium to examine the inhibitory rate. The inhibitory effect of acetone on mycelial growth of *B. cinerea* was 0.00001% in the sensitivity analysis of iprodione content which is ignorable. The mycelial plug of 5 mm diameter of 3d old *B. cinerea* colony was placed in the center of the 90 mm petri plate that contains iprodione amended media. These plates incubated at 23 °C for 3 days and radial mycelial growth (colony diameter) of each isolate measured in Petri plate by using a scale in the perpendicular direction and 5 mm original plug subtracted from the whole measurement. The experiment preliminarily repeated thrice, and each treatment had 3 replicates with control (only PDA medium). Those isolates that grew successfully on iprodione amended PDA were considered as resistant and failed as sensitive. Data processing system statistics (DPS version 7.05, Zhejiang, China) was used to analyze the effect of different concentrations of fungicides on *B. cinerea* mycelial growth inhibition rate to inhibit the rate of probability, the value of the ordinate (y), the concentration of the agent to the value of the abscissa (x), obtained virulence linear regression equation $y = a + bx$, effective medium concentration (EC_{50}) value, and correlation coefficient (r).

2.3. Evaluation of B. cinerea Mutants Resistant to Iprodione

After measuring the colony diameter of *B. cinerea*, the plates were further analyzed to obtain resistant mutants induced by iprodione. The method of UV mutagenesis induced resistance into drug sensitive strains. The mycelial plug of two resistant strains was placed in 1 μg/mL PDA medium containing iprodione and incubated at 23 °C. After 3 days incubation, these plates put in preheated, 20W UV lamp, 25 cm irradiation. After applying the treatment, the plates were kept at the incubator in the dark for ten days. After observation, each isolate was transferred to a higher concentration of drug-containing culture plates and repeat the above UV-induced colonies taken at edge cake. The resistant strains were plugged into the PDA medium and continuously cultured for 15 generations and measured as described above. Only stable resistant mutants selected for further analysis (Supplementary Table S2).

2.4. Characteristics of Iprodione Mutants and Sensitive Isolates

Mycelial growth, sporulation, virulence, cell membrane permeability, osmotic potential, and enzymatic activities were performed to compare the fitness characteristics between the sensitive and resistant strains. Mycelial colony diameter and sporulation were assessed with or without iprodione amended PDA medium. The mycelial colony diameter measured perpendicularly after 24, 48, 96, and 72 h and sporulation after 12, 13, 14, and 15 days of incubation at 23 °C.

Mycelium growth assay was conducted on fungicide-free PDA. Mycelial plugs were cut from the borders of 3-day-old colony and transferred to the center of PDA plates. Four plates for each isolate were incubated at 23 °C in the dark and colony diameter was measured at two perpendicular directions after 60 h of incubation.

The virulence was assessed on detached tomato leaves as previously described by Fan et al. [32]. Fresh tomato leaves were washed with double distilled water then disinfected by dipping in 75% ethanol for 1–2 min followed by three washings of double distilled water and allow them to dry on filter paper at room temperature. Leaves were placed in a 15 cm petri dish and cover the petioles with a wet cotton ball for moisture. Each leaf was punctured with a sterile lancet (Yangzhou Shuangling Medical Appliance Co., Ltd., Shuangling, China) in the middle as previously applied by Fan et al. [32] than placed in 5 mm mycelial plug on top of the wounds. Lesion diameter recorded with the help of measuring tape from each leaf after 4 days of incubation in the dark at 23 °C.

Osmotic sensitivity was measured to evaluate the cell wall elasticity by adding 10, 20, 40, and 80 mg/mL NaCL in PDA medium after 3 days of incubation at 23 °C. Mycelial growth inhibition rate (MGIR) calculated by the formula MGIR (%) = (CK-N)/(CK-5) × 100, whereas CK (mm) is the control plate colony diameter, N (mm) is that of a plate containing NaCl amendment. To determine cell membrane permeability, the wild type and its mutant strains were first incubated in 100 mL potato dextrose broth in a conical flask. These conical flasks were placed in a continuous shaker at a temperature 23 °C for 3 days. A (5 mm) eight mycelial plug of *B. cinerea* was added in each 250 mL conical flask contains 20 mL solution of iprodione with a concentration of 0, 1, 5, and 10 µg/mL. The conductivities were detected after 0 h, 0.5 h, 1 h, 2 h, and 3 h with the help of DDS-11A conductivity meter (Nanjing T-Bota Scietech Instruments & Equipment Co., Ltd., Nanjing, China) after each treatment. The final conductivities were measured via boiling the mycelium in water for 5 min. Each treatment has three replications. The relative permeability was calculated by using this formula: Relative permeability (%) = (Ct − C0)/C × 100 whereas C0: initial conductivity value; Ct: conductivity value at a certain moment; C: After boiling treatment [14].

2.5. Enzymatic Activities of Iprodione Resistant Mutant and Its Corresponding Wild Type

Polyglacturonase (PG), Polymethylglacturonase (PMG) and Cellulase (CE) Performed by DNS (3,5-dinitrosalicylic acid) Method Previously Described by Jiang et al. [33]. All three isolates were grown on potato dextrose broth media (PDB) at 120 r/min Shaker culture at 23 °C for 3 days. 0.1 g mycelium was grounded in liquid nitrogen in precooled pestle and mortar. Add 5 mL of sodium acetate buffer (pH 5.5) and centrifuge at 16,000 rpm for 20 min. The supernatant was collected and stored at 4 °C for further enzyme analysis. The substrate used to estimate the PG activity was 1% polygalacturonic acid in 50 mM sodium acetate buffer. The reaction mixture contains 0.5 mL sample volume, 0.5 mL substrate and 1 mL sodium acetate buffer in an eppendorf tube. The mixture was incubated in water bath at 37 °C for 1 h. After incubation add 1.6 mL DNS and boil for 5 min. The absorbance was measured at 540 nm by using a spectrophotometer (Multiskan GO, ThermoFisher Scientific, Boston, MA, USA). The boiled enzyme was used as a control. The standard curve was drawn by taking different concentrations of galacturonic acid. The activity of CE and PMG was measured by using the above mentioned procedure except substrate. For PMG and CE 1% pectin added in 50 mM sodium acetate buffer and 1% CMC dissolved in 50 mM citric acid-sodium citrate buffer respectively. Total proteins were determined by coomassie brilliant blue method [34]. Standard solutions of proteins were prepared by Bovine serum mg/mL also used as control without sample (Supplementary Tables S3 and S4).

2.6. Transcriptome Analysis of Wild and Mutant Strain

To explore whole sequence analysis, highly iprodione resistant mutant M0 and its wild type W0 were selected according to their fitness stability. Three biological repeats of wild type and mutant were grown in 100 mL of PDB under 0 µg/mL iprodione treatment grouped as (W0, M0). Similarly, three biological repeats of wild type and mutant were grown in 100 mL of PDB amended with 1 µg/mL iprodione and cultured at 220 rpm for 48 hr at 28 °C in the dark grouped as (W-ipro, M-ipro). After 2 days mycelia of all treatments were collected, washed with double distilled water, frozen in liquid nitrogen and stored in freezer at −80 °C. Untreated wild type (W0), mutant (M0) samples used as a control.

Total RNA was extracted according to the RNA isolation kit (TRIzole reagent, Invitrogen, Carlsbad, California, CA, USA). The purification of RNA was estimated with the Nano Photometer spectrophotometer at 260/280 nm (IMPLEN, California, CA, USA) and integrity was evaluated by assay kit (Nano 6000, California, CA, USA) using Bioanalyzer 2100 system (Agilent Technologies, California, CA, USA). The library generation and RNA sequencing was carried out by staff at I-sanger cloud platform. The further library was prepared by using an NEB-Next Ultra RNA illumina platform [35]. The Illumina platform converts the sequenced image signal to a text signal via CASAVA base calling and stores it in fastq format as raw data. Quality assessment was performed on raw data of each sample including base quality, base error rate and base level distribution statistics to obtain high quality clean reads by using FASTQ for subsequent analysis. The clean reads were mapped into a transcript and compared to a reference genome using Tophat2 alignment software. Some transcripts without annotation to the reference genome were called new transcripts.

Differential Gene Expression

We identified differentially express genes of two phenotypical groups of strains (Mutant and wild type, with or without fungicide) by using DEseq2 [36]. To estimate the gene expression level fragment per Kilobase of exon model per million fragments mapped (FPKM) tool was used. The statistical difference among genes was analyzed using the recommended Benjamini-Hochberg correction method (p-value ≤ 0.05) for controlling the false discovery rate (FDR). Eventually, the fold change (log fold$_2$) and FDR values used as a key indicator the expression amount of different genes among samples and represent a heat map. Functional annotation of genes was performed as described by Wang et al. [14] and Cai et al. [37]. GO enrichment of differential expression of genes was implemented by the GO seq R packages (1.10.1) based on Wallenius' non-central hyper-geometric distribution. KEGG pathways enrichment statistical analysis was performed by KOBAS software.

2.7. Quantitative Real Time PCR Analysis

For the confirmation of differential gene expression levels attained from the RNA sequencing data analysis, the qRT-PCR investigation was carried out. Total RNA was extracted from mutant and wild type with and without treatment of EC$_{50}$ concentration of iprodione 1 µg/mL according to the kit instructions (TaKaRa Biotechnol. Co., Ltd., Dalian, China). For the preparation of reverse transcription first Single stranded cDNA was synthesized according to labelled kit instructions (TaKaRa Biotechnol. Co., Ltd., Dalian, China). The expression of seven genes were studied. Moreover, one control gene UBQ used as a reference gene. The primers were designed by using oligo software v7.37 and the specificity was confirmed by blast against *B. cinerea*. B.010 genome. The sequence of all primers were listed in (Supplementary Table S5). The length of primers fragment were between 19–23 base pair with melting temperature 80 °C. 1 µg RNA of each sample was first treated with RNase Free dHH$_2$O and 4 × gDNA wiper Mix (Nanjing Nuo Weizan Biotechnology Co., Ltd., Nanjing, China) for removal of contaminated DNA in the extract. For the preparation of reverse transcription reaction system, the reaction mixture consisted of template cDNA 2 µL, reverse primer 0.8 µL (5 µM), forward primer 0.8 µL (5 µM) and ChamQ SYBR Color qPCR Master Mix 16.5 µL (TaKaRa Biotechnol. Co., Ltd., Dalian, China) of total volume 20 µL. The qRT-PCR reaction was conducted in a thermal cycler (ABI 7500, Hangzhou Langji Scientific Instruments Co, Ltd., Hangzhou, China) with initial temperature 95 °C for 5 min, 40 cycles include melting at 95 °C for 5 s annealing for 30 s and finally extension at 72 °C for 40 s. Three biological repeats of each treatment were performed with triplicate of each gene reaction vs. reference gene. Then changing in fold expression of different genes of the mutant and wild type was evaluated by using algorithm $2^{-\Delta\Delta CT}$ value. All qRT-PCR data were analyzed by using Light Cycler® 480 software version 1.5.1 (Roche Diagnostics Corporation, Indianapolis, IN, USA).

2.8. DNA Extraction, Cloning and Sequence Analysis of the Tubulin Genes

For DNA extraction wild type W0 and its mutant M0, mycelium was cultured on potato dextrose broth (PDB) and incubated for 48 h at 28 °C under shaking condition (200 rpm). Mycelia was harvested and washed with sterilized water and ethylenediamine tetra acetic acid (EDTA). The DNA was extracted by cetyl trimethylammonium bromide (CTAB) method [6]. The specific primers β-TUB (F-5′-TGAAGGTATGGACGAGAT-3′) (R-5′-GCATCCTGGTATTGTTGA-3′) under accession number (XM_001560987.1) and α-TUB (F-5′GTTGGAGTTCTGTGTCTA-3′) (R-5′GTGGTCAAGATGGAGTTA-3′) under accession number (XM_001555875.1) were used to amplify the complete coding sequence (CDS) of two Tubulin genes *bctubA* and *bctubB*. Three biological replicates of each strain used for DNA extraction and the PCR reactions were conducted three times independently for each sample. The amplified PCR products were purified using a PCR Purification Kit (TIANGEN, Beijing, China), ligated into the pMD18-T Vector (TaKaRa Biotechnol. Co., Ltd., Dalian, China), and then sequenced by Sangon (Guangzhou, China). The exon sequences of the *bctubA* and *bctubB* genes were translated into amino acid sequences and aligned using DNAMAN8.0 software (Lynnon Biosoft, Quebec, Canada) to check the mutation point.

2.9. Statistical Data Analysis

All values related sensitivity, osmotic potential, enzymatic activities are mean of three replicates was analyzed using statistical software (DPS version 7.05, Zhejiang, China). The LSD test was used to determine significant differences ($\alpha = 0.05$). Pearson's correlation coefficients were calculated to evaluate the correlation of gene expression obtained by RNA-seq and qRT-PCR using Origin 9.0 software (Origin Lab, Newyork, USA). In the SAM method, the delta value was set to obtain an average. A false discovery rate (FDR) of 5% and the fold change cut-off value was established as 1.5. In LIMMA analysis, genes with a fold change >1.5 and $p < 0.05$ were considered as differentially expressed. Only the genes identified as differentially expressed by both SAM and LIMMA were considered.

3. Results

3.1. Sensitivity of B. cinerea to Iprodione

One hundred and twelve samples of *B. cinerea* were collected from different locations in tomato production area and tomato fields, Baise City, Guangxi Province, China. The sensitivity of B. *cinerea* was checked on PDA medium amended with iprodione at 0.01 µg/mL. The inhibition rate of all samples ranged from 7.69–74.35% with an average of 50.70% (Figure 1). The EC$_{50}$ of iprodione against all samples ranged from 0.07 to 0.87 µg/mL with an average of 0.47 µg/mL, indicating these *B. cinerea* isolates were susceptible to iprodione. The value of EC$_{50}$ of all isolates indicated that 0.47 µg/mL was an appropriate threshold concentration to assess iprodione resistance in the consequent experiments. Among them 5 isolates were highly sensitive to EC_{50} for 0.134 µg/mL iprodione and five isolates were moderately sensitive to EC_{50} for 0.434 µg/mL iprodione (Supplementary Tables S6–S9).

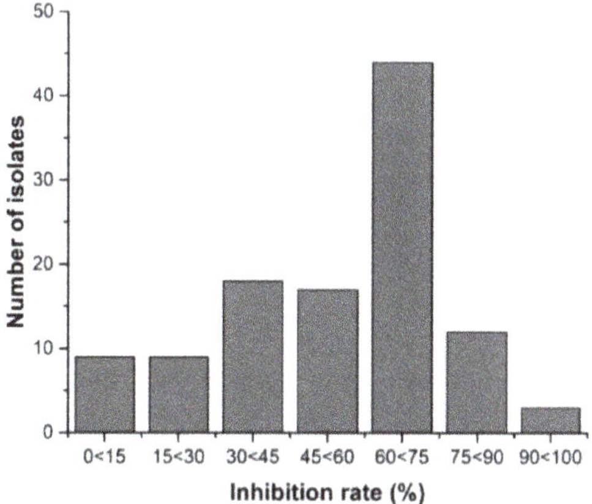

Figure 1. The inhibition rate of *B. cinerea* isolates against iprodione (0.01 µg/mL) collected from different areas of Guangxi Province China.

3.2. In Vitro Iprodione-Induced B. cinerea Mutants

UV radiation is a toxic mutagen and was expected to decrease the viability of the cells but also increase the probability of the emergence of mutants of drug sensitive strains. Five iprodione sensitive isolates of *B. cinerea* exposed to different concentrations of iprodione fungicide with 20W UV lamp were continuously cultured at 28 °C to induce rapid growth of mutants. The only strain B67 showed two mutants M1 and M0 respectively. The EC_{50} of these mutants were 674.48 µg/mL and 1025.74 µg/mL, respectively, and 597.63 and 906.94 times than that of wild isolate (EC_{50} was 1.12 µg/mL). However, other isolates showed higher sensitivity to iprodione and did not produce any mutant used as a control. These two mutants were continuously sub-cultured for 1, 5, 10, and 15 generations on drug free PDA medium for stability test (Supplementary Table S2).

3.3. Morphology and Physiology of Mutants

3.3.1. Iprodione Resistant Mutant's Mycelium Growth Rate and Sclerotia Formation

The results showed that mycelial growth of wild type on PDA medium is significantly higher than both mutants after 5 days at 28 °C (Figure 2A). Wild type (W0) showed maximum sclerotia formation on PDA after 8 days; in petri dish edges are produced around a small contiguous black sclerotia; did not spread throughout the surface of the medium. M0 produced dark grey to black sclerotia after 12 days and spread over the medium and M1 after 14 days produced sclerotia over the surface of the medium (Figure 2B). The mutant M1 after 10 days began to produce small contiguous black sclerotia circles at the edge of the dish and after 12 days spread on the petri dish. When PDA medium was amended with 100, 500, 600, and 1000 µg/mL iprodione, mutant M0 showed significantly high mycelial growth than M1 and wild type W0 (Figure 2C). Mutant M0 after 12 days began to produce fewer black sclerotia circles at the edge of the dish with 5 µg/mL iprodione containing media. In contrast, wild type failed to produce any (Supplementary Figure S1).

Wild type (W0)   Mutant (M0)   Mutant (M1)

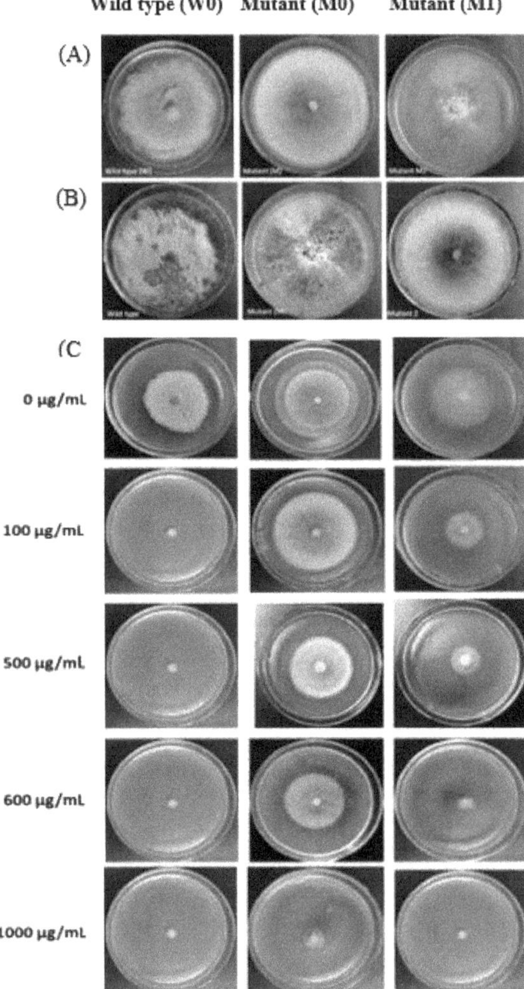

Figure 2. Colony morphology of *B. cinerea* Wild type (W0) and its mutant (M0, M2); (**A**) Mycelial growth of mutants (M0, M1) and its wild type on PDA medium after 5 days; (**B**) Sclerotia formation of wild type (W0) and its mutants (M0,M1) on PDA medium after 8, 12 and 14 days respectively; Sporulation: mean number ($\times10^6$) seclerotia per square centimeter, (**C**) Mycelial growth of wild type (W0) and its mutants (M0, M1) after exposed 100, 500, 600 and 1000 µg/mL to iprodione on PDA medium at 28 °C for 3 days.

3.3.2. Cell Membrane Permeability Osmotic Sensitivity Pathogenicity and Enzymatic Activity of Mutants and its Wild Type

Cell membrane permeability at four different concentrations (0, 1, 5, 10 µg/mL) of iprodione were measured, the relative rate of infiltration with the extension of the processing time increases and gradually stabilized in M1, indicating that the cell membrane permeability is significantly higher in M1 and wild strain than M0 (Supplementary Figure S2). Whereas, there was a no significant difference in osmotic potential of wild type and its mutants (Supplementary Figure S3). Protein concentration was calculated according to the sample suction photometric method. The remarkably highest protein contents were in W0 (1023.97 µg/mL) than M0 and M1 mutant (941.38, 908.02 µg/mL respectively (Supplementary Table S10). Enzymatic activities of both mutants and wild type were also measured. Although PG and CE activity was higher in M0 than M1 and wild type W0. (Supplementary Table S11). In pathogenicity assays, detached tomato leaves inoculated with both mutants (M0, M1) or wild type strain showed typical symptoms and lesions by W0 and M0, while the only PDA or control plants and M1 remained asymptomatic after 24 h of inoculation (Figure 3). These results showed that mutant M0 pathogenicity and enzymatic activity is more vulnerable than M1.

Figure 3. Virulence of *B. cinerea* Wild type (W0) and its mutnats (M0, M1) on detached tomato leaves after 24 h of inoculation.

3.3.3. Cross Resistance

The sensitivity of both *B. cinerea* mutants (M0, M1) were also determined against tebuconazole and fludioxinil using a discriminatory dose. Both iprodione mutants showed positive cross resistance against these fungicides.

3.4. Transcriptomic Data Analysis of Iprodione-Resistant Mutant and Its Wild Type after Exposed to Iprodione In Vitro

The results revealed that transcriptomic sequencing of twelve iprodine resistant mutant and wild type samples generating 187,062,138 raw reads after the screening and filtration of raw reads, a total of 186,026,964 clean reads were obtained. The percentage nucleotide quality score of more than 20 (Q20) was noted as high as 97.75%, and the percentage of Guanine and Cytosine GC (%) among all nucleotides was obtained as 47.33% (Table 1). According to Illuminia platform, the contrast efficiencies of mapped reads, uniquely mapped reads and multi mapped reads were 95.91%, 0.52%, and 95.2%, respectively, as compared to the reference genome (Table 1). A total number of transcripts was 20,375 and its length varied from 201 to1800 bp with an average 1000 bp (Supplementary Figure S4). On the basis of existing reference genome based assembly is performed by using compare software. To check the presence of novel transcripts, we combined the RNA-seq data of 4 samples with 3 biological repeats to identify novel transcripts, which are not assembled in the database. New transcript is obtained by comparing it with known transcript and further classified into 12 different class codes (Supplementary Figures S5 and S12). Out of 20,375 transcripts, 13,639 complete matches of intron chain, 3574 potentially novel isoforms, 1229 unknown intergenic transcript, and 372 generic exonic overlaps with a reference transcript were obtained during analysis. The raw data is submitted to NCBI under SRA number SRP254522.

Table 1. Details of raw and clean data of twelve transcriptomes of *B. cinerea* and the reference genomes.

Strain	Raw Reads	Clean Reads	Clean Bases	Mapped Reads	Q 20 Avg (%)	GC Avg (%)
M0	47,684,718	47,218,908	7,057,947,521	44,704,018 (95.8%)		
M-ipro	45,648,745	46,083,059	6,885,980,894	44,251,477 (96%)		
W0	47,109,858	46,628,979	6,970,197,563	44,597,601 (95.62%)	97.76	47.23
W-ipro	46,618,817	46,096,018	6,885,797,287	44,052,162 (96.22%)		
Total	187,062,138	186,026,964				

3.4.1. Functional Annotation of Transcripts and Unigenes

Almost all transcripts (13,703) and unigenes (11,698) sequences were alligned to the NCBI and annotated at least one of these six databases, Non Reductase (NR), SwissProt, Protein family (Pfam), Gene ontology (GO), Clustre of Orthologous Groups of Proteins (COG) and Kyoto Encyclopedia of Genes and Genomes (KEGG) (Supplementary Figure S6). The database indicated that maximum transcripts (13,687; 99%) and unigenes (11,685; 99%) were alligned by NR, whereas more than 50% of transcripts and unigenes were aligned to the COG, Pfam, and Swissprot databases. The minimum number of transcripts vs. unigenes was annotated by KEGG 33%. Moreover, 14 transcripts and 12 unigenes remained unannotated (Supplementary Table S13 Excel sheet).

3.4.2. Discovery of New Genes

A total of 1024 new genes were discovered according to the above mentioned six databases. COG annotated 150 new genes in 14 different categories, GO 160 new genes in 19 different compartments (Figure 4), and KEGG 17 in 12 different disciplines. Furthermore, GO database is secondarily classified into three categories, namely molecular functions (47), cellular components (57), and biological processes (56) of new genes were explored. The highest number of new genes were involved in binding, metabolism, and cellular processes. COG annotated a maximum 84 new genes, which were poorly characterized, and 44 genes were involved in the repair, replication, and recombination of RNA. KEGG aligned the highest number of genes in the biosynthesis of the secondary metabolism process.

Figure 4. Histogram of KEGG pathways.

3.4.3. Differential Genes Expression of Iprodione Resistant Mutant and its Wild Type After Exposure to Iprodione

After obtaining the clean reads, the differential expression of unigenes was analyzed by using software DESeq2. A total of 281 unigenes were expressed in mutant type (M0), including 166 up-regulated and 115 downregulated with or without iprodione treatment. Meanwhile, wild type (W0) showed 99 unigenes expressions in which 85 were up-regulated and 14 downregulated (Table 2).

Table 2. The total number of DEG'S in wild type and its mutant with or without iprodione.

Strain	All DEG	DEG Upregulated	DEG Downregulated
M-ipro vs. W-ipro	1897	886	1011
M0 vs. W0	1707	890	817
M-ipro vs. M0	281	166	115
W-ipro vs. W0	99	85	14

The analysis showed that mutant and wild type shared 19 DEGs and 262 and 80 unique DEGs were detected in mutant and wild type respectively when exposed to iprodione. Overall, 1897 DEGs were detected after iprodione exposure and 1707 without iprodione between mutant and wild type. Furthermore, M-ipro vs. W-ipr0 and M0 vs. Wo share 1192 common unigenes and 720 unigenes are upregulated in mutant corresponding to wild type after iprodione treatment (Supplementary Figure S7).

These results demonstrated that the DEGs pattern significantly changed in mutant and wild type with or without iprodione exposure, suggesting that some compounds may be specific to produce resistance in mutants against iprodione treatment.To understand the mechanism of resistance in *B cinerea* against iprodione, the gene function, expression level, and expression difference were analyzed in gene set analysis. Genes of the same function were located on one transcript, particularly within the three loci easy to annotate by gene ontology (GO) rather than those situated on different transcripts. The genes related to metabolic process, localization, ATP binding, transmembrane transport antibiotic activity, and the cellular process were most abundant in mutant type (M-ipro vs. M0) relative to wild type (W-ipro vs. W0) (Figure 5).

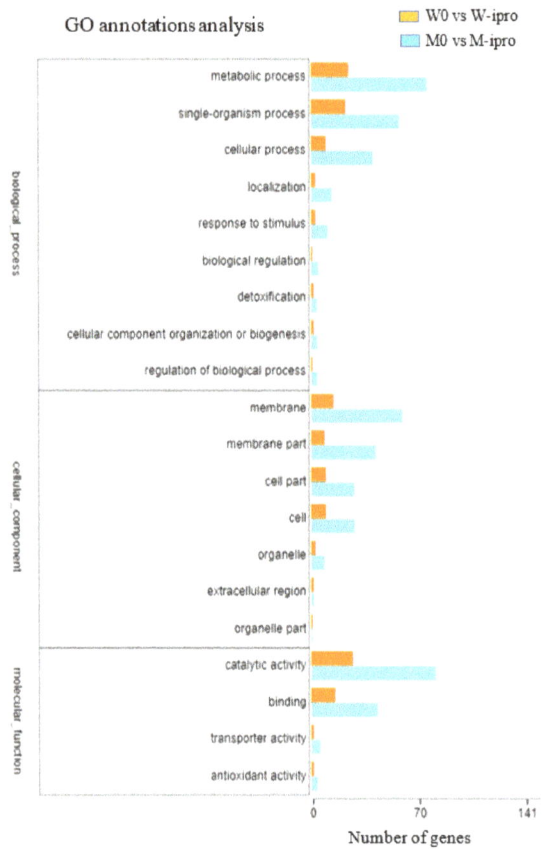

Figure 5. Histogram showed a significant difference (*p*-value < 0.001) in W0 vs. W-ipro and M0 vs. M-ipro by GO annotation. The X-axis represents the number of genes and Y-axis represents the three GO terms under biological processes, cellular component and molecular function.

To elucidate the difference between mutant and wild type, the expression pattern of four treatments were divided into a hierrachial clustering analysis (Supplementary Figure S8). Out of 1912 genes, we focused our attention on highly expressed genes in M-ipro that were more or less related to iprodione resistance and assembled them into 12 small clusters according to their functions (Figure 6). In clusture I, MFO (multifunctional genes) were analyzed (*BCIN_06g07150, BCIN_09g01190, BCIN_07g01720, BCIN_02g04800, Bcape1*) highly expressed in M-ipro and involved in molecular and biological functions of *B. cinerea*. Among the five genes of aspartic proteinase family (*bcap1, bcap4, bcap6, bcap8, bcap10*), only *bcap8* log2 value was significantly high in M-ipro, while the remaining genes did not show any significant difference in both mutant and wild type with or without iprodione treatment. In cluster II genes, set ABC transporter genes (*BcatrD*, *Bcbfr1*) were highly expressed in mutant type M-ipro than M0 and downregulated in wild type (W-ipro, W0). Various cytochrome p450 coding genes were expressed in a comprehensive data base and assembled in cluster III. Almost all genes were depressed in wild type (W-ipro, W0). *BccpoA90*, *Bccyp51*, and *BCIN_15g04350* expression was high in M0 and M-ipro (Figure 6) except *BCIN_02g00240*. Genes that were involved in amino acid metabolism exhibited high variability among mutant (M-ipro, M0) and wild type (W1, W0). Cluster IV has transmembrane transporter genes which were also expressed in both wild and mutant type. Only *BCIN_14g04470* and *BCIN_12g02430* were highly downregulated in W-ipro and W0 and upregulated in M0 and M-ipro.

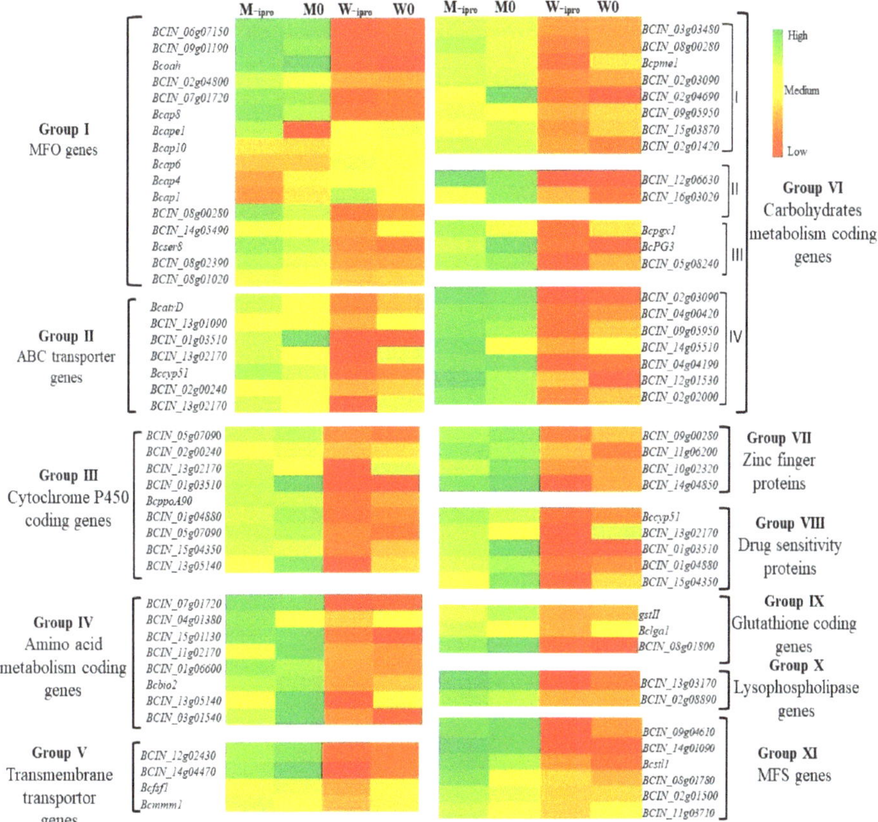

Figure 6. Variability in altered genes of wild type and its corresponding mutant with or without iprodione application represented in the form of Heat map. Genes with the same annotation modulated in a similar group. Twelve major groups are displayed in the heat map (Group I, II, III, IV, V, VI, VII, VIII, IX, X, XI, XII). Group VII have four subgroups I (Glycosyl family), II (Cellulase genes). III (Polygalacturose genes) and IV (carbohydrate metabolism).

In contrast, many more genes had no significant difference among all treatments (Supplementary Table S14 Excel sheet). Moreover, the maximum high expression was detected in carbohydrate metabolism, glycosyl family, polyglacturose family and cellulose related genes attributed in cluster VI. *BCIN_02g04690* log fold 2 values significantly high in M0 and low in W0 related to carbohydrate binding. In cluster VII several genes were upregulated belongs to zinc finger proteins in mutant type (M-ipro, M0). Intriguingly *BCIN_10g0230*, *BCIN_09g00280*, and *BCIN_14g04850* genes were highly upregulated in M0 rather than M-ipro and downregulated in wild type (W-ipro, W0). Drug sensitive proteins were highly downregulated in wild type after exposure to iprodione pooled in cluster VIII. *BCIN_15g04850* and *BCIN_13g05140*, highly upregulated in mutant without iprodione treatment instead of its application (Figure 6). Glutathione-S transferase encoding genes, i.e., GSt enzymes play an important role to detoxify the chemicals. *BCIN_08g01800* gene regulating glutathione enzyme was more expressed in mutant without iprodione application than wild type. Clusture X, XI represents the lysophospholipase and super family genes (MFS). Two sugar transporter genes (*BCIN_09g04610*, *BCIN_14g01090*) of super family upregulated after exposure to iprodione fungicide in the resistant mutant. *BCIN_13g03170* lysophospholipase gene expression log fold 2 value was −0.629 in W0, showed decrease −1.072 in W-ipro while expression level was increased in mutant after exposure to iprodione (Table 3). On the basis of these results, the presence of resistance in mutant strain may not be due to particular resistant gene against chemical or inactivation of enzymes and metabolic process. These data base also suggested that synergistic and combination of several genes belong to different functions or families generate resistance in mutants against a particular drug or multi drugs.

Table 3. Major genes related to resistance in *B. cinerea* wild type and its mutant with or without iprodione application.

Gene ID	Log$_2$ Fold Change Value				Function Annotation
	M-ipro vs. W-ipro	M0 vs. W0	M-ipro vs. M0	W-ipro vs. W0	
BCIN_02g08890	9.104	8.746	0.565	0	Integral cellular component domain
BCIN_13g03170	8.222	5.529	0.378	−0.231	Glycerophospholipid metabolism
BCIN_06g07150	6.213	5.503	0.12	−0.599	Energy production and conservation
BCIN_09g01190	4.878470612	6.006	−0.511	1.2044	Zinc finger proteins
BCIN_13g04220	8.618	−7.08119	8.576	−21.308	Chaperones proteins
BCIN_08g06520	7.317	4.154955	1.07079	−4.216	Oxygenase super family
Bcpg2	4.978	−0.967	2.896255	−2.978	Glycosyl hydrolase family
Bcoah	4.878	6.600	−0.511	1.204	Isocitrtae lyase family
BcatrD	2.529	1.66	0.266	−0.249	ABC transporter proteins
Bclga1	2.266	0.745	0.865	−0.705	Glutathione S transferase proteins
BCIN_12g0243	2.597	3.064	−0.407	−0.046	Catalyze aminotransferase acid
Bcrds2	1.229	1.092	−0.062	−0.192	Drug sensitivity proteins
Bcltf1	0.531	1.292	−1.108	−0.3241	Sexual development transcription factor NsdD
BCIN_06g02680	2.00	0.835	1.292	−1.108	Phenylalanine aminomutase enzyme
BCIN_08g02390	2.837039	2.628415	0.436	0.223	Heat shock binding proteins
BCIN_02g01500	1.929	0.199	0.318	1.779	Carbohydrate transport and metabolism
BCIN_04g01200	3.721	2.744	1.035	0.061	Steroid biosynthesis proteins
BCIN_08g01540	2.119	1.637	0.0715	−0.411	Steroid biosynthesis proteins
BCIN_03g03480	4.555	4.296	0.285	0.019	Xylanase proteins
Bcpgx1	2.536	1.212	0.496	−0.823	Polygalacturose proteins
BCIN_02g04800	4.861	4.796	0.611	0.536	Amino acid metabolism proteins
BCIN_01g01440	2.811	−1.449	2.354	−1.871	Dioxygenase TDA family
BCIN_12g04510	1.087	0.736	0.193	−0.165	Histidine kinase activity proteins
BCIN_01g03510	4.531	3.321	−2.06396	0.7321	Cytochrome P450
BCIN_02g08880	9.495	8.718	0.024	−0.752	Unknown function hypothetical proteins
BCIN_02g08890	9.104	8.745	0.565	0	Unknown function hypothetical proteins

3.5. qRT-PCR Amplification of Some Specific Genes

In order to verify the transcriptomic analysis of mutant and wild type *B. cinerea*, a total of eight highly expressed encoding genes of lysophospholipase (*BCIN_13g03170*), drug sensitivity (*BCIN_04g01200*), cytochrome p450 (*Bccyp51*), cellulase (*BCIN_12g06630*), glutathione-S- transferase (*BCIN_08g01800*), oxaloacetate acetyl hydrolase (*bcoah*), cellulase (*BCIN_12g06630*), glycosyl family (*BCIN_12g01530*), and reference gene *UBQ* have been selected for expression analysis of RT-qPCR. Among these genes gluthathione-S transferase gene (*BCIN_18g01800*) highly upregulated in mutant type with or without iprodione application (Figure 7C). Moreover, two lysophopholipase genes (*BCIN_13g01370*, *bcoah*) expression in the mutant (M-ipro) were highly upregulated as compared to M0 and wild type (W-ipro, W0) (Figure 7A,B). The drug sensitivity and cytochrome familyP450 genes (*Bccyp51*, *BCIN_14g01200*) were downregulated in wild type after iprodione treatment (Figure 7E,F). The cellulase gene (*BCIN_08g01800*) was highly expressed in all treatments (Figure 7G).

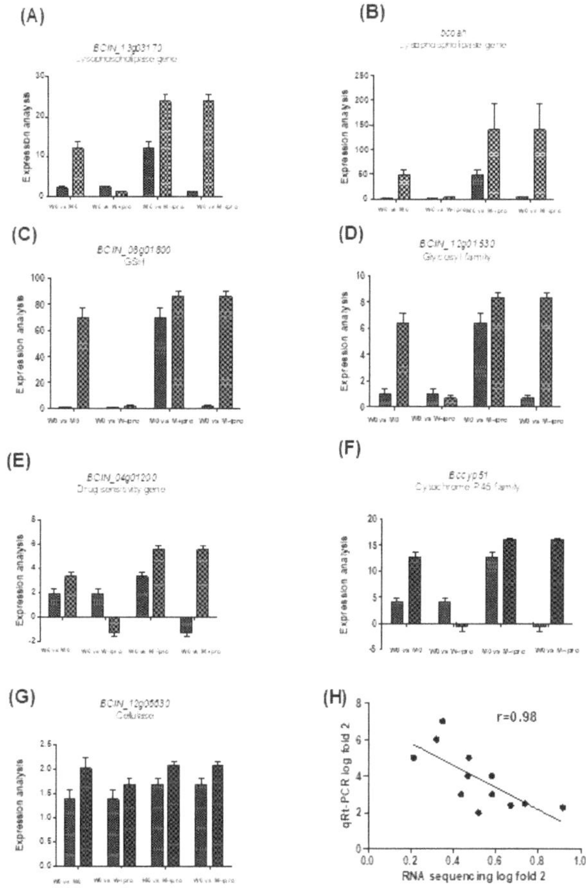

Figure 7. Validation of seven DEG's log fold 2 value of RNA seq by qRT-PCR of wild type and its corresponding mutant with or without iprodione treatment. (**A,B**) lysopholipase genes, (**C**) Glutathione-S transferase gene, (**D**) Glycosyl family carbohydrate metabolism, (**E**) Drug sensitivity genes, (**F**) Cytochrome P450 family, (**G**) Cellulase genes, (**H**) Pearson correlation of log fold 2 value of qRT-PCR and RNA sequencing of wild type and its mutant after iprodione application. The mRNA abundance was normalized by using the reference gene UBQ and relative expression (log fold 2) was valued as $2^{-\Delta\Delta CT}$. All values of qRT-PCR represents as mean $\pm$ SD ($n = 7$).

3.6. Detection of Mutations in Tubulin Genes in Iprodione Mutants

Genome sequencing of wild type (W0) and its mutant (M0) showed two tubulin genes encoding *bctubA* and *bctubB*. The coding region of *bctubA* had 1985 nucleotides encoding 661 amino acids, which was 65% match with *bctubA* B05.10 strain of *B. cinerea* (Gene Bank accession number: XP_024546500.1). While the coding region of *bctubB* gene had 1919 nucleotides encoding 639 aminoacids was match with *bctubB* (*B. cinerea* B05.10) (GeneBank accession number: XP_024546928.1). The *bctubA* gene in mutant (M0) replaced TTC and GAT codon at position 593 and 599 by TTA and GAA, resulting in replacement of phenyl alanine into leucine (transversion C/A) and aspartic acid into glutamic acid (transversion T/C) respectively. Whereas, in *bctubB* gene GAT codon at position 646 replaced by AAT and aspartic acid converted into asparagine (transition G/A).(Figure 8) No point mutation was found in wild type as compared to control (*B. cinerea* B05.10 strain).

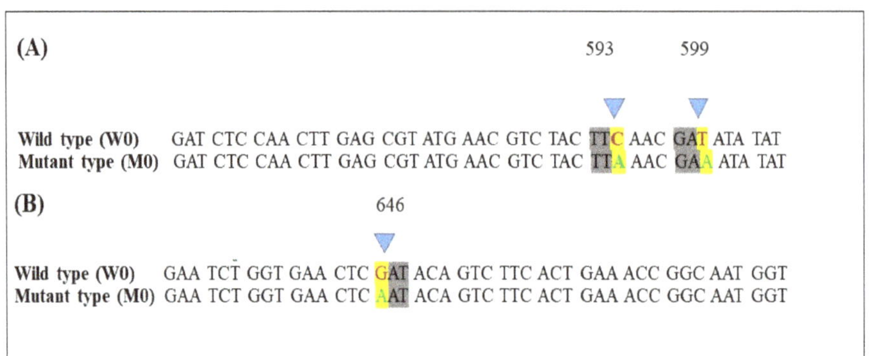

Figure 8. Amino acid sequence alignment of *B. cinerea* tubulin genes in wild type (W0) and its mutant (M0), (**A**): Mutation variation in *bctubA* gene at position 593 and 599, (**B**): Mutation variation in *bctubB* gene at position 646.

4. Discussion

Fungicide resistance in different pathogens has been a major problem in crop protection worldwide in two decades. The extensive use of fungicides to control pathogens in same area for several seasons creating this problem; as a result the efficacy of fungicides decreasing and resistance induced into the pathogens against a specific or multiple fungicides is increasing. Therefore, it is necessary to evaluate the resistance risk in the lab before new fungicides are widely practice into the field. In Guangxi province, grey mould, caused by the fungal pathogen *B. cinerea*, is one of the most devastating tomato diseases, and the control of this disease is mainly by the application of chemicals. In the present study, 112 isolates of *B. cinerea* collected from major tomato production area of Guangxi province, China and screened with different concentrations of iprodione and found that toxicity steadily increased in mutants. This is considered as the first report to assess iprodione sensitivity of *B. cinerea* collected from Guangxi Province China. Previously sensitive isolates of *B. cinerea* on tomato were detected in Germany [38]. All strains were sensitive to iprodione from 0.07 to 0.87 µg/mL with an average of 0.47 µg/mL. These results are similar to *B. cinerea* collected from strawberry, a procymidone and zoxamide sensitive strain from Hubei province having EC_{50} value of 0.25 µg/mL and 0.360 µg/mL, respectively [22]. No 100% sensitive strain was discovered in our collected samples because iprodione is a site specific fungicide with high efficacy, less toxicity and lower application rate. The high dozes or sustained application of fungicides leads to put selection pressure to develop fungicide resistance [39]. In order to solve this problem, there was an urgent need to develop effective resistance management strategies.

High frequencies of *B. cinerea* resistant isolates have been recorded not only iprodione but also to various groups of fungicides including DC, MBC, and PP across the globe [40–42]. Although field resistant mutants of iprodione was documented in Northeatern part (21%), Henan (8%), and also in Anhuai (5%) of China [18,43]. This study was showed low frequency of resistant mutants as compared to the previous one, but with the passage of time resistance level will increase with rapid and continuous application of DC's fungicides. In the present work, UV irradiation used to detect specific iprodione resistant mutants to assess the risk from emergence strain. After stability and sensitivity testing the EC_{50} of UV mutants (M0, M1) exceeded 1 µg/mL which was higher 1025.74 µg/mL and 674.48 µg/mL respectively than corresponding wild strain (W0). The cell membrane permeability was significantly increased in M1 and wild type than M0 which indicate M0 is more stable. In contrast CE and PG enzyme activity were higher in M0. Polygalactronase family genes (*Bcpgx1* and *Bcpg3*) involved in pathogenicity expression were higher in M-ipro than wild type. These findings were distinction with the Guo et al. [44] found that resistant strains when continuously dealt with fungicides, pathogens lose its potential and viability resulted in failure of infection. The virulence of M0 like as wild type while M1 results supported the previous findings of Chen et al. [45] mutant strain of *Verticillium dahliae* lose its pathogenicity by the repressed of cytochrome p450 gene expression. Comparative genomic studies have been frequently conducted to understand the expression of selected genes against different chemicals. However, inadequate information is available due to a lack of whole genomic sequencing (RNA seq data) and suitable methodologies for comparative transcriptome. In recent studies, transcriptome analysis has been widely applied on fungicidal resistant plant pathogens, including Fusarium spp. [40], *B. cinerea* [37], and *Penecillium didgitatum* [46]. According to our transcriptome analysis the expression of lysophospholipase genes, transmembrane transporter genes, MF (multifunctional genes), MFS (super family genes) encoding, amino acid and carbohydrate metabolism genes were clearly upregulated in field mutant (M0, M-ipro) than wild type (W0, W-ipro) with or without application of iprodione (Supplementary Table S15 Excel sheet). Many microorganisms produce phospholipases heterogeneous groups of enzymes, either secreted or induced intracellularly by physical disruption of the cellular membrane [47]. Among them lysophospholipases are key enzymes that hydrolyze the esters linkages in glycephosopholipids and contribute to detoxification of potentially cellular lysophospholipids that facilitate the survival of fungi in vivo, cell wall integrity, proliferation expression of virulence, fungal cell signaling and immunomodulatory pathways [48]. Here, the expression of lysophopholipase genes (*BCIN_13g03170, BCIN_02g08890*) were superiorly upregulated in M-ipro vs. W-ipro than M0 vs. W0 (Figure 7) contribute to iprodione resistance. *BCIN_02g08890* is highly upregulated in mutant strain with or without fungicide and absent in wild type.

Plant cell wall mainly composed of polysaccharides with less amount of glycoproteins esters, mineral contents, phenolic compounds and enzymes [49]. The predominant polysaccharides are cellulose, hemicellulose galacto (mannans, xylans, and xyloglucans), and pectin. Carbohydrate-degrading enzymes of pathogens constitute a key factor involved in the metabolic breakdown of glycoconjugates, oligosaccharides, and polysaccharides of host plant cell wall components during infection or invasion [50]. Cell wall degrading enzymes are abundantly found in *B. cinerea* [51]. Recent studies revealed that cellulase, xylanase, and pectinase (glucanase) enzymes functioned as a virulence factor in phytopathogens and were recognized as PAMPS by plants to trigger the PTI responses, during host plant–pathogen interactions [52].

In this study, we analyzed the enzymatic activity of cellulase, polygalactrose, and polymethylgalactrose in mutant and wild type *B.cinerea*. Enzymatic assays and gene expressions of cellulase (*BCIN_12g06630, BCIN_16g03020*) and xylanase A (*BCIN_03g03480, Bcxyn11A*) were more upregulated in mutant than wild type (Figure 7). Cellulase catalyzes the degradation of the β-1,4-glycosidic bonds in cellulose [53]. Cellulase is an elicitor in plant–pathogen interactions but its enzymatic activity is independent of its elicitor. In contrast, it was previously reported that xylanase activity promotes the necrotic infection of *B. cinerea* into plant tissues [54]. *Bcxyn11A*, an endo-β-1,4-xylanase degrades plant cell wall xylan contents, and is required for successful infection. Furthermore, a small nanogram of xyn A was sufficient as an elicitor in *S.lycopersci* and *N. benthamiana* [55]. In contrast, not all fungal xylanases have been conclusively involved in pathogenicity and virulence [51]. The *B. cinerea* equipped with different patterns of endopolygalactrase (*Bcpg*) genes and exopolygalactrase (*Bcpgx*) genes to degrade pectate machinery of host cell. In the current study, the expression of *Bcpg2* and *Bcpgx1* genes were highly upregulated in (M-ipro vs. W-ipro) and there is no change in *Bcpg1* and *Bcpg4* in all treatments. *Bcpg2* is a necessary gene during primary infection and lesion expansion in tomato [56]. A major result of our work is reported that there is no diversification in virulence genes of mutant and wild type before and after iprodione application. *B. cinerea* secretes several genes of the aspartic proteinase (AP) family to perform proteolytic activity. A functional analysis of our results showed that there is no change in the expression of *Bcap* genes of wild and mutant type after iprodione application (Figure 6). No significant difference was found in *Bcap1-5* genes mutants and the wild type strain of B05.10.

The resistance of fungi is sturdily associated with multiple mechanisms, including (1) the nonsynonyms mutation in the target protein encoding genes, (2) the upregulation of the target proteins, and (3) the overexpression of transporting and membrane encoding genes. Fungal efflux pumps such as cytochrome P450 are the most versatile natural bio-catalyst genes and constitute a large superfamily related to the detoxification of fungicides, insecticides, and xenobiotics under mild conditions [57,58].

The mechanism of resistance to iprodione was associated with point mutations in the tubulin gene that changes the structure of the fungicide binding site to decrease sensitivity according to Grabake et al. [28]. According to the genome sequencing information, two tubulin genes showed mutation. A point mutation at codons 593 (A198E) and 599 (F200Y) in *bctubA* gene and at 646 codon (R216C) in *bctubB* were detected from resistant strain M0, and a similar mutation was reported in a field isolate of *B. cinerea* resistant to benzimidazole that had a mutation at points A198E and F200Y [59]. A novel point mutation at codon 646 (R216C) was detected in *bctubB* in mutant (M0). Contrarily, point mutation variations at single codon I365S/N/R of the *Bos1* gene were responsible for dicarboximide (iprodione) low resistance, as reported from France, England, Israel, Japan, New Zealand, Italy, Switzerland, and the United States. Iprodione reduced DNA, RNA synthesis in the germinating fungal spore and inhibited the enzymatic acivity of NADH cytochrome c reductase, thereby preventing lipid and membrane synthesis and ultimately mycelium growth [60]. Wang et al. [14] reported that the resistance in *B. cinerea* to fenhexamid mainly relied on the mutation of *BCIN_16062* encoding P450 gene. In a hypersensitive strain of *Candidas albicans*, *CaALK8* gene was promoted and confers multidrug resistance [61]. The expression of *BCIN_01g03510* and *BCIN_13g05140* were upregulated in mutant (M0) relative to wild type (W0, W-ipro) (Figure 6). Glutathione S-transferases are multifunctional detoxification enzymes that regulate the cell functions, countering oxidative stress and signal transduction with several resistance mechanisms [62,63]. Our findings revealed that *BCIN_08g01800* was highly downregulated in wild type (W1-ipro) after iprodione application while other genes showed similar expression. Remarkably, transmembrane proteins in fungal efflux systems (ABC and MFS transporter genes) have been reported to provide protection for fungal cells against antibiotics and fungicides found in the environment [64]. Furthermore, these transporters determine the baseline of sensitivity or resistance to fungicides [65]. Several genes encoding transmembrane transporters were identified from the RNA sequencing data (Figure 6). Most of them showed higher expression in mutant strain than wild type. Particularly, two MFS encoding genes *Bcstl1* and *BCIN_08g01780* were highly upregulated in mutant (M-ipro) after iprodione treatment than M0 and wild type (W-ipro, W0). Intriguingly, *BCIN_14g01090* gene logfold2 value were intensely upregulated before and after iprodione application in mutant (M-ipro, M0) and downregulated in (W-ipro, W0). The high expression of drug efflux transporter has been reported in different isolates of *B. cinerea* against different classes of fungicides [66,67]. On the other hand, Grabke et al. [17] reported the overexpression of MFS encoding genes of *B. cinerea* conferring a low level of resistance to iprodione in strawberry. Contrarily, MFS genes contributed to resistance in *Fusarium* spp against prochloraz and carbendazim fungicides [40,46]. Likewise, *BCIN_14g04470* and *BCIN_12g02430* in fungal efflux were highly upregulated in mutant strain (M0) relative to wild type (W0).

5. Conclusions

The risk of resistance selection in *B. cinerea* due to the extensive application of DCs (iprodione) may be a severe problem in Guangxi Province China. In the current study, the results showed that *B. cinerea* isolates were sensitive to iprodione The laboratory UV induced resistant mutants were obtained through continuous iprodione treatment, which indicated that *B. cinerea* quickly adopt resistance to iprodione. The resistant mutant demonstrated an EC_{50} value 1025.74 µg/mL higher than curative dosage. The resistant mutant had a high level of fitness (mycelial growth, sclerotia formation and aggressiveness) as compared to sensitive isolates. We identified two mutation points at codon 593 (A198E) and 599 (F200Y) in the *bctubA* gene, including one novel point mutation in *bctubB* gene at position 646 codon. Our research work also provided a comprehensive molecular mechanism involved in the *B. cinerea* resistant mutant. By analyzing RNA sequencing data for wild type and mutant with or without iprodione application, genes related to iprodione resistance were highlighted and identified. These DEGs were involved in the production of detoxification enzymes, metabolism, catalytic activity, MAPK signaling pathway, drug efflux, and transporter functions to resist the chemicals. Resistant management strategies should be implemented to delay the spread of iprodione resistant mutants in the field. Furthermore, integrated disease management strategies should be practiced followed, including the use of biological control and agricultural practices.

Supplementary Materials: The following are available online at http://www.mdpi.com/1422-0067/21/14/4865/s1.

Author Contributions: Conceptualization, A.M., C.W., and H.W.; methodology, A.M.; software, A.M.; validation, S.A., C.W., and A.M.; formal analysis, A.M., C.W., and S.A.; investigation, A.M. and H.W.; resources, H.W.; data curation, A.M. and S.A.; writing—original draft preparation, A.M.; writing—review and editing, A.M., C.W., S.A., and H.W.; supervision, H.W. All authors have read and agreed to the published version of the manuscript.

Funding: We express our sincere gratitude to Guangxi Innovation Team of National Modern Agricultural Technology System (nycytxgxcxtd-10-04), the National Natural Science Foundation of China (31660511), Construction project of characteristic specialty and experimental training teaching base (Center)—Characteristic specialty—Plant Protection (2018–2020).

Conflicts of Interest: The authors declare that the research was conducted in the absence of any commercial or financial relationship that could be construed as a potential conflict of interest.

References

1. Williamson, B.; Tudzynski, B.; Tudzynski, P.; van Kan, J.A. *Botrytis cinerea*: The cause of grey mould disease. *Mol. Plant Pathol.* **2007**, *8*, 561–580. [CrossRef] [PubMed]

2. Dean, R.; Van Kan, J.A.; Pretorius, Z.A.; Hammond-Kosack, K.E.; Di Pietro, A.; Spanu, P.D.; Rudd, J.J.; Dickman, M.; Kahmann, R.; Ellis, J. The Top 10 fungal pathogens in molecular plant pathology. *Mol. Plant Pathol.* **2012**, *13*, 414–430. [CrossRef] [PubMed]

3. Hua, L.; Yong, C.; Zhanquan, Z.; Boqiang, L.; Guozheng, Q.; Shiping, T. Pathogenic mechanisms and control strategies of *Botrytis cinerea* causing post-harvest decay in fruits and vegetables. *Food Qual. Saf.* **2018**, *2*, 111–119. [CrossRef]

4. Amselem, J.; Cuomo, C.A.; Van Kan, J.A.; Viaud, M.; Benito, E.P.; Couloux, A.; Coutinho, P.M.; de Vries, R.P.; Dyer, P.S.; Fillinger, S. Genomic analysis of the necrotrophic fungal pathogens *Sclerotinia sclerotiorum* and *Botrytis cinerea*. *PLoS Genet.* **2011**, *7*, e1002230. [CrossRef] [PubMed]

5. Tiedemann, B.; Have, A.; Hansen, M.E.; Tenberge, K. Infection Strategies of *Botrytis cinerea* and related necrotrophic pathogens. *Fungal Pathol.* **2013**, 33–64.

6. Maqsood, A.; Rehman, A.; Ahmad, I.; Nafees, M.; Ashraf, I.; Qureshi, R.; Jamil, M.; Rafay, M.; Hussain, T. Physiological attributes of fungi associated with stem end rot of mango (*Mangifera indica* L.) cultivars in postharvest fruit losses. *Pak. J. Bot.* **2014**, *46*, 1915–1920.

7. Munhuweyi, K.; Lennox, C.L.; Meitz-Hopkins, J.C.; Caleb, O.J.; Opara, U.L. Major diseases of pomegranate (*Punica granatum* L.), their causes and management—A review. *Sci. Horticult.* **2016**, *211*, 126–139. [CrossRef]

8. Tsitsigiannis, D.I.; Antoniou, P.P.; Tjamos, S.E.; Paplomatas, E.J. Major diseases of tomato, pepper and egg plant in green houses. *Eur. J. Plant Sci. Biotechnol.* **2008**, *2*, 106–124.

9. Leroux, P.; Fritz, R.; Debieu, D.; Albertini, C.; Lanen, C.; Bach, J.; Gredt, M.; Chapeland, F. Mechanisms of resistance to fungicides in field strains of *Botrytis cinerea*. *Pest Manag. Sci.* **2002**, *58*, 876–888. [CrossRef]

10. Xiang, F.; Han, Y.; Zeng, X.; Gu, Y.; Chen, F. The development situation and prospect of strawberry industry in western Hubei mountain districts. *Hubei Agric. Sci.* **2015**, *54*, 5930–5932.

11. Rosslenbroich, H.J.; Stuebler, D. *Botrytis cinerea*—History of chemical control and novel fungicides for its management. *Crop Prot.* **2000**, *19*, 557–561. [CrossRef]

12. Leroux, P. Chemical control of Botrytis and its resistance to chemical fungicides. In *Botrytis: Biology, Pathology and Control*; Springer: Berlin, Germany, 2007; pp. 195–222.

13. Kroes, R.; Galli, C.; Munro, I.; Schilter, B.; Tran, L.A.; Walker, R.; Würtzen, G. Threshold of toxicological concern for chemical substances present in the diet: A practical tool for assessing the need for toxicity testing. *Food Chem. Toxicol.* **2000**, *38*, 255–312. [CrossRef]

14. Wang, X.; Gong, C.; Zhao, Y.; Shen, L. Transcriptome and resistance-related genes analysis of *Botrytis cinerea* B05. 10 strain to different selective pressures of cyprodinil and fenhexamid. *Front. Microbiol.* **2018**, *9*, 2591. [CrossRef] [PubMed]

15. Hirooka, T.; Ishii, H. Chemical control of plant diseases. *J. Gen. Plant Pathol.* **2013**, *79*, 390–401. [CrossRef]

16. Fernandez-Cornejo, J.; Nehring, R.F.; Osteen, C.; Wechsler, S.; Martin, A.; Vialou, A. Pesticide use in US agriculture: 21 selected crops, 1960–2008. *USDA-ERS Econ. Inf. Bull.* **2014**, *124*. [CrossRef]

17. Grabke, A.; Fernández-Ortuño, D.; Amiri, A.; Li, X.; Peres, N.A.; Smith, P.; Schnabel, G. Characterization of iprodione resistance in *Botrytis cinerea* from strawberry and blackberry. *Phytopathology* **2014**, *104*, 396–402. [CrossRef]

18. Hamada, M.S.; Yin, Y.; Ma, Z. Sensitivity to iprodione, difenoconazole and fludioxonil of *Rhizoctonia cerealis* isolates collected from wheat in China. *Crop Prot.* **2011**, *30*, 1028–1033. [CrossRef]

19. Ma, Z.; Luo, Y.; Michailides, T.J. Resistance of *Botryosphaeria dothidea* from pistachio to iprodione. *Plant Dis.* **2001**, *85*, 183–188. [CrossRef] [PubMed]

20. Ma, Z.; Michailides, T.J. Characterization of iprodione-resistant *Alternaria* isolates from pistachio in California. *Pestic. Biochem. Physiol.* **2004**, *80*, 75–84. [CrossRef]

21. Adnan, M.; Hamada, M.; Li, G.; Luo, C. Detection and molecular characterization of resistance to the dicarboximide and benzamide fungicides in *Botrytis cinerea* from tomato in Hubei Province, China. *Plant Dis.* **2018**, *102*, 1299–1306. [CrossRef]

22. Adnan, M.; Hamada, M.S.; Hahn, M.; Li, G.Q.; Luo, C.X. Fungicide resistance of *Botrytis cinerea* from strawberry to procymidone and zoxamide in Hubei, China. *Phytopathol. Res.* **2019**, *1*, 17. [CrossRef]

23. Sun, H.Y.; Wang, H.C.; Chen, Y.; Li, H.X.; Chen, C.J.; Zhou, M.G. Multiple resistance of *Botrytis cinerea* from vegetable crops to carbendazim, diethofencarb, procymidone, and pyrimethanil in China. *Plant Dis.* **2010**, *94*, 551–556. [CrossRef]

24. Ma, Z.; Yan, L.; Luo, Y.; Michailides, T.J. Sequence variation in the two-component histidine kinase gene of *Botrytis cinerea* associated with resistance to dicarboximide fungicides. *Pestic. Biochem. Physiol.* **2007**, *88*, 300–306. [CrossRef]

25. Marschall, R.; Schumacher, J.; Siegmund, U.; Tudzynski, P. Chasing stress signals–exposure to extracellular stimuli differentially affects the redox state of cell compartments in the wild type and signaling mutants of *Botrytis cinerea*. *Fungal Genet. Biol.* **2016**, *90*, 12–22. [CrossRef] [PubMed]

26. Avenot, H.F.; Quattrini, J.; Puckett, R.; Michailides, T.J. Different levels of resistance to cyprodinil and iprodione and lack of fludioxonil resistance in *Botrytis cinerea* isolates collected from pistachio, grape, and pomegranate fields in California. *Crop Prot.* **2018**, *112*, 274–281. [CrossRef]

27. Panebianco, A.; Castello, I.; Cirvilleri, G.; Perrone, G.; Epifani, F.; Ferrara, M.; Polizzi, G.; Walters, D.R.; Vitale, A. Detection of *Botrytis cinerea* field isolates with multiple fungicide resistance from table grape in Sicily. *Crop Prot.* **2015**, *77*, 65–73. [CrossRef]

28. Marie, C.; White, T.C. Genetic basis of antifungal drug resistance. *Curr. Fungal Infect. Rep.* **2009**, *3*, 163–169. [CrossRef] [PubMed]

29. Maqsood, A.; Wu, H.; Kamran, M.; Altaf, H.; Mustafa, A.; Ahmar, S.; Hong, N.T.T.; Tariq, K.; He, Q.; Chen, J.T. Variations in growth, physiology, and antioxidative defense responses of two tomato (*Solanum lycopersicum* L.) cultivars after co-infection of *Fusarium oxysporum* and *Meloidogyne incognita*. *Agronomy* **2020**, *10*, 159. [CrossRef]

30. Fernández-Ortuño, D.; Pérez-García, A.; Chamorro, M.; de la Peña, E.; de Vicente, A.; Torés, J.A. Resistance to the SDHI fungicides boscalid, fluopyram, fluxapyroxad, and penthiopyrad in *Botrytis cinerea* from commercial strawberry fields in Spain. *Plant Dis.* **2017**, *101*, 1306–1313. [CrossRef]

31. Luo, B.; Young, V.G. Jr.; Gladfelter, W.L. Si-C bond cleavage in the reaction of gallium chloride with lithium bis (trimethylsilyl) amide and thermolysis of base adducts of dichloro (trimethylsilyl) amido gallium compounds. *J. Organomet. Chem.* **2002**, *649*, 268–275. [CrossRef]

32. Fan, H.; Yu, G.; Liu, Y.; Zhang, X.; Liu, J.; Zhang, Y.; Rollins, J.A.; Sun, F.; Pan, H. An atypical forkhead-containing transcription factor SsFKH1 is involved in sclerotial formation and is essential for pathogenicity in *Sclerotinia sclerotiorum*. *Mol. Plant Pathol.* **2017**, *18*, 963–975. [CrossRef] [PubMed]

33. Jiang, M.; Cao, Y.; Guo, Z.F.; Chen, M.; Chen, X.; Guo, Z. Menaquinone biosynthesis in Escherichia coli: Identification of 2-succinyl-5-enolpyruvyl-6-hydroxy-3-cyclohexene-1-carboxylate as a novel intermediate and re-evaluation of MenD activity. *Biochemistry* **2007**, *46*, 10979–10989. [CrossRef] [PubMed]

34. Georgiou, C.; Grintzalis, K.; Zervoudakis, G.; Papapostolou, I. Mechanism of coomassie brilliant blue G-250 binding to proteins: A hydrophobic assay for nanogram quantities of proteins. *Anal. Bioanal. Chem.* **2008**, *391*, 391–403. [CrossRef] [PubMed]

35. Yang, L.; Xu, Y.; Zhang, R.; Wang, X.; Yang, C. Comprehensive transcriptome profiling of soybean leaves in response to simulated acid rain. *Ecotoxicol. Environ. Saf.* **2018**, *158*, 18–27. [CrossRef]

36. Feng, J.; Meyer, C.A.; Wang, Q.; Liu, J.S.; Shirley Liu, X.; Zhang, Y. GFOLD: A generalized fold change for ranking differentially expressed genes from RNA-seq data. *Bioinformatics* **2012**, *28*, 2782–2788. [CrossRef] [PubMed]

37. Cai, N.; Liu, C.; Feng, Z.; Li, X.; Qi, Z.; Ji, M.; Qin, P.; Ahmed, W.; Cui, Z. Design, synthesis, and SAR of novel 2-glycinamide cyclohexyl sulfonamide derivatives against *Botrytis cinerea*. *Molecules* **2018**, *23*, 740. [CrossRef]

38. Leroch, M.; Kretschmer, M.; Hahn, M. Fungicide resistance phenotypes of *Botrytis cinerea* isolates from commercial vineyards in South West Germany. *J. Phytopathol.* **2011**, *159*, 63–65. [CrossRef]

39. Beckerman, J.L.; Sundin, G.W.; Rosenberger, D.A. Do some IPM concepts contribute to the development of fungicide resistance? Lessons learned from the apple scab pathosystem in the United States. *Pest Manag. Sci.* **2015**, *71*, 331–342. [CrossRef]

40. Xu, S.; Wang, J.; Wang, H.; Bao, Y.; Li, Y.; Govindaraju, M.; Yao, W.; Chen, B.; Zhang, M. Molecular characterization of carbendazim resistance of *Fusarium* species complex that causes sugarcane pokkah boeng disease. *BMC Genom.* **2019**, *20*, 115. [CrossRef]

41. Cai, M.; Lin, D.; Chen, L.; Bi, Y.; Xiao, L.; Liu, X.L. M233I mutation in the β-tubulin of *Botrytis cinerea* confers resistance to zoxamide. *Sci. Rep.* **2015**, *5*, 16881. [CrossRef]

42. Veloukas, T.; Kalogeropoulou, P.; Markoglou, A.; Karaoglanidis, G. Fitness and competitive ability of *Botrytis cinerea* field isolates with dual resistance to SDHI and QoI fungicides, associated with several sdh B and the cyt b G143A mutations. *Phytopathology* **2014**, *104*, 347–356. [CrossRef] [PubMed]

43. Lu, X.H.; Jiao, X.L.; Hao, J.J.; Chen, A.J.; Gao, W.W. Characterization of resistance to multiple fungicides in *Botrytis cinerea* populations from Asian ginseng in northeastern China. *Eur. J. Plant Pathol.* **2016**, *144*, 467–476. [CrossRef]

44. Guo, L.N.; Xiao, M.; Cao, B.; Qu, F.; Zhan, Y.L.; Hu, Y.J.; Wang, X.R.; Liang, G.W.; Gu, H.T.; Qi, J. Epidemiology and antifungal susceptibilities of yeast isolates causing invasive infections across urban Beijing, China. *Future Microbiol.* **2017**, *12*, 1075–1086. [CrossRef]

45. Chen, J.Y.; Xiao, H.L.; Gui, Y.J.; Zhang, D.D.; Li, L.; Bao, Y.M.; Dai, X.F. Characterization of the Verticillium dahliae exoproteome involves in pathogenicity from cotton-containing medium. *Front. Microbiol.* **2016**, *7*, 1709. [CrossRef] [PubMed]

46. Liu, J.; Wang, S.; Qin, T.; Li, N.; Niu, Y.; Li, D.; Yuan, Y.; Geng, H.; Xiong, L.; Liu, D. Whole transcriptome analysis of *Penicillium digitatum* strains treatmented with prochloraz reveals their drug-resistant mechanisms. *BMC Genom.* **2015**, *16*, 855. [CrossRef] [PubMed]

47. Djordjevic, J. Role of phospholipases in fungal fitness, pathogenicity, and drug development–lessons from *Cryptococcus neoformans*. *Front. Microbiol.* **2010**, *1*, 125. [CrossRef]

48. Gilbert, N.M.; Donlin, M.J.; Gerik, K.J.; Specht, C.A.; Djordjevic, J.T.; Wilson, C.F.; Sorrell, T.C.; Lodge, J.K. KRE genes are required for β-1, 6-glucan synthesis, maintenance of capsule architecture and cell wall protein anchoring in Cryptococcus neoformans. *Mol. Microbiol.* **2010**, *76*, 517–534. [CrossRef]

49. Luczaj, J.; Leavitt, S.; Csank, A.; Panyushkina, I.; Wright, W. Comment on "Non-Mineralized Fossil Wood" by George E. Mustoe (Geosciences, 2018). *Geosciences* **2018**, *8*, 462. [CrossRef]

50. Zerillo, M.M.; Adhikari, B.N.; Hamilton, J.P.; Buell, C.R.; Lévesque, C.A.; Tisserat, N. Carbohydrate-active enzymes in Pythium and their role in plant cell wall and storage polysaccharide degradation. *PLoS ONE* **2013**, *8*, e72572. [CrossRef]

51. Yang, Y.; Yang, X.; Dong, Y.; Qiu, D. The *Botrytis cinerea* xylanase BcXyl1 modulates plant immunity. *Front. Microbiol.* **2018**, *9*, 2535. [CrossRef]

52. Tang, D.; Wang, G.; Zhou, J.M. Receptor kinases in plant-pathogen interactions: More than pattern recognition. *Plant Cell* **2017**, *29*, 618–637. [CrossRef] [PubMed]

53. Ma, Y.; Han, C.; Chen, J.; Li, H.; He, K.; Liu, A.; Li, D. Fungal cellulase is an elicitor but its enzymatic activity is not required for its elicitor activity. *Mol. Plant Pathol.* **2015**, *16*, 14–26. [CrossRef] [PubMed]

54. Noda, J.; Brito, N.; González, C. The *Botrytis cinerea* xylanase Xyn11A contributes to virulence with its necrotizing activity, not with its catalytic activity. *BMC Plant Biol.* **2010**, *10*, 38. [CrossRef] [PubMed]

55. Brito, N.; Espino, J.J.; González, C. The endo-β-1, 4-xylanase Xyn11A is required for virulence in *Botrytis cinerea*. *Mol. Plant-Microbe Interact.* **2006**, *19*, 25–32. [CrossRef]

56. Zhang, L.; Van Kan, J.A. *Botrytis cinerea* mutants deficient in D-galacturonic acid catabolism have a perturbed virulence on *Nicotiana benthamiana* and *Arabidopsis*, but not on tomato. *Mol. Plant Pathol.* **2013**, *14*, 19–29. [CrossRef]

57. Črešnar, B.; Petrič, Š. Cytochrome P450 enzymes in the fungal kingdom. *BBA Proteins Proteom.* **2011**, *1814*, 29–35. [CrossRef]

58. Durairaj, P.; Hur, J.S.; Yun, H. Versatile biocatalysis of fungal cytochrome P450 monooxygenases. *Microb. Cell Fact.* **2016**, *15*, 125. [CrossRef]

59. Muñoz, C.; Talquenca, S.G.; Volpe, M.L. Tetra primer ARMS-PCR for identification of SNP in β-tubulin of *Botrytis cinerea*, responsible of resistance to benzimidazole. *J. Microbiol. Methods* **2009**, *78*, 245–246. [CrossRef]

60. Sameer, W.M. Control of the root rot diseases of wheat using fungicides alone and in combinations with selenium. *Al-Azhar Bull. Sci.* **2018**, *29*, 27–37.

61. Lata Panwar, S.; Krishnamurthy, S.; Gupta, V.; Alarco, A.M.; Raymond, M.; Sanglard, D.; Prasad, R. CaALK8, an alkane assimilating cytochrome P450, confers multidrug resistance when expressed in a hypersensitive strain of Candida albicans. *Yeast* **2001**, *18*, 1117–1129. [CrossRef]

62. Alias, Z. The role of glutathione transferases in the development of insecticide resistance. *Insectic. Resist.* **2016**, 315. [CrossRef]

63. Kolawole, A.O.; Ajele, J.O.; Sirdeshmuhk, R. Studies on glutathione transferase of cowpea storage bruchid, *Callosobrochus maculatus* F. *Pestic. Biochem. Physiol.* **2011**, *100*, 212–220. [CrossRef]

64. Holmes, A.R.; Keniya, M.V.; Ivnitski-Steele, I.; Monk, B.C.; Lamping, E.; Sklar, L.A.; Cannon, R.D. The monoamine oxidase A inhibitor clorgyline is a broad-spectrum inhibitor of fungal ABC and MFS transporter efflux pump activities which reverses the azole resistance of *Candida albicans* and *Candida glabrata* clinical isolates. *Antimicrob. Agents Chemother.* **2012**, *56*, 1508–1515. [CrossRef] [PubMed]

65. Stergiopoulos, I.; Zwiers, L.H.; De Waard, M.A. Secretion of natural and synthetic toxic compounds from filamentous fungi by membrane transporters of the ATP-binding cassette and major facilitator superfamily. *Eur. J. Plant Pathol.* **2002**, *108*, 719–734. [CrossRef]

66. Mernke, D.; Dahm, S.; Walker, A.S.; Lalève, A.; Fillinger, S.; Leroch, M.; Hahn, M. Two promoter rearrangements in a drug efflux transporter gene are responsible for the appearance and spread of multidrug resistance phenotype MDR2 in *Botrytis cinerea* isolates in French and German vineyards. *Phytopathology* **2011**, *101*, 1176–1183. [CrossRef]

67. De Waard, M.A.; Andrade, A.C.; Hayashi, K.; Schoonbeek, H.j.; Stergiopoulos, I.; Zwiers, L.H. Impact of fungal drug transporters on fungicide sensitivity, multidrug resistance and virulence. *Pest Manag. Sci.* **2006**, *62*, 195–207. [CrossRef]

International Journal of
Molecular Sciences

Article

Ethylene Is Not Essential for R-Gene Mediated Resistance but Negatively Regulates Moderate Resistance to Some Aphids in *Medicago truncatula*

Lijun Zhang [1,2,†], Lars G. Kamphuis [1,3], Yanqiong Guo [1,2,†], Silke Jacques [1,2,‡],
Karam B. Singh [1,3,*] and Ling-Ling Gao [1,*]

[1] CSIRO Agriculture and Food, Wembley, WA 6014, Australia; butterflycoco@163.com (L.Z.);
     lars.kamphuis@csiro.au (L.G.K.); guoyq1979@163.com (Y.G.); silke.jacques@curtin.edu.au (S.J.)
[2] College of Plant Protection, Shanxi Agricultural University, Taigu 030801, China
[3] The UWA Institute of Agriculture, The University of Western Australia, Crawley, WA 6009, Australia
*   Correspondence: karam.singh@csiro.au (K.B.S.); lingling.gao@csiro.au (L.-L.G.);
     Tel.:+61-8-9333-6320 (K.B.S.); Fax: +61-8-9387-8991 (K.B.S.)
†   Current addresses: College of Plant Protection, Shanxi Agricultural University, Taigu 030801, China.
‡   Current addresses: Centre for Crop and Disease Management, Curtin University, Bentley,
     WA 6102, Australia.

Received: 30 May 2020; Accepted: 27 June 2020; Published: 30 June 2020

Abstract: Ethylene is important for plant responses to environmental factors. However, little is known about its role in aphid resistance. Several types of genetic resistance against multiple aphid species, including both moderate and strong resistance mediated by R genes, have been identified in *Medicago truncatula*. To investigate the potential role of ethylene, a *M. truncatula* ethylene- insensitive mutant, *sickle*, was analysed. The *sickle* mutant occurs in the accession A17 that has moderate resistance to *Acyrthosiphon kondoi*, *A. pisum* and *Therioaphis trifolii*. The *sickle* mutant resulted in increased antibiosis-mediated resistance against *A. kondoi* and *T. trifolii* but had no effect on *A. pisum*. When *sickle* was introduced into a genetic background carrying resistance genes, *AKR* (*A. kondoi* resistance), *APR* (*A. pisum* resistance) and *TTR* (*T. trifolii* resistance), it had no effect on the strong aphid resistance mediated by these genes, suggesting that ethylene signaling is not essential for their function. Interestingly, for the moderate aphid resistant accession, the *sickle* mutant delayed leaf senescence following aphid infestation and reduced the plant biomass losses caused by both *A. kondoi* and *T. trifolii*. These results suggest manipulation of the ethylene signaling pathway could provide aphid resistance and enhance plant tolerance against aphid feeding.

Keywords: disease resistance; plant defense; herbivore; phytohormone; plant biotic stress; plant signalling; *Medicago truncatula*

---

1. Introduction

Aphids (Hemiptera: Aphidoidea) constitute a large group of sap-sucking insect pests that cause substantial losses to agriculture worldwide by draining plant nutrients and transmitting pathogenic viruses [1–3]. The plant-aphid interaction is distinctive from plant interactions with microbial pathogens and chewing insects mainly because aphid infestation instigates very little physical damage to the plant. With their stylets, aphids penetrate plant tissues by piercing intercellularly through epidermal and mesophyll cell layers and ultimately feed specifically from the phloem sieve element [1–5]. A number of plant genes or loci, including Resistance (R) genes, that modulate plant defenses against aphids have been identified in a range of plant species against various aphid species ([6–8]. Molecular studies have revealed that plant phytohormones are also involved in the regulation of plant interactions with aphids. However, in many cases, the specific roles that phytohormone pathways play in basal and *R* gene-mediated aphid resistance remains largely unknown.

Ethylene (ET) is a gaseous plant hormone which is involved in the regulation of various developmental as well as abiotic and biotic stress responses. Studies using mutants impaired in ET biosynthesis and signalling demonstrated a direct role for ET in plant defence against microbial pathogens and insect pests and in the control of plant association with beneficial microbes, such as rhizobia and mycorrhizas. ET signalling has also been shown to regulate plant interactions with insect herbivores [4].

A number of studies have been conducted to investigate the role that ET plays in both compatible and incompatible plant–aphid interactions. For compatible interactions, most studies were carried out with the generalist aphid *Myzus persicae* (Sulzer) for which there is no natural genetic resistance. However, the results from these studies were inconsistent. Divol et al. [5] and Moran et al. [6]showed that genes involved in ET biosynthesis and signalling were induced in both celery and *Arabidopsis* following infestation by *M. persicae*. Other studies showed that ET accumulation remained unchanged in both *M. persicae*-infested *Arabidopsis* and *Nicotiana attenuate* compared to the un-infested control plants [7,8] When the performance of *M. persicae* was compared between the *Arabidopsis* wild-type and ET-insensitive *etr* or *ein2* mutant plants, Kettles et al. [9]. found that the aphid fecundity did not differ between the *Arabidopsis* wild-type and *etr1* mutant, whilst the *ein2* mutant did show higher *M. persicae* fecundity than wild-type plants. In contrast, Mewis et al. [10] found that the fecundity of both *M. persicae* and *Brevicoryne brassicae* was reduced on the *etr1* mutant compared to wild-type *Arabidopsis* plants. These contradictory results in the *Arabidopsis–M. persicae* interaction highlight the need to further study the role of ET in plant-aphid interactions.

Studies also indicated that ET may modulate *R* gene-mediated plant defence against aphids. In aphid-resistant barley plants, ET production was significantly induced following the infestation by *Schizaphis graminum*, *Rhopalosiphum padi*, and *Diuraphis noxia* [11,12]. Upon feeding by *D. noxia*, transcript levels of ET-related genes increased in aphid resistant wheat plants [13]. Furthermore, the induction of genes involved in ET signalling and downstream responses was also found in both susceptible and resistant interactions of tomato with *Macrosiphum euphorbiae* and in melon with *Aphis gossypii* [14]. However, in melon with *A. gossypii* stronger induction of ET pathway genes was shown in the resistant variety than the susceptible plants, but this was not the case in tomato with *M. euphorbiae*.

Medicago truncatula is a model legume species for studying plant interaction with aphids [15,16]. *M. truncatula* is a host to several important aphid species including *Acyrthosiphon kondoi* (bluegreen aphid), *Therioaphis trifolii* (spotted alfalfa aphid) and *A. pisum* (pea aphid). In *M. truncatula*, various types of resistance against these aphid species have been identified and the resistance is controlled either through major dominant resistance genes and/or quantitative loci [17–24]. In *M. truncatula* cv. Jester, three single dominant resistance genes, named *AKR* (*A. kondoi* resistance), *APR* (*A. pisum* resistance) and *TTR* (*T. trifolii* resistance), provide strong resistance to *A. kondoi*, *A. pisum* and *T. trifolii*, respectively [17,21,22]. In addition to the major dominant resistance genes for *A. kondoi* and *A. pisum*, a second semi-dominant resistance gene termed *AIN* (*Acyrthosiphon induced necrosis*) has been identified [23]. *AIN* confers a moderate level of resistance to both aphid species and forms hypersensitive response (HR-like) necrotic lesions at the site of infestation by both *A. kondoi* and *A. pisum* [23,25]. This locus is present in both Jester and the reference *M. truncatula* accession, Jemalong (A17) which lacks the three major resistance genes, *AKR*, *APR* and *TTR*. The *M. truncatula* cv. Jester is closely related to A17 [21]. When compared with the highly susceptible *M. truncatula* accession A20, A17 shows moderate resistance to all three aphid species [15]. In A17, in addition to the antibiosis resistance conferred by the *AIN* locus, two distinct quantitative trait loci (QTLs) have been identified for tolerance to *A. kondoi* and *A. pisum*, respectively [18]. Furthermore, three QTL involved in the moderate antibiosis and tolerance to *T. trifolii* have also been identified in A17 [20].

The molecular mechanisms underlying the various types of aphid resistance in *M. truncatula* are largely unknown. Expression analysis of genes involved in defence signalling pathways indicated that salicylic acid (SA)-related genes were induced in both A17 and Jester following the infestation by *A. kondoi* and *A. pisum* [17,26,27]. However, jasmonic acid (JA)-related genes were highly induced only in Jester when infested by *A. kondoi* suggesting that JA might be involved in the *AKR*-mediated resistance to *A. kondoi*. In the interactions between *M. truncatula* and European biotypes of *A. pisum*, the induction of phytohormones in *Medicago truncatula* was dependent upon the genotypes of both plant and insect as well the time post-infestation and aphid density [28]. There was some induction of hormones in the compatible interaction but higher concentration of JA, SA and medicarpin exhibited during the incompatible interaction. Although Gao et al. [26] showed an induction of ET associated genes following *A. kondoi* infestation of both Jester and A17, little is known about the role of the ET signalling pathway in moderate or *R*-gene mediated responses in *M. truncatula* following aphid predation.

The primary aim of this study was to determine the role of ET signalling in the different modes of resistance in *M. truncatula* against the three aphid species, *A. kondoi*, *A. pisum* and *T. trifolii*. The ET insensitive mutant *sickle* in the A17 background, provides a useful genetic tool to decipher the function of ET signalling in the control of different plant-aphid interactions [16,29,30]. Therefore, the role of ET in the moderate resistance to aphids found in A17[18,20,23], was tested by comparing aphid performance and plant tolerance in *sickle* to A17 wild-type plants. To examine the role of *R* gene mediated resistance, crosses were made between Jester, which harbours the major resistance genes and *sickle*. Offspring that carry both the homozygous *sickle* mutation and homozygous resistance genes for the respective aphid species were then tested for aphid performance and plant damage caused by aphid feeding. We found that ET is a negative regulator of moderate resistance to *A. kondoi* and *T. trifolii* but not to the Australian, *A. pisum* biotype. Our results also showed that ET is not essential for *R* gene mediated resistance against the three aphid species or for the *AIN* mediated HR-like response to both *A. kondoi* and *A. pisum*.

2. Results

2.1. The M. truncatula Sickle Mutant Modulates Aphid Resistance

To examine if the ethylene-insensitive, *M. truncatula sickle* mutant affects the moderate aphid resistance observed in A17, the performance of three aphid species, *A. kondoi*, *T. trifolii* and *A. pisum* was measured. The highly resistant accession, Jester, which carries single dominant resistance genes to all three aphid species and is near isogenic to A17, was included and aphid performance was measured with single trifoliate leaves and subsequently with whole plant assays. Although the durations of aphid infestation were different, seven days in single leaf experiments and 14 days in the whole plant assays, for each aphid species and each *M. truncatula* accession, the results were consistent between these two experiments. As shown in Figures 1 and 2, with all three aphid species, the aphid population weight were significantly higher on the moderately resistant A17 than the highly resistant Jester, which was consistent with our previous reports [17,21,31]. Interestingly, for both *A. kondoi* and *T. trifolii*, the aphid weight was significantly lower ($p < 0.05$) on *sickle* than its wild-type parent, A17, but significantly higher ($p < 0.05$) than on Jester plants. The reduction of aphid population on the *sickle* plants was most pronounced for *T. trifolii* (Figures 1 and 3). At seven days following aphid infestation on single trifoliate leaves, the average aphid population weight per trifoliate leaf on *sickle* was reduced to one third that on A17. In the whole plant experiments, on A17, aphids were able to feed but performed poorly with low aphid weight per plant dry weight (2.36 mg/g) (Figure 2). In contrast, no aphids were observed on *sickle* or on Jester plants after 14 days of aphid infestation. However, for *A. pisum*, the aphid weight did not differ significantly between *sickle* and A17 but was significantly higher than on Jester. Consistent results were obtained in repeat experiments.

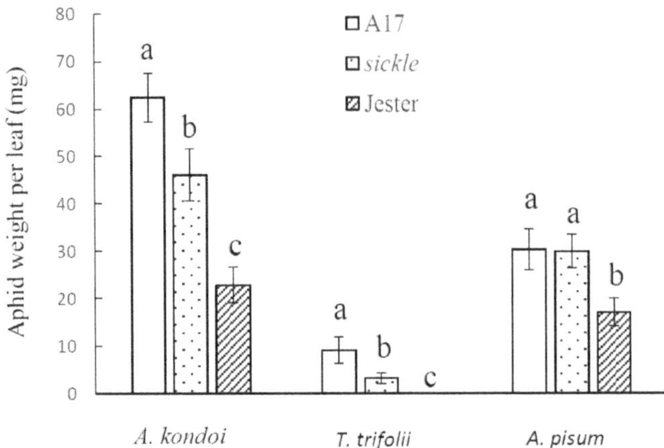

Figure 1. Aphid performance on single intact trifoliate leaves of *Medicago truncatula* genotypes A17, *sickle* and Jester. The aphid performance was shown as aphid fresh weight per trifoliate leaf at seven days following infestation with *Acyrthosiphon kondoi* (five aphids), *Therioaphis trifolii* (seven aphids) and *A. pisum* (four aphids). The values depict the mean and standard error of six biological replicates. The means were only compared among *M. truncatula* genotypes within each aphid species and means with different letters indicate the differences are significant as determined by ANOVA, GenStat ($p < 0.05$).

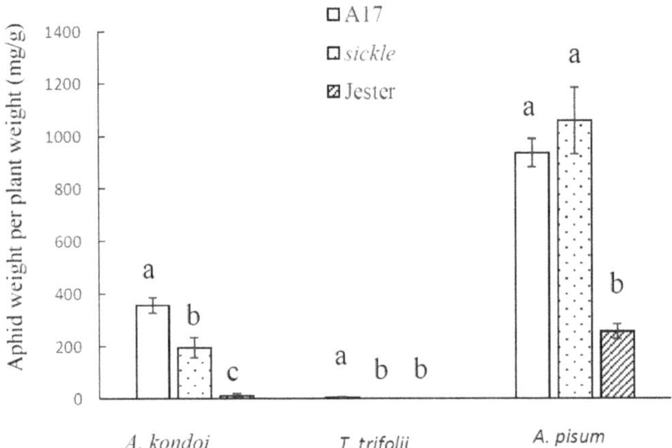

Figure 2. Aphid performance on *Medicago truncatula* genotypes A17, *sickle* and Jester with whole plant assays. The aphid performance was shown as aphid fresh weight per plant dry weight at 14 days following infestation with *Acyrthosiphon kondoi* (five aphids), *Therioaphis trifolii* (seven aphids) and *A. pisum* (four aphids). The values depict the mean and standard error of six biological replicates. The means were only compared among *M. truncatula* genotypes within each aphid species and means with different letters indicate the differences are significant as determined by ANOVA, GenStat ($p < 0.05$).

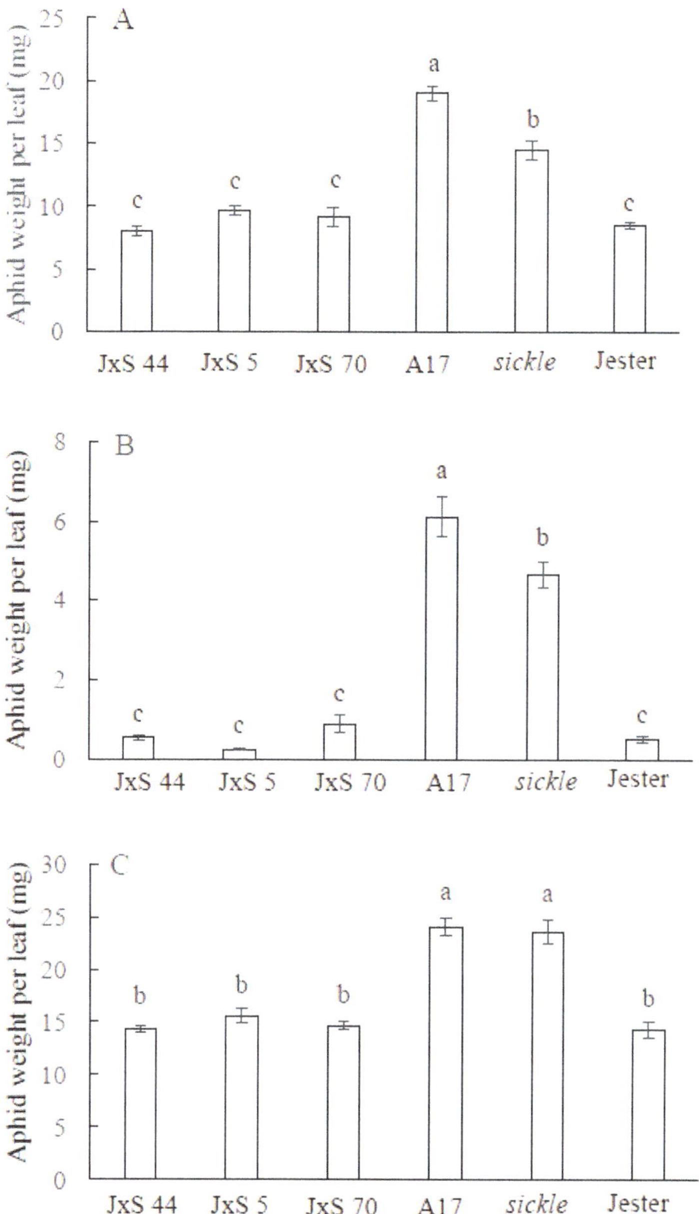

Figure 3. Aphid performance on *Medicago truncatula* accession A17, *sickle*, Jester and three independent F3 lines of Jester x *sickle* crosses, JxS 44, JxS 5 and JxS 70, containing the homozygous *sickle* mutation and all three aphid resistance genes, *AKR* (*Acyrthosiphon kondoi* resistance), *APR* (*Acyrthosiphon pisum* resistance) and *TTR* (*Therioaphis trifolii* resistance), seven days after the infestation by *A. kondoi* (five aphids) (**A**), *T. trifolii* (seven aphids) (**B**) and *A. pisum* (four aphids) (**C**). The values depict the mean and standard error of six biological replicates. For each aphid species, different letters indicate significant differences in the aphid weight between *M. truncatula* genotypes as determined by ANOVA, GenStat ($p < 0.05$).

2.2. Effect of M. truncatula Sickle Mutant on Plant Tolerance to Aphid Feeding

Prior to aphid infestation, the leaves of A17, *sickle* and Jester plants (three weeks after planting) showed no noticeable phenotypical difference. Experiments with single trifoliate leaves demonstrated that following aphid feeding for seven days all three aphid species caused significant damage to the leaves of A17, resulting in reduced leaf size and leaf senescence (Figure 4). The infestation of *A. kondoi* and *A. pisum* also caused leaf senescence on the resistant Jester plants despite much lower aphid population weight on Jester than on A17. Interestingly, with all three aphid species, the leaf senescence was not noticeable on the leaves of *sickle*, even though higher aphid populations were observed on *sickle* than on the resistant Jester plants. This was most apparent with *A. pisum* where on *sickle* leaves the aphid population levels were comparable to A17, and significantly higher than on Jester (Figures 1 and 4). Consistent outcomes were obtained under both growth cabinet and glasshouse conditions in three repeat experiments.

Figure 4. The damage symptoms on single intact trifoliate leaves of *Medicago truncatula* genotypes A17, *sickle* and Jester (left panels). The necrotic flecks (indicated by arrows) on each *M. truncatula* genotype are depicted on the right. The photos were taken at seven days following infestation with *Acyrthosiphon kondoi* (five aphids) (**A**), *Therioaphis trifolii* (seven aphids) (**B**) and *A. pisum* (four aphids) (**C**).

The tolerance responses of *sickle*, A17 and Jester plants to aphids when measured with whole plants were consistent with the results on single trifoliate leaves. As illustrated in Figure 5, for each aphid species, the degree of plant biomass reduction caused by aphid infestation correlated with the aphid population levels shown in Figure 2. A17 showed the highest plant biomass reduction relative to the un-infested control plants. In response to the infestation by both *A. kondoi* and *T. trifolii*, *sickle* demonstrated significantly ($p < 0.05$) lower plant biomass reduction than A17. With *A. kondoi*, the biomass reduction in *sickle* was significantly ($p < 0.05$) higher than Jester; however, with *T. trifolii*, the plant biomass of *sickle* or Jester was not significantly different between aphid-infested and control plants. In contrast, the plant biomass reduction caused by *A. pisum* infestation did not differ between *sickle* and A17, which was significantly ($p < 0.05$) higher than that of Jester (Figure 5).

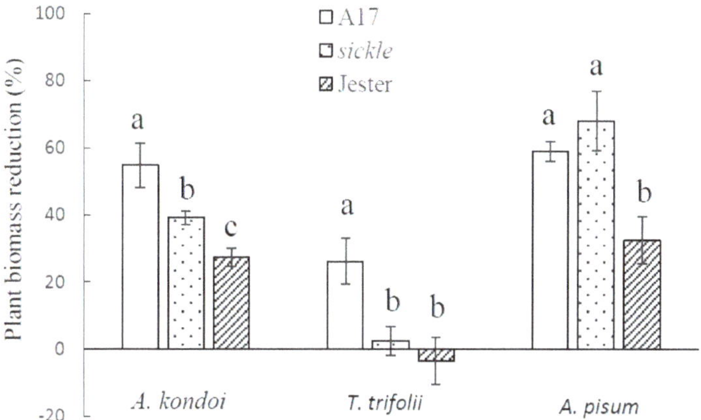

Figure 5. Plant tolerance of *Medicago truncatula* genotypes A17, *sickle* and Jester with whole plant assays. The plant tolerance was measured as the percentage of plant biomass reduction caused by the infestation of *Acyrthosiphon kondoi* (five aphids), *Therioaphis trifolii* (seven aphids) and *A. pisum* (four aphids) for 14 days. The values depict the mean and standard error of six biological replicates. The means were only compared among *M. truncatula* genotypes within each aphid species and means with different letters indicate the differences are significant as determined by ANOVA, GenStat ($p < 0.05$).

2.3. Effect of M. truncatula Sickle Mutant on R Gene Mediated Resistance to Aphids

To examine if the *sickle* mutation affects *R* gene mediated aphid resistance, we crossed the *sickle* locus (located on chromosome 7) to the Jester background. Jester contains all three single dominant aphid resistance genes *AKR*, *APR* and *TTR*, which are closely linked and located on chromosome 3 [17,22,32,33]. The F_2 plants from these crosses were first screened for the *sickle* locus using 1-aminocyclopropane-1-carboxylic acid (ACC) (see Materials and Methods). Out of 512 F_2 seedlings screened, 163 showed normal embryonic root growth similar to ACC treated *sickle* mutants and untreated controls, which is consistent with a 1:3 segregation ratio (Chi-square = 2.34, $p = 0.125$) for the 163 lines containing the homozygous *sickle* allele. A subset (90) of 163 pre-selected *sickle* mutant lines were further assessed using high throughput, Multiplex-Ready marker technology and molecular markers linked to these resistance gene loci. Nine homozygous *sickle* plants also contained homozyogous alleles of all three aphid resistance genes (Supplementary Table S2).

To investigate if the *sickle* mutation affects *R* gene mediated resistance to the three aphid species, aphid performance and leaf tolerance were first measured on single trifoliate leaves of individual F_2 plants with the *sickle* mutation and the three aphid resistance loci and the results compared to *sickle*, A17 and Jester. A follow-up experiment with three randomly selected F_3 lines, each containing homozygous *sickle*, *AKR*, *APR* and *TTR* alleles was conducted. The results were consistent in both studies using F_2 or F_3 plants, but only the results using the F_3 homozygous lines are presented in Figure 3. With all three aphid species, aphid weights on the control plants of *sickle*, A17 and Jester were consistent to the results shown in Figure 1. For each aphid species, the aphid weights on the three independent F_3 lines were not significantly different ($p > 0.05$) from each other and did not differ significantly from the aphid weight on the Jester plants (Figure 3). The results demonstrate that the ET insensitive *sickle* mutation has no impact on the antibiosis effect conferred by the three aphid *R* genes; *AKR*, *TTR* and *APR*, against *A. kondoi*, *T. trifolii* and *A. pisum*, respectively.

2.4. The Role of Ethylene Insensitivity in the AIN-Mediated Hypersensitive Response to A. kondoi and A. pisum Infestation

Both *M. truncatula* A17 and Jester carry the semi-dominant *AIN* gene (*Acyrthosiphon*-induced necrosis) which causes HR-like necrotic flecks upon feeding by both *A. kondoi* and *A. pisum* [23]. To evaluate if the *sickle* mutant interacts with AIN-mediated necrosis, the number of necrotic flecks per trifoliate leaf were recorded. *A. kondoi* and *A. pisum* both induced necrotic like spots on A17 and *sickle* (Figures 4 and 6). As shown in Figure 4, when infested with *A. kondoi*, the average number of necrotic flecks per leaf varied significantly ($p < 0.05$) between A17 and *sickle* with 19.5 and 12 per leaf, respectively (Figure 6). In contrast, upon feeding by *A. kondoi* there were no macroscopic lesions observed on Jester nor on leaves of F_2 and F_3 lines containing both homozygous *sickle* and *AKR* loci (Figures 4 and 6). When infested with *A. pisum*, necrotic spots were observed on the leaves of all *M. truncatula* accessions examined. The number of necrotic spots per leaf did not differ significantly ($p > 0.05$) between A17 and *sickle* or between Jester and lines carrying both the *sickle* and *APR* loci (Figure 6).

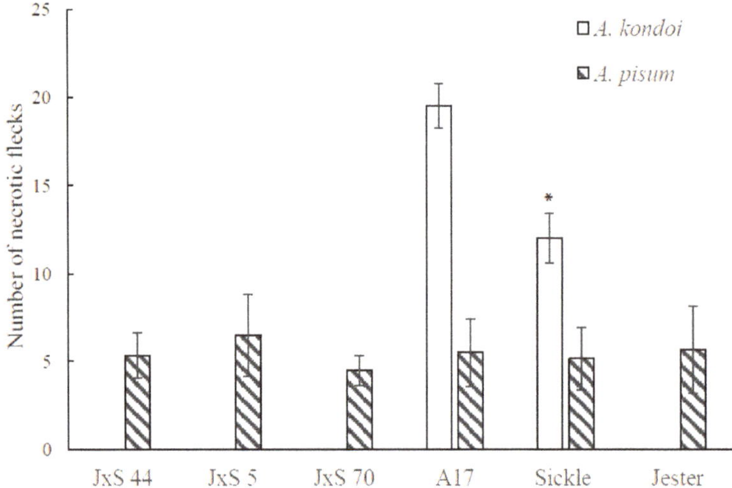

Figure 6. The numbers of necrotic flecks on *Medicago truncatula* accessions A17, *sickle*, Jester and three independent F3 lines of Jester x *sickle* crosses, JxS 44, JxS 5 and JxS 70, containing the homozygous *sickle* mutation and all three aphid resistance genes, *AKR* (*Acyrthosiphon kondoi* resistance) and *APR* (*A. pisum* resistance), seven days after the infestation by *A. kondoi* (five aphids) and *A. pisum* (four aphids). The values depict the mean and standard error of six biological replicates. The means were only compared among *M. truncatula* accessions within each aphid species. The differences in the number of necrotic flecks caused by *A. pisum* are not significant as determined by ANOVA, GenStat ($p > 0.05$). With *A. kondoi*, the necrotic flecks were not observed on *M. truncatula* accession, Jester, JxS 44, JxS 5 and JxS 70, containing homozygous *sickle* and *AKR* alleles. * indicates that with *A. kondoi* the numbers of necrotic flecks on A17 and *sickle* are significantly different as determined by ANOVA, GenStat ($p < 0.05$).

3. Discussion

The model legume *M. truncatula* provides a great opportunity to decipher the molecular mechanisms underlying plant defence against sap-sucking insects, where various types of interactions have been identified to multiple aphid species [16]. The well characterised, ET insensitive, *M. truncatula sickle* mutant allowed us to determine the specific roles that ET plays in plant-aphid interactions. We have found that the *sickle* mutant enhanced the antibiosis effect on *A. kondoi* and *T. trifolii* but not *A. pisum* (Figures 1 and 2), and delayed leaf senescence caused by the feeding of all three aphid species.

We show that the ET signalling pathway is not essential for the function of the major aphid resistance genes, *AKR*, *APR* or *TTR* against the three aphid species and is also not required for the *AIN*-mediated hypersensitive response to *A. kondoi* or *A. pisum* infestation.

Our results showed that on *sickle*, the growth of both *A. kondoi* and *T. trifolii* colonies was significantly reduced compared to its wild-type parent, A17. The results suggest that ET is a negative regulator of the moderate resistance in *M. truncatula* against these two aphid species. The ET signalling pathway has previously been demonstrated as a negative regulator in other plant species against sap-sucking insects. For instance, in rice, the suppression of ET biosynthesis enhanced resistance against a piercing-sucking insect, the brown planthopper (*N. lugens*) but reduced plant resistance against a chewing insect, striped stem borer (*C. suppressalis*) [34]. In *Arabidopsis*, several studies suggested that ET is a negative regulator of aphid defense responses. The fecundities of the generalist *M. persicae* and the specialist *Brevicoryne brassicae* were reduced on the ET-insensitive *etr1* mutant compared to wild-type plants [10,35]. The overexpression of a transcription factor gene, *MYB102*, which promotes ET biosynthesis by upregulation of some 1-aminocyclopropane-1-carboxylate synthase (ACS) genes in the ET-synthetic pathway led to an increase in aphid performance [36]. Furthermore, in tomato, ET signalling contributes to the susceptibility of potato aphid, *M. euphorbiae*, in the absence of the *Mi-1.2* gene. In choice assays, potato aphids preferred wild-type plants to the ET-insensitive, *Neverripe* mutant [37]. Our results with *sickle* together with other results discussed, suggest that ET can benefit the feeding for some aphid species, and the impediment of ET pathway impairs the infestation of these aphids.

In contrast to *A. kondoi* and *T. trifolii*, the *sickle* mutant did not affect the growth of the Australian biotype of *A. pisum* (Figures 1 and 2). These findings were consistent with studies with the European *A. pisum* biotype (PS01), which is distinct from the Australian *A. pisum* biotype, and where ET was also found to not be involved in the aphid susceptible or resistant interactions [24,28]. However, *sickle* was found to promote the growth of a Chinese biotype of *A. pisum* [38–40]. These differences between the *A. pisum* biotypes and with *A. kondoi* and *T. trifolii* suggest that the role of ethylene in the *M. truncatula*-aphid interactions is both biotype- and species-dependent.

How ET signalling modulates the moderate resistance in A17 against *A. kondoi* and *T. trifolii* is unknown. A17 carries multiple QTLs conferring antibiosis factors against these two aphid species [18,20]. The direct link between the ET signalling pathway and a specific QTL(s) in A17 is yet to be investigated. It is possible that the suppression of the ET pathway in *sickle* led to upregulation of other signalling pathways, such as for SA and JA, which might increase plant defence mechanisms against these two aphid species, as these signalling pathways are often inter-linked and work synergistically or antagonistically [41–43]. Further research on the interactions between the ET insensitivity in *M. truncatula sickle* and other defense signalling pathways in plant-aphid interactions will facilitate the understanding of the function of ET in plant resistance to aphids.

Our results also showed that the *sickle* mutant delayed the leaf senescence caused by the feeding of all three aphid species. There was no noticeable phenotypical difference among the leaves of A17, *sickle* and Jester control plants prior to aphid infestation at three weeks after planting though *sickle* could demonstrate concomitant alternation of some ethylene related phenotypes, including delayed petal and leaf senescence and decreased abscission of seed pod and leaves at the later stage of plant growth [29]. As all our experiments were carried out with young plants of three to five weeks old, it is unlikely these concomitant ethylene related phenotypes have a direct impact on aphid performance or plant symptom in response to aphid infestation. Aphid infestation causes changes in source allocation in the host plant to direct nutrients to the insect infested tissues [44]. Premature leaf senescence has been suggested to be a plant defence mechanism used to counteract aphid feeding by redirecting the nutrients to the un-infested source tissues [45]. In *Arabidopsis*, infestation by *M. persicae* induced the transcription level of *SENESCENCE ASSOCIATED GENES* (*SAGs*). Silencing of the *SAGs* delayed plant senescence which led to an increase *M. persicae* levels [45]. Here we observed the opposite with the *M. truncatula sickle* mutant. Whether the delay in leaf senescence directly relates to the increased antibiosis resistance to *A. kondoi* and *T. trifolii* remains unknown. Further research would help elucidate the relationship between leaf senescence and aphid feeding processes.

We have determined that ET is not essential for *AKR*, *APR* or *TTR* mediated resistance against *A. kondoi*, *A. pisum* or *T. trifolii*, respectively. ET has also been found to be dispensable for the *RAP* gene mediated resistance against the European *A. pisum* biotype [28]. However, upon infestation by *A. kondoi* or *T. trifolii*, some ET related genes were found to be induced in both A17 and Jester, with higher induction in Jester than A17 [26,27]. The lack of difference in aphid performance between Jester and Jester with the homozygous *sickle* mutation suggests that induction of the ET related genes previously found in Jester may be insufficient in limiting aphid feeding from the plant [26]. This might also be the case in other plant-aphid systems, such as in resistant barley plants with *Schizaphis graminum*, *Rhopalosiphum padi*, and *Diuraphis noxia* [11,12], wheat with *D. noxia* [13], tomato (*Mi1.2*) with *Macrosiphum euphorbiae* and melon (*Vat*) with *Aphis gossypii* [14]. Although in these plant-aphid systems, ET production or ET related genes were shown to be highly induced in the resistant interactions, whether ET contributes to the resistance outcome is still a question.

Both A17 and Jester carry a semi-dominant locus called *AIN* which mediates necrotic lesions resembling a hypersensitive response at the site of infestation by both *A. kondoi* and *A. pisum* [23]. It is unlikely that ET signalling is negatively regulating the activity of *AIN* in A17 for the following reasons: firstly, while *AIN* is important for resistance to both *A. kondoi* and *A. pisum* in A17, the *sickle* mutant only results in an increase in resistance *to A. kondoi*. Secondly, the *sickle* mutant still displays the same HR-like symptoms conferred by *AIN* following infestation with either aphid. While the overall number of necrotic lesions were less in *sickle* than A17 following *A. kondoi* infestation, this is most likely a reflection of the lower number of aphids feeding on *sickle*, due to the increase in aphid resistance. Importantly, the size of the necrotic lesions remained similar between *sickle* and A17. Collectively, these data suggest that ET signalling is not a negative regulator of *AIN* activity. As discussed earlier, there are other QTLs that have been identified in A17 as being important for the moderate resistance to *A. kondoi* but not *A. pisum* [18], which may be the target of ET negative regulation, or the target(s) may be an unidentified loci.

In conclusion, the *M. truncatula sickle* mutant has previously been shown to be defective in the control of root infecting micro-organisms including beneficial rhizobia, mycorrhizal fungi, as well as infection by the fungal and oomycete pathogens, *R. solani* and *P. medicaginis* [46,47]. Here we observed a positive effect of the *sickle* mutant on the control of the infestation by insect herbivory, by *A. kondoi* and *T. trifolii*. While ET signaling is not essential for the activity of three R genes for resistance against *A. kondoi*, *A. pisum* and *T. trifolii*, it is also not involved in the *AIN* mediated hypersensitive response to *A. kondoi* and *A. pisum*. However, the *sickle* mutant delayed leaf senescence by all three aphid species, but enhanced tolerance only to infestation by *A. kondoi* and *T. trifolii*, The results suggest that manipulation of the ET signaling pathway could also help provide resistance to certain aphid species and enhance plant tolerance against aphid feeding.

4. Materials and Methods

4.1. Plant Materials and Growth Conditions

Included in this study were *M. truncatula* A17 (referred to as "wild-type"), which is the reference *M. truncatula* accession, an ethylene insensitive *sickle* mutant, a *tnt1* retrotransposon mutant that arose from a single *tnt1* insertion in the genetic background of A17 [29,30] and Jester which is closely related to A17 sharing 89% genome identity [21]. In addition to ethylene insensitivity, *sickle* also demonstrates delayed petal senescence and decreased abscission of seed pod and leaves. In addition F_2 and F_3 progenies were generated by reciprocal crossing between Jester and *sickle* accession plants according to the methods described by [48].

Prior to planting, seeds were scarified and germinated on moist filter paper in the dark at room temperature for two days. Plants were grown in a growth chamber with 16 h light (22 °C)/8 h dark (20 °C) under metal halide and incandescent lamps producing 240 to 260 µE m^{-2} s^{-1} or in a glasshouse with controlled temperature around 22 °C and ambient light condition. In both glasshouse and growth chamber experiments, plants were grown in individual 0.9 L pots with *Arabidopsis* soil mix (Richgrow company, Perth, WA, Australia). Plants were fertilized with liquid Nitrosol fertilizer (Amgrow Australia, Perth, WA, Australia) once planted and watered two times per week throughout the experiments.

4.2. Aphid Species and Rearing Conditions

The aphid species used were *A. kondoi* (bluegreen aphid), *A. pisum* (pea aphid.) and *T. trifolii* f. *maculate* (spotted alfalfa aphid). Aphids of each species were obtained from colonies initiated from single aphid clones collected in Western Australia and were reared on Subterranean clover (*Trifolium subterraneum*) for *A. kondoi*, faba bean (*Vicia faba* L.) for *A. pisum* and alfalfa (*M. sativa*) for *T. trifolii* with 14 h light (23 °C)/10 h dark (20 °C) under high pressure sodium lamps and fluorescent light at 280 µE m^{-2} s^{-1}. Aphids were transferred to experimental plants with a fine paintbrush.

4.3. Screening for F$_2$ Plants Containing the Sickle Homozygous Allele

To obtain lines with the homozygous *sickle* locus and all three aphid resistance genes, *AKR*, *APR* and *TTR* crosses between *sickle* and Jester were performed F$_2$ plants were first screened for the *sickle* mutation using 1-aminocyclopropane-1-carboxylic acid (ACC) [30]. To establish an effective and reliable condition for the screening, five concentrations of ACC, 20, 40, 60, 80 and 100 ppm were initially evaluated with the *sickle* mutant, A17 and Jester with water as the control. While the *sickle* mutant showed no response to all ACC concentrations, 20 ppm ACC started to show impact on the embryonic root of A17 and Jester, while 80 ppm or 100 ppm of ACC resulted in severe stunting of root radicles of both A17 and Jester (Supplementary Figure S1). The 100 ppm was used for the screening of the Jester x *sickle* F$_2$ population (Supplementary Figure S2).

4.4. Genotyping F$_2$ Plants with AKR, APR and TTR Loci

In order to obtain F$_2$ lines that combined the homozygous *sickle* allele and the three aphid resistance gene loci, 90 of the 163 pre-selected F$_2$ *sickle* mutant plants were randomly selected and analysed using high throughput Multiplex-Ready marker technology (MRT) and molecular markers linked to these loci (Supplementary Table S1). The plants were grown in the growth room conditions as described above. Two weeks after planting, a single trifoliate leaf from each plant was collected and DNA isolated using the CTAB method as described previously [49]. DNA was subsequently diluted to a concentration of 50 ng/µL in a 96-well plate and multiplex ready PCRs were setup using the primers in Supplementary Table S1 and the protocol described by Hayden et al. [50]. The multiplexed PCR products were subjected to fragment analysis on an ABI3730 DNA analyser (Applied Biosystems, Melbourne, Victoria, Australia) according to Hayden et al. [50] and marker allele sizing determined using the Genemarker software (SoftGenetics LLC, State College, PA, USA).. After the genotyping, seeds from the F$_2$ plants with both the homozygous *sickle* and the three homozygous aphid resistance gene alleles, *AKR*, *APR* and *TTR*, were harvested to obtain F$_3$ seeds for the subsequent aphid infestation experiments.

4.5. Aphid Performance and Plant Damage on Single Trifoliate Leaves

To assess the aphid performance and leaf tolerance of *sickle* in comparison with its wild-type, parent A17 and Jester against three aphid species, three experiments were conducted, one in the glasshouse and two in growth chambers. Six replicate plants of each *M. truncatula* accession were randomly arranged. For all three experiments, three weeks after sowing, a single trifoliate leaf of similar age (fourth or fifth trifoliate leaf to emerge on the primary stem) of each plant was infested with four, five or seven adults of *A. pisum*, *A. kondoi* or *T. trifolii*, respectively. The number for each aphid species was determined based on our previous experiences with regards to the aphid size, the speed of aphid reproduction and degree of leaf damage caused to create a condition that allowed the aphid growth and leaf damage to be fully expressed to make comparison between the three *M. truncatula* accessions [17,22,23,26,31–33]. The aphids were caged on a single trifoliate leaf in a linen mesh cage (35 × 200 mm) per plant. A wooden stake supported the stem and cage [26]. Seven days after aphid infestation, the aphids on each leaf were collected and weighed. The damage on each leaf was visually assessed after the removal of the aphids.

The effect of the *sickle* mutant on *R* gene mediated aphid resistance was first measured using the F_2 plants after the genotyping. This was followed by an experiment using three independent F_3 lines which contained the *sickle*, *AKR*, *APR* and *TTR* homozygous alleles. A17, Jester and the *sickle* mutant were included for comparison. For each accession/F3 line, six replicate plants were set up for the aphid infestation as described above with single trifoliate leaves. The aphid population weight and leaf damage symptoms, such as leaf senescence and necrosis, were assessed. To examine if the *sickle* mutant affects the *AIN* -mediated hypersensitive response to *A. kondoi* and *A. pisum*, the numbers of macroscopic necrotic flecks per single trifoliate leaves were also recorded.

4.6. Aphid Performance and Plant Tolerance Experiments on Whole Plants

To assess the aphid performance and plant tolerance on whole plants, non-choice experiments were conducted under glasshouse conditions. Plants of *M. truncatula sickle* mutant, A17 and Jester were grown as described above. Two weeks after planting, each plant was infested with four, five or seven adults of *A. pisum*, *A. kondoi* or *T. trifolii*, respectively. Six replicate plants were set up for each *M. truncatula* accession with or without aphid infestation. Plants were randomly arranged. Fourteen days after the aphid infestation, aphids were collected from each plant and weighed immediately. After the removal of the aphids, the aerial part of all the plants including the non-infested control plants were dried in the oven at 50°C for two days. The dried weight of each plant was recorded. For each plant, aphid fresh weight per plant dry weight was calculated to determine aphid performance on the plant. For each *M. truncatula* accession, the tolerance of individual plants to aphid infestation was measured as the percentage of plant biomass reduction (PBR) relative to mean biomass of the control plants of the same *M. truncatula* accession using the formula: PBR = [(A − B)/A] × 100, in which A: average of the non-infested plant dry weight; B: dry weight of individual aphid-infested plant.

With each experiment, the aphid weight, the number of necrotic flecks or plant biomass reduction were analyzed by one-way ANOVA and compared by the LSD test at a 5% significance level using GenStat (VSN International, Rothamsted Research, Hertfordshire, UK).

Supplementary Materials: The following are available online at http://www.mdpi.com/1422-0067/21/13/4657/s1. Table S1. Overview of the PCR primers used; Table S2. Overview of the genotyping data. Supplementary Figure S1. The radicle root growth of *Medicago truncatula* A17, Jester and *sickle* mutant. Supplementary Figure S2. Screening of *Medicago truncatula* F_2 plants of crosses between Jester and *sickle*

Author Contributions: Conceptualization, L.-L.G. and K.B.S.; investigation, L.Z., L.-L.G., L.G.K., and Y.G.; formal analysis, L.Z., L.-L.G., L.G.K..; resources, K.B.S.; writing—original draft preparation, L.-L.G.; review and editing, L.Z., L.G.K., S.J. and K.B.S.; supervision L.-L.G. and K.B.S. All authors have read and agreed to the published version of the manuscript.

Funding: This research received no external funding

Acknowledgments: The authors thank Elaine Smith for technical assistance and Jonathan Anderson, Louise Thatcher and Kemal Kazan CSIRO for critical review of the manuscript. This research was financially supported by CSIRO A&F. The authors also would like to acknowledge the Shanxi Agricultural University and Scholarship Council of Shanxi Province, China to support L.J. Zhang and Y.Q. Guo during their visit to CSIRO.

Conflicts of Interest: The authors declare no conflict of interest.

References

1. Ng, J.C.K.; Perry, K.L. Transmission of plant viruses by aphid vectors. *Mol. Plant Pathol.* **2004**, *5*, 505–511. [CrossRef] [PubMed]

2. Whitfield, A.E.; Falk, B.W.; Rotenberg, D. Insect vector-mediated transmission of plant viruses. *Virology* **2015**, *479–480*, 278–289. [CrossRef] [PubMed]

3. Carr, J.P.; Murphy, A.M.; Tungadi, T.; Yoon, J.Y. Plant defense signals: Players and pawns in plant-virus-vector interactions. *Plant Sci.* **2019**, *279*, 87–95. [CrossRef] [PubMed]

4. Erb, M.; Meldau, S.; Howe, G.A. Role of phytohormones in insect-specific plant reactions. *Trends Plant Sci.* **2012**, *17*, 250–259. [CrossRef]

5. Divol, F.; Vilaine, F.; Thibivilliers, S.; Amselem, J.; Palauqui, J.C.; Kusiak, C.; Dinant, S. Systemic response to aphid infestation by Myzus persicae in the phloem of Apium graveolens. *Plant Mol. Biol.* **2005**, *57*, 517–540. [CrossRef]

6. Moran, P.J.; Cheng, Y.F.; Cassell, J.L.; Thompson, G.A. Gene expression profiling of Arabidopsis thaliana in compatible plant-aphid interactions. *Arch. Insect Biochem. Physiol.* **2002**, *51*, 182–203. [CrossRef]

7. Heidel, A.J.; Baldwin, I.T. Microarray analysis of salicylic acid- and jasmonic acid-signalling in responses of Nicotiana attenuata to attack by insects from multiple feeding guilds. *Plant Cell Environ.* **2004**, *27*, 1362–1373. [CrossRef]

8. De Vos, M.; Van Oosten, V.R.; Van Poecke, R.M.P.; Van Pelt, J.A.; Pozo, M.J.; Mueller, M.J.; Buchala, A.J.; Metraux, J.P.; Van Loon, L.C.; Dicke, M.; et al. Signal signature and transcriptome changes of Arabidopsis during pathogen and insect attack. *Mol. Plant-Microbe Interact.* **2005**, *18*, 923–937. [CrossRef]

9. Kettles, G.J.; Drurey, C.; Schoonbeek, H.-j.; Maule, A.J.; Hogenhout, S.A. Resistance of Arabidopsis thaliana to the green peach aphid, Myzus persicae, involves camalexin and is regulated by microRNAs. *New Phytol.* **2013**, *198*, 1178–1190. [CrossRef]

10. Mewis, I.; Tokuhisa, J.G.; Schultz, J.C.; Appel, H.M.; Ulrichs, C.; Gershenzon, J. Gene expression and glucosinolate accumulation in Arabidopsis thaliana in response to generalist and specialist herbivores of different feeding guilds and the role of defense signaling pathways. *Phytochemistry* **2006**, *67*, 2450–2462. [CrossRef]

11. Miller, H.L.; Neese, P.A.; Ketring, D.L.; Dillwith, J.W. Involvement of ethylene in aphid infestation of barley. *J. Plant Growth Regul.* **1994**, *13*, 167–171. [CrossRef]

12. Argandona, V.H.; Chaman, M.; Cardemil, L.; Munoz, O.; Zuniga, G.E.; Corcuera, L.J. Ethylene production and peroxidase activity in aphid-infested barley. *J. Chem. Ecol.* **2001**, *27*, 53–68. [CrossRef] [PubMed]

13. Boyko, E.V.; Smith, C.M.; Thara, V.K.; Bruno, J.M.; Deng, Y.P.; Starkey, S.R.; Klaahsen, D.L. Molecular basis of plant gene expression during aphid invasion: Wheat *Pto-* and *Pti*-like sequences are involved in interactions between wheat and Russian wheat aphid (Homoptera: Aphididae). *J. Econon. Entomol.* **2006**, *99*, 1430–1445. [CrossRef] [PubMed]

14. Anstead, J.; Samuel, P.; Song, N.; Wu, C.J.; Thompson, G.A.; Goggin, F. Activation of ethylene-related genes in response to aphid feeding on resistant and susceptible melon and tomato plants. *Entomol. Exp. Appl.* **2010**, *134*, 170–181. [CrossRef]

15. Kamphuis, L.; Gao, L.; Turnbull, C.; Karam, S. Medicago—Aphid interactions. In *The Model Legume Medicago Truncatula*; de Bruijn, F.J., Ed.; Wiley & Sons, Inc.: New York, NY, USA, 2019; pp. 363–368.

16. Kamphuis, L.G.; Zulak, K.; Gao, L.-L.; Anderson, J.; Singh, K.B. Plant–aphid interactions with a focus on legumes. *Funct. Plant Biol.* **2013**. [CrossRef] [PubMed]

17. Gao, L.L.; Klingler, J.P.; Anderson, J.P.; Edwards, O.R.; Singh, K.B. Characterization of pea aphid resistance in *Medicago truncatula*. *Plant Physiol.* **2008**, *146*, 996–1009. [CrossRef]

18. Guo, S.M.; Kamphuis, L.G.; Gao, L.L.; Klingler, J.P.; Lichtenzveig, J.; Edwards, O.; Singh, K.B. Identification of distinct quantitative trait loci associated with defence against the closely related aphids *Acyrthosiphon pisum* and *A. kondoi* in *Medicago truncatula*. *J. Exp. Bot.* **2012**, *63*, 3913–3922. [CrossRef]

19. Kamphuis, L.G.; Guo, S.M.; Gao, L.L.; Singh, K.B. Genetic Mapping of a Major Resistance Gene to Pea Aphid (*Acyrthosipon pisum*) in the Model Legume *Medicago truncatula*. *Int. J. Mol. Sci.* **2016**, *17*, 1224. [CrossRef]

20. Kamphuis, L.G.; Lichtenzveig, J.; Peng, K.F.; Guo, S.M.; Klingler, J.P.; Siddique, K.H.M.; Gao, L.L.; Singh, K.B. Characterization and genetic dissection of resistance to spotted alfalfa aphid (*Therioaphis trifolii*) in *Medicago truncatula*. *J. Exp. Bot.* **2013**, *64*, 5157–5172. [CrossRef]

21. Klingler, J.; Creasy, R.; Gao, L.L.; Nair, R.M.; Calix, A.S.; Jacob, H.S.; Edwards, O.R.; Singh, K.B. Aphid resistance in *Medicago truncatula* involves antixenosis and phloem-specific, inducible antibiosis, and maps to a single locus flanked by NBS-LRR resistance gene analogs. *Plant Physiol.* **2005**, *137*, 1445–1455. [CrossRef]

22. Klingler, J.P.; Edwards, O.R.; Singh, K.B. Independent action and contrasting phenotypes of resistance genes against spotted alfalfa aphid and bluegreen aphid in *Medicago truncatula*. *New Phytol.* **2007**, *173*, 630–640. [CrossRef] [PubMed]

23. Klingler, J.P.; Nair, R.M.; Edwards, O.R.; Singh, K.B. A single gene, *AIN*, in *Medicago truncatula* mediates a hypersensitive response to both bluegreen aphid and pea aphid, but confers resistance only to bluegreen aphid. *J. Exp. Bot.* **2009**, *60*, 4115–4127. [CrossRef] [PubMed]

24. Stewart, S.A.; Hodge, S.; Ismail, N.; Mansfield, J.W.; Feys, B.J.; Prosperi, J.M.; Huguet, T.; Ben, C.; Gentzbittel, L.; Powell, G. The RAP1 Gene Confers Effective, Race-Specific Resistance to the Pea Aphid in Medicago truncatula Independent of the Hypersensitive Reaction. *Mol. Plant-Microbe Interact.* **2009**, *22*, 1645–1655. [CrossRef] [PubMed]

25. Kamphuis, L.G.; Klingler, J.P.; Jacques, S.; Gao, L.-L.; Edwards, O.R.; Singh, K.B. Additive and epistatic interactions between *AKR* and *AIN* loci conferring bluegreen aphid resistance and hypersensitivity in *Medicago truncatula*. *J. Exp. Bot.* **2019**, *70*, 4887–4902. [CrossRef] [PubMed]

26. Gao, L.L.; Anderson, J.P.; Klingler, J.P.; Nair, R.M.; Edwards, O.R.; Singh, K.B. Involvement of the octadecanoid pathway in bluegreen aphid resistance in *Medicago truncatula*. *Mol. Plant-Microbe Interact.* **2007**, *20*, 82–93. [CrossRef]

27. Gao, L.L.; Kamphuis, L.G.; Kakar, K.; Edwards, O.R.; Udvardi, M.K.; Singh, K.B. Identification of potential early regulators of aphid resistance in *Medicago truncatula* via transcription factor expression profiling. *New Phytol.* **2010**, *186*, 980–994. [CrossRef]

28. Stewart, S.A.; Hodge, S.; Bennett, M.; Mansfield, J.W.; Powell, G. Aphid induction of phytohormones in Medicago truncatula is dependent upon time post-infestation, aphid density and the genotypes of both plant and insect. *Arthropod-Plant Interact.* **2016**, *10*, 41–53. [CrossRef]

29. Penmetsa, R.V.; Cook, D.R. A legume ethylene-insensitive mutant hyperinfected by its rhizobial symbiont. *Science* **1997**, *275*, 527–530. [CrossRef]

30. Penmetsa, R.V.; Uribe, P.; Anderson, J.; Lichtenzveig, J.; Gish, J.C.; Nam, Y.W.; Engstrom, E.; Xu, K.; Sckisel, G.; Pereira, M.; et al. The Medicago truncatula ortholog of Arabidopsis EIN2, sickle, is a negative regulator of symbiotic and pathogenic microbial associations. *Plant J.* **2008**, *55*, 580–595. [CrossRef]

31. Gao, L.L.; Horbury, R.; Nair, R.M.; Singh, K.B.; Edwards, O.R. Characterization of resistance to multiple aphid species (Hemiptera: Aphididae) in *Medicago truncatula*. *Bull. Entomol. Res.* **2007**, *97*, 41–48. [CrossRef]

32. Guo, S.M.; Kamphuis, L.G.; Gao, L.L.; Edwards, O.R.; Singh, K.B. Two independent resistance genes in the *Medicago truncatula* cultivar jester confer resistance to two different aphid species of the genus Acyrthosiphon. *Plant Signal. Behav.* **2009**, *4*, 328–331. [CrossRef]

33. Klingler, J.; Creasy, R.; Gao, L.L.; Nair, R.M.; Jacob, H.S.; Edwards, O.R.; Singh, K.B. Genetics of phloem-aphid interactions in *Medicago truncatula*. *Comp. Biochem. Physiol. A-Mol. Integr. Physiol.* **2005**, *141*, S230.

34. Lu, J.; Li, J.; Ju, H.; Liu, X.; Erb, M.; Wang, X.; Lou, Y. Contrasting Effects of Ethylene Biosynthesis on Induced Plant Resistance against a Chewing and a Piercing-Sucking Herbivore in Rice. *Mol. Plant* **2014**, *7*, 1670–1682. [CrossRef] [PubMed]

35. Mewis, I.; Appel, H.M.; Hom, A.; Raina, R.; Schultz, J.C. Major signaling pathways modulate Arabidopsis glucosinolate accumulation and response to both phloem-feeding and chewing insects. *Plant Physiol.* **2005**, *138*, 1149–1162. [CrossRef] [PubMed]

36. Zhu, L.; Guo, J.; Ma, Z.; Wang, J.; Zhou, C. Arabidopsis Transcription Factor *MYB102* Increases Plant Susceptibility to Aphids by Substantial Activation of Ethylene Biosynthesis. *Biomolecules* **2018**, *8*, 39. [CrossRef] [PubMed]

37. Mantelin, S.; Bhattarai, K.K.; Kaloshian, I. Ethylene contributes to potato aphid susceptibility in a compatible tomato host. *New Phytol.* **2009**, *183*, 444–456. [CrossRef] [PubMed]

38. Guo, H.J.; Sun, Y.C.; Li, Y.F.; Liu, X.H.; Zhang, W.H.; Ge, F. Elevated CO_2 decreases the response of the ethylene signaling pathway in *Medicago truncatula* and increases the abundance of the pea aphid. *New Phytol.* **2014**, *201*, 279–291. [CrossRef]

39. Sun, Y.; Guo, H.; Zhu-Salzman, K.; Ge, F. Elevated CO_2 increases the abundance of the peach aphid on Arabidopsis by reducing jasmonic acid defenses. *Plant Sci.* **2013**, *210*, 128–140. [CrossRef]

40. Sun, Y.C.; Guo, H.J.; Ge, F. Plant-Aphid Interactions Under Elevated CO_2: Some Cues from Aphid Feeding Behavior. *Front. Plant Sci.* **2016**, *7*. [CrossRef]

41. Buscaill, P.; Rivas, S. Transcriptional control of plant defence responses. *Curr. Opin. Plant Biol.* **2014**, *20*, 35–46. [CrossRef]

42. Faulkner, C.; Robatzek, S. Plants and pathogens: Putting infection strategies and defence mechanisms on the map. *Curr. Opin. Plant Biol.* **2012**, *15*, 699–707. [CrossRef] [PubMed]

43. Shah, J. Plants under attack: Systemic signals in defence. *Curr. Opin. Plant Biol.* **2009**, *12*, 459–464. [CrossRef] [PubMed]

44. Girousse, C.; Moulia, B.; Silk, W.; Bonnemain, J.-L. Aphid Infestation Causes Different Changes in Carbon and Nitrogen Allocation in Alfalfa Stems as Well as Different Inhibitions of Longitudinal and Radial Expansion. *Plant Physiol.* **2005**, *137*, 1474–1484. [CrossRef]

45. Pegadaraju, V.; Knepper, C.; Reese, J.; Shah, J. Premature leaf senescence modulated by the Arabidopsis PHYTOALEXIN DEFICIENT4 gene is associated with defense against the phloem-feeding green peach aphid. *Plant Physiol.* **2005**, *139*, 1927–1934. [CrossRef]

46. Anderson, J.P.; Lichtenzveig, J.; Gleason, C.; Oliver, R.P.; Singh, K.B. The B-3 Ethylene Response Factor *MtERF1-1* Mediates Resistance to a Subset of Root Pathogens in *Medicago truncatula* without Adversely Affecting Symbiosis with Rhizobia. *Plant Physiol.* **2010**, *154*, 861–873. [CrossRef] [PubMed]

47. Liu, Y.; Hassan, S.; Kidd, B.N.; Garg, G.; Mathesius, U.; Singh, K.B.; Anderson, J.P. Ethylene Signaling Is Important for Isoflavonoid-Mediated Resistance to *Rhizoctonia solani* in Roots of *Medicago truncatula*. *Mol. Plant-Microbe Interact.* **2017**, *30*, 691–700. [CrossRef]

48. Veerappan, V.; Kadel, K.; Alexis, N.; Scott, A.; Kryvoruchko, I.; Sinharoy, S.; Taylor, M.; Udvardi, M.; Dickstein, R. Keel petal incision: A simple and efficient method for genetic crossing in Medicago truncatula. *Plant Methods* **2014**, *10*, 11. [CrossRef]

49. Ellwood, S.R.; D'Souza, N.K.; Kamphuis, L.G.; Burgess, T.I.; Nair, R.M.; Oliver, R.P. SSR analysis of the *Medicago truncatula* SARDI core collection reveals substantial diversity and unusual genotype dispersal throughout the Mediterranean basin. *Theor. Appl. Genet.* **2006**, *112*, 977–983. [CrossRef]

50. Hayden, M.J.; Nguyen, T.M.; Waterman, A.; Chalmers, K.J. Multiplex-Ready PCR: A new method for multiplexed SSR and SNP genotyping. *BMC Genom.* **2008**, *9*, 80. [CrossRef]

International Journal of
Molecular Sciences

Review

Crosstalk between Hydrogen Sulfide and Other Signal Molecules Regulates Plant Growth and Development

Lijuan Xuan [†], Jian Li [†], Xinyu Wang and Chongying Wang *

Ministry of Education Key Laboratory of Cell Activities and Stress Adaptations, School of Life Sciences, Lanzhou University, Lanzhou 730000, China; xuanlj17@lzu.edu.cn (L.X.); Lijian18@lzu.edu.cn (J.L.); wangxy@lzu.edu.cn (X.W.)
* Correspondence: wangcy@lzu.edu.cn; Tel./Fax: +86-093-1891-4155
† These authors contributed equally to this work.

Received: 31 May 2020; Accepted: 24 June 2020; Published: 28 June 2020

Abstract: Hydrogen sulfide (H_2S), once recognized only as a poisonous gas, is now considered the third endogenous gaseous transmitter, along with nitric oxide (NO) and carbon monoxide (CO). Multiple lines of emerging evidence suggest that H_2S plays positive roles in plant growth and development when at appropriate concentrations, including seed germination, root development, photosynthesis, stomatal movement, and organ abscission under both normal and stress conditions. H_2S influences these processes by altering gene expression and enzyme activities, as well as regulating the contents of some secondary metabolites. In its regulatory roles, H_2S always interacts with either plant hormones, other gasotransmitters, or ionic signals, such as abscisic acid (ABA), ethylene, auxin, CO, NO, and Ca^{2+}. Remarkably, H_2S also contributes to the post-translational modification of proteins to affect protein activities, structures, and sub-cellular localization. Here, we review the functions of H_2S at different stages of plant development, focusing on the S-sulfhydration of proteins mediated by H_2S and the crosstalk between H_2S and other signaling molecules.

Keywords: hydrogen sulfide; reactive oxygen species; S-sulfhydration; plant hormone; gasotransmitter

---

1. Introduction

Sulfur (S) is an essential element and is involved in the synthesis and metabolism of the sulfur-containing amino acids cysteine (Cys) and methionine (Met), as well as co-enzyme A, thiamine, biotin, iron-sulfur clusters, and nitrogenase. Only plants, algae, fungi, and some prokaryotes can take advantage of the inorganic sulfur (sulfate, SO_4^{2-}) naturally found in soils and incorporate it into organic forms [1]. During sulfur assimilation in plants, the SO_4^{2-} absorbed by roots is first reduced to hydrogen sulfide (H_2S) under the catalysis of adenosine-5′-phosphoryl sulfate reductase (APSR) and sulfite reductase (SIR) and then transformed into Cys under the catalysis of O-Acetylserine (thiol) lyase (OAS-TL). Therefore, H_2S is an extremely important intermediate in the thio-metabolism pathway. H_2S can also be generated from chloroplasts and mitochondria through the reduction of Cys by β-cyanoalanine synthase (CAS) and cysteine desulfhydrase (CDes) [2–5]. CAS can transform cyanide (CN^-) and L-Cys into β–cyanoalanine and H_2S to degrade the toxin cyanogen (Figure 1) [3,6,7]. CDes, such as L-cysteine desulfhydrase (LCD, at3g62130) [2], L-cysteine desulfhydrase 1 (DES1, at5g28030) [4], D-cysteine desulfhydrase 1 (DCD1, at1g48420), and D-cysteine desulfhydrase 2 (DCD2, at3g26115) in Arabidopsis, catalyze both L-Cys and D-Cys into H_2S, pyruvate, and ammonia. LCD and DES1 use L-Cys as substrate and are the two pivotal enzymes in the process of endogenous H_2S production [2].

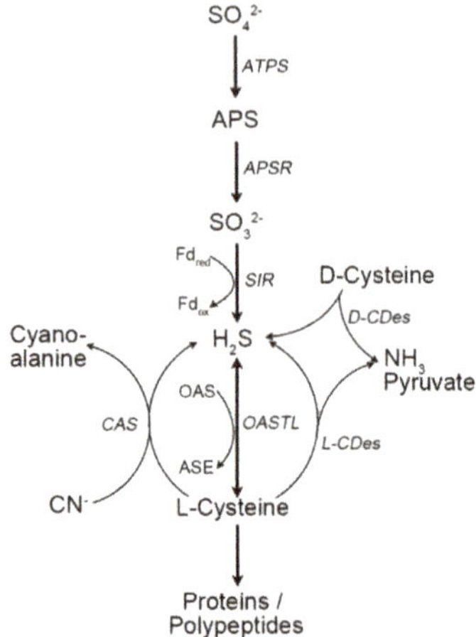

Figure 1. The synthesis and metabolism of H$_2$S in higher plants. H$_2$S is generated coincident with sulfate reduction in the plant cell. The key enzymes in H$_2$S biosynthesis and metabolism include sulfite reductase (SIR), L-cysteine desulfhydrase (L-CDes), D-cysteine desulfhydrase (D-CDes), β-cyanoalanine synthase (CAS), and O-acetylserine(thiol)lyase (OAS-TL). Plants are capable of reducing activated sulfate (SO$_4^{2-}$) to sulfite (SO$_3^{2-}$), after that SIR catalyzes SO$_3^{2-}$ to H$_2$S, with ferredoxin (Fd$_{red}$) as the electron donor. In the presence of OAS-TL, the generated H$_2$S is reversibly reduced to L-cysteine by reacting with O-acetylserine (OAS). L-CDes and D-CDes catalyze the degradation of L/D-cysteine to produce H$_2$S, amine (NH$_3$) and pyruvate to maintain H$_2$S homeostasis. CAS, located in the mitochondria, can also catalyze the production of H$_2$S, using cyanide (CN$^-$) and cysteine as substrates, removing the toxin cyanogen.

H$_2$S is a toxic gaseous molecule with the pungent odor of rotten eggs and has serious impacts on animals and plants [8]. Just 30 μmol/L H$_2$S can inhibit the activity of mitochondrial cytochrome c oxidase and reduce the intensity of mitochondrial respiration by 50% [9]. Surprisingly, H$_2$S also functions as a gasotransmitter, with essential roles at different stages of plant development. H$_2$S interacts with other signals, such as plant hormones, other gasotransmitters, and ionic signals. H$_2$S can also post-translationally modify proteins or affect secondary metabolism [10]. During seed imbibition, the endogenous H$_2$S level increases in Arabidopsis under normal growth conditions [11]. When the germination of wheat seed is inhibited under copper (Cu) stress, treatment with an appropriate concentration of exogenous H$_2$S (1.4 mM) promotes germination by reducing oxidative damage [12]. In addition, H$_2$S induces stomatal movement of guard cells and serves as a switch in stomatal opening [13]. The application of an exogenous H$_2$S donor (sodium hydrosulfide (NaHS) or GYY4137 (morpholine-4-4-methoxyphenyl)) reduces the nitric oxide (NO) accumulation induced by abscisic acid (ABA) and promotes guard cell movement to allow stomatal opening in light or darkness [14]. In Arabidopsis, mutation of *des1* leads to premature senescence of leaves [15]. In many fruits and vegetables, H$_2$S treatment delays premature leaf senescence and the decay of fruits after harvest via reducing the accumulation of reactive oxygen species (ROS) [16,17] and inhibits the abscission of plant organs via increasing the content of auxin in abscission zone tissues [18]. A recent report has shown that there is a significant increase in the S-sulfhydration level of the actin proteins in an H$_2$S-overproducing line, created by the over-expression of *LCD* in the Arabidopsis *O-acetylserine(thiol)lyase isoform a1* (*oasa1*) mutant (*OE LCD-5/oas-a1*). This increase in S-sulfhydration decreased the distribution of the actin cytoskeleton, which directly weakened actin polymerization and impaired root hair growth [19].

Here, we comprehensively review the functions of H$_2$S in plant growth and development under normal or adverse environmental conditions and the mechanisms by which H$_2$S influences different processes. The review focuses on both the crosstalk of H_2S with other signals and the H_2S-mediated S-sulfhydration of proteins.

2. Roles of H_2S at Different Stages of Plant Development

2.1. H_2S Promotes Seed Germination

Seed germination, the first step of the plant life cycle, is quite vulnerable to unfavorable environmental conditions [20], and several studies have addressed the concentration of H_2S that contributes to seed germination under normal or stress conditions. For instance, when the seeds and later roots of bean, pea, wheat, and corn were exposed to 10–100 mM H_2S solutions, their germination rate and seedling size were increased, and their germination times were shortened. After growing to maturity in soil, the total mass, roots and fruits of all H_2S-pretreated plants were greater than the controls [21]. In imbibed seeds, the activities of L/D-CDes were stimulated and the content of H_2S increased slightly compared with the dry seeds [11]. In the presence of hypotaurine (HT, an H_2S scavenger) or DL propargylglycine (PAG, a DES1 inhibitor), seed germination was delayed [11], suggesting that H_2S is indispensable in seed germination. Furthermore, metal, osmotic, and heat stresses often cause oxidative damage during seed germination. In wheat seeds inhibited by Cu, aluminum (Al), or osmotic stresses, treatment with the H_2S donor, 1.4 mM NaHS, not only increased the content of endogenous H_2S but also improved germination, with increased activities of amylase and esterase. Meanwhile, NaHS treatment prevented the absorption of Cu and maintained lower levels of malondialdehyde (MDA) and hydrogen peroxide (H_2O_2) [12,22,23]. It was concluded that H_2S plays an important role in promoting seed germination during ionic stress by reducing oxidative damage and preventing the absorption of metal ions.

Heat stress generally suppresses seed germination by enhancing the contents of ABA, which acts through ABA-INSENSITIVE 5 (ABI5) and ELONGATED HYPCOTYL 5 (HY5), positive regulators of ABA inhibition of seed germination [24]. In maize seeds under high temperature, pre-soaking with 0.5 mM NaHS enhanced seed germination rates, sprout length, root length, and fresh weight [25]. In Arabidopsis seeds under heat stress, 0.1 mM H_2S treatment broke the ABA inhibition on seed germination. This was shown to be due to decreased translocation of the E3 ligase CONSTITUTIVE PHOTOMORPHOGENESIS 1 (COP1) from the nucleus to the cytoplasm, causing continued degradation of ELONGATED HYPCOTYL 5 (HY5) in the nucleus. Degradation of HY5 in the nucleus inhibits ABA signaling since the transcription of *ABI5* could not be activated by HY5. H_2S is thus potentially important in the modulation of thermotolerance of seed germination [24].

Li et al. (2012) found that soaking *Jatropha curcas* seeds with H_2O_2 could greatly improve the germination rate by stimulating LCD activity and H_2S accumulation [26]. Interestingly, germination was enhanced by exogenous H_2S but was reduced by pretreatment with an H_2S biosynthesis inhibitor (aminooxyacetic acid, AOA). Thus, H_2S plays a vital role in H_2O_2-induced seed germination in *Jatropha curcas* [26].

Together, these reports show that the content of endogenous H_2S increases during seed germination and that exogenous NaHS treatment enhances the production of endogenous H_2S, which in turn protects seed germination from damage by enhancing the activities of amylase and esterase, by reducing oxidative damage, by preventing the absorption of metal ions, and by repressing ABA signaling.

However, there are also reports of confounding roles for H_2S during seed germination. In the *des1* mutant of Arabidopsis, the content of H_2S remained unchanged after imbibition, and there was no significant difference in seed germination between wild type (WT) and *des1* under a range of temperatures (15–25 °C) and either 1 µM or 5 µM ABA [11]. In another study, when wheat TaD-CDes was ectopically overexpressed in Arabidopsis, both the transcription level and enzyme activity of D-CDes were increased, but the seed germination of TaD-CDes-expressing plants was more sensitive to ABA [27]. Therefore, the above results indicate that appropriate increase in the content of H_2S aids seed germination under both normal and stress conditions but that endogenous H_2S has an incompatible role with the exogenous application of H_2S under ABA treatment, and the mechanism is not clear absolutely.

2.2. H_2S Affects Formation of Lateral Roots

The development of plant root is primarily regulated by indoleacetic acid (IAA) [28]. However, recent studies have shown that H_2S plays a significant role in the development of lateral roots by interacting with IAA, NO or H_2O_2 [29–31]. For instance, Zhang et al. (2009) reported that the application of 0.2 mM NaHS on cuttings of *Ipomoea batatas* seedlings promoted the number and the length of adventitious roots in a dose-dependent manner with increases in IAA and NO [29]. Further research showed that 1 mM NaHS pretreatment induced the up-regulation of an auxin-dependent *Cyclin Dependent Kinases (CDK)* gene (*SlCDKA1*) and a cell cycle regulatory gene (*SlCYCA2*) and the down-regulation of the *Kip-Related Protein 2* (*SlKRP2*), which is dependent on NO signaling [32]. The gene expression induced by H_2S could be blocked by an IAA transport inhibitor (N-1-naphthylphthalamic acid; NPA) or a NO scavenger [2-(4-carboxyphenyl)-4,4,5,5 -tetramethylimidazoline-1- oxyl-3- oxide; cPTIO], indicating that the lateral root development promoted by H_2S is also dependent on NO and IAA signaling through regulation of *SlCDKA1*, *SlCYCA2* and *SlKRP2* [29]. Moreover, 1 mM NaHS treatment upregulated the *respiration burst oxidase homologous (RBOH1)* transcript, resulting in the overproduction of H_2O_2, contributing to lateral root formation in tomato. However, when plants were co-treated with an H_2O_2 scavenger (dimethylthiourea; DMTU) and an inhibitor of NADPH oxidase (diphenylene iodonium; DPI), the lateral root formation induced by NaHS was impaired, and the up-regulation of *SlCYCA2;1*, *SlCYCA3;1*, and *SlCDKA1* and the down-regulation of *SlKRP2* induced by H_2S was suppressed [31]. Therefore, it can be concluded that H_2S treatment upregulated the *RBOH1* transcript and promoted the production of H_2O_2, which stimulated NO and IAA signaling through regulation of the expression of *SlCDKA1*, *SlCYCA2, SlCDKA1* and *SlKRP2*, leading to lateral root formation.

Methane (CH_4) is another gaseous compound that may transmit signals. Recent studies have shown that CH_4 plays an important role in some plant physiological processes, such as responses to water, salt and heavy metal stressors [33–35]. In addition, CH_4 was also found to participate in root organogenesis, an activity that might be related to NO [36]. Kou et al. (2018) discovered that CH_4 treatment increased the expression levels of *L-CDes* genes and the endogenous H_2S content, which then promoted adventitious root development in cucumber with the up-regulation of genes related to cell division, namely *CsDNAJ-1, CsCDPK, CsCDPK5*, and *CsCDC6*, to auxin signaling, namely *CsAux22D-like* and *CsAux22B-like*, and that this response was disrupted by the presence of the H_2S scavenger HT or the DES1 inhibitor PAG [37]. Similar results were reported in tomatoes [38]. Further, the lateral roots of the *Atdes1* mutant showed defects in the presence of CH_4 in Arabidopsis [38]. All these data demonstrated that *DES*-dependent H_2S signaling plays a major role in CH_4-triggered lateral root formation.

Chen et al. (2014) found that selenium (Se) stress inhibited root growth in *Brassica napus* by suppressing the expression of most of the LCD and DCD homologues [30]. Pretreatment with 0.5 mM NaHS alleviated the inhibitory effect of Se on root growth by partly restoring the endogenous H_2S content in roots and reducing the accumulation of ROS by increasing the content of glutathione (GSH), suggesting that both H_2S and GSH are involved in the regulation of lateral root growth under stress through antioxidation [30].

However, in some other studies, high levels of H_2S (100–500 µM) changed root development by inhibiting auxin transport and thus altering the polar subcellular distribution of the PIN proteins, which is an actin-dependent process [39]. NaHS treatment (100 µM) and the overproduction of endogenous H_2S in the *OE LCD-5/oasa1* line significantly increased the S-sulfhydration level of actin-2 and decreased the distribution of actin cytoskeleton in root cells, which directly weakened the aggregation of actin and reduced the root hair density of Arabidopsis [19]. Overexpression of D-CDes further inhibited root growth under ABA treatment [27]. Based on the above studies, it is speculated that the concentration of H_2S may vary in different plants. The appropriate concentration of H_2S promotes the formation of adventitious roots by affecting the expression of cell division-related genes and auxin signaling-related genes or by reducing the accumulation of ROS induced by stress. In Arabidopsis, increasing the endogenous H_2S levels, either through treatment with high concentration of NaHS (100 μM) or through overexpression of a *CDes* gene, to a harmful level that affects lateral root development through the S-sulfhydration of actin-2, a posttranslational modification.

2.3. H_2S Regulates Plant Stomata Movement and Photosynthesis

Stomata are the channels that allow the exchange of gas and water between plants and the environment, and their opening and closing regulate the important physiological processes of photosynthesis and transpiration, thus affecting the growth and development of plants [40]. Stomatal movement and photosynthesis are most often influenced by environmental factors—including light, temperature, and water-and regulated by plant hormones, such as ABA, jasmonic acid (JA), or ethylene (ET). Other important signaling components influencing stomatal movement are Ca^{2+}, NO, and H_2O_2 [41–45].

At present, a number of studies have confirmed that H_2S, as a gaseous signaling molecule, also regulates stomatal movement of guard cells [13,46]. For instance, under normal conditions, 0.01 mM H_2S treatment improved photosynthesis by increasing stomatal aperture and density and reducing photorespiration in rice and *Spinacia oleracea* [16,47]. In tall fescue, 500 μM H_2S increased photochemical efficiency and antioxidant enzyme activities while reducing the levels of H_2O_2 and MDA under low-light stress conditions [48]. In blueberry seedlings, exogenous 500 μM H_2S alleviated low temperature stress by maintaining the content of chlorophyll, carotenoids, and the osmotic regulator proline and by reducing photosynthetic inhibition and membrane peroxidation [49].

In guard cells, other studies have shown that both ET and ABA could increase L/D-CDes activity, resulting in an increase of H_2S content [50,51]. In *Vicia faba* L. and Arabidopsis, the application of an H_2S synthesis inhibitor (AOA), NO scavenger (cPTIO), or NO synthesis inhibitor (Na_2WO_4) suggested that H_2S was located downstream of the NO signal that regulates ET-induced stomatal closure [51–53]. In addition, D-CDes overexpression accelerated ABA-induced stomatal closure by up-regulating the expression of ABA-responsive genes [27], while the mutation of *des1* blocked ABA-induced stomatal closure through the signaling pathway of LONG HYPOCOTYL1 (HY1, a member of the heme oxygenase family) [54]. Further investigation revealed that, under the induction of ABA, the Cys44 and Cys205 residues of DES1 were persulfidated by H_2S, and DES1 activity was also rapidly activated, resulting in a large amount of intracellular H_2S accumulation in a short period of time. Furthermore, this sustainable H_2S accumulation contributed to the S-sulfhydration of the NADPH oxidase RBOHD at Cys825 and Cys890, which could then stimulate a large amount of ROS production. Simultaneously, excessive intracellular production of ROS could induce stomatal closure and negatively regulate the degree of S-sulfhydration of DES1 and RBOHD, and thus played a role in feedback inhibition of ABA signaling [55]. Additionally, the accumulation of H_2S induced by ABA could also mediate the S-sulfhydration of SNF1-RELATED PROTEIN KINASE 2.6 (SnRK2.6), which in turn positively regulates ABA signaling to induce stomatal closure [56]. Therefore, during ABA-induced stomatal closure, H_2S, on the one hand, activates ABA signaling via the S-sulfhydration of SnRK2.6 and, on the other hand, is a feed-back regulator of ABA signaling via the S-sulfhydration of RBOHD, which then induces stomatal movement.

Inconsistence with the above results, the application of an exogenous H_2S donor, 200 μM NaHS or 200 μM GYY4137, caused guard cell to open stomata in light or darkness by reducing the NO accumulation induced by ABA in *Capsium anuum* and Arabidopsis [14,57]. As discussed above, it is speculated that a low concentration of H_2S participates in regulating stomatal closure induced by drought, ABA, or ET and enhances photosynthesis by acting with NO, H_2O_2 or persulfidation-based modification of proteins. However, a high concentration of H_2S can prevent stomatal closure. This paradox is emblematic of the double-sided effect of H_2S.

2.4. H₂S Delays Plant Senescence

Plant senescence is an actively programmed cell death (PCD), which not only occurs naturally in the plant life cycle during times such as leaf senescence, fruit ripening and abscission, but also when a plant is subjected to darkness, drought, disease, low temperature and other stresses [58]. At the molecular level, plant senescence is mainly regulated by plant hormones—including cytokinin (CTK), gibberellin (GA), ET, brassinolide (BR), salicylic acid (SA), and JA—by *senescence-associated genes* (*SAGs*) and by WRKY family transcription factors [59]. However, recent research revealed that H_2S also participates in the regulation of plant senescence.

2.4.1. H₂S Delays Leaf Senescence

Leaf senescence is an important developmental process, which involves a variety of metabolic changes related to macromolecular degradation, recycling nutrients back to the main plant body [60]. In *S. oleracea* seedlings, the senescent leaves had higher H_2S levels than the new leaves, indicating that H_2S may also be involved in the regulation of plant senescence [16]. Zhang et al. (2011) showed that the flower and shoot explants from *Gossypium* and *Salix*, treated with 0.6 mM and 0.2 mM NaHS, respectively, increased the activities of catalase (CAT), superoxide dismutase (SOD), and APX and kept the low levels of MDA, H_2O_2 and superoxide anion ($\bullet O_2^-$), which resulted in prolonging fresh cut flowers and one-year-old shoots [61]. In detached leaves of Arabidopsis, 0.5 mM H_2S inhibited chlorophyll degradation by regulating the dark-dependent response, and actively regulated the expression of *SAGs*, such as *SAG1* and *SAG21*, in a manner dependent on *S-nitrosoglutathione reductase 1* (*GSNOR1*) under long dark condition [62]. The leaves in the Arabidopsis *des1* mutant showed premature senescence and higher expression of *SAG1*, *SAG21* and related transcription factors compared to WT. Remarkably, senescence-associated vesicles, related to cell autophagy, were detected in mesophyll protoplasts in the *des1* mutant, and *DES1* deficiency stimulated the accumulation and lipidation of autophagy related protein-8 (ATG8) [15]. Moreover, treatment with an H_2S donor, NaHS, or sodium sulfide (Na₂S), negatively regulated autophagy in Arabidopsis in a way that was unrelated to ROS or nutrient deficiency [63,64]. Thus, H_2S might regulate plant senescence by reducing ROS accumulation and chlorophyll degradation, positively regulating *SAG* genes expression, and negatively regulating autophagy. Nevertheless, the mechanism by which H_2S regulates autophagy is unclear. There is some evidence that H_2S regulation of autophagy might be related to the persulfidation of autophagy-related proteins (ATGs), such as ATG18a, ATG3, ATG5, ATG4, or ATG7 [15,63,65,66].

2.4.2. H₂S Delays the Postharvest Maturation of Fruits

Postharvest maturation of fruits and vegetables is also a type of senescence. H_2S treatment positively regulates certain physiological aspects of ripening, such as color metabolism, softening, and postharvest decay during storage, suggesting that H_2S might regulate aging to protect the ripening and quality changes in various fruits and vegetables [17]. In addition, H_2S can eliminate ROS in harvested produce by promoting the activities of antioxidant enzymes, through synergism (NO) or antagonism (ET) with other molecules, and by regulating the expression of *SAGs* related to protein and chlorophyll degradation in order to maintain the integrity of membranes and to slow senescence [67]. In softening kiwifruit, 45–90 µM NaHS treatment up-regulated the activities of protective enzymes, such as SOD and CAT, and down-regulated the levels of ROS and ET during storage [68]. Moreover, H_2S contributed to the maintenance of firmness and the soluble solids content, affecting the expression of related genes, and to the protection of the integrity of the cell wall and modulation of ET signal transduction [69]. During postharvest storage of tomato fruit, H_2S acts as an antagonist to ET, coordinates antioxidative enzymes, and reduces the production of $\bullet O_2^-$, MDA, and H_2O_2 [70]. H_2S has a significant role in postharvest fruit biology, through establishing crosstalk with ET, ROS, NO, oxidative stress signaling, sulfate metabolism, and post-translational modification of proteins [71]. Therefore, all of the above studies indicate that H_2S delays postharvest maturation of fruits mainly by enhancing their antioxidant capacity to reduce the production of $\bullet O_2^-$, MDA, and H_2O_2 and by establishing crosstalk with NO and ET signaling pathways. It is speculated that H_2S can be used to delay crop aging for increasing crop yield and for keeping fruits and vegetables fresh during storage and transport.

2.4.3. H_2S Inhibits Organ Abscission in Plants

Abscission in plants refers to the process by which some organs, including leaves, flowers, fruits, seeds, and petioles, grow to a certain extent and then are removed naturally from the plant itself. Normal organ abscission is often associated with maturation and senescence [72]. For instance, most fruits undergo abscission during ripening, and petals wither and fall from flowers after pollination and fertilization [73]. Abnormal organ abscission also occurs when plants encounter unfavorable environmental conditions or are damaged by diseases or insects [74,75]. Numerous experiments have shown that plant hormones, such as auxin, ET and SA, are involved in regulating organ abscission in plants [76]. ET is a pivotal abscission inducer and has an indispensable role at different stages of abscission, such as the initiation and progression of floral and organ abscission [77,78]. Furthermore, ET is associated with INFLORESCENCE DEFICIENT IN ABSCISSION (IDA)-mediated floral organ abscission through regulation of the transcription of DNA binding with one finger 4.7 (AtDOF4.7), which can directly impair the expression of the abscission-related gene *ARABIDOPSIS DEHISCENCE ZONE POLYGALACTURONASE 2 (ADPG2)* in Arabidopsis [79]. Liu et al. (2020) demonstrated that H_2S also participates in ET-induced petiole abscission of tomato [18]. The research showed that H_2S treatment could delay abscission of the tomato petiole, but the situation was reversed when the plants were exposed to an H_2S scavenger. Moreover, H_2S treatment reduced the enzymatic activities that modify the cell wall. Along with the expression levels of IAA/AUX family genes (*SlIAA3* and *SlIAA4*), the transcription of genes in the IAA-amino acid conjugate hydrolase (ILR) family (*ILR-L3* and *ILR-L4*) were found to be up-regulated and down-regulated, respectively, in the abscission process, suggesting that H_2S prevented ET-induced petiole abscission by increasing the content of auxin in abscission zone tissues [18]. Additionally, Hideo et al. (2019) reported that D-Cys, as a physiologically relevant substrate, participates in the process of root abscission and that exogenous application of H_2S chemical donors or polysulfides can positively induce abscission to cope with environmental stimuli in the water fern *Azolla* [80]. Therefore, H_2S also plays a positive role in ET-induced organ abscission by regulating the transcription of IAA-related genes and by promoting the accumulation of auxin in abscission zone tissues.

3. Mechanism by which H_2S Regulates Plant Growth and Development

3.1. Crosstalk of H_2S with Plant Hormones

Phytohormonea are indispensable regulators of plant growth and development. A large number of studies have showed that H_2S closely interacts with the plant hormones ABA, ET, auxin, SA, GA, and JA during plant growth and development under normal or stress conditions.

3.1.1. Crosstalk of H_2S with Abscisic Acid

ABA plays important roles in many physiological processes of plants, such as maintaining seed dormancy, promoting plant senescence, and even responding to drought stress [81]. A recent study showed that ABA could activate the gene expression and enzyme activities of *LCD/DES1*, which are responsible for the synthesis of H_2S [13]. On the other hand, exogenous H_2S treatment accelerated stomatal closure induced by ABA in *Vicia faba*, *Arabidopsis thaliana*, and *Impatiens walleriana*, suggesting that H_2S may participate in the ABA-induced stomatal closure [46]. Further analysis showed that ectopic expression of *D-CDes* from wheat (*TaD-CDes*) in Arabidopsis makes plants more sensitive to ABA, which means that ectopic expression of *TaD-CDe* amplifies the stomatal closure and root shortening and further delays the seed germination and cotyledon greening induced by ABA. Simultaneously, TaD-CDe plants showed up-regulation of the ABA receptor PYR1; the ABA responsive element-binding factors ABF2 and ABF4; and the ABA negative regulators ABI1, ABI2, HAB1, and HAB2, and down-regulation of ABA-induced SNF1-related protein kinases (SnRK2.2, SnRK2.3, and SnRK2.6) [27]. Moreover, the accumulation of H_2S induced by ABA in turn activates the activity of SnRK2.6 by the S-sulfhydration of SnRK2.6 at Cys131 and Cys137, which enhances the interaction of SnRK2.6 with ABF2. Thereby, H_2S plays a positive role in the regulation of ABA-induced stomatal closure through mediating the S-sulfhydration of SnRK2.6 [56]. Another study demonstrated that H_2S mediated the S-sulfhydration of DES1 at Cys44 and Cys205, which is stimulated by ABA and positively activates DES1 activity, leading to further accumulation of H_2S [55]. However, excessive production of ROS in turn inhibits the S-sulfhydration of DES1 and RBOHD, forming a feedback regulation mechanism to control ABA signaling [55]. On the other hand, pretreatment with an ATP-binding cassette (ABC)-transporter inhibitor (glibenclamide), an H_2S scavenger (HT) or an H_2S synthesis inhibitor (PAG), blocks ABA signaling, suggesting that the regulation of ABC transporters play a critical role in the signaling transduction of ABA-dependent stomatal closure mediated by H_2S [46]. Taken together, we can conclude that H_2S activates ABA signaling through mediating the S-sulfhydration of SnRK2.6 and that higher levels of H_2S tamps down ABA signaling by mediating the S-sulfhydration of RBOHD, leading to an increase in ROS, thereby balancing the ABA signal. ATP also plays an important role in the cross-talk between H_2S and ABA [46].

In order to study the close relationship between H_2S and ABA under drought stress, the mutants *lcd*, *aba3*, and *abi1* were studied. Compared with WT, the *lcd* mutant showed a weakened response to ABA-induced stomatal closure and was more sensitive to drought stress with the decrease of expression of ion-channel coding genes for Ca^{2+} and outward-rectifying K^+ channels, and, conversely, an increase of inward-rectifying K^+ and anion channels. In both the *aba3* and *abi1* mutants, the stomatal aperture was increased with the decrease of *LCD* expression and H_2S production rate. Remarkably, NaHS treatment rescues all the above defects, implying that H_2S is an important mediator in the ABA-regulated stomatal response to drought through ion channels [82]. In addition, Li et al. (2016) found that ABA treatment increased the activity of LCD in tobacco cells under high temperature and that application of NaHS enhanced the heat tolerance induced by ABA by alleviating the increase in MDA content and electrolyte leakage [83]. This effect of exogenous H_2S or ABA treatment was weakened by the addition of an H_2S scavenger or a specific inhibitor of H_2S biosynthesis, suggesting that there is a synergistic effect between H_2S- and ABA-mediated heat resistances of tobacco suspension-cultured cells [83]. More research discovered that application of H_2S promoted the accumulation of the E3 ligase COP1 in the nucleus, resulting in the degradation of HY5 and a decrease in *ABI5* expression, which lead to a decrease of ABA content and enhanced seed germination under high temperatures [24]. Therefore, it is speculated that H_2S may cooperate with ABA signaling to enhance the tolerance of plants to drought stress by activating Ca^{2+} signaling and inward-rectifying K^+ channels. Under heat stress, H_2S cooperates with ABA signaling to promote seed germination and growth by reducing oxidative damage and regulating the expression of ABA-related genes.

3.1.2. Crosstalk of H_2S with Ethylene

ET has many roles, including inducing stomatal closure. In Arabidopsis, ET significantly affects the transcription of *AtD-CDes*. Similarly,1-aminocyclopropane-1-carboxylic acid (ACC), a precursor of ET, treatment increases the content of H_2S and the activities of D/L-CDes [51]. Although inhibitors of D/L-CDes alone cannot inhibit stomatal closure, they do significantly inhibit ACC-induced stomatal closure. Furthermore, *L/D-Cdes* overexpression plants are more sensitive to ET. Thus, H_2S may be located downstream of ET and work synergistically with ET to induce stomatal closure, similar to its interaction with ABA [51,52]. However, further research revealed that when the NO content decreased, the ET induction of H_2S and of L/D-CDes activities was reduced. The inhibition of H_2S synthesis had no effect on the accumulation of NO and the activity of nitrate reductase (NR). Furthermore, ET induced NO synthesis but failed to enhance stomatal closure in the NO-related mutants *atnia1* and *nia2*, indicating that H_2S enhances ET-induced stomatal closure under the guidance of NO [52].

Equally, in *Vicia faba*, H_2S is a key participant in ET-induced stomatal closure downstream of NO [53], but their interaction mechanism is not clear.

ET-fumigation promotes the ripening of fruits with increases in the content of ROS and MDA. Li et al. (2017) found that H_2S treatment could effectively alleviate ET-induced fruit softening when fumigated kiwifruit with both ET and H_2S while increasing the levels of ascorbic acid, titratable acid, starch, and soluble protein and reducing sugar [84]. In addition, ET and H_2S treatment enhanced the activities of antioxidant enzymes (CAT, APX) and reduced the oxidative stress of the fruits. Further research showed that H_2S inhibited the expression of ET synthesis-related genes and decreased the expression of Cys protease genes [84]. In addition to fruit ripening, ET positively regulates organ abscission. Liu et al. (2020) recently showed through H_2S-ET co-treatment that H_2S inhibited the up-regulation of ET synthesis and signal transduction genes, including *ACS6*, *ACO1*, *ACO4*, *ERF1*, and *ETR4*, eventually resulting in the suppression of ET-induced petiole abscission in tomato [18]. Together, these experiments show that, during fruits ripening, senescence and organ abscission, H_2S antagonizes the effects of ET by reducing oxidative stress and reducing the expression of ET-related genes and ET synthesis, thereby suppressing the ET signaling.

What is the relationship between H_2S and ET under stress condition? Jia et al. (2018) revealed that an H_2S scavenger (HT) or synthesis inhibitor (PAG) could eliminate the effect of ET or osmotic stress on stomatal closure, indicating that H_2S is a necessary downstream factor of ET-induced stomatal closure under osmotic stress [50]. However, under hypoxia, NaHS pretreatment inhibited the activity of ACC oxidase (ACO), a key enzyme in ET biosynthesis [85]. Moreover, it was documented that H_2S reduced ethylene synthesis by inhibiting the transcription of *LeACO* genes and restraining the activities of LeACO1 and LeACO2 by inducing the S-sulfhydration of LeACO1 at Cys60 in a dose-dependent manner [50]. In short, these data show that the ET-induced H_2S signal has a negative regulatory effect on ET biosynthesis through mediating S-sulfhydration of ACO.

3.1.3. Crosstalk of H_2S with Auxin

Auxin affects many stages of plant growth and development, coordinating the adaptation of plant growth and morphology to environmental conditions [86]. During lateral root development, NaHS treatment rapidly increases the content of auxin and promotes the number and length of adventitious roots, showing that there may be also a close cross-talk between H_2S and auxin [29]. Auxin normally inhibits organ abscission, and further investigation showed that the IAA/auxin family genes (*IAA3* and *IAA4*) are often up-regulated by H_2S [18]. In cuttings from sweet potato seedlings and excised willow shoots and soybean seedlings, both the IAA polar transport inhibitor NPA and the NO scavenger (cPTIO) can disturb the formation of root system mediated by H_2S. It is speculated that H_2S acts as upstream of NO and IAA to promote root hair development or to restrain organ abscission [29]. However, auxin-insufficiency weakened DES1 activity and reduced the content of H_2S in tomato. Both NAA and NaHS can counteract the effects of auxin deficiency on *SlDES1* transcription, DES1 activity and endogenous H_2S content and can rescue the stimulation of lateral roots induced by auxin depletion [32]. Simultaneously, NaHS- or NAA-induced up-regulation of the cell cycle regulatory genes *SlCDKA;1* and *SlCYCA2;1* and down-regulation of *SlKRP2* were reversed after exposure to the scavenger HT, suggesting that H_2S might be downstream of auxin to promote the formation of lateral roots [32]. These data suggest that there may also be feedback regulation between H_2S and auxin during plant growth and development, in which H_2S can up-regulate the transcription of IAA family genes, and IAA can also affect the *DES1* expression and DES1 activity.

During the plant response to pathogen, the expression of *auxin signaling F-box protein 1 (AFB1)*, *AFB2*, and *AFB3* are negatively regulated by H_2S [87]. Furthermore, cold stress promoted the accumulation of H_2S and also triggered the endogenous IAA system. Application of NaHS significantly increased the activity of favin monooxygenase (FMO) and the relative expression of the FMO-like protein *YUCCA2* in cucumber seedlings, which in turn increased the level of endogenous IAA and improved cold tolerance, seen as decreases in electrolyte leakage and accumulation of ROS and increases in expression of genes and enzyme activities related to photosynthesis. Application of IAA or removal of H_2S had little effect on the signaling of the other molecule, but the IAA polar transport inhibitor NPA inhibited H_2S-induced cold tolerance and defense gene expression [88]. IAA participates in H_2S-induced stress tolerance in plants as a downstream signaling molecule, while H_2S promotes auxin signal transduction by regulating the expression of auxin-related genes and the synthesis of auxin, thereby enhancing the plant tolerance to adverse environmental conditions.

3.1.4. Crosstalk of H_2S with Gibberellin

GA can regulate many aspects of plant growth and development, such as seed germination, leaf expansion, and flowering [89]. During seed germination, GA can stimulate the synthesis of α- amylase and some secreted hydrolases to break seed dormancy. H_2S significantly enhances the activity of β-amylase and accelerates the germination of barley seeds with or without GA, although the survival rate of cells without GA is higher than those with GA. It is speculated that at the early stage of seed germination, the activation of β-amylase by H_2S is ahead of the activation of α-amylase by GA, both of which can then degrade starch and provide sugar for seedling growth and development [90]. In the wheat aleurone layer, GA accelerates PCD, and during these, both the activity of LCD and the production of H_2S are reduced [91]. Interestingly, application of NaHS not only inhibits the production of endogenous H_2S, but also alleviates the PCD induced by GA. It was speculated that this reversal is related to GSH because NaHS causes an increase of endogenous GSH content, and the alleviation of NaHS-mediated PCD is eliminated by an inhibitor of GSH synthesis [91]. Therefore, the interaction between H_2S and GA is likely indirect through the regulation of GSH homeostasis.

3.1.5. Crosstalk of H_2S with Salicylic Acid

The phenolic compound SA widely exists in plants, can be transported in the phloem, and plays multiple roles, such as improving disease resistance, drought resistance and heat resistance [92]. Li et al. (2015) discovered that SA pretreatment enhances the activity of LCD and contributes to the accumulation of endogenous H_2S during heat tolerance response of maize seedlings [93]. The heat resistance induced by SA is enhanced by the addition of NaHS and decreased by the addition of an H_2S-synthesis inhibitor (PAG) or scavenger (HT). However, there was no significant effect on key enzymes of SA biosynthesis and endogenous SA content. In addition, pretreatment with SA-biosynthesis inhibitors (paclobutrazol, PAC and 2-aminoindan-2-phosphonic acid, AIP) do not affect the heat tolerance induced by NaHS [93]. These results indicate that H_2S is located downstream of SA and works with SA to induce plant resistance to heat stress.

3.1.6. Crosstalk of H_2S with Jasmonate

JA is an important endogenous regulator in higher plants, especially as an environmental signaling molecule, and both regulates plant growth and development and mediates plant defense response to biotic and abiotic stresses [94,95]. JA and JASMONATE INSENSITIVE (JIN/MYC) transcription factors are key factors in regulating stomatal development in Arabidopsis [96]. A recent experiment suggested that the removal of H_2S increased the number of stomata inhibited by JA, while the application of NaHS alleviated the stomatal inhibition in the JA-signaling-deficient *myc234* mutant. H_2S reduces the expression of stomate-associated genes and blocks key components of the stomatal signaling pathway, such as TOO MANY MOUTHS (TMM), STOMATAL DENSITY AND DISTRIBUTION1 (SDD1), and SPEECHLESS (SPCH). Interestingly, mutation of *LCD* increased stomatal density and index values, and an H_2S synthesis inhibitor (HT) counteracts the JA-mediated reduction of stomatal density [97]. All of these data confirm that H_2S is located downstream of JA and cooperates with JA to negatively regulate stomatal development.

3.2. Crosstalk between H₂S and Other Gasotransmitters

H_2S is the third known gaseous signaling molecule, along with carbon monoxide (CO) and NO. There are many similarities between these three molecules in their physiological functions in plants, such as regulating growth, enhancing the response of plants to various adversities, and improving the antioxidation capacity, and many close interactions between their signaling pathways.

3.2.1. Crosstalk between H₂S and NO

During root organogenesis, IAA, H_2S, and NO all promote root hair growth in *Ipomoea batatas* in a dose-dependent manner, as shown by the application of the H_2S donor NaHS, the NO donor sodium nitroprusside (SNP) and IAA [29]. Furthermore, both the NO scavenger cPTIO and the IAA transport inhibitor NPA could inhibit H_2S-induced root hair growth. Interestingly, an H_2S scavenger also inhibits the lateral root formation induced by NO, but not by IAA, indicating that only H_2S and NO might be interdependent, although both NO and IAA are involved in the adventitious root formation induced by H_2S [29]. Moreover, Zhang et al. (2017) found that a high level of NaHS treatment inhibited the growth of the primary root, which was accompanied by the accumulation of ROS and NO and activation of MITOGEN-ACTIVATED PROTEIN KINASE 6 (MPK6) [98]. Further studies showed that ROS was required for the generation of NO inducted by H_2S, and that this induction was mediated by MPK6. Moreover, the *respiration burst oxidase homologous (rbohd/f)* mutant and NO biosynthesis-related mutants (*nia1-2/2-5* double mutant and *noa1*) were less sensitive to NaHS, and the inhibition of NaHS on the growth of root was reduced by the NO scavenger cPTIO. These results indicate that ROS-MPK6-NO signaling mediates the inhibitory effect of high levels of H_2S on root growth [98].

From the previous discussion of crosstalk between H_2S and ET, we know that H_2S may be a signaling molecule downstream of NO in ET-induced stomatal closure [52,53]. However, Lisjak et al. found that H_2S causes stomatal opening in Arabidopsis and *Capsicum anuum*, when plants are treated with an H_2S donor (NaHS) and a slow-release H_2S donor molecule (GYY4137) [14,57]. Moreover, both donor molecules reduced NO accumulation caused by ABA treatment of leaf tissue [14,57]. These results suggest that the adjustment of both H_2S and NO affects the sensitivity of stomatal movement. In the *gsnor1* mutant (which normally clears SNO to prevent NO signal transmission), the positive effect of H_2S on *SAGs* was weakened in the dark [62], indicating that H_2S signaling during the regulation of plant senescence depends on NO signaling [99]. Proteomic studies have also found that sites in proteins that can be S-nitrosylated by NO can also be S-sulfhydrated by H_2S [100]. Therefore, NO and H_2S, may compete with each other through the post-translational modification of proteins to regulate plant growth and development.

Under adverse conditions, both H_2S and NO are important signaling molecules, but their crosstalk relationship needs to be sorted out. Recent research revealed that both application of the H_2S donor NaHS and the NO donor SNP improved the survival rate of plants under heat stress because of reduced electron leakage accumulation of MDA, and improved antioxidant capacity [101,102]. In maize under heat stress, SNP pretreatment increases the activity of LCD, inducing the accumulation of endogenous H_2S [101]. The application of NaHS and GYY4137 enhances the heat resistance induced by SNP, but this is eliminated by an H_2S scavenger. Therefore, H_2S might be a downstream signaling molecule during NO-induced heat tolerance in maize seedlings [101]. However, in strawberry during the early stage of exposure to high temperature, the application of NaHS reduced NO content, enhancing the tolerance to the heat stress [103].

During Al stress, NO is also a negative regulator [104]. H_2S alleviates the inhibition of Al on Arabidopsis elongation by enhancing the activity of antioxidant enzymes and reducing ROS damage. In rice, H_2S increases Al transport into vacuoles and reduces the content of NO in roots [105,106]. Therefore, it is hypothesized that H_2S interacts with NO signaling to improve Al and heat tolerance of plants by reducing the content of NO and oxidative damage.

Hypoxic conditions, when O_2 is lacking, often cause a ROS burst. Group VII ET-responsive factors (ERFVII) sense hypoxia and then initiate the hypoxia response. NO is required for the destabilization of ERFVII [107]. H_2S can also enhance tolerance to hypoxia by removing the accumulated ROS and increasing the transcription of hypoxia-responsive genes (*ADH, CRT1, GS,* and *CYP51*) [85,108]. In maize seedling root tips, pretreatment with SNP enhanced the activity of key H_2S metabolic enzymes (LCD, CAS, OAS-TL) and the accumulation of endogenous H_2S under hypoxia, but these effects were reversed by cPTIO. Application of an H_2S synthesis inhibitor (HA) and an H_2S scavenger (HT) canceled out the increased survival rate induced by SNP [108]. Therefore, under adverse conditions, NO and H_2S work interdependently to remove accumulated ROS and enhance the stress tolerance of plants.

3.2.2. Crosstalk between H_2S and CO

Although CO is also an important signaling molecule, there are relatively few studies on any crosstalk between CO and H_2S. During root development, Heme Oxygenase-1 (HO-1), which catalyzes the production of CO, acts downstream of the auxin signaling pathway, leading to the formation of adventitious roots of cucumber [109]. Further analysis found that the addition of CO and H_2S could also promote adventitious root formation in cucumber [110]. In pepper, NaHS induced both the *CsHO-1* gene and CsHO-1 protein expression in a time-dependent manner. The application of ZnPPIX, a specific inhibitor of HO-1, could reverse the formation of adventitious roots induced by NaHS. However, the addition of an H_2S scavenger (HT) could not alter the effect of CO on adventitious root formation [110]. This indicates that H_2S may play a specific role upstream of CO in the formation of adventitious roots and may promote the production of CO, which then stimulates the formation of lateral roots.

3.3. Crosstalk of H_2S with Ionic Signals

3.3.1. Crosstalk of H_2S with Ca^{2+}

The Ca^{2+} is one of the most important nutrient elements in plants. Ca^{2+} functions to maintain the stability of the cell wall, cell membrane and membrane binding proteins, but is also an important signaling molecule and participates in the regulation of cell homeostasis, plant growth and stress responses.

The application of exogenous NaHS increases the intracellular Ca^{2+} content under both hypoxia and heat stress [108,111]. In the suspension culture cells of tobacco, exogenous Ca^{2+} and its ionophore A23187 significantly enhances the high temperature tolerance induced by NaHS. On the other hand, the heat tolerance induced by H_2S could be weakened by a Ca^{2+} chelating agent, the plasma membrane channel blocker La^{3+}, or the calmodulin antagonist chlorpromazine or trifluoperazine. This suggests that the H_2S-induced thermostability requires the participation of Ca^{2+}, which acts as a downstream molecule, at least in tobacco suspension cells [111]. However, the application of Ca^{2+} or calmodulin (CaM), a calcium ion receptor, activates the activity of DES1 and induces the accumulation of endogenous H_2S in tobacco suspension culture cells, and the application of a Ca^{2+} chelator or CaM antagonists reduces DES1 enzyme activity and H_2S content. All of these increases induced by Ca^{2+}/CaM, in DES1 activity, H_2S content, and heat tolerance are enhanced by the H_2S donor NaHS or weakened by H_2S synthesis inhibitors or an H_2S scavenger. Therefore, during the heat stress response process, the H_2S and Ca^{2+} signals may be interdependent [112].

Similarly, chromium (Cr^{6+}) stress activates endogenous H_2S synthesis and Ca^{2+} signaling transduction. The damage caused by Cr^{6+} stress is greatly alleviated by application of H_2S and Ca^{2+} alone or in combination, with the combined addition more effective. In contrast, the induced stress was intensified by treatment with an H_2S synthesis inhibitor or Ca^{2+} chelators. This illustrated the synergistic effect of H_2S and Ca^{2+} under Cr^{6+} stress [113]. Furthermore, during Cr^{6+} stress, the metallothionein (encoded by *MT3A*) and phytochelatin (synthesized by phytochelatin synthase, PCS) bind the heavy metal to provide protection to the plant cells. The upregulation of *MT3A* and *PCS*, regulated by Ca^{2+}, is dependent on H_2S signaling [113].

Calcium dependent protein kinases (CDPK) are important protein kinases in plant signal transduction. CDPK can be activated directly by combination with Ca^{2+}. An activated CDPK protein can be phosphorylated to amplify Ca^{2+} signaling. Experiments in Arabidopsis revealed that both H_2S and CDPK are involved in the cadmium (Cd) stress response through the alleviation of the oxidative stress. Moreover, mutation of *CDPK* or treatment with the CDPK inhibitor TFP reduces LCD enzyme activity and H_2S content. In the *cdpk3* mutant, H_2S increases the transcription of Cd stress-responsive genes, such as *MYB107*, *CAX3*, *POX1*, *MT3*, and *PCS1*, suggesting that H_2S and CDPK are linked under Cd stress [114].

All of these results show that H_2S and Ca^{2+} signaling, especially under adverse conditions, are interrelated. Ca^{2+} signaling can activate LCD enzyme activity, thereby promoting the accumulation of H_2S. In turn, H_2S regulates the expression of stress response-related genes by stimulating the Ca^{2+} signal. Together, these two signals enhance the tolerance of the plant to stress.

3.3.2. Crosstalk of H_2S with Na^+ and K^+

Salt stress invariably causes a rapid increase in the intracellular Na^+ level and leads to an imbalance of Na^+/K^+, which in turn represses plant growth. Therefore, maintaining the balance of Na^+/K^+ is a crucial factor in conquering salt stress [115]. Several studies have proclaimed that H_2S can reduce the sensitivity of plants to salt stress mainly by preventing both uptake of Na^+ and K^+ efflux and by promoting Na^+ efflux and uptake of K^+ and thus mediating the balance of Na^+/K^+ [116–118], which have begun to reveal the regulatory mechanisms by which H_2S helps to mediate the balance of Na^+/K^+. In wheat, the addition of $CaCl_2$ (an inhibitor of nonselective cation channels (NSCCs)) or amiloride (an inhibitor of salt overly sensitive 1 (SOS1), a Na^+/H^+ antiporter) disrupts the Na^+/K^+ balance promoted by H_2S, indicating that NSCC and SOS1 may be the main pathway of reducing Na^+ by H_2S [116]. In *Populus popularis*, NaCl induces K^+ loss mainly due to the activation of H^+-ATPase on the plasma membrane. Application of Na^+/H^+ antiporter inhibitors, sodium orthovanadate and amiloride effectively inhibited the Na^+ efflux, but NaHS enhanced it. Thus, the Na^+/K^+ balance maintained by H_2S may be achieved by regulating the Na^+/H^+ antiport system in *Populus popularis* [117]. In Arabidopsis, application of NaHS alleviates the suppression of salt stress on root growth and promotes the accumulation of H_2O_2, while exogenous application of H_2O_2 reduces the ratio of Na^+/H^+ and strengthens the role of H_2S. Application of a ROS scavenger (DMTU), a plasma membrane (PM) NADPH oxidase inhibitor (DPI) or a glucose-6-phosphate dehydrogenase (G6PDH) inhibitor (glycerol) all eliminate the effect of H_2S, further indicating that H_2O_2 may be involved in the H_2S-mediated tolerance to salt stress via the regulation of G6PDH and PM NADPH oxidase [119]. In conclusion, under salt stress, H_2S works to maintain ion homeostasis within plant cells by regulating the Na^+/H^+ antiport system in the way that is H_2O_2-dependent and that uses the enzymes NSCCs and the SOS1 antiporter to reduce Na^+ levels.

3.4. S-sulfhydration Modification of Proteins Mediated by H_2S

At present, many studies have proved that H_2S can regulate the spatial structure of certain target proteins via the post-translational modification named S-sulfhydration. S-sulfhydration affects protein structure, subcellular localization, and function, in a way that can regulate plant growth and development and responses to stress [65]. S-sulfhydration occurs when H_2S reacts with Cys residues (-SH, -S-S-, -S-OH or S-NO) in target proteins to form a persulfide group (-SSH) [120]. In a persulfidation proteome in Arabidopsis treated with NaHS, a total of 106 persulfidated proteins were identified, which were mainly involved in photosynthesis, protein synthesis, cell organization, and primary metabolism [100,121]. Using a different technique, proteome analysis of endogenous persulfidated proteins in leaves of WT Arabidopsis and the *des1* mutant identified 2015 persulfidated proteins, which were mainly involved in regulating primary metabolism, responses to abiotic and biotic stress, plant growth and development, and RNA translation [65]. At least 5% of proteins in Arabidopsis may be persulfidated under normal growth conditions [65], which is consistent with the persulfidation proteome with application of NaHS [100]. Further analysis found that the activities of APX, glyceraldehyde 3-phosphate dehydrogenase (GAPDH) and glyceraldehyde 3-phosphate dehydrogenase, isoform C1 (GAPC1) were increased by S-sulfhydration, indicating that S-sulfhydration may be a mechanism that promotes reduction of oxidative stress in plants [100]. Physiological research further confirms that S-sulfhydration, mediated by H_2S, plays key roles in plant growth, development, and stress response. For example, Li et al. found that ACTIN2 (ACT2) can be S-sulfhydrated by H_2S at Cys287. This S-sulfhydration interrupts actin-2 polymerization, resulting in root hair dysplasia in Arabidopsis [19]. Furthermore, Shen et al. (2020) and Chen et al. (2020) found that ABA-induced stomatal closure was also related to H_2S-mediated S-sulfhydration. In Arabidopsis, ABA addition stimulates the S-sulfhydration of DES1 at Cys44 and Cys205 to activate DES1, which catalyzes the accumulation of H_2S [55,56]. This higher levels of H_2S then mediates the S-sulfhydration of SnRK2.6 at Cys131 and Cys137, promote its activity and the interaction between SnRK2.6 and ABF2, which in turn positively regulates ABA signaling [56]. On the other hand, the produced H_2S also drives the S-sulfhydration of RBOHD at Cys825 and Cys890, enhancing the production of ROS. Physiologically, ROS is the rate-limiting messenger in ABA-mediated stomatal closure and is part of the negative feedback loop for inhibiting ABA signal [55].

In cucumber, H_2S improves cold tolerance via actively modifying the synthesis of Cucurbitacin C (CuC) by driving S-sulfhydration of the His-Csa5G156220 and His-Csa5G157230 proteins, transcription factors that activate the CuC synthetase gene [122]. In tomato, H_2S, as a downstream component of ET-induced stomatal closure, reduces ET content by impairing the activity of ACOs through persulfidation, which in turn enhances the osmotic stress response [50]. Consequently, H_2S-mediated S-sulfhydration occurs during many aspects of plant growth and S-sulfhydration of proteins may be an essential mechanism by which H_2S affects plant growth and development under both normal and stress conditions.

Based on the above descriptions, it can be clearly seen that H_2S does not function independently in plants, but interacts with plant hormones and other signaling molecules, such as Ca^{2+}, NO, H_2O_2, and even proteins, form a complex signaling network that finely regulates plant growth, development, and stress responses. In the future, we can make full use of advanced proteomics to further explore the mechanisms by which H_2S influences signaling pathways in plants.

4. Conclusions and Perspectives

Continuing investigation into H_2S has revealed its numerous and varied regulatory roles in biology and has brought more attention to this gasotransmitter. It is now recognized that H_2S promotes seed germination, root development, photosynthesis, stomatal movement, and plant senescence. H_2S also regulates plant responses to stress by activating antioxidant defenses, improving expression of genes encoding resistance-related enzymes, and interacting with different signaling molecules. Additionally, S-sulfhydration of proteins induced by H_2S is an essential mediator (Figure 2).

However, there remain numerous issues to be explored. For example, it has been confirmed that an appropriate concentration of H_2S produces a marked effect on plant development and responses to stress, but different plants have different tolerances to H_2S. This means it is particularly important to monitor the concentration of H_2S in cells. Second, most of the existing research has focused on how exogenous H_2S improves plant resistance to stress, but the mechanism(s) by which endogenous H_2S functions is barely clear. In some studies, the *des1* mutant exhibited stronger tolerance to Cd and pathogen stress, which differs from the theory that increases in H_2S could improve the stress resistance of plants. Therefore, it is not clear whether H_2S enhances the antioxidant capacity of plants through the homeostasis of H_2S-Cys or as an antioxidant signaling molecule itself. Furthermore, it is unclear if endogenous and exogenous H_2S have different function mechanisms. It is also unknown how environmental stimulation triggers the accumulation of H_2S, how cells perceive the H_2S signal and what are the direct targets and downstream cascades of H_2S plant signal transduction. Numerous reports have documented that H_2S can crosstalk with the signaling pathways of plant hormones, other gasotransmitters, and ions to form a complex regulatory network for all aspects of plant growth and development, but the interactional mechanisms of H_2S with other signals remain to be elucidated. It is also unknown whether H_2S plays important roles through its receptor. Therefore, the functions of H_2S in plant growth and development need to be deeply studied by transcriptomics, proteomics, metabolomics, and functional genomics, in combination with more genetic materials and H_2S donors, scavengers, and synthetic inhibitors in the future.

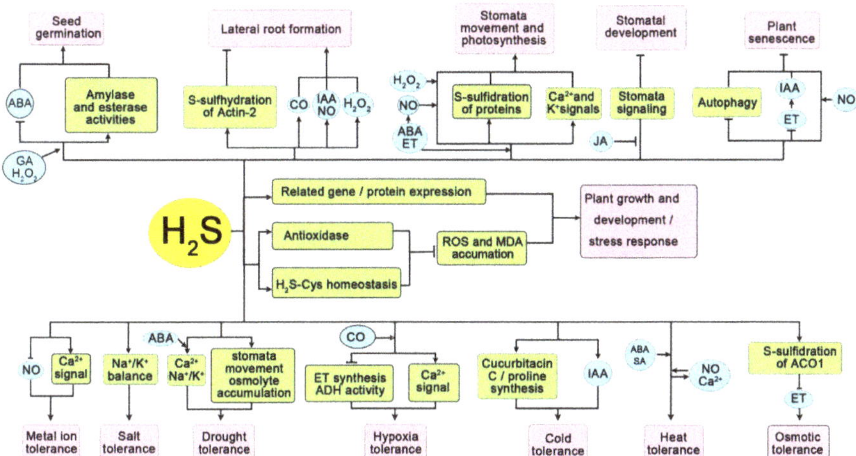

Figure 2. A model of the roles of H_2S in plant development and stress responses. H_2S has recently been recognized as a novel gaseous signaling molecule with various functions during plant development at different stages and during stress responses. H_2S functions by promoting the expression of specific genes, enhancing the activity of the antioxidant system and maintaining H_2S-Cys homeostasis. Growing evidence suggests that H_2S is involved in seed germination, by increasing amylase and esterase content for greater energy efficiency. H_2S can also fine-tune lateral root formation, stomatal movement, photosynthesis, and plant senescence by regulating protein S-sulfhydration and by establishing crosstalk with CO, NO, IAA, ABA, ET, and other signaling pathways. In addition, H_2S may also be involved in plant senescence by inhibiting autophagy. Both exogenous and endogenous H_2S are able to optimize plant adaptation to various stresses (e.g., metal ion, drought, hypoxia, temperature, salt, and osmotic stress) through positively regulating ionic equilibrium, stomatal movement, osmolyte accumulation, ethylene synthesis, related enzyme activity, interaction with other reactive species, and plant hormones. H_2S can also regulate the expression of related genes and proteins, reduce the oxidative stress caused by various stresses by enhancing the activities of antioxidant enzymes and the accumulation of antioxidants, so as to improve the stress resistance and promote plant development.

Author Contributions: C.W. and X.W. conceived and designed the main content. L.X. designed and produced the figures. C.W., L.X. and J.L. wrote the article. All authors have read and agreed to the published version of the manuscript.

Funding: This work was funded by the National Natural Science Foundation of China (NSFC) (Grant Nos. 31670254, 31770199 and 31700215), and the Foundation of the Key Laboratory of Cell Activities and Stress Adaptations, Ministry of Education of China (lzujbky-2019-kb05).

Conflicts of Interest: The authors declare no conflict of interest.

Abbreviations

S	Sulfur
Cys	Cysteine
L-Cys	L-cysteine
Met	Methionine
H_2S	Hydrogen sulfide
SIR	Sulfite reductase
OAS-TL	O-acetylserine (thiol) lyase
OAS	O-acetylserine
APS	Adenosine 5′-phosphosulfate
ATPS	ATP sulfurylase
APSR	Adenosine- 5′- phosphoryl sulfate reductase
CDes	Cysteine desulfhydrase
LCD	L-cysteine desulfhydrase
DES1	L-cysteine desulfhydrase 1
DCD1	D-cysteine desulfhydrase1
DCD2	D-cysteine desulfhydrase2
CAS	β-cyanoalanine synthase
CN^-	Cyanide
GSH	Glutathione
AOA	Aminooxyacetic acid, an H_2S synthesis inhibitor
HT	Hypotaurine, an H_2S scavenger
PAG	Propargylglycine, a DES1 inhibitor
NPA	N-1-naphthylphthalamic acid, IAA transport inhibitor
Cptio	2-(4-carboxyphenyl)-4,4,5,5-tetramethylimidazoline-1-oxyl-3-oxide, NO scavenger
GSNOR1	S-nitrosoglutathione reductase 1
GAPDH	Glyceraldehyde phosphate dehydrogenase
SNP	Sodium nitroprusside, NO donor
HA	Hydroxylamine, H_2S synthesis inhibitor

References

1. Takahashi, H.; Kopriva, S.; Giordano, M.; Saito, K.; Hell, R. Sulfur assimilation in photosynthetic organisms: Molecular functions and regulations of transporters and assimilatory enzymes. *Annu. Rev. Plant Biol.* **2011**, *62*, 157–184. [CrossRef] [PubMed]

2. Rennenberg, H.; Arabatzis, N.; Grundel, I.J.P. Cysteine desulphydrase activity in higher plants: Evidence for the action of L- and D-cysteine specific enzymes. *Phytochemistry* **1987**, *26*, 1583–1589. [CrossRef]

3. Li, Z.G. Analysis of some enzymes activities of hydrogen sulfide metabolism in plants. *Method Enzymol.* **2015**, *555*, 253–269.

4. Romero, L.C.; Garcia, I.; Gotor, C. L-Cysteine Desulfhydrase 1 modulates the generation of the signaling molecule sulfide in plant cytosol. *Plant Signal. Behav.* **2013**, *8*, e24007. [CrossRef]

5. Leon, S.; Touraine, B.; Briat, J.F.; Lobreaux, S. The AtNFS2 gene from *Arabidopsis thaliana* encodes a NifS-like plastidial cysteine desulphurase. *Biochem. J.* **2002**, *366*, 557–564. [CrossRef]

6. Hatzfeld, Y.; Maruyama, A.; Schmidt, A.; Noji, M.; Ishizawa, K.; Saito, K. beta-cyanoalanine synthase is a mitochondrial cysteine synthase-like protein in spinach and *Arabidopsis*. *Plant Physiol.* **2000**, *123*, 1163–1171. [CrossRef]

7. Yamaguchi, Y.; Nakamura, T.; Kusano, T.; Sano, H. Three *Arabidopsis* genes encoding proteins with differential activities for cysteine synthase and beta-cyanoalanine synthase. *Plant Cell Physiol.* **2000**, *41*, 465–476. [CrossRef]

8. Hancock, J.T. Harnessing evolutionary toxins for signaling: Reactive oxygen species, nitric oxide and hydrogen sulfide in plant cell regulation. *Front. Plant Sci.* **2017**, *8*, 189. [CrossRef]

9. Mancardi, D.; Penna, C.; Merlino, A.; Del Soldato, P.; Wink, D.A.; Pagliaro, P. Physiological and pharmacological features of the novel gasotransmitter: Hydrogen sulfide. *BBA Bioenergetics* **2009**, *1787*, 864–872. [CrossRef]

10. Li, Z.G.; Min, X.; Zhou, Z.H. Hydrogen sulfide: A signal molecule in plant cross-adaptation. *Front. Plant Sci.* **2016**, *7*, 1621. [CrossRef]

11. Baudouin, E.; Poilevey, A.; Hewage, N.I.; Cochet, F.; Puyaubert, J.; Bailly, C. The significance of hydrogen sulfide for *Arabidopsis* seed germination. *Front. Plant Sci.* **2016**, *7*, 930. [CrossRef] [PubMed]

12. Zhang, H.; Hu, L.Y.; Hu, K.D.; He, Y.D.; Wang, S.H.; Luo, J.P. Hydrogen sulfide promotes wheat seed germination and alleviates oxidative damage against copper stress. *J. Integr. Plant Biol.* **2008**, *50*, 1518–1529. [CrossRef] [PubMed]

13. Jin, Z.P.; Pei, Y.X. Hydrogen sulfide: The shutter button of stomata in plants. *Sci. China Life Sci.* **2016**, *59*, 1187–1188. [CrossRef] [PubMed]

14. Lisjak, M.; Srivastava, N.; Teklic, T.; Civale, L.; Lewandowski, K.; Wilson, I.; Wood, M.E.; Whiteman, M.; Hancock, J.T. A novel hydrogen sulfide donor causes stomatal opening and reduces nitric oxide accumulation. *Plant Physiol. Biochem.* **2010**, *48*, 931–935. [CrossRef]

15. Alvarez, C.; Garcia, I.; Moreno, I.; Perez-Perez, M.E.; Crespo, J.L.; Romero, L.C.; Gotor, C. Cysteine-generated sulfide in the cytosol negatively regulates autophagy and modulates the transcriptional profile in *Arabidopsis*. *Plant Cell* **2012**, *24*, 4621–4634. [CrossRef]

16. Chen, J.; Wu, F.H.; Wang, W.H.; Zheng, C.J.; Lin, G.H.; Dong, X.J.; He, J.X.; Pei, Z.M.; Zheng, H.L. Hydrogen sulphide enhances photosynthesis through promoting chloroplast biogenesis, photosynthetic enzyme expression, and thiol redox modification in *Spinacia oleracea* seedlings. *J. Exp. Bot.* **2011**, *62*, 4481–4493. [CrossRef]

17. Ali, S.; Nawaz, A.; Ejaz, S.; Haider, S.T.A.; Alam, M.W.; Javed, H.U. Effects of hydrogen sulfide on postharvest physiology of fruits and vegetables: An overview. *Sci. Hortic.* **2019**, *243*, 290–299. [CrossRef]

18. Liu, D.M.; Li, J.N.; Li, Z.W.; Pei, Y.X. Hydrogen sulfide inhibits ethylene-induced petiole abscission in tomato (*Solanum lycopersicum* L.). *Hortic. Res. England* **2020**, *7*, 14. [CrossRef]

19. Li, J.S.; Chen, S.S.; Wang, X.F.; Shi, C.; Liu, H.X.; Yang, J.; Shi, W.; Guo, J.K.; Jia, H.L. Hydrogen sulfide disturbs actin polymerization via S-sulfhydration resulting in stunted root hair growth. *Plant Physiol.* **2018**, *178*, 936–949. [CrossRef]

20. Yuan, X.; Wen, B. Seed germination response to high temperature and water stress in three invasive Asteraceae weeds from Xishuangbanna, SW China. *PLoS ONE* **2018**, *13*, e0191710. [CrossRef]

21. Dooley, F.D.; Nair, S.P.; Ward, P.D. Increased growth and germination success in plants following hydrogen sulfide administration. *PLoS ONE* **2013**, *8*, e62048. [CrossRef]

22. Zhang, H.; Wang, M.J.; Hu, L.Y.; Wang, S.H.; Hu, K.D.; Bao, L.J.; Luo, J.P. Hydrogen sulfide promotes wheat seed germination under osmotic stress. *Russ. J. Plant Physl.* **2010**, *57*, 532–539. [CrossRef]

23. Zhang, H.; Tan, Z.Q.; Hu, L.Y.; Wang, S.H.; Luo, J.P.; Jones, R.L. Hydrogen sulfide alleviates aluminum toxicity in germinating wheat seedlings. *J. Integr. Plant Biol.* **2010**, *52*, 556–567. [CrossRef]

24. Chen, Z.; Huang, Y.W.; Yang, W.J.; Chang, G.X.; Li, P.; Wei, J.L.; Yuan, X.J.; Huang, J.L.; Hu, X.Y. The hydrogen sulfide signal enhances seed germination tolerance to high temperatures by retaining nuclear COP1 for HY5 degradation. *Plant Sci.* **2019**, *285*, 34–43. [CrossRef] [PubMed]

25. Zhou, Z.H.; Wang, Y.; Ye, X.Y.; Li, Z.G. Signaling molecule hydrogen sulfide improves seed germination and seedling growth of maize (*Zea mays* L.) under high temperature by inducing antioxidant system and osmolyte biosynthesis. *Front. Plant Sci.* **2018**, *9*, 1288. [CrossRef] [PubMed]

26. Li, Z.G.; Gong, M.; Liu, P. Hydrogen sulfide is a mediator in H_2O_2-induced seed germination in *Jatropha Curcas*. *Acta Physiol. Plant* **2012**, *34*, 2207–2213. [CrossRef]

27. Li, H.; Zhang, Y.; Li, X.; Xue, R.L.; Zhao, H.J. Identification of wheat d-cysteine desulfhydrase (TaD-CDes) required for abscisic acid regulation of seed germination, root growth, and stomatal closure in *Arabidopsis*. *J. Plant Growth Regul.* **2018**, *37*, 1175–1184. [CrossRef]

28. Casimiro, I.; Marchant, A.; Bhalerao, R.P.; Beeckman, T.; Dhooge, S.; Swarup, R.; Graham, N.; Inze, D.; Sandberg, G.; Casero, P.J.; et al. Auxin transport promotes *Arabidopsis* lateral root initiation. *Plant Cell* **2001**, *13*, 843–852. [CrossRef]

29. Zhang, H.; Tang, J.; Liu, X.P.; Wang, Y.; Yu, W.; Peng, W.Y.; Fang, F.; Ma, D.F.; Wei, Z.J.; Hu, L.Y. Hydrogen sulfide promotes root organogenesis in *Ipomoea batatas*, *Salix matsudana* and *Glycine max*. *J. Integr. Plant Biol.* **2009**, *51*, 1086–1094. [CrossRef]

30. Chen, Y.; Mo, H.Z.; Zheng, M.Y.; Xian, M.; Qi, Z.Q.; Li, Y.Q.; Hu, L.B.; Chen, J.; Yang, L.F. Selenium inhibits root elongation by repressing the generation of endogenous hydrogen sulfide in *Brassica rapa*. *PLoS ONE* **2014**, *9*, e110904. [CrossRef]

31. Mei, Y.D.; Chen, H.T.; Shen, W.B.; Shen, W.; Huang, L.Q. Hydrogen peroxide is involved in hydrogen sulfide-induced lateral root formation in tomato seedlings. *BMC Plant Biol.* **2017**, *17*, 1–12. [CrossRef] [PubMed]

32. Fang, T.; Cao, Z.Y.; Li, J.L.; Shen, W.B.; Huang, L.Q. Auxin-induced hydrogen sulfide generation is involved in lateral root formation in tomato. *Plant Physiol. Biochem.* **2014**, *76*, 44–51. [CrossRef] [PubMed]

33. Gu, Q.; Chen, Z.P.; Cui, W.T.; Zhang, Y.H.; Hu, H.L.; Yu, X.L.; Wang, Q.Y.; Shen, W.B. Methane alleviates alfalfa cadmium toxicity via decreasing cadmium accumulation and reestablishing glutathione homeostasis. *Ecotoxicol. Environ. Saf.* **2018**, *147*, 861–871. [CrossRef]

34. Zhu, K.K.; Cui, W.T.; Dai, C.; Wu, M.Z.; Zhang, J.; Zhang, Y.H.; Xie, Y.J.; Shen, W.B. Methane-rich water alleviates NaCl toxicity during alfalfa seed germination. *Environ. Exp. Bot.* **2016**, *129*, 37–47. [CrossRef]

35. Han, B.; Duan, X.L.; Wang, Y.; Zhu, K.K.; Zhang, J.; Wang, R.; Hu, H.L.; Qi, F.; Pan, J.C.; Yan, Y.X.; et al. Methane protects against polyethylene glycol-induced osmotic stress in maize by improving sugar and ascorbic acid metabolism. *Sci. Rep. UK* **2017**, *7*, 7–14. [CrossRef] [PubMed]

36. Qi, F.; Xiang, Z.X.; Kou, N.H.; Cui, W.T.; Xu, D.K.; Wang, R.; Zhu, D.; Shen, W.B. Nitric oxide is involved in methane-induced adventitious root formation in cucumber. *Physiol. Plantarum.* **2017**, *159*, 366–377. [CrossRef]

37. Kou, N.H.; Xiang, Z.X.; Cui, W.T.; Li, L.N.; Shen, W.B. Hydrogen sulfide acts downstream of methane to induce cucumber adventitious root development. *J. Plant Physiol.* **2018**, *228*, 113–120. [CrossRef]

38. Mei, Y.D.; Zhao, Y.Y.; Jin, X.X.; Wang, R.; Xu, N.; Hu, J.W.; Huang, L.Q.; Guan, R.Z.; Shen, W.B. L-Cysteine desulfhydrase-dependent hydrogen sulfide is required for methane-induced lateral root formation. *Plant Mol. Biol.* **2019**, *99*, 283–298. [CrossRef]

39. Jia, H.; Hu, Y.; Fan, T.; Li, J. Hydrogen sulfide modulates actin-dependent auxin transport via regulating ABPs results in changing of root development in *Arabidopsis*. *Sci. Rep. UK* **2015**, *5*, 8251. [CrossRef]

40. Nunes, T.D.G.; Zhang, D.; Raissig, M.T. Form, development and function of grass stomata. *Plant J.* **2020**, *101*, 780–799. [CrossRef]

41. Schachtman, D.P.; Goodger, J.Q.D. Chemical root to shoot signaling under drought. *Trends Plant Sci.* **2008**, *13*, 281–287. [CrossRef] [PubMed]

42. Hossain, M.A.; Munemasa, S.; Uraji, M.; Nakamura, Y.; Mori, I.C.; Murata, Y. Involvement of endogenous abscisic acid in methyl jasmonate-induced stomatal closure in *Arabidopsis*. *Plant Physiol.* **2011**, *156*, 430–438. [CrossRef] [PubMed]

43. Xie, Y.J.; Mao, Y.; Zhang, W.; Lai, D.W.; Wang, Q.Y.; Shen, W.B. Reactive oxygen species-dependent nitric oxide production contributes to hydrogen-promoted stomatal closure in *Arabidopsis*. *Plant Physiol.* **2014**, *165*, 759–773. [CrossRef]

44. Wilkinson, S.; Davies, W.J. Ozone suppresses soil drying- and abscisic acid (ABA)-induced stomatal closure via an ethylene-dependent mechanism. *Plant Cell Environ.* **2009**, *32*, 949–959. [CrossRef]

45. Hoque, T.S.; Uraji, M.; Ye, W.X.; Hossain, M.A.; Nakamura, Y.; Murata, Y. Methylglyoxal-induced stomatal closure accompanied by peroxidase-mediated ROS production in *Arabidopsis*. *J. Plant Physiol.* **2012**, *169*, 979–986. [CrossRef]

46. Garcia-Mata, C.; Lamattina, L. Hydrogen sulphide, a novel gasotransmitter involved in guard cell signalling. *New Phytol.* **2010**, *188*, 977–984. [CrossRef] [PubMed]

47. Duan, B.B.; Ma, Y.H.; Jiang, M.R.; Yang, F.; Ni, L.; Lu, W. Improvement of photosynthesis in rice (*Oryza sativa* L.) as a result of an increase in stomatal aperture and density by exogenous hydrogen sulfide treatment. *Plant Growth Regul.* **2015**, *75*, 33–44. [CrossRef]

48. Liu, Y.H.; Zhang, X.H.; Liu, B.W.; Ao, B.; Liu, Q.; Wen, S.Y.; Xu, Y.F. Hydrogen sulfide regulates photosynthesis of tall fescue under low-light stress. *Photosynthetica* **2019**, *57*, 714–723. [CrossRef]

49. Tang, X.D.; An, B.Y.; Cao, D.M.; Xu, R.; Wang, S.Y.; Zhang, Z.D.; Liu, X.J.; Sun, X.G. Improving photosynthetic capacity, alleviating photosynthetic inhibition and oxidative stress under low temperature stress with exogenous hydrogen sulfide in blueberry seedlings. *Front. Plant Sci.* **2020**, *11*, 108. [CrossRef] [PubMed]

50. Jia, H.L.; Chen, S.S.; Liu, D.; Liesche, J.; Shi, C.; Wang, J.; Ren, M.J.; Wang, X.F.; Yang, J.; Shi, W.; et al. Ethylene-induced hydrogen sulfide negatively regulates ethylene biosynthesis by persulfidation of ACO in tomato under osmotic stress. *Front. Plant Sci.* **2018**, *9*, 1517. [CrossRef]

51. Hou, Z.H.; Wang, L.X.; Liu, J.; Hou, L.X.; Liu, X. Hydrogen sulfide regulates ethylene-induced stomatal closure in *Arabidopsis thaliana*. *J. Integr. Plant Biol.* **2013**, *55*, 277–289. [CrossRef] [PubMed]

52. Liu, J.; Hou, L.X.; Liu, G.H.; Liu, X.; Wang, X.C. Hydrogen sulfide induced by nitric oxide mediates ethylene-induced stomatal closure of *Arabidopsis thaliana*. *Chin. Sci. Bull.* **2011**, *56*, 3547–3553. [CrossRef]

53. Liu, J.; Hou, Z.H.; Liu, G.H.; Hou, L.X.; Liu, X. Hydrogen sulfide may function downstream of nitric oxide in ethylene-induced stomatal closure in *Vicia faba* L. *J. Integr. Agric.* **2012**, *11*, 1644–1653. [CrossRef]

54. Zhang, J.; Zhou, M.J.; Ge, Z.L.; Shen, J.; Zhou, C.; Gotor, C.; Romero, L.C.; Duan, X.L.; Liu, X.; Wu, D.L.; et al. Abscisic acid-triggered guard cell l-cysteine desulfhydrase function and in situ hydrogen sulfide production contributes to heme oxygenase-modulated stomatal closure. *Plant Cell Environ.* **2020**, *43*, 624–636. [CrossRef] [PubMed]

55. Shen, J.; Zhang, J.; Zhou, M.J.; Zhou, H.; Cui, B.M.; Gotor, C.; Romero, L.C.; Fu, L.; Yang, J.; Foyer, C.H.; et al. Persulfidation-based modification of cysteine desulfhydrase and the NADPH oxidase RBOHD controls guard cell abscisic acid signaling. *Plant Cell* **2020**, *32*, 1000–1017. [CrossRef] [PubMed]

56. Chen, S.S.; Jia, H.L.; Wang, X.F.; Shi, C.; Wang, X.; Ma, P.Y.; Wang, J.; Ren, M.J.; Li, J.S. Hydrogen sulfide positively regulates abscisic acid signaling through persulfidation of SnRK2.6 in guard cells. *Mol. Plant* **2020**, *13*, 732–744. [CrossRef] [PubMed]

57. Lisjak, M.; Teklic, T.; Wilson, I.D.; Wood, M.; Whiteman, M.; Hancock, J.T. Hydrogen sulfide effects on stomatal apertures. *Plant Signal. Behav.* **2011**, *6*, 1444–1446. [CrossRef] [PubMed]

58. Gregersen, P.L.; Culetic, A.; Boschian, L.; Krupinska, K. Plant senescence and crop productivity. *Plant Mol. Biol.* **2013**, *82*, 603–622. [CrossRef]

59. Woo, H.R.; Masclaux-Daubresse, C.; Lim, P.O. Plant senescence: How plants know when and how to die. *J. Exp. Bot.* **2018**, *69*, 715–718. [CrossRef]

60. Watanabe, M.; Tohge, T.; Balazadeh, S.; Erban, A.; Giavalisco, P.; Kopka, J.; Mueller-Roeber, B.; Fernie, A.R.; Hoefgen, R. Comprehensive metabolomics studies of plant developmental senescence. *Methods Mol. Biol.* **2018**, *1744*, 339–358.

61. Zhang, H.; Hu, S.L.; Zhang, Z.J.; Hu, L.Y.; Jiang, C.X.; Wei, Z.J.; Liu, J.A.; Wang, H.L.; Jiang, S.T. Hydrogen sulfide acts as a regulator of flower senescence in plants. *Postharvest Biol. Technol.* **2011**, *60*, 251–257. [CrossRef]

62. Wei, B.; Zhang, W.; Chao, J.; Zhang, T.R.; Zhao, T.T.; Noctor, G.; Liu, Y.S.; Han, Y. Functional analysis of the role of hydrogen sulfide in the regulation of dark-induced leaf senescence in *Arabidopsis*. *Sci. Rep. UK* **2017**, *7*, 2615. [CrossRef] [PubMed]

63. Gotor, C.; Garcia, I.; Crespo, J.L.; Romero, L.C. Sulfide as a signaling molecule in autophagy. *Autophagy* **2013**, *9*, 609–611. [CrossRef]

64. Laureano-Marin, A.M.; Moreno, I.; Romero, L.C.; Gotor, C. Negative regulation of autophagy by sulfide is independent of reactive oxygen species. *Plant Physiol.* **2016**, *171*, 1378–1391. [PubMed]

65. Aroca, A.; Benito, J.M.; Gotor, C.; Romero, L.C. Persulfidation proteome reveals the regulation of protein function by hydrogen sulfide in diverse biological processes in *Arabidopsis*. *J. Exp. Bot.* **2017**, *68*, 4915–4927. [CrossRef] [PubMed]

66. Filomeni, G.; Desideri, E.; Cardaci, S.; Rotilio, G.; Ciriolo, M.R. Under the ROS: Thiol network is the principal suspect for autophagy commitment. *Autophagy* **2010**, *6*, 999–1005. [CrossRef]

67. Huo, J.Q.; Huang, D.J.; Zhang, J.; Fang, H.; Wang, B.; Wang, C.L.; Liao, W.B. Hydrogen sulfide: A gaseous molecule in postharvest freshness. *Front. Plant Sci.* **2018**, *9*, 1172. [CrossRef]

68. Zhu, L.Q.; Wang, W.; Shi, J.Y.; Zhang, W.; Shen, Y.G.; Du, H.Y.; Wu, S.F. Hydrogen sulfide extends the postharvest life and enhances antioxidant activity of kiwifruit during storage. *J. Sci. Food Agric.* **2014**, *94*, 2699–2704. [CrossRef]

69. Lin, X.C.; Yang, R.; Dou, Y.; Zhang, W.; Du, H.Y.; Zhu, L.Q.; Chen, J.Y. Transcriptome analysis reveals delaying of the ripening and cell-wall degradation of kiwifruit by hydrogen sulfide. *J. Sci. Food Agric.* **2020**, *100*, 2280–2287. [CrossRef]

70. Gai-Fang, Y.; Zeng-Zheng, W.; Ting-Ting, L.; Jun, T.; Zhong-Qin, H.; Feng, Y.; Yan-Hong, L.; Zhuo, H.; Fan, H.; Lan-Ying, H.; et al. Enhanced antioxidant activity is modulated by hydrogen sulfide antagonizing ethylene in tomato fruit ripening. *J. Agric. Food Chem.* **2018**, *66*, 10380–10387.

71. Ziogas, V.; Molassiotis, A.; Fotopoulos, V.; Tanou, G. Hydrogen sulfide: A potent tool in postharvest fruit biology and possible mechanism of action. *Front. Plant Sci.* **2018**, *9*, 1375. [CrossRef] [PubMed]

72. Gulfishan, M.; Jahan, A.; Bhat, T.A.; Sahab, D. Plant senescence and organ abscission. In *Senescence Signalling and Control in Plants*; Bhat, T.A., Ed.; Academic Press: Aligarh, India, 2019; pp. 255–272.

73. Yasong, C.; Qingfu, C.; Dongao, H.; Guihaia, L.H.J. Research progress on molecular mechanism of plant falling flowers and fruits. *Guihaia* **2018**, *38*, 1234–1247.

74. Olsson, V.; Butenko, M.A. Abscission in plants. *Curr. Biol.* **2018**, *28*, 329–341. [CrossRef] [PubMed]

75. Patharkar, O.R.; Walker, J.C. Advances in abscission signaling. *J. Exp. Bot* **2018**, *69*, 733–740. [CrossRef]

76. Taylor, J.E.; Whitelaw, C.A. Signals in abscission. *New Phytol.* **2001**, *151*, 323–339. [CrossRef]

77. Botton, A.; Ruperti, B. The yes and no of the ethylene involvement in abscission. *Plants* **2019**, *8*, 187. [CrossRef] [PubMed]

78. Meir, S.; Philosoph-Hadas, S.; Riov, J.; Tucker, M.L.; Patterson, S.E.; Roberts, J.A. Re-evaluation of the ethylene-dependent and -independent pathways in the regulation of floral and organ abscission. *J. Exp. Bot.* **2019**, *70*, 1461–1467. [CrossRef] [PubMed]

79. Wang, G.Q.; Wei, P.C.; Tan, F.; Yu, M.; Zhang, X.Y.; Chen, Q.J.; Wang, X.C. The transcription factor AtDOF4.7 is involved in ethylene- and IDA-mediated organ abscission in *Arabidopsis*. *Front. Plant Sci.* **2016**, *7*, 863. [CrossRef]

80. Yamasaki, H.; Ogura, M.P.; Kingjoe, K.A.; Cohen, M.F. d-cysteine-induced rapid root abscission in the water fern *Azolla Pinnata*: Implications for the linkage between d-amino acid and reactive sulfur species (RSS) in plant environmental responses. *Antioxidants* **2019**, *8*, 411. [CrossRef]

81. Cutler, S.R.; Rodriguez, P.L.; Finkelstein, R.R.; Abrams, S.R. Abscisic acid: Emergence of a core signaling network. *Annu. Rev. Plant Biol.* **2010**, *61*, 651–679. [CrossRef]

82. Jin, Z.P.; Xue, S.W.; Luo, Y.A.; Tian, B.H.; Fang, H.H.; Li, H.; Pei, Y.X. Hydrogen sulfide interacting with abscisic acid in stomatal regulation responses to drought stress in *Arabidopsis*. *Plant Physiol. Biochem.* **2013**, *62*, 41–46. [CrossRef]

83. Li, Z.G.; Jin, J.Z. Hydrogen sulfide partly mediates abscisic acid-induced heat tolerance in tobacco (*Nicotiana tabacum* L.) suspension cultured cells. *Plant Cell Tiss. Org.* **2016**, *125*, 207–214. [CrossRef]

84. Li, T.T.; Li, Z.R.; Hu, K.D.; Hu, L.Y.; Chen, X.Y.; Li, Y.H.; Yang, Y.; Yang, F.; Zhang, H. Hydrogen sulfide alleviates kiwifruit ripening and senescence by antagonizing effect of ethylene. *Hortscience* **2017**, *52*, 1556–1562. [CrossRef]

85. Cheng, W.; Zhang, L.; Jiao, C.J.; Su, M.; Yang, T.; Zhou, L.; Peng, R.Y.; Wang, R.; Wang, C.Y. Hydrogen sulfide alleviates hypoxia-induced root tip death in *Pisum sativum*. *Plant Physiol. Biochem.* **2013**, *70*, 278–286. [CrossRef] [PubMed]

86. Zhao, Y.D. Auxin biosynthesis and its role in plant development. *Annu. Rev. Plant Biol.* **2010**, *61*, 49–64. [CrossRef]

87. Shi, H.T.; Ye, T.T.; Han, N.; Bian, H.W.; Liu, X.D.; Chan, Z.L. Hydrogen sulfide regulates abiotic stress tolerance and biotic stress resistance in *Arabidopsis*. *J. Integr. Plant Biol.* **2015**, *57*, 628–640. [CrossRef]

88. Zhang, X.W.; Liu, F.J.; Zhai, J.; Li, F.D.; Bi, H.G.; Ai, X.Z. Auxin acts as a downstream signaling molecule involved in hydrogen sulfide-induced chilling tolerance in cucumber. *Planta* **2020**, *251*, 1–19. [CrossRef]

89. Achard, P.; Genschik, P. Releasing the brakes of plant growth: How GAs shutdown DELLA proteins. *J. Exp. Bot.* **2009**, *60*, 1085–1092. [CrossRef]

90. Zhang, H.; Dou, W.; Jiang, C.X.; Wei, Z.J.; Liu, J.; Jones, R.L. Hydrogen sulfide stimulates β-amylase activity during early stages of wheat grain germination. *Plant Signal. Behav.* **2010**, *5*, 1559–2324. [CrossRef]

91. Xie, Y.J.; Zhang, C.; Lai, D.W.; Sun, Y.; Samma, M.K.; Zhang, J.; Shen, W.B. Hydrogen sulfide delays GA-triggered programmed cell death in wheat aleurone layers by the modulation of glutathione homeostasis and heme oxygenase-1 expression. *J. Plant Physiol.* **2014**, *171*, 53–62. [CrossRef]

92. Raskin, I. Role of salicylic-acid in plants. *Annu. Rev. Plant Phys.* **1992**, *43*, 439–463. [CrossRef]

93. Li, Z.G.; Xie, L.R.; Li, X.J. Hydrogen sulfide acts as a downstream signal molecule in salicylic acid-induced heat tolerance in maize (*Zea mays* L.) seedlings. *J. Plant Physiol.* **2015**, *177*, 121–127. [CrossRef] [PubMed]

94. Devoto, A.; Turner, J.G. Regulation of jasmonate-mediated plant responses in *Arabidopsis*. *Ann. Bot. London* **2003**, *92*, 329–337. [CrossRef] [PubMed]

95. Balbi, V.; Devoto, A. Jasmonate signalling network in *Arabidopsis thaliana*: Crucial regulatory nodes and new physiological scenarios. *New Phytol.* **2008**, *177*, 301–318. [CrossRef] [PubMed]

96. Han, X.; Hu, Y.R.; Zhang, G.S.; Jiang, Y.J.; Chen, X.L.; Yu, D.Q. Jasmonate negatively regulates stomatal development in *Arabidopsis* cotyledons. *Plant Physiol.* **2018**, *176*, 2871–2885. [CrossRef] [PubMed]

97. Deng, G.B.; Zhou, L.J.; Wang, Y.Y.; Zhang, G.S.; Chen, X.L. Hydrogen sulfide acts downstream of jasmonic acid to inhibit stomatal development in *Arabidopsis*. *Planta* **2020**, *251*, 42. [CrossRef]

98. Zhang, P.; Luo, Q.; Wang, R.L.; Xu, J. Hydrogen sulfide toxicity inhibits primary root growth through the ROS-NO pathway. *Sci. Rep. UK* **2017**, *7*, 868. [CrossRef]

99. Jaffrey, S.R.; Erdjument-Bromage, H.; Ferris, C.D.; Tempst, P.; Snyder, S.H. Protein S-nitrosylation: A physiological signal for neuronal nitric oxide. *Nat. Cell Biol.* **2001**, *3*, 193–197. [CrossRef]

100. Aroca, A.; Serna, A.; Gotor, C.; Romero, L.C. S-Sulfhydration: A cysteine posttranslational modification in plant systems. *Plant Physiol.* **2015**, *168*, 334–342. [CrossRef]

101. Li, Z.G.; Yi, X.Y.; Li, Y.T. Effect of pretreatment with hydrogen sulfide donor sodium hydrosulfide on heat tolerance in relation to antioxidant system in maize (*Zea mays*) seedlings. *Biologia* **2014**, *69*, 1001–1009. [CrossRef]

102. Uchida, A.; Jagendorf, A.T.; Hibino, T.; Takabe, T.; Takabe, T. Effects of hydrogen peroxide and nitric oxide on both salt and heat stress tolerance in rice. *Plant Sci.* **2002**, *163*, 515–523. [CrossRef]

103. Christou, A.; Filippou, P.; Manganaris, G.A.; Fotopoulos, V. Sodium hydrosulfide induces systemic thermotolerance to strawberry plants through transcriptional regulation of heat shock proteins and aquaporin. *BMC Plant Biol.* **2014**, *14*, 42. [CrossRef]

104. Zhou, Y.; Xu, X.Y.; Chen, L.Q.; Yang, J.L.; Zheng, S.J. Nitric oxide exacerbates Al-induced inhibition of root elongation in rice bean by affecting cell wall and plasma membrane properties. *Phytochemistry* **2012**, *76*, 46–51. [CrossRef] [PubMed]

105. Zhu, C.Q.; Zhang, J.H.; Sun, L.M.; Zhu, L.F.; Abliz, B.; Hu, W.J.; Zhong, C.; Bai, Z.G.; Sajid, H.; Cao, X.C.; et al. Hydrogen sulfide alleviates aluminum toxicity via decreasing apoplast and symplast al contents in rice. *Front. Plant Sci.* **2018**, *9*, 294. [CrossRef]

106. Sun, C.L.; Lu, L.L.; Yu, Y.; Liu, L.J.; Hu, Y.; Ye, Y.Q.; Jin, C.W.; Lin, X.Y. Decreasing methylation of pectin caused by nitric oxide leads to higher aluminium binding in cell walls and greater aluminium sensitivity of wheat roots. *J. Exp. Bot.* **2016**, *67*, 979–989. [CrossRef]

107. Pucciariello, C.; Perata, P. New insights into reactive oxygen species and nitric oxide signalling under low oxygen in plants. *Plant Cell Environ.* **2017**, *40*, 473–482. [CrossRef] [PubMed]

108. Peng, R.Y.; Bian, Z.Y.; Zhou, L.N.; Cheng, W.; Hai, N.; Yang, C.Q.; Yang, T.; Wang, X.Y.; Wang, C.Y. Hydrogen sulfide enhances nitric oxide-induced tolerance of hypoxia in maize (*Zea mays* L.). *Plant Cell Rep.* **2016**, *35*, 2325–2340. [CrossRef] [PubMed]

109. Xuan, W.; Zhu, F.Y.; Xu, S.; Huang, B.K.; Ling, T.F.; Qi, J.Y.; Ye, M.B.; Shen, W.B. The heme oxygenase/carbon monoxide system is involved in the auxin-induced cucumber adventitious rooting process. *Plant Physiol.* **2008**, *148*, 881–893. [CrossRef]

110. Lin, Y.T.; Li, M.Y.; Cui, W.T.; Lu, W.; Shen, W.B. Haem oxygenase-1 is involved in hydrogen sulfide-induced cucumber adventitious root formation. *J. Plant Growth Regul.* **2012**, *31*, 519–528. [CrossRef]

111. Li, Z.G.; Gong, M.; Xie, H.; Yang, L.; Li, J. Hydrogen sulfide donor sodium hydrosulfide-induced heat tolerance in tobacco (*Nicotiana tabacum* L.) suspension cultured cells and involvement of Ca^{2+} and calmodulin. *Plant Sci.* **2012**, *185*, 185–189. [CrossRef]

112. Li, Z.G.; Long, W.B.; Yang, S.Z.; Wang, Y.C.; Tang, J.H.; Wen, L.; Zhu, B.Y.; Min, X. Endogenous hydrogen sulfide regulated by calcium is involved in thermotolerance in tobacco *Nicotiana tabacum* L. suspension cell cultures. *Acta Physiol. Plant* **2015**, *37*, 1–11. [CrossRef]

113. Fang, H.H.; Jing, T.; Liu, Z.Q.; Zhang, L.P.; Jin, Z.P.; Pei, Y.X. Hydrogen sulfide interacts with calcium signaling to enhance the chromium tolerance in *Setaria italica*. *Cell Calcium* **2014**, *56*, 472–481. [CrossRef] [PubMed]

114. Qiao, Z.J.; Jing, T.; Jin, Z.P.; Liang, Y.L.; Zhang, L.P.; Liu, Z.Q.; Liu, D.M.; Pei, Y.X. CDPKs enhance Cd tolerance through intensifying H_2S signal in *Arabidopsis thaliana*. *Plant Soil* **2016**, *398*, 99–110. [CrossRef]

115. Zhu, J.K. Regulation of ion homeostasis under salt stress. *Curr. Opin. Plant Biol.* **2003**, *6*, 441–445. [CrossRef]

116. Deng, Y.Q.; Bao, J.; Yuan, F.; Liang, X.; Feng, Z.T.; Wang, B.S. Exogenous hydrogen sulfide alleviates salt stress in wheat seedlings by decreasing Na$^+$ content. *Plant Growth Regul.* **2016**, *79*, 391–399. [CrossRef]
117. Zhao, N.; Zhu, H.P.; Zhang, H.P.; Sun, J.; Zhou, J.C.; Deng, C.; Zhang, Y.H.; Zhao, F.; Zhou, X.Y.; Lu, C.F.; et al. Hydrogen sulfide mediates K$^+$ and Na$^+$ homeostasis in the roots of salt-resistant and salt-sensitive poplar species subjected to NaCl stress. *Front. Plant Sci.* **2018**, *9*, 1366. [CrossRef] [PubMed]
118. Mostofa, M.G.; Saegusa, D.; Fujita, M.; Tran, L.S.P. Hydrogen sulfide regulates salt tolerance in rice by maintaining Na$^+$/K$^+$ balance, mineral homeostasis and oxidative metabolism under excessive salt stress. *Front. Plant Sci.* **2015**, *6*, 1055. [CrossRef]
119. Li, J.S.; Jia, H.L.; Wang, J.; Cao, Q.H.; Wen, Z.C. Hydrogen sulfide is involved in maintaining ion homeostasis via regulating plasma membrane Na$^+$/H$^+$ antiporter system in the hydrogen peroxide-dependent manner in salt-stress *Arabidopsis thaliana* root. *Protoplasma* **2014**, *251*, 899–912. [CrossRef]
120. Paul, B.D.; Snyder, S.H. H$_2$S signalling through protein sulfhydration and beyond. *Nat. Rev. Mol. Cell Biol.* **2012**, *13*, 499–507. [CrossRef]
121. Aroca, A.; Gotor, C.; Romero, L.C. Hydrogen sulfide signaling in plants: Emerging roles of protein persulfidation. *Front. Plant Sci.* **2018**, *9*, 1369. [CrossRef]
122. Liu, Z.Q.; Li, Y.W.; Cao, C.Y.; Liang, S.; Ma, Y.S.; Liu, X.; Pei, Y.X. The role of H$_2$S in low temperature-induced cucurbitacin C increases in cucumber. *Plant Mol. Biol.* **2019**, *99*, 535–544. [CrossRef] [PubMed]

 International Journal of
Molecular Sciences

Article

Characterization of *FLOWERING LOCUS C* Homologs in Apple as a Model for Fruit Trees

Hidenao Kagaya [1], Naoko Ito [1], Tomoki Shibuya [2], Sadao Komori [3], Kazuhisa Kato [1,* and Yoshinori Kanayama [1,*

[1]   Graduate School of Agricultural Science, Tohoku University, Aoba-ku, Sendai 980-8572, Japan
[2]   Faculty of Life and Environmental Science, Shimane University, Matsue 690-8504, Japan
[3]   Faculty of Agriculture, Iwate University, Morioka 020-8550, Japan
*   Correspondence: kazuhisa.kato.d8@tohoku.ac.jp (K.K.); yoshinori.kanayama.a7@tohoku.ac.jp (Y.K.)

Received: 28 April 2020; Accepted: 24 June 2020; Published: 26 June 2020

Abstract: To elucidate the molecular mechanism of juvenility and annual flowering of fruit trees, *FLOWERING LOCUS C (FLC)*, an integrator of flowering signals, was investigated in apple as a model. We performed sequence and expression analyses and transgenic experiments related to juvenility with annual flowering to characterize the apple *FLC* homologs *MdFLC*. The phylogenetic tree analysis, which included other MADS-box genes, showed that both MdFLC1 and MdFLC3 belong to the same FLC group. MdFLC1c from one of the *MdFLC1* splice variants and MdFLC3 contain the four conserved motives of an MIKC-type MADS protein. The mRNA of variants *MdFLC1a* and *MdFLC1b* contain intron sequences, and their deduced amino acid sequences lack K- and C-domains. The expression levels of *MdFLC1a*, *MdFLC1b*, and *MdFLC1c* decreased during the flowering induction period in a seasonal expression pattern in the adult trees, whereas the expression level of *MdFLC3* did not decrease during that period. This suggests that *MdFLC1* is involved in flowering induction in the annual growth cycle of adult trees. In apple seedlings, because phase change can be observed in individuals, seedlings can be used for analysis of expression during phase transition. The expression levels of *MdFLC1b*, *MdFLC1c*, and *MdFLC3* were high during the juvenile phase and low during the transitional and adult phases. Because the expression pattern of *MdFLC3* suggests that it plays a specific role in juvenility, *MdFLC3* was subjected to functional analysis by transformation of *Arabidopsis*. The results revealed the function of *MdFLC3* as a floral repressor. In addition, *MdFT* had CArG box-like sequences, putative targets for the suppression of flowering by MdFLC binding, in the introns and promoter regions. These results indicate that apple homologs of FLC, which might play a role upstream of the flowering signals, could be involved in juvenility as well as in annual flowering. Apples with sufficient genome-related information are useful as a model for studying phenomena unique to woody plants such as juvenility and annual flowering.

Keywords: *Malus domestica*; Rosaceae; juvenility; FLOWERING LOCUS C; flowering

1. Introduction

Fruit trees, which are perennial woody plants, have a long juvenile period after germination and before flowering and fruit set. Because fruit quality cannot be evaluated during this juvenile phase, it is necessary to address the long juvenile period of fruit trees with regard to fruit tree breeding. Lengths of juvenile periods differ among species, and apple requires six to eight years of juvenility [1]. Although some cultivation techniques to shorten the juvenile period have been proposed, such as plant hormone treatment, grafting to dwarf rootstocks, suppression of dormancy, and adjustment of cultivation conditions [2–4], little is known about the molecular mechanisms of juvenility. It is more difficult to elucidate the mechanism underlying the control of juvenility in fruit trees than in herbaceous plants because fruit trees take longer to grow and are larger in size. Apple is one of the earliest plants whose genome sequences have been reported as woody plants and fruit tree [5], and is utilized as a model because of its extensive genome information [6,7]. Therefore, apples are useful for studying phenomena unique to woody plants such as juvenility and annual flowering.

In general, juvenility is stronger in young seedlings and gradually weakens as age progresses. There is also known to be a gradient of juvenility in individual trees; that is, juvenility is stronger at the base of trunks and branches and becomes weaker approaching the tip [8]. Genes that are potentially related to juvenility have been reported in studies of homologs of flowering-related genes from *Arabidopsis*, which is a model herbal plant. Previous studies focused on the relationship between juvenility and the apple homologs of *TERMINAL FLOWER 1* (*TFL1*), *FLOWERING LOCUS T* (*FT*), *LEAFY* (*LFY*), and *APETALA1* (*AP1*), which are floral meristem identity-related genes in *Arabidopsis*. The expression of apple *TFL1* homolog *MdTFL* is high during the juvenile phase, and the expression of apple *FT* homolog *MdFT* is high during the adult phase [9]. The overexpression of apple homologs of *LFY* and *AP1* in *Arabidopsis* promotes flowering, and the overexpression of *MdTFL* delays flowering [10–12]. In citrus, the overexpression of *LFY* or *AP1* of *Arabidopsis* reportedly decreases the juvenile period from seven years to two years [13], and the methylation of the *LEAFY* homolog is thought to be involved in juvenility [14]. However, most of these genes are located relatively downstream of the flowering pathway, and in order to clarify the molecular mechanism of juvenility, it is necessary to analyze genes farther upstream.

There are several flowering pathways, including the photoperiod, vernalization, gibberellin, and autonomous pathways [15]. The photoperiod and vernalization-dependent pathways are controlled by environmental factors, and the gibberellin-dependent pathway is comprised of a group of genes related to the synthesis and signal transduction of gibberellin. In contrast, the autonomous pathway is dependent on endogenous growth-related factors. Juvenility is not reduced by environmental factors such as temperature and day length, but is reduced by growth over several years, which suggests that juvenility is controlled by endogenous growth-related factors. Some genes in the autonomous pathway are known to induce flowering by suppressing the expression of *FLOWERING LOCUS C* (*FLC*) [16]. *FLC* plays a role as a key regulator of the autonomous and vernalization pathways and inhibits flowering by suppressing the expression of floral induction genes *SOC1* and *FT* in *Arabidopsis* [17–19]. Since *FLC* is a key gene for the flowering pathways, including the autonomous pathway, it can be expected to play a role in the suppression of flowering in the juvenile phase.

There have been some reports on *FLC* homologs in fruit trees. *FLC* homologs have been identified in apple, and divergent functions have been suggested based on their nucleotide sequences [20]. Two of these exhibit increased expression during dormancy and decreased expression with dormancy release, which suggests that they repress flowering as described in *Arabidopsis*. In fact, the possibility that the *FLC*-like gene is a candidate gene is mentioned in the genome-wide association mapping of the flowering period in apple [21]. In other apple *FLC*-like genes, the repression of bud outgrowth during dormancy and the promotion of flowering under non-chilling conditions have been reported [22,23]. Among other Rosaceae fruit trees, the *FLC* homolog in peach (*Prunus persica*) has not been reported to be associated with dormancy, and its role has not been elucidated [24]. No *FLC* homolog has been found in Japanese apricot (*Prunus mume*) either [25]. In fruit trees other than Rosaceae, changes in the expression and splicing of an *FLC*-like gene were reported in trifoliate orange (*Poncirus trifoliata*) [26]. Although these reports provide some information about the role of *FLC* in fruit trees, information about the relationship between juvenility and *FLC* homologs is still lacking. Therefore, to investigate the physiological roles of apple *FLC* homolog *MdFLC* at the molecular level, sequence and expression analyses and transgenic experiments were performed.

2. Results

2.1. cDNA Isolation and Phylogenetic Tree Analysis of MdFLC

Nine apple expressed sequence tag (EST) sequences were obtained by a BLAST search in DNA Data Bank of Japan (DDBJ, http://blast.ddbj.nig.ac.jp) using the amino acid sequence of the MADS region of *Arabidopsis FLC*. The contig sequences corresponding to these EST sequences were searched in GDR (https://www.rosaceae.org), a Rosaceae genome database, to obtain four contig sequences. Six EST or contig sequences remained after excluding duplicate sequences. A phylogenetic tree was prepared based on the amino acid sequences of these six genes and MADS-box genes from *Arabidopsis* and apple, and three of the six sequences were assigned to the same group as *Arabidopsis FLC*. Next, one of the three sequences was detected by PCR using cDNA from the juvenile phase of apple seedlings with primers specific to the three sequences and designated as *MdFLC1* (MD05G1037100) [23]. Three kinds of mRNA sequence were obtained by the RACE method and designated as splice variants *MdFLC1a* (accession number LC550081), *MdFLC1b* (LC550082), and *MdFLC1c* (LC550083).

PCR was performed using cDNA from the juvenile phase of apple seedlings with degenerate primers in the MADS region, and 10 kinds of MADS box-like sequences were obtained. One sequence among them was found to be juvenile phase-specific and homologous to *Arabidopsis FLC*; it was designated as *MdFLC3* (MD10G1041100) [23].

Figure 1 shows the amino acid sequence alignment of MdFLC1, MdFLC3, and AtFLC (AF537203) from *Arabidopsis*. MdFLC1c, which was the longest of the *MdFLC1* mRNA variants, and MdFLC3 contained MADS-, K-, I-, and C-domains with AtFLC. A phylogenetic tree of the amino acid sequence of MdFLC1c, MdFLC3, and FLC homologous proteins from other plants was prepared with other MADS-box proteins based on [27] (Figure 2). FLC, SVP, SOC1, AP1, and SEP groups were formed, and MdFLC1c and MdFLC3 were included in the FLC group.

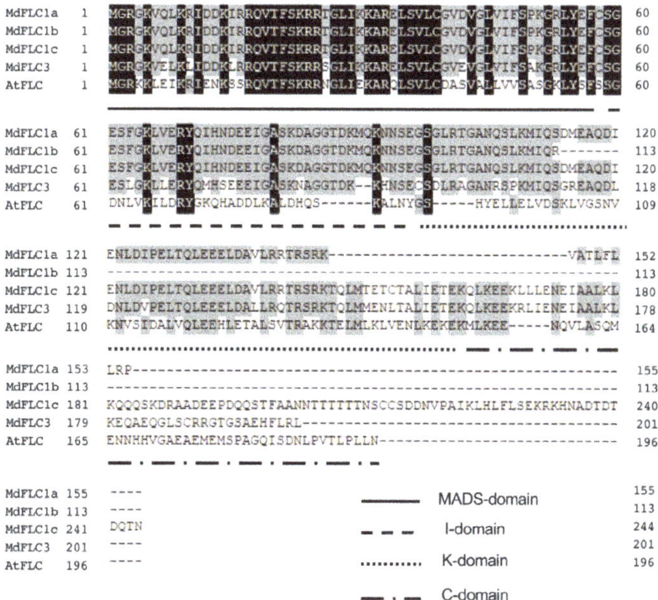

Figure 1. Amino acid sequence alignment of MdFLC and AtFLC. The MADS-, K-, I-, and C-domains are underlined. Identical amino acids for five and less proteins are shown in black and gray boxes, respectively.

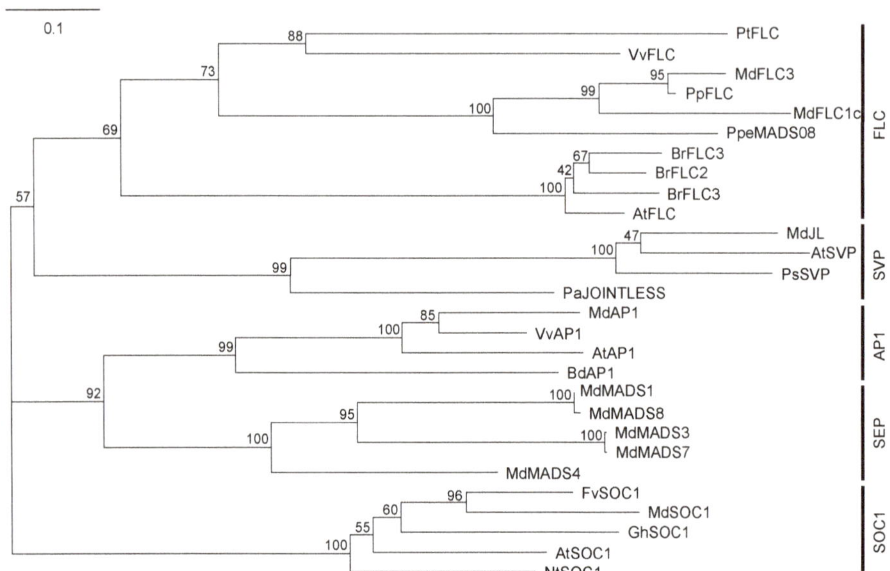

Figure 2. A phylogenetic tree based on the amino acid sequences of FLC, SVP, AP1, SEP, and SOC1 homologs from various species. The tree was constructed by the neighbor-joining method after sequence alignment using the ClustalW program. Branch numbers refer to percentage of replicates that support the branch using the bootstrap method (1000 replicates). The scale bar corresponds to 0.1 amino acid substitutions per residue. The accession numbers of the proteins added to construct the phylogenetic tree are as follows: PtFLC (EU497676), VvFLC (GU133630), PpFLC (KP164015), BrFLC1 (DQ866874), BrFLC2, (DQ866875), BrFLC3 (DQ866876), MdJOINTLESS (DQ402055), AtSVP (AF211171), PsSVP (AY830919), PaJOINTLESS (EU332978), VvAP1 (GU133634), MdAP1 (EU672877), AtAP1 (BT004113), BdAP1 (HQ588324), MdMADS1 (U78947), MdMADS3 (U78949), MdMADS4 (U78950), MdMADS7 (AJ000760), MdMADS8 (AJ001681), FvSOC1 (FJ531999), MdSOC1 (DQ846833), GhSOC1 (JF701982), NtSOC1 (JQ686938), AtSOC1 (AY093967).

2.2. Expression Analysis of MdFLC in the Adult Trees

Seasonal changes in the expression of *MdFLC* were examined in the adult trees. Flowering induction occurs between late June and mid-July [27]. The expression levels of *MdFLC1a*, *MdFLC1b*, and *MdFLC1c* were high in early June and decreased in early July during the period of flowering induction (Figure 3a–c). On the other hand, the expression level of *MdFLC3* did not change from early June to early July, but increased in August (Figure 3d). FLC suppresses the expression of *FT* in leaves as mentioned above [17]. *FT* generally produces mobile floral signals in leaves [28], and FT signal movement is also reported in Rosaceae fruit trees [29]. Therefore, leaves were used for analysis in the present study.

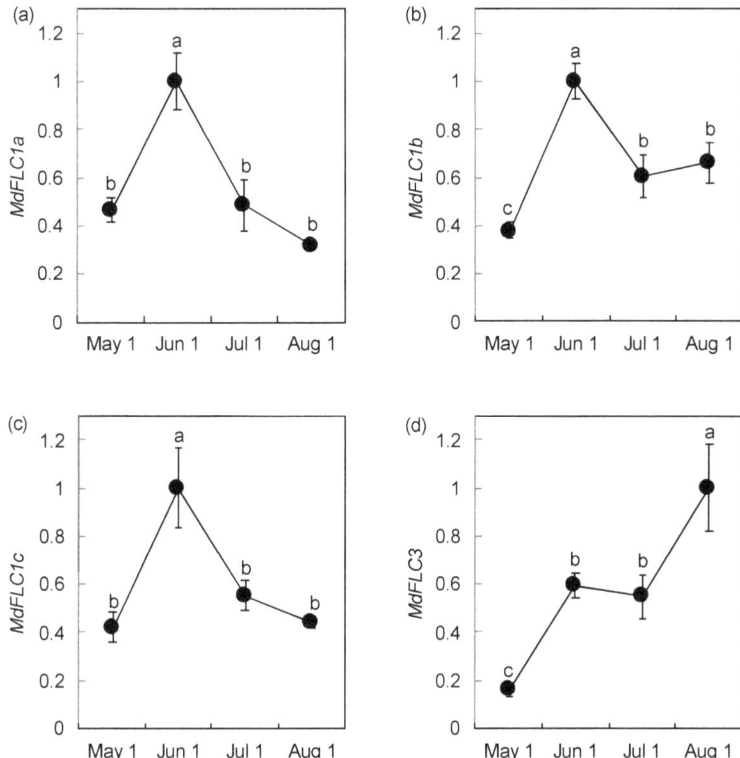

Figure 3. Seasonal changes in the expression levels of *MdFLC1a* (**a**), *MdFLC1b* (**b**), *MdFLC1c* (**c**), and *MdFLC3* (**d**) in the leaves of adult trees. Total RNA was prepared for the expression analysis of *MdFLC1* and *MdFLC3* on May 1, June 1, July 1, and August 1, 2016. Relative expression was determined using triplicate measurements taken from three independent biological replicates. The relative expression levels were normalized against *MdACTIN* with standard errors, and the maximum level of the transcripts was set at 1.0. The values with different letters for each gene significantly differed between days at $p < 0.05$, according to a Tukey test.

2.3. Expression Analysis of MdFLC During Phase Transition

The expression level of *MdFLC* during phase transition was performed in apple seedlings. While *MdFLC1a* expression was not detectable, the expression levels of *MdFLC1b* and *MdFLC1c* were high during the juvenile phase and low during the transitional and adult phases (Figure 4). The expression level of *MdFLC3* during the juvenile phase was also high compared to that in the transitional and adult phases, and it was 7.4 times the level in the adult phase.

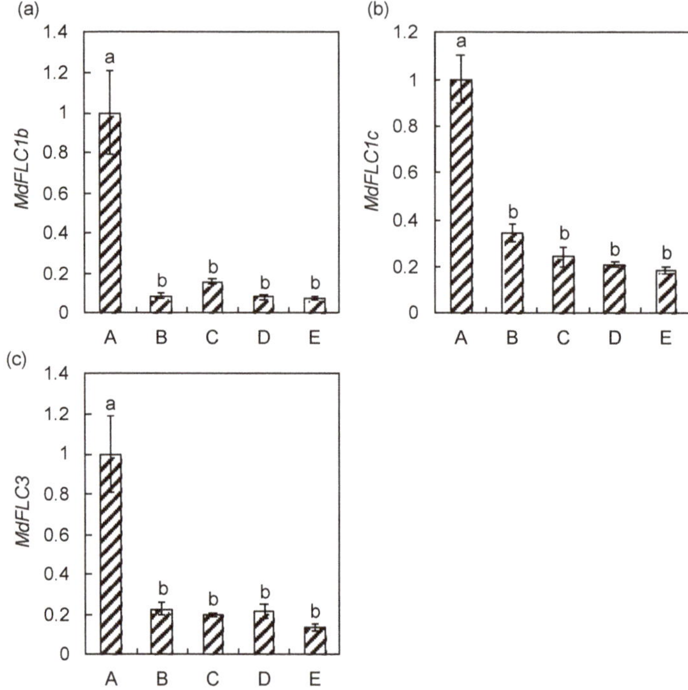

Figure 4. Changes in the expression levels of *MdFLC1b* (**a**), *MdFLC1c* (**b**), and *MdFLC3* (**c**) during phase transition in apple seedlings. The expression level of *MdFLC1a* was undetectable. Total RNA was prepared from the juvenile phase (A), transition phase (B and C), and adult phase (D and E) in early July 2009 as described in [27]. Relative expression was determined in triplicate measurements taken from three independent biological replicates. The relative expression levels were normalized against *MdACTIN* with standard errors, and the maximum level of the transcripts was set at 1.0. The values with different letters for each gene significantly differed between positions at $p < 0.05$ according to a Tukey test.

2.4. Transformation of Arabidopsis with MdFLC3 cDNA

Delays in bolting were observed in more than three transgenic plants from individual seeds (T$_1$) obtained by *Agrobacterium* in planta vacuum infiltration transformation. Detailed analysis was performed on their progenies, lines FOX1 and FOX2. The growth of FOX1 and FOX2 lines was observed and expression analysis of *AtFT* was then performed (Figure 5; Figure S1a,b, Supplementary Materials). The number of days from sowing to bolting was 31.8 in the wild-type (WT) control, and that in FOX1 and FOX2 was 35.5 and 35.7 days, respectively. The number of rosette leaves at the time of bolting was 14.4 in the wild-type and 17.5 and 18.2 in FOX1 and FOX2, respectively. A lower expression level of *AtFT* was observed with late flowering in the FOX lines. A very high expression level of *MdFLC3* was confirmed in the FOX lines, whereas only a trace expression of endogenous *FLC* (*AtFLC*) was found in the FOX lines and wild-type (Figure S2, Supplementary Materials). The average value of transformants derived from individual seeds, which are different from FOX1 and FOX1 lines, at the beginning of transformation showed delayed bolting, supporting the above result (Figure S1c,d).

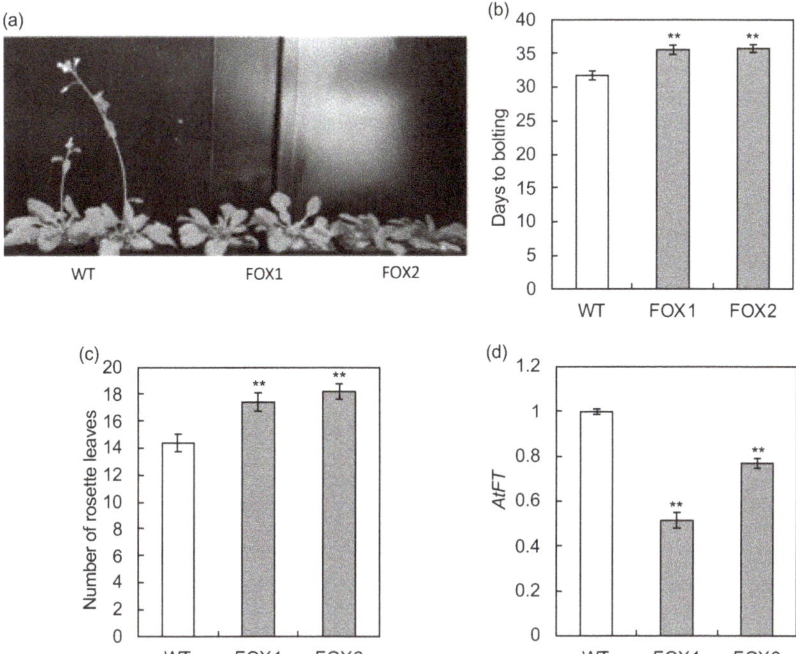

Figure 5. The overexpression of *MdFLC3* and phenotypic analysis in *Arabidopsis*. Flowering phenotypes 32 days after sowing (**a**), days to bolting from sowing (**b**), the number of rosette leaves at bolting time (**c**), and the expression levels of *AtFT* (**d**) in the *MdFLC3* transgenic (FOX1, 2) and wild-type (WT) plants grown under a 16 h photoperiod. The values with ** significantly differed between FOX and WT plants at $p < 0.01$, according to the Dunnett test. The values in (**b**,**c**) indicate means and standard errors (n = 30 or 31). The relative expression of *AtFT* was determined from triplicate measurements of three independent biological replicates 15 days after sowing (**d**). The relative expression levels were normalized against *AtACTIN* with standard errors and the maximum level of the transcripts was set at 1.0.

3. Discussion

A phylogenetic tree analysis, including other MADS-box genes such as SVP and SOC1, revealed that both MdFLC1 and MdFLC3 belong to the same FLC group as VvFLC [30] and PtFLC [26], which has been reported to function as a floral repressor (Figure 2). Therefore, MdFLC1 and MdFLC3 were further investigated as apple FLCs in this study. FLC is one of the MADS-box proteins, which are transcription factors having a highly conserved region of approximately 60 amino acids, MADS-box, that is involved in DNA binding and dimer formation. Many MADS-box proteins in plants are classified as MIKC-type, contain MADS-, I-, K-, and C-domains [31], and form dimers and higher multimers to function [32,33]. MdFLC1c and MdFLC3 contained these four conserved domains.

Three cDNA sequences of *MdFLC1* were searched in the GDR database, and *MdFLC1* was found to correspond to MD05G1037100. *MdFLC1c* mRNA contained all exons, whereas *MdFLC1a* and *MdFLC1b* mRNA contained the sequences of the fourth and third introns, which were not removed in splicing, respectively (Figure 6). Therefore, the deduced amino acid sequence of MdFLC1c contained MADS-, I-, K-, and C-domains, whereas the sequences of MdFLC1a and MdFLC1b lacked K- and C- domains because of the stop codons in the intron sequences of their mRNA (Figure 1). K- and C-domains are important for protein–protein interactions and other functions in MIKC-type MADS proteins [34]. These results suggest that *MdFLC1c* plays the role of *MdFLC1* and that *MdFLC1a* and *MdFLC1b* are expected to be non-functional. However, since the regulation of expression by selective splicing has been reported in plant response to environmental stress [35], expression analysis was performed in the three splicing variants of *MdFLC1*. The *MdFLC3* cDNA sequence was consistent with that of MD10G1041100 in the GDR database, and its deduced amino acid sequence contained the four domains of an MIKC-type MADS protein (Figure 1), suggesting that MdFLC3 functions as an MIKC-type MADS protein.

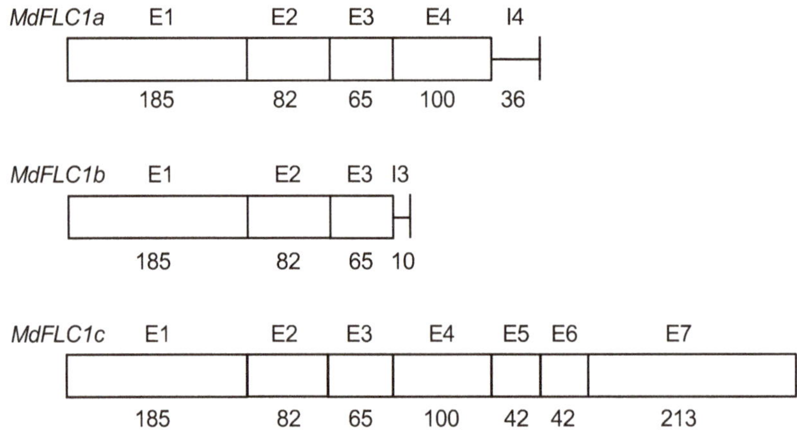

Figure 6. Transcript structures of *MdFLC1* splice variants, *MdFLC1a*, *MdFLC1b*, and *MdFLC1c*. E and I with numbers indicate exons and introns, respectively. Numbers under the bar correspond to their sizes, in base pairs.

In the present study, the expression levels of *MdFLC1a*, *MdFLC1b*, and *MdFLC1c* decreased during the period of flowering induction in a seasonal expression pattern in the adult trees. In the annual growth cycle of apple, *MdFT1*, which is a floral integrator, shows high expression, and the expression level of *MdTFL*, which is a floral repressor, decreases during the period of flowering induction [9,27]. Therefore, the expression of *MdFLC1a*, *MdFLC1b*, and particularly *MdFLC1c*, which is expected to translate functional proteins, is likely involved in suppression of flowering in the annual growth cycle of adult apple trees, as is *MdTFL*. In contrast, the expression level of *MdFLC3* did not decrease during the period of flowering induction, suggesting that *MdFLC3* is not involved in flowering induction in the annual growth cycle of adult trees.

Because phase change is observable in individuals, seedlings can be used for expression analysis of phase transition [36]. Five sites in the seedlings used in this study show phase transition based on juvenile characteristics such as flower bud formation, leaf size, and leaf serration, and were used for *MdFLC* expression analysis [27]. Although expression of *MdFLC1a* was not detected, the expression levels of *MdFLC1b* and *MdFLC1c* were high in the juvenile phase and low in the transitional and adult phases. The expression level of *MdFLC3* was similarly high in the juvenile phase. The expression pattern of *MdFLC3* suggests that it does not play a role in flowering in the annual growth cycle but acts specifically in juvenility. A similar seasonal expression pattern was observed in *MdFLC3* within buds in the adult trees and confirmed the specific role of MdFLC3 (Figure S3, Supplementary Materials). Furthermore, the expression of *MdFLC1c* in the juvenile phase of seedlings is significantly lower than the expression in the adult trees (approximately 1/6, data not shown). Therefore, we focused on this specific role of *MdFLC3* and subjected *MdFLC3* to functional analysis by the transformation of *Arabidopsis*. The results showed that *MdFLC3* is a floral repressor, confirming its role in the juvenility of apple. Since the result for phase transition was obtained in crossed seedlings, a genotype-specific effect could not be excluded. Further study using other several genotype combinations and apomixis will be necessary to confirm the findings of the present study.

FLC suppresses the expression of *FT* containing the CArG box, which is an FLC-binding sequence, in the promoter region and first intron in *Arabidopsis* leaves [17]. Apples also have sufficient genomic information and such binding sites can be analyzed. Since the *MdFTs* in the database have CArG box-like sequences in the intron and promoter region (Table 1), it can be expected that MdFLC1 and MdFLC3 bind to these sequences to control annual flowering and juvenility. Regarding the molecular mechanism of juvenility of woody plants, including fruit trees, little is known about genes located upstream of the flowering pathway. Our results suggest that the apple homolog of *FLC*, which appears to play a central role relatively upstream of the flowering pathway, could be involved in juvenility as well as in annual flowering. Testing the correlation between MdFLC characteristics and the length of the juvenile phase could provide valuable insight into the function of MdFLC in the regulation of this process in the near future. In addition, if this gene could be used as a marker, it would be possible to breed cultivars with a short juvenile period as well as various useful traits by marker-assisted selection [21,37]. As apples have advanced in genome information and genome editing technology [38], accumulating results as a model will be utilized for other fruit trees.

Table 1. Putative CArG boxes in *MdFT* [a].

Gene	Accession Number	Putative CArG Box	Strand	Position of 1st C from ATG [b]
MdFT1	AB458506	AACT**CC**ATTAATTGCAGG	Top	+289
		TACT**CC**TTATTTTGTCAA	Top	+846
MdFT2	AB458504	CTAA**CC**ATTAATTGTGTT	Top	+1001
		AGAT**CC**TAAAAAAGTATA	Bottom	+994
		GTAT**CC**AAATAAGTTGC	Bottom	−159
		GTCT**CC**TAATTTGTTGT	Bottom	−745

[a] The boxes in the introns and promoter regions, the upstream 1500-bp *MdFT* promoter sequences of the start codon were checked in both strands according to Helliwell et al. [17]. [b] The boxes in the introns and the promoter regions are shown with the positive and negative number of positions, respectively.

4. Materials and Methods

4.1. Plant Materials

Mature leaves in the juvenile phase of 8-year-old apple (*Malus domestica*) seedlings of a cross between 'Fuji' and 'Himekami' were used for cDNA isolation and sequence analysis of *MdFLC*. Mature leaves from 11-year-old apple trees ('Fuji' grafted onto M.9, *M. prunifolia*) were used as adult trees. For expression analysis in adult trees, mature leaves were sampled from 20-year-old apple trees ('Fuji' grafted onto M.9, *M. prunifolia*) on May 1, June 1, July 1, and August 1, 2016. For expression analysis of phase transition, mature leaves were collected from the juvenile to adult phase in 8-year-old apple seedlings of a cross between 'Fuji' and 'Himekami' [27]. All samples were collected in the Tohoku University (Sendai, Japan) experimental field at 38°16′ N and 140°52′ E.

4.2. cDNA Isolation, Sequence Analysis, and Phylogenetic Tree Analysis of MdFLC

To isolate *MdFLC1* cDNA, the apple EST sequences were searched using DDBJ and GDR. DDBJ was also used to search homologous genes in other plants. The cetyltrimethylammonium bromide (CTAB) method [27] and a TaKaRa RNA PCR kit (AMV) (Takara-Bio, Kusatsu, Japan) were used for RNA extraction and reverse transcription, respectively. Cloning of *MdFLC3* cDNA was performed by PCR with degenerate primers MdMADSF and MdMADSR (Table S1, Supplementary Materials) based on highly conserved sequences in the MADS-box protein using PROSITE (https://prosite.expasy.org). The restriction sites of *Eco*RI and *Bam*HI were added to the primers in advance. PCR was performed with cDNA from leaves in the juvenile and adult phases. PCR products were electrophoresed on agarose gel, and amplified fragments of the expected size were collected using TaKaRa RECOCHIP (TaKaRa). pUC18 was digested with *Eco*RI and *Bam*HI, and the PCR fragments were ligated into this vector and then transformed into *Escherichia coli*. The plasmid was purified for sequence analysis. Sequence Alignment by ClustalW (http://align.genome.jp) was used to prepare the sequence alignments and phylogenetic trees.

4.3. Expression Analysis by Real-Time PCR

For expression analysis in the adult trees, RNA extraction and reverse transcription were performed using a Cica Geneus RNA prep kit for Plant (Kanto Kagaku, Tokyo, Japan) and ReverTra Ase qPCR RT Master Mix with gDNA Remover (TOYOBO, Osaka, Japan), respectively. THUNDERBIRD SYBR qPCR Mix (TOYOBO) was used for the subsequent PCR. For expression analysis with phase transition, RNA extraction and reverse transcription were performed using the CTAB method [27] and QuantiTect Reverse Transcription Kit (Qiagen, Hilden, Germany), respectively. A QuantiTect SYBR Green PCR Kit (Qiagen) was used for the subsequent PCR. Real-time PCR was performed as described in Ikeda et al. [39], using the primer sets listed in Table S1. Coefficients of variant for quantification cycle of the reference genes among samples were 1.85%, 2.75%, and 1.01% in Figures 3–5, respectively.

4.4. Transformation of Arabidopsis with MdFLC3 cDNA

The translated region of *MdFLC3* cDNA was amplified by PCR using Q5 DNA polymerase (New England Biolabs, Ipswich,, MA, USA) electrophoresed in an agarose gel. It was then collected from the gel and used as an insert. The primers used were MdFLC3InsertF and MdFLC3InsertR, in which *Bam*HI and *Sac*I sites were added to the 5′ and 3′ ends, respectively (Table S2, Supplementary Materials). The *GUS* sequence in the binary vector pBI121 was excised using *Bam*HI and *Sac*I, and the *MdFLC3* insert was introduced into the vector instead. The resultant vector containing *MdFLC3* cDNA with CaMV *35S* promoter was used for the transformation of *Arabidopsis thaliana* ecotype Columbia (Inplanta Innovation Inc., Yokohama, Japan). Transformed seeds (T_1) were selected in a medium containing kanamycin, and a transgene check was performed by PCR using the primer set for *MdFLC3* (Table S2). Expression analysis was performed by real-time PCR using the primer sets for *AtFT*, *MdFLC3*, and *AtFLC*, and the *Arabidopsis* actin primer was set as a reference (Table S2). Seeds (T_2) were collected separately from individuals (T_1) derived from individual seeds initially obtained by *Agrobacterium* in planta vacuum infiltration transformation and each of them was sown as a line. In the T_2 generation, segregation of the transgene was checked in each line and seeds (T_3) were collected. Homozygous seeds that did not segregate in T_3 were used. The homozygous seeds (T_3) were planted and grown in vermiculite/pearlite (1:1) at 22 °C under a 16 h photoperiod, and the number of rosette leaves and days after sowing was measured at bolting [40].

Supplementary Materials: Supplementary materials can be found at http://www.mdpi.com/1422-0067/21/12/4562/s1. Table S1. Primers used for cloning of cDNA encoding *MdFLC* and real-time PCR in apple; Table S2. Primers used for transformation of *Arabidopsis* with *MdFLC3* cDNA and real-time PCR in transgenic *Arabidopsis*; Figure S1. Phenotype analysis in the *MdFLC3* transgenic (FOX) and wild-type (WT) plants as support data for Figure 5; Figure S2. Expression levels of *MdFLC3* and *AtFLC* in the *MdFLC3* transgenic (FOX1, 2) and wild-type (WT) plants; Figure S3. Seasonal changes in the expression levels of *MdFLC3* in the buds of adult trees.

Author Contributions: Conceptualization, S.K. and Y.K.; methodology, H.K., T.S., and S.K.; investigation, H.K. and N.I.; formal analysis, H.K., N.I., and T.S.; resources, K.K.; data curation, K.K.; writing—original draft preparation, H.K.; writing—review and editing, K.K. and Y.K.; visualization, H.K.; supervision, S.K. and Y.K.; project administration, Y.K.; funding acquisition, K.K. and Y.K. All authors have read and agreed to the published version of the manuscript.

Funding: This research was supported by Grants-in-Aid for Scientific Research (24248006), Japan.

Acknowledgments: The authors thank Shohei Hano and Mai Asakawa for preliminary experiments and materials used for experiments.

Conflicts of Interest: The authors declare no conflict of interest.

Abbreviations

FLC	FLOWERING LOCUS C
TFL1	*TERMINAL FLOWER 1*
FT	FLOWERING LOCUS T
LFY	LEAFY
AP1	APETALA1
EST	expressed sequence tag
DDBJ	DNA Data Bank of Japan
WT	wild-type

References

1. Zimmerman, R.H. Juvenility and flowering in woody plants: A review. *HortScience* **1972**, *7*, 447–455.
2. Visser, T. Juvenile phase and growth of apple and pear seedlings. *Euphytica* **1964**, *13*, 119–129.
3. Visser, T. The effect of rootstocks on growth and flowering of apple seedlings. *J. Am. Soc. Hortic. Sci.* **1973**, *98*, 26–28.
4. Hackett, W.P. Juvenility, maturation, and rejuvenation in woody plants. *Hortic. Rev.* **1985**, *7*, 109–155.
5. Velasco, R.; Zharkikh, A.; Affourtit, J.; Dhingra, A.; Cestaro, A.; Kalyanaraman, A.; Fontana, P.; Bhatnagar, S.K.; Troggio, M.; Pruss, D.; et al. The genome of the domesticated apple (*Malus* × *domestica* Borkh.). *Nat Genet.* **2010**, *42*, 833–839. [CrossRef]
6. Eccher, G.; Ferrero, S.; Populin, F.; Colombo, L.; Botton, A. Apple (*Malus domestica* L. Borkh) as an emerging model for fruit development. *Plant Biosyst.* **2014**, *148*, 157–168. [CrossRef]
7. He, P.; Li, L.G.; Li, H.F.; Wang, H.B.; Yang, J.M. Development and application S-SAP molecular markers for the identification of apple (*Malus Domestica* Borkh.) sports. *J. Hortic. Sci. Biotechnol.* **2015**, *90*, 297–302. [CrossRef]
8. Poething, R.S. Phase change and the regulation of shoot morphogenesis in plants. *Science* **1990**, *250*, 923–930. [CrossRef]
9. Mimida, N.; Komori, S.; Suzuki, A.; Wada, M. Functions of the apple *TFL1/FT* orthologs in phase transition. *Sci. Hortic.* **2013**, *156*, 106–112. [CrossRef]
10. Kotoda, N.; Wada, M.; Kusaba, S.; Kano-Murakami, Y.; Masuda, T.; Soejima, J. Overexpression of *MdMADS5*, an *APETALA1*-like gene of apple, causes early flowering in transgenic *Arabidopsis*. *Plant Sci.* **2002**, *162*, 679–687. [CrossRef]
11. Kotoda, N.; Wada, M. *MdTFL1*, a *TFL1*-like gene of apple retards the transition from the vegetative to reproductive phase in transgenic *Arabidopsis*. *Plant Sci.* **2005**, *168*, 95–104. [CrossRef]
12. Wada, M.; Cao, Q.; Kotoda, N.; Soejima, J.; Masuda, T. Apple has two orthologues of *FLORICAULA/LEAFY* involved in flowering. *Plant Mol. Biol.* **2002**, *49*, 567–577. [CrossRef] [PubMed]
13. Pena, L.; Martin-Trillo, M.; Juarez, J.; Pina, J.A.; Navarro, L.; Martinez-Zapater, J.M. Constitutive expression of *Arabidopsis LFAFY* or *APETALA1* gene in citrus reduces their generation time. *Nat. Biotech.* **2001**, *19*, 263–267. [CrossRef]
14. Zhang, J.Z.; Mei, L.; Liu, R.; Khan, M.R.G.; Hu, C.G. Possible involvement of locus-specific methylation on expression regulation of *LEAFY* homologous gene (*CiLFY*) during precocious trifoliate orange phase change process. *PLoS ONE* **2014**, *9*, e88558. [CrossRef] [PubMed]
15. Srikanth, A.; Schmid, M. Regulation of flowering time: All roads lead to Rome. *Cell Mol. Life Sci.* **2011**, *68*, 2013–2037. [CrossRef] [PubMed]
16. Michaels, S.D.; Amasino, R.M. *FLOWERING LOCUS C* encodes a novel MADS domain protein that acts as a repressor of flowering. *Plant Cell* **1990**, *11*, 949–956. [CrossRef] [PubMed]
17. Helliwell, C.A.; Wood, C.C.; Robertson, M.; James, P.W.; Dennis, E.S. The *Arabidopsis* FLC protein interacts directly in vivo with *SOC1* and *FT* chromatin and is part of a high-molecular-weight protein complex. *Plant J.* **2006**, *46*, 183–192. [CrossRef]

Int. J. Mol. Sci. **2020**, *21*, 4562

18. Hepworth, S.R.; Valverde, F.; Ravenscroft, D.; Mouradov, A.; Coupland, G. Antagonistic regulation of flowering-time gene *SOC1* by *CONSTANS* and *FLC* via separate promoter motifs. *EMBO J.* **2002**, *21*, 4327–4337. [CrossRef]

19. Searle, I.; He, Y.; Turck, F.; Vincent, C.; Fornara, F.; Krober, S.; Amasino, R.A.; Coupland, G. The transcription factor *FLC* confers a flowering response to vernalization by repressing meristem competence and systemic signaling in *Arabidopsis*. *Genes Dev.* **2006**, *20*, 898–912. [CrossRef]

20. Kumar, G.; Arya, P.; Gupta, K.; Randhawa, V.; Achaya, V.; Singh, A.K. Comparative phylogenetic analysis and transcriptional profiling of MADS-box gene family identified *DAM* and *FLC*-like genes in apple (*Malus* x *domestica*). *Sci. Rep.* **2016**, *6*, 20695. [CrossRef]

21. Urrestarazu, J.; Muranty, H.; Denancé, C.; Leforestier, D.; Ravon, E.; Guyader, A.; Guisnel, R.; Feugey, L.; Aubourg, S.; Celton, J.-M.; et al. Genome-wide association mapping of flowering and ripening periods in apple. *Front. Plant Sci.* **2017**, *8*, 1923. [CrossRef] [PubMed]

22. Porto, D.D.; Bruneau, M.; Perini, P.; Anzanello, R.; Jean-Pierre, R.; Dos Santos, H.P.; Fialho, F.B.; Revers, L.F. Transcription profiling of the chilling requirement for bud break in apples: A putative role for *FLC*-like genes. *J. Exp. Bot.* **2015**, *66*, 2659–2672. [CrossRef] [PubMed]

23. Zong, X.; Zhang, Y.; Walworth, A.; Tomaszewski, E.M.; Callow, P.; Zhong, G.Y.; Song, G.Q. Constitutive expression of an apple FLC3-like gene promotes flowering in transgenic blueberry under nonchilling conditions. *Int. J. Mol. Sci.* **2019**, *20*, 2775. [CrossRef] [PubMed]

24. Wells, E.C.; Vendramin, E.; Tarodo, J.S.; Verge, I.; Bielenberg, G.D. A genome-wide analysis of MADS-box genes in peach [*Prunus persica* (L.) Batsch]. *BMC Plant Biol.* **2015**, *15*, 41. [CrossRef]

25. Xu, Z.; Zhang, Q.; Sun, L.; Du, D.; Cheng, T.; Pan, H.; Yang, W.; Wang, J. Genome-wide identification, characterization and expression analysis of the MADS-box gene family in *Prunus mume*. *Mol. Genet. Genom.* **2014**, *289*, 903–920. [CrossRef]

26. Zhang, J.Z.; Li, Z.M.; Mei, L.; Yao, J.L.; Hu, C.G. *PtFLC* homolog from trifoliate orange (*Poncirus trifoliate*) is regulated by alternative splicing and experiences seasonal fluctuation in expression level. *Planta* **2009**, *229*, 847–859. [CrossRef]

27. Hatsuda, Y.; Nishio, S.; Komori, S.; Nishiyama, M.; Kanahama, K.; Kanayama, Y. Relationship between *MdMADS11* gene expression and juvenility in apple. *J. Soc. Hortic. Sci.* **2011**, *80*, 396–403. [CrossRef]

28. Zeevaart, J.A.D. Leaf-produced floral signals. *Curr. Opin. Plant Biol.* **2008**, *11*, 541–547. [CrossRef]

29. Freiman, A.; Golobovitch, S.; Yablovitz, Z.; Belausov, E.; Dahan, Y.; Peer, R.; Avraham, L.; Freiman, Z.; Evenor, D.; Reuveni, M.; et al. Expression of *flowering locus T2* transgene from *Pyrus communis* L. delays dormancy and leaf senescence in *Malus* x *domestica* Borkh, and causes early flowering in tobacco. *Plant Sci.* **2015**, *241*, 164–176. [CrossRef]

30. Díaz-Riquelme, J.; Martínez-Zapater, J.M.; Carmona, M.J. Transcriptional analysis of tendril and inflorescence development in grapevine. *PLoS ONE* **2014**, *9*, e92339. [CrossRef]

31. De Bodt, S.; Raes, J.; Florquin, K.; Rombauts, S.; Rouzé, P.; Theissen, G.; Van de Peer, Y. Genomewide structural annotation and evolutionary analysis of the type I MADS-box genes in plants. *J. Mol. Evol.* **2003**, *56*, 573–586. [CrossRef]

32. Theissen, G.; Saedler, H. Floral quartets. *Nature* **2001**, *409*, 469–471. [CrossRef]

33. Honma, T.; Goto, K. Complexes of MADS-box proteins are sufficient to convert leaves into floral organs. *Nature* **2001**, *409*, 525–529. [CrossRef] [PubMed]

34. Kaufmann, K.; Melzer, R.; Theissen, G. MIKC-type MADS-domain proteins: Structural modularity, protein interactions and network evolution in land plants. *Gene* **2005**, *347*, 183–198. [CrossRef] [PubMed]

35. Shang, X.; Cao, Y.; Ma, L. Alternative splicing in plant genes: A means of regulating the environmental fitness of plants. *Int. J. Mol. Sci.* **2017**, *18*, 432. [CrossRef] [PubMed]

36. Carlsbecker, A.; Tandre, K.; Johanson, U.; Englund, M.; Engstrom, P. The MADS-box gene DAL1 is a potential mediator of the juvenile-to-adult transition in Norway spruce (*Picea abies*). *Plant J.* **2004**, *40*, 546–557. [CrossRef]

37. Moriya, S.; Terakami, S.; Okada, K.; Shimizu, T.; Adachi, Y.; Katayose, Y.; Fujisawa, H.; Wu, J.; Kanamori, H.; Yamamoto, T.; et al. Identification of candidate genes responsible for the susceptibility of apple (*Malus* × *domestica* Borkh.) to *Alternaria blotch*. *BMC Plant Biol.* **2019**, *19*, 132. [CrossRef]

38. Zhou, J.; Li, D.; Wang, G.; Wang, F.; Kunjal, M.; Joldersma, D.; Liu, Z. Application and future perspective of CRISPR/Cas9 genome editing in fruit crops. *J. Integr. Plant Biol.* **2020**, *62*, 269–286. [CrossRef]

39. Ikeda, H.; Shibuya, T.; Imanishi, S.; Aso, H.; Nishiyama, M.; Kanayama, Y. Dynamic Metabolic regulation by a chromosome segment from a wild relative during fruit development in a tomato introgression line, IL8-3. *Plant Cell Physiol.* **2016**, *57*, 1257–1270. [CrossRef]

40. Kanayama, Y.; Mizutani, R.; Yaguchi, S.; Hojo, A.; Ikeda, H.; Nishiyama, M.; Kanahama, K. Characterization of an uncharacterized aldo-keto reductase gene from peach and its role in abiotic stress tolerance. *Phytochemistry* **2014**, *104*, 30–36. [CrossRef]

International Journal of
Molecular Sciences

Review

Regulation of Three Key Kinases of Brassinosteroid Signaling Pathway

Juan Mao [1,2] and Jianming Li [1,2,3,*]

1   State Key Laboratory for Conservation and Utilization of Subtropical Agro-Bioresources, South China Agriculture University, Guangzhou 510642, China; maojuan@scau.edu.cn
2   Guangdong Key Laboratory for Innovative Development and Utilization of Forest Plant Germplasm, College of Forestry and Landscape Architecture, South China Agricultural University, Guangzhou 510642, China
3   Department of Molecular, Cellular, and Developmental Biology, University of Michigan, Ann Arbor, MI 48109, USA
*   Correspondence: jmlumaa@scau.edu.cn

Received: 1 June 2020; Accepted: 16 June 2020; Published: 18 June 2020

Abstract: Brassinosteroids (BRs) are important plant growth hormones that regulate a wide range of plant growth and developmental processes. The BR signals are perceived by two cell surface-localized receptor kinases, Brassinosteroid-Insensitive1 (BRI1) and BRI1-Associated receptor Kinase (BAK1), and reach the nucleus through two master transcription factors, bri1-EMS suppressor1 (BES1) and Brassinazole-resistant1 (BZR1). The intracellular transmission of the BR signals from BRI1/BAK1 to BES1/BZR1 is inhibited by a constitutively active kinase Brassinosteroid-Insensitive2 (BIN2) that phosphorylates and negatively regulates BES1/BZR1. Since their initial discoveries, further studies have revealed a plethora of biochemical and cellular mechanisms that regulate their protein abundance, subcellular localizations, and signaling activities. In this review, we provide a critical analysis of the current literature concerning activation, inactivation, and other regulatory mechanisms of three key kinases of the BR signaling cascade, BRI1, BAK1, and BIN2, and discuss some unresolved controversies and outstanding questions that require further investigation.

Keywords: brassinosteroids; receptor-like kinases; GSK3-like kinases; somatic embryogenesis receptor-like kinases; protein phosphatases

---

1. Introduction

Brassinosteroids (BRs) are plant-specific steroid hormones and play essential roles in a broad range of plant growth and developmental processes, including cell elongation, cell division, and differentiation, seed germination, stomata formation, root development, vascular differentiation, plant architecture, flowering, male fertility, and senescence [1–4]. BRs are also involved in responding to various abiotic and biotic stresses, such as drought, flooding, salinity, extreme temperatures, microbial pathogens, and insect herbivores [5–7]. Plants with defects in BR biosynthesis or signaling show a characteristic set of developmental defects, including dwarfed statue, male sterility, delayed senescence and flowering, and photomorphogenesis in the dark [8,9].

Using various experimental and theoretical approaches, including genetics, biochemistry, cell biology, chemical biology, structural biology, proteomics, transcriptomics, genomics, mathematical modeling, and computational dynamics simulation, a series of important BR signaling components have been well established and intensively studied, revealing a protein phosphorylation-mediated BR signaling cascade [10–12] (Figure 1). BRs are perceived at the plasma membrane (PM) by the extracellular domains of BRI1 (Brassinosteroid-Insensitive1) receptor, a leucine-rich repeat receptor-like kinase (LRR-RLK) [9,13], and its co-receptor BAK1 (BRI1-Associated receptor Kinase1, also known as SERK3 for Somatic Embryogenesis Receptor Kinase3) [14–16], a versatile LRR-RLK involved in many signaling processes [17]. BR binding to the BR-binding pocket formed by the extracellular domains of BRI1 and BAK1 is thought to trigger conformational changes of their cytoplasmic domains [18,19]. BRI1 subsequently phosphorylates its inhibitor BKI1 (BRI1 Kinase Inhibitor1) and induces its dissociation from the PM [14,18,20,21], thus enabling heterodimerization, reciprocal phosphorylation, and full activation of the kinase activities of BRI1 and BAK1 [14,18,20–23]. The fully activated BRI1 triggers a series of phosphorylation or dephosphorylation events to transduce the extracellular BR signals into the cytosol. BRI1 phosphorylates BSK1 (BR-Signaling Kinase1), CDG1 (Constitutive Differential Growth1), and some of their homologs, leading to phosphorylation and subsequent activation of members of a unique family of protein phosphatases with Kelch-like domain (PPKLs) that include BSU1 (bri1 suppressor1) and BSU1-Like1-3 (BSL1-3) [24–26]. It is generally believed that the phosphorylated BSU1/BSLs inactivate BIN2 (Brassinosteroid-Insensitive2), which is a member of the plant GSK3 (Glycogen Synthase Kinase3)-like kinase family, via dephosphorylation of a phosphorylated tyrosine (Tyr) residue in the activation loop of BIN2 [27]. The dephosphorylated BIN2 also interacts with KIB1 (Kink suppressed in *bzr1-1D*1), an F-box E3 ubiquitin ligase, leading to BIN2 ubiquitination and proteasome-mediated degradation [28]. Upon BIN2 inactivation and degradation, two highly similar BIN2 substrates, BZR1 (Brassinazole-resistant1) and BES1 (bri1-EMS suppressor1) [29,30] are rapidly dephosphorylated by certain nuclear-localized members of the PP2A (protein phosphatase 2A) family [31], leading to their nuclear accumulation. The dephosphorylated BZR1 and BES1 bind to their target promoters containing BRRE (BR-response element) (CGTGC/TG) and/or E-box (CANNTG) motif to regulate expression of thousands of BR-responsive genes that are crucial for plant growth and development [32,33].

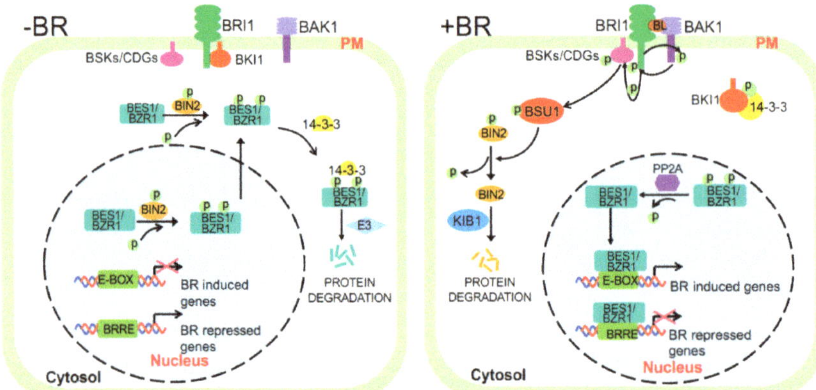

Figure 1. A current model of BR signaling. When BRs (brassinosteroids) are absent (left), BRI1 (Brassinosteroid-Insensitive1) is kept inactive by its autoinhibitory C-terminus and BKI1 (BRI1 Kinase Inhibitor1) association. BIN2 (Brassinosteroid-Insensitive2) is constitutively active and phosphorylates BES1 (bri1-EMS suppressor1)/BZR1 (Brassinazole-resistant1) transcription factors to promote their degradation and 14-3-3-mediated cytosolic retention, and to directly inhibit their DNA-binding activities. When BRs are present (right) and bind to the extracellular domains of BRI1 and its co-receptor BAK1 (BRI1 Associated receptor Kinase1) to activate the two receptor kinases, leading to dissociation of BKI1 from BRI1, phosphorylation and activation of BSKs (BR-signaling kinases)/CDGs (Constitutive Differential Growth) and BSU1 (bri1 suppressor1). The activated BSU1 dephosphorylates and inactivates BIN2 while KIB1 (Kink suppressed in *bzr1-1D*1) promotes BIN2 degradation, causing nuclear accumulation of PP2A (protein phosphatase 2A)-dephosphorylated BES1/BZR1 that bind BRRE (BR response element)/E-box-containing promoters to regulate expression of thousands of BR-responsive genes important for plant growth and development.

In the past few years, several excellent reviews were published that summarized the significant progresses in understanding how the BR signal is perceived at the cell surface and how the extracellular BR signal is transduced into the nucleus to regulate a wide range of plant developmental and physiological processes [10–12,34]. In this review, we present a critical analysis of currently available data on three early BR signaling components, including the BR receptor BRI1, its coreceptor BAK1, and the crucial negative regulator BIN2, highlighting recent findings and discussing controversies and unanswered questions on the regulatory mechanisms that control their protein abundance, subcellular localizations, and signaling activities.

2. BRI1, the BR Receptor

BRI1 localizes to the PM and belongs to a large and plant-specific family of LRR-RLKs, which is composed of 223 members in *Arabidopsis thaliana* [35]. BRI1 consists of an extracellular LRR domain of 25 LRRs, a single-pass transmembrane segment, and a cytoplasmic kinase domain [9]. The 25 LRRs are interrupted by a 70-residue island domain (ID, from amino acid Lys^{586} to Met^{657}), which constitutes the BR binding domain with the 22nd LRR [36]. Its cytoplasmic kinase domain can be subdivided into the JM (juxtamembrane) region, a canonical serine/threonine (Ser/Thr) kinase domain, and a short C-terminal tail (CT) of 36 amino acids (AAs). The Arabidopsis genome encodes three BRI1 homologs: BRL1 (BRI1-Like1), BRL2 (also known as VH1 for Vascular Highway1 [37]), and BRL3. BRL1 and BRL3 can also bind BRs and function as BR receptors [38,39], whereas BRL2/VH1 does not bind BR but functions in a BR-independent manner to regulate vascular development [37,40]. Gene expression analysis and genetic studies revealed that BRI1 is expressed in most plant tissues/organs, whereas BRLs are found only in the vascular tissues and stem cell niches [41]. BRLs have been shown to be involved in vascular development [42] and plant tolerance to abiotic stresses, such as hypoxia and drought [43,44].

2.1. Maintaining the Inactive State in the Absence of BR

In the absence of BR ligands, BRI1 is maintained at its inactive state via its inhibitory CT and its binding to BKI1 at the PM [20,23]. Removal of the BRI1 CT not only increased the in vitro kinase activity of a recombinant BRI1 kinase protein, but also led to a hyperactive BR receptor in vivo [23]. Further investigation will likely be required to fully understand the mechanism of this autoinhibitory activity of the BRI1 kinase activity given the missing CT in several available crystal structures of the BRI1 kinase domain [45,46]. The 36-AA CT could block the substrate binding site to interfere with the in vitro homodimerization for the auto(trans)phosphorylation activity or the in vivo binding of the kinase domains of BRI1 and BAK1. In addition to the *"cis"* autoinhibition, the binding of the PM-associated BKI1 to the BRI1 cytoplasmic domain demonstrates a *"trans"* inhibitory mechanism to prevent association and subsequent cross phosphorylation of the kinase domains of BRI1 and BAK1, which is likely triggered by BR-independent heterodimerization of their extracellular domains known to occur in plant cells [15,47,48]. A previous genetic demonstration of the absolute requirement of BAK1/SERKs for the BRI1's activation [49] implies that BRI1 can be kept in its inactive state via competitive binding of BAK1 with certain BAK1-interacting proteins. Indeed, recent studies revealed that ligand-independent interaction of BAK1 with BIR3 (BAK1-Interacting Receptor-like kinase3) could inhibit BR-stimulated BAK1-BRI1 heterodimerization and BRI1 activation (see below for more discussion on the BAK1-BIR interaction) [50].

2.2. BRI1 Activation

It was well known that many kinases are activated via phosphorylation of crucial Ser/Thr residues in the activation segment catalyzed by upstream kinases [51]. In animal cells, many receptor kinases are activated by homodimerization and auto(trans)phosphorylation [52,53]. Despite several reports of BRI1 homodimerization in vitro or in vivo [15,23,54,55], BRI1 activation requires its heterodimerization and subsequent transphosphorylation with its coreceptor BAK1/SERKs. An earlier in vitro kinase assay with dephosphorylated recombinant kinases of BRI1 and BAK1 showed that neither BRI1 nor BAK1 was active when incubated alone with the ATP-containing kinase reaction buffer but became active when incubated together [45]. Similarly, the full-length BRI1 or BAK1 was inactive when expressed alone in yeast cells but became active kinases when they were coexpressed. Neither BRI1 nor BAK1 was active when coexpressed in yeast with a kinase-dead partner [18]. More important, a BRI1-FLAG fusion protein could not be activated by exogenously applied BR in a transgenic Arabidopsis line that lacks BAK1 and its two close homologs [49]. Furthermore, a recent study showed that ligand-independent heterodimerization of a chimeric LRR-RLK, which was composed of the extracellular domain of BIR3 and the intracellular domain of BRI1, and BAK1 led to constitutive activation of the BR signaling pathway [56]. All these biochemical and genetic/transgenic experiments argue against the widely accepted sequential-phosphorylation model. This model posits that BR binding-triggered conformational changes result in weak activation of BRI1 and its subsequent phosphorylation and activation of BAK1, which can then transphosphorylate BRI1 for its full activation [21]. Further studies, especially structural analysis of the ligand-bound full-length BRI1-BAK1 complexes and mass spectrometry (MS)-based quantitative phosphorylation analysis of BRI1 in *bak1/serk* mutants, are needed to fully understand the activation mechanism of BRI1 and the genetic requirement of BAK1/SERKs for its activation.

Structural analyses of the BRI1-BAK1 extracellular heterodimers and molecular dynamics simulations suggested that BR binding to the extracellular domain of BRI1 causes conformational changes in BRI1 [57,58]. Such subtle structural changes not only stabilize the ID to create a docking platform for the BRI1-BAK1 association, which is further stabilized by a bound active BR (functioning as "molecular glue" to interact with the N-terminal cap of BAK1), but also potentially create a secondary interface involving the N-terminal 12 LRRs of BRI1 to interact with BAK1 [16,19] (Moffett and Shukla, 2020 bioRxiv: http://doi.org/10.1101/630640). Despite lack of experimental evidence or structural information, it has been widely accepted that stable association of the extracellular domains of BRI1 and BAK1 brings their cytoplasmic domains into close proximity, thus permitting cross phosphorylation, especially at Ser^{1044} of BRI1 and Thr^{450} of BAK1 within their respective activation loop, and subsequent activation of both kinases [45,46,59]. A recent structural modeling study suggested that the cytoplasmic domains of BRI1 and BAK1 could weakly interact independently of BR binding to their extracellular domains [60]. However, it is important to point out that the structural models of the kinase domains of BRI1 and BAK1 used for the modeling study were derived from autophosphorylated and activated kinase domains with stabilized αC-helixes and activation segments.

Structural models of the activated BRI1 kinase domain [45,46] suggested that the phosphorylated Ser^{1044} is likely trapped in a positively-charged phosphate-binding pocket consisting of Arg^{922} (at the beginning of the αC-helix), Arg^{1008} [within the HRD (His-Arg-Asp) motif], and Arg^{1032} (at the beginning of the β9-strand), thus stabilizing the activation loop and establishing an active kinase conformation (Figure 2). Specifically, the interaction of the phosphorylated Ser^{1044} (pSer^{1044}) and Arg^{922} of the αC-helix is likely critical for creating the so-called regulatory-spine (R-spine) consisting of 5 residues [the anchoring Asp^{1068} residue of the αF-helix, His^{1007} of the HRD motif, Phe^{1028} of the DFG (Asp-Phe-Gly) motif, Ile^{931} of the αC-helix, and Leu^{942} of the β4-strand] (Figure 2). The assembly of the R-spine has been considered as a crucial indicator of an active conformation of eukaryotic protein kinases [61,62]. As expected, a Ser^{1044}-Ala mutation resulted in a strong loss of kinase and signaling activity of BRI1 in vitro and in vivo [22].

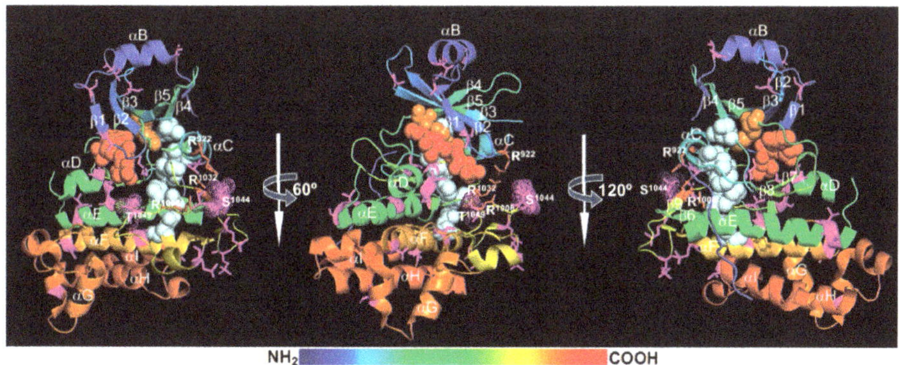

Figure 2. A structural model of the activated BRI1 kinase domain. Shown here are three (0°, 60°, and 180°) rotational views of a rainbow-colored ribbon model of the crystal structure of the BRI1 kinase domain (Protein Data Bank No. 4oh4). Individual α-helices (αB-αI) and β-strands (β1-β5) were labelled. The magenta-colored sticks indicate phosphorylated Ser/Thr residues with the magenta dots surrounding Ser^{1044} and Thr^{1049} of the activation segment, the green sticks denote the Lys^{911} and Glu^{927} residues that form the salt bridge between the β3-strand and αC-helix, the red sticks mark the three positively charged residues of the phosphate-binding pocket, and the orange sticks represent the phosphorylated Tyr residue. The red spheres show adenylyl-imidodiphosphate (a non-hydrolysable ATP analog), the orange spheres indicate the gatekeeper Tyr^{956} residue that is also phosphorylated in in vitro assays, and the light-blue spheres denote the five regulatory-spine (R-spine) residues (from lower to upper: Asp^{1068}, His^{1077}, Phe^{1028}, Ile^{931}, and Leu^{942}). The rainbow bar indicates the order of amino acids (AAs) from the N-terminus (blue) to the C-terminus (red).

In addition to the pSer^{1044} residue within the activation loop, previous studies revealed an essential role of a conserved Ser/Thr residue (Thr^{1049}) in the "P+1 loop" [21,22], which is thought to be the docking site for the backbone of the substrate P-site and the side-chain of the P+1 residue [61]. Previous bioinformatic analysis indicated that this position is occupied by a Ser/Thr residue in >99% of the so-called "RD"-type RLKs containing Arg-Asp residues within the conserved HRD motif [21]. By contrast, only 30% of "non-RD"-type RLKs contain a Ser/Thr residue at this position. Mutating Thr^{1049} to Ala completely inhibited the in vitro phosphorylation activity of a recombinant BRI1 kinase domain and greatly reduced its signaling activity in transgenic Arabidopsis plants [23], demonstrating its crucial role in BRI1 kinase activity. Mutating the equivalent Ser/Thr residue in several other "RD"-type LRR-RLKs resulted in loss of in vitro phosphorylation activity [63]. Further studies will be needed to determine whether the phosphorylation of this residue is truly required for BRI1 activation as the phosphorylation-mimic Thr^{1049}-Asp mutation also led to a strong loss of the in vitro phosphorylation activity of a recombinant BRI1 kinase [21]. Based on the structural model (Figure 2), the phosphorylation of this residue could potentially interfere with ATP binding and/or substrate binding.

The activated BRI1 is thought to autophosphorylate additional Ser, Thr, and Tyr residues in both N-lobe and C-lobe (Figure 2). The phosphorylated N-lobe residues include Ser^{838}, Thr^{842}, Thr^{846}, Thr^{851}, and Ser^{858} of the JM (not shown in Figure 2), Thr^{872} and Thr^{880} on the αB-helix that caps the N-lobe, Ser^{887} and Ser^{891} on both ends of the β1-strand, Ser^{906} in the β2-β3 loop, Ser^{917} in the β3-αC loop (missing in the BRI1 crystal structure), and Thr^{930} on the αC-helix. The phosphorylated C-lobe residues include Ser^{963} at the start of the αD-helix, Ser^{981}, Thr^{982}, and Ser^{990} of the αE-helix, Ser^{1012} and Ser^{1013} of a half helical twist before the β7-strand, Ser^{1026} located between the β8-strand and the DFG motif, Thr^{1039}, Ser^{1042}, and Thr^{1045} of the activation segment, Ser^{1071} and Thr^{1081} of the αF-helix, Ser^{1109} located at the start of the twisted helical αG-αH linker, Thr^{1147} at the start of the αI-helix, and at least 5 residues in the CT (Ser^{1166}, Ser^{1168}, Thr^{1169}, Ser^{1179}, and Ser^{1187}) [21,22,63–66] (Figure 2). Among those phosphorylated residues, at least 11 were identified from immunoprecipitated BRI1 fusion protein from transgenic Arabidopsis plants, including 6 uniquely identified (Ser^{838}, Ser^{858}, Thr^{872}, Thr^{880}, Thr^{982}, and Ser^{1168}), 2 additional Thr sites (most likely Thr^{842} and Thr^{846}) and 3 ambiguous sites within the activation fragment [22]. In addition, at least 3 Tyr residues were reported to be autophosphorylated, including Tyr^{831} (in the JM), Tyr^{956} (the gatekeeper residue on the β5-strand), and Tyr^{1072} (on the αF-helix) [67,68] (Figure 2). It should be noted that the published MS data from several in vitro and in vivo phosphorylation site-mapping experiments of BRI1 did not identify any phosphorylated Tyr residue [21,22,63–66]. The 3 phosphorylated Tyr residues (Tyr^{831}, Tyr^{956}, and Tyr^{1072}) that were discussed in the current literature were deduced from immunoblot assays with generic and site-specific anti-phosphorylated Tyr (anti-pTyr) antibodies [67,68]. It is generally believed that these phosphorylated Ser/Thr/Tyr sites could regulate the abundance and/or signaling activity of BRI1 and create docking sites for binding downstream signaling components or regulators to build a BRI1 signaling complex, which likely include scaffolding proteins such as TTLs (Tetratricopeptide Thioredoxin-Like proteins) and BSK3 [69,70]. The activated BRI1 can then transphosphorylate these regulators and downstream targets to transduce the extracellular BR signal into the cytosol, ultimately reaching the nucleus to alter gene expression.

One question that remains to be answered is how binding of BR to BRI1 triggers phosphorylation of BKI1. As discussed above, the activation of BRI1's kinase activity requires its heterodimerization and transphosphorylation with BAK1, a process that is prevented by the PM-localized BRI1-binding BKI1, while BKI1 dissociation from BRI1 and the PM absolutely requires phosphorylation by activated BRI1 [71,72]. Previous studies showed that BRI1-catalyzed Tyr phosphorylation at the conserved Tyr^{211}, which likely requires BRI1-catalyzed phosphorylation at Ser^{270} and Ser^{274} [72], is necessary and sufficient for its dissociation from the PM [71]. It might be possible that BR binding to the BRI1's extracellular domain triggers yet undefined conformational changes in its cytoplasmic domain. Such structural changes could weaken the BRI1-BKI1 binding and permit weak BRI1-BAK1 association to allow the BAK1-catalyzed transphosphorylation of Thr^{1044} in the activation loop of BRI1, leading to further conformational changes and full activation of BRI1. Activated BRI1 can subsequently phosphorylate BKI1 at Ser$^{270/274}$ and Tyr^{211}, resulting in BKI1 dissociation from BRI1 and the PM and a stronger BRI1-BAK1 binding. A better understanding of the BRI1-BKI1 and BRI1-BAK1 interaction requires further structural analyses of the protein complexes of full-length proteins in the absence or presence of active BRs.

2.3. Attenuation and Deactivation

The magnitude and duration of the BR signaling is dynamically regulated by activation and deactivation of the BR receptor. One simple mechanism of receptor inactivation or attenuation is mediated by autophosphorylation. Previous studies suggested that autophosphorylation at certain residues inhibited the in vitro kinase activity of an *E. coli* expressed BRI1 kinase domain and reduced the physiological activity of BRI1 in transgenic Arabidopsis plants [21,23,67,73]. For example, mutating Thr^{872} to Ala, which is located at the start of the αB-helix that caps the N-lobe of 5 antiparallel β-strands (Figure 2), could significantly enhance the in vitro autophosphorylation activity or transphosphorylation activity towards an artificial peptide substrate [22]. It is interesting to note that Thr^{872} and Thr^{880}, known to be phosphorylated and located at the start and end of the αB-helix (Figure 2), are highly conserved among 213 Arabidopsis LRR-RLK [22], suggesting that phosphorylation of these residues might be a conserved attenuation mechanism for many plant LRR-RLKs. Similarly, Ser^{891}, located in the β1-β2 ATP-binding loop (better known as "glycine-rich loop" or "G-loop") containing the GXGXXG-motif (X indicating any AA), was previously shown to deactivate BRI1 [73], likely by interfering with ATP binding, as the "G-loop" is known for positioning the three phosphate groups and the adenine ring of ATP [61]. An earlier study also suggested that phosphorylation of two tyrosine residues, Tyr^{831} (located in the JM) and Tyr^{956} (a critical gatekeeper of the ATP binding pocket), might also inhibit the BRI1 kinase and signaling activity [67]. However, the lack of appropriate phosphorylation-mimic

AA for a pTyr residue makes it complicated to interpret the published experimental results on the regulatory roles of BRI1's Tyr phosphorylation.

In addition to the "*cis*" attenuation by autophosphorylation, the signaling activity of BRI1 could be "*trans*" attenuated by dephosphorylation of certain phosphorylated Ser/Thr residues through PP2A [66,74]. A PP2A holoenzyme is composed of three subunits: a catalytic C subunit, a regulatory B subunit, and a scaffolding A subunit. A B subunit from one of the 3 subfamilies in Arabidopsis: B, B′, and B″, determines the substrate specificity, while an A subunit brings the B and C subunits together to form an active protein phosphatase complex [75]. An earlier study implicated the Arabidopsis SBI1 (suppressor of bri1), a leucine carboxylmethyltransferase, in methylating the C subunits of PP2A to enhance the PM-association of certain PP2A holoenzymes [74]. As a result, PP2A could bind and dephosphorylate activated BRI1, leading to increased BRI1 degradation and attenuated BR signaling. Further support for the involvement of PP2A in attenuating BR signaling came from a recent study showing that at least four cytoplasm-localized PP2A B′ subunits, including B′γ, B′η, B′θ, and B′ζ [75], interact with BRI1 and mediate its dephosphorylation and inactivation [66]. Overexpression of any of these *PP2A-B′* genes in a weak *bri1-5* mutant significantly enhanced its dwarf phenotype and further reduced its BR signaling activity. By contrast, simultaneously eliminating these four *PP2A-B′* genes could enhance BR signaling and partially suppressed the dwarf phenotype of a weak BR-deficient mutant *det2* (*de-etiolation2*) [66]. Quantitative MS assays of autophosphorylated BRI1 kinase domain incubated with immunoprecipitated PP2A complexes identified several autophosphorylated Ser/Thr residues as the potential dephosphorylation sites of PP2A, including a phosphorylated residue (Thr^{872}, Thr^{880}, or Ser^{887} on the αB-helix and the αB-β1 loop), Ser^{917} (in the β3-αC loop), Ser^{981} (on the αD helix), and four phosphorylated Ser/Thr residues in the CT (Ser^{1166}, Ser^{1168}, Thr^{1169}, or Ser^{1172}, and S^{1179}/Thr^{1180}) [66]. An earlier mutagenesis experiment indicated that individual mutations of these Ser/Thr residues (except Thr^{872}) had little impact on the in vitro autophosphorylation activity of the BRI1 kinase but did reduce its transphosphorylation activity towards a peptide substrate [22]. In addition, the dephosphorylated CT was previously shown to exert inhibitory effects on BRI1 signaling activity [23]. Further MS analyses of the endogenous BRI1 in *PP2A B′*-overexpressing transgenic Arabidopsis lines and the Arabidopsis quadruple mutant lacking B′γ, B′η, B′θ, and B′ζ are needed to pinpoint the exact Ser/Thr residues that are dephosphorylated by PP2A. Given the importance of Tyr phosphorylation in BR signaling [67,68,71] and a previous report of Tyr dephosphorylation of an LRR-RLK immunity receptor EFR (Elongation Factor-Tu Receptor) by a bacterial tyrosine phosphatase [76], it will be interesting to investigate if the BRI1 signaling activity could also be regulated by a member of the Arabidopsis PTP/DSPP (protein tyrosine phosphatase/dual specificity protein phosphatase) family [77].

2.4. Regulating the Abundance of BRI1 on the PM

2.4.1. Trafficking from the ER to the PM

The BR signaling activity at the PM is also controlled by the amount of the PM-localized BRI1 receptor, which starts its secretory journey in the endoplasmic reticulum (ER) (Figure 3). It was well known that the ER houses several stringent quality control (QC) systems that permit export of only correctly folded proteins into the Golgi apparatus but retain incompletely or incorrectly folded proteins in the ER for chaperone-assisted folding repair or degradation via ER-associated degradation mechanism (ERAD) [78,79]. Two structurally defect but biochemically competent mutant variants of BRI1, bri1-5 with Cys^{69}-Tyr mutation and bri1-9 carrying a Ser^{662}-Phe mutation, are retained in the ER and degraded by ERAD [80–82] (Figure 3). Loss-of-function mutations in the ER quality control (ERQC) or ERAD system resulted in increased amounts of mutant bri1 receptors on the PM, thus partially suppressing the dwarfism phenotype of these *bri1* mutants as the corresponding mutant BRI1 proteins still retain partial signaling activity after being correctly targeted to the PM [83–86]. A recent study also suggested the presence of a yet to be defined QC system at the PM to remove misfolded/damaged PM-localized proteins, such as bri1-301 (with a Gly^{989}-Ile mutation) that escapes from ERQC/ERAD [87,88]. TWD1 (Twisted Dwarf1), a well-studied cochaperone protein that was previously shown to be localized in the ER and at the PM, where it regulates auxin transport [89], also interacts with both BRI1 and BAK1 to enhance BR signaling [90,91]. Given its demonstrated interaction with HSP90 (heat shock protein 90) [92,93], it is tempting to speculate that TWD1 could promote optimal folding of BRI1 and its coreceptor, thus maximizing their BR signaling activities at the PM.

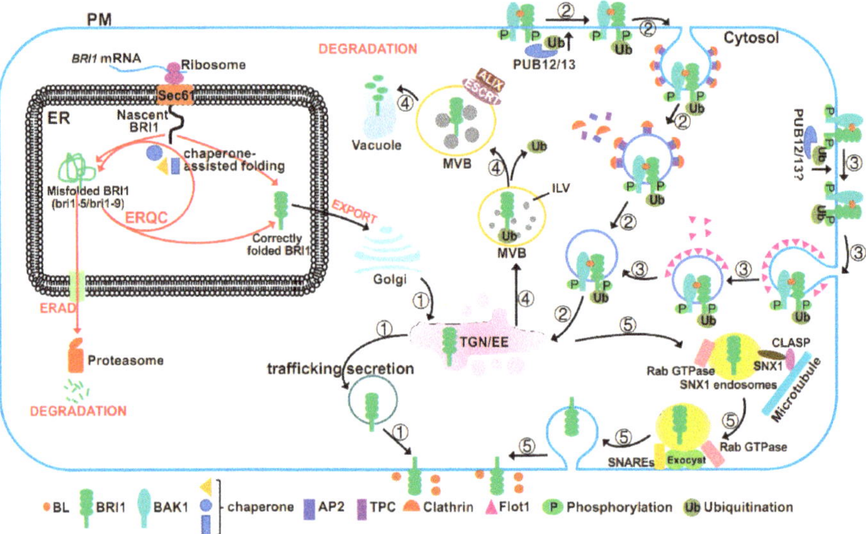

Figure 3. A current model of regulating the BRI1 abundance on the plasma membrane (PM). Newly synthesized BRI1 is translocated through the Sec61 translocon from the ER (endoplasmic reticulum)-associated ribosomes into the ER where it undergoes chaperone-assisted protein folding. Correctly folded BRI1 is exported into the Golgi to continue its secretory journey through the TGN (*trans*-Golgi network)/EE (early endosome) to be targeted on the PM (1), whereas the incorrectly/misfolded BRI1, such as its mutant variants bri1-5 and bri1-9, are retained in the ER by ERQC (ER quality control) for refolding and degradation via ERAD (ER-associated degradation) that involves retrotranslocation and cytosolic proteasome. The PM-localized BRI1 or BRI1/BAK1 heterodimer is presumably ubiquitinated by the PUB12/13 (plant U-box protein12/13) E3 ligases and undergoes constitutive CME (clathrin-mediated endocytosis)-mediated internalization (2) or ligand-induced CIE (clathrin-independent endocytosis)-mediated internalization (3). The endocytosed BRI1 at the TGN/EE could be packaged into ILVs (intraluminal vesicles) and delivered to the vacuole for degradation via ESCRT (endosomal sorting complex required for transport)-mediated biogenesis of MVBs (multivesicular bodies) and eventual MVB-vacuole fusion (4). Alternatively, the TGN/EE-localized BRI1 can be recycled back to the PM via retromer-mediated cargo selection, microtubule-assisted vesicle trafficking, and exocyst/SNARE (soluble N-ethylmaleimide-sensitive factor-attachment protein receptor)-involved exocytosis (5). The circled numbers indicate different secretion/trafficking routes. SNX1 (sorting nexin1) is one of the core retromer subunits, CLASP (cytoplasmic linker-associated protein) is a microtubule-associated protein, and ALIX (the Arabidopsis homolog of apoptosis-linked gene 2-interacting protein X) is a cytosolic ESCRT-associated protein.

2.4.2. Targeting to Unique PM Nano/Microdomains

Recent fluorescence microscopy studies coupled with live-cell imaging techniques have revealed that plant LRR-RLKs, including BRI1 and BAK1, are not uniformly distributed on the PM but are rather localized to specific PM subcompartments known as lipid rafts, nanodomains, or microdomains with unique lipid and protein compositions [48,94,95]. It was thought that members of the two protein families: Flotillins (Flot) and plant-specific remorins, might function as organization centers to form protein nanoclusters that are important for receptor signaling [55,95]. Interestingly, AtFlot1, one of the two Arabidopsis Flot homologs, was shown to coexist with BRI1 in a PM microdomain to influence BRI1 endocytosis [55] while OsREM4.1 (*Oryza sativa* remorin4.1), a member of the rice remorin family, interacted with OsSERK1, a rice homolog of BAK1, to regulate the heterodimerization of OsSERK1 with a rice homolog of BRI1, OsBRI1 [96]. It remains to be determined how AtFlot1 and OsREM4.1 recruit BRI1/OsBRI1 to unique nano/microdomains on the PM given the presence of >600 RLKs, but only 2 Flots and 16 remorins in Arabidopsis [97,98].

2.4.3. Endocytosis

Like many PM-localized proteins, the wild-type BRI1 is also known to dynamically undergo clathrin-mediated endocytosis (CME) or clathrin-independent endocytosis (CIE) from the PM to the trans-Golgi network or early endosomes (TGN/EE) [15,55,99], where BRI1 can then be recycled back to the PM or sorted into the vacuole for degradation via multiple multimeric ESCRT (endosomal sorting complex required for transport) complexes [100–104]. Thus, BRI1 endocytosis serves to attenuate BR signaling despite an early report that suggested endosomal initiation of BR signaling [99,105]. CME is a highly conserved cellular process that requires coordination of several groups of proteins, including clathrin triskelia consisting of clathrin heavy chains and light chains, small GTPases and their regulators such as ADP-ribosylation factor-guanine nucleotide exchange factors (ARF-GEFs), and adapter complexes, including the canonical heterotetrameric adapter protein 2 (AP2)-complex and the heterooctomeric TPLATE complex (TPC) [106]. Consistent with this, inhibition of BRI1 endocytosis by tyrphostin A23, which is a widely-used chemical that specifically blocks the cargo recruitment step of CME [107], could inhibit BRI1 endocytosis and enhance BR signaling [55,104]. Similarly, genetic mutations or transgenic interferences (dominant-negative and/or gene silencing) of the CME components, including clathrins, Rab GTPases, two ARF-GEFs (GNOM and GNOM-LIKE1), the AP2-complex, and the TPC complex, impaired BRI1 endocytosis, leading to increased BRI1 abundance at the PM and enhanced BR signaling [104,108,109]. It was generally thought that the CME-mediated BRI1 internalization is a constitutive process; however, a recent study suggested that BRI1 internalization could be stimulated by BR treatment, which is mediated by a CIE pathway involving a membrane microdomain-associated protein AtFlot1 (*Arabidopsis thaliana* Flot1 [110]. AtFlot1 directly associates with BRI1 and interference with AtFlot1 protein affects BRI1 endocytosis, leading to increased BRI1 on the PM and enhanced BR signaling [55]. It remains to be determined to what degree coordination of the CME and CIE pathways regulate BRI1 endocytosis and downstream BR signaling.

2.4.4. The Endocytic Pathway to the Vacuole for Degradation

Endocytosed BRI1 proteins accumulate in the TGN/EE compartments where they are sorted into the vacuole for degradation, an endocytic pathway that is mediated through sorting/packaging of BRI1 into ILVs (intraluminal vesicles) of the LE/MVBs (late endosomes/multivesicular bodies) and eventual MVB-vacuole fusion [111]. Previous studies suggested that BRI1 endocytosis and its subsequent sorting at TGN/EE into MVBs is likely regulated by the E3 ligases PUB12/13 (plant U-box protein12/13)-catalyzed ubiquitination of BRI1 [112] while its packaging into MVBs is regulated by ESCRT protein complexes and their associated proteins [113]. Interestingly, BR treatment stimulated the BRI1-PUB13 interaction and PUB13 phosphorylation and simultaneous elimination of PUB12 and PUB13 inhibited BRI1 internalization, leading to increased BRI1 abundance and enhanced BR sensitivity [112]. It was known that K63-linked polyubiquitination is involved in endocytosis and subsequent ESCRT-mediated vacuolar delivery, whereas K48-linked polyubiquitination is the signal for proteasome-mediated degradation [114,115]. A previous study indicated that PUB12/13-catalyzed ubiquitination of FLS2 (Flagellin Sensing2), another LRR-RLK plant immunity receptor that senses bacterial flagellin [116], is involved in proteasome-mediated FLS2 degradation [117], implying that PUB12/13 likely catalyzes the K48-linked polyubiquitination of FLS2. Thus, it will be interesting to determine the exact type of the ubiquitin-linkage and the exact site(s) of the PUB12/13-catalyzed BRI1 ubiquitination given an earlier report indicating that the K63-linked polyubiquitination at Lys^{866} of an immunoprecipitated BRI1 is the likely signal to drive the BRI1 endocytosis [115]. Experiments are also needed to fully understand the discrepancy between the BR-dependent BRI1-PUB12/13 interaction [112] and the kinase-dependent but ligand-independent BRI1 ubiquitination [115].

Interestingly, a partial loss-of-function mutation in Arabidopsis ALIX (apoptosis-linked gene 2-interacting protein X), which is required for localizing an Arabidopsis deubiquitinating enzyme in LE/MVBs and associates with an ESCRT complex to mediate packaging of cargos into ILVs [118], was found to be defective in the vacuolar delivery of BRI1 [113]. It remains unknown whether the failure to package endocytosed BRI1 into ILVs is functionally related to the mislocalization of the deubiquitinating enzyme in the *alix* mutant. A recent study also implicated BIL4 (Brassinazole-Insensitive-Long hypocotyl4), a 7-transmembrane protein localized in the TGN/EE, LE/MVB, and vacuolar membrane, in regulating the endocytic trafficking of BRI1 to the vacuole [119]. RNAi-mediated *BIL4* silencing resulted in increased BRI1 localization in the vacuole, whereas *BIL4* overexpression reduced the BRI1 trafficking from the TGN/EE to the vacuole. It remains to be investigated to fully understand the biochemical function of BIL4 in the plant endocytic pathway.

2.4.5. Endocytic Recycling

Most of the endocytosed BRI1 are thought to be recycled back to the PM to replenish the PM pool of the BR receptor, thus enhancing BR signaling [99,101,102,104]. It was thought that the TGN/EE-PM recycling involves retromer complex-mediated cargo retrieval, microtubule-assisted trafficking of recycling endosomes, and SNARE (soluble N-ethylmaleimide-sensitive factor attachment proteins receptor)-mediated exocytosis [111]. A hypomorphic allele of *DET3* that encodes the cytosolic C subunit of the Arabidopsis vacuolar ATPase (V-ATPase) reduces the ability of V-ATPase to acidify the TGN/EE but not the Golgi or the vacuole, leading to compromised secretion and recycling of BRI1 and reduced BR sensitivity [120,121]. An Arabidopsis microtubule-associated protein CLASP (cytoplasmic linker protein-associated protein) was previously known to interact with SNX1 (sorting nexin1), a component of the Arabidopsis retromer complex, to mediate the endosome-microtubule association [122]. Interestingly, the expression of CLASP is regulated by BR in a BRI1-dependent manner and CLASP also mediates the BR-induced microtubule reorganization [123]. Importantly, a loss-of-function mutation in CLASP compromised the TGN/EE-PM trafficking of the constitutive cycle of BRI1 endocytosis-exocytosis, leading to reduced BRI1 abundance on the PM and dampened BR sensitivity [123]. The exocytosis is known to be coordinated by Rab GTPase, the vesicle tethering complex known as exocyst, and SNAREs [124]. Mutations in EXO70A1 (exocyst subunit 70A1, a component of the Arabidopsis exocyst complex) or components of the SNARE complex reduced BRI1 recycling back to the PM [125,126]. A recent study also implicated BIG3 (brefeldin A-inhibited guanine nucleotide-exchange protein3) and BIG5, two members of the BIG subfamily of the Arabidopsis ARF-GEFs, in BRI1 recycling. Simultaneous elimination of BIG3 and BIG5 resulted in dwarfed plants with reduced BR sensitivity [127], but their exact cellular functions remain to be defined. Given the widely accepted model of constitutive endocytosis of BRI1, it will be interesting to investigate how plant cells integrate developmental cues and environmental signals to balance the endocytic vacuolar degradation and the endocytic recycling process to control the abundance of signaling competent BRI1 on the PM.

3. BAK1, the Coreceptor

BAK1/SERK3 and three other members of the SERK family, SERK1, SERK4, and SERK5 (nonfunction in the Col-0 ecotype of *Arabidopsis thaliana* but remains function in the L*er*-0 ecotype [128]), are required to function as the BRI1 coreceptor to initiate the BR signaling at the PM [14,18,129,130].

The 5 members of the SERK family share the same structural organization with an extracellular domain of 5 LRRs plus the N-terminal cap, a single transmembrane helix, a cytoplasmic kinase domain, and a CT with a conserved Ser-Gly-Pro-Arg motif at their C-terminal end known to be important for their kinase activity [131]. Structural analysis of the BRI1-BAK1 heterodimer of their extracellular domains identified key residues involved in forming and stabilizing the BAK1-BRI1 dimer [19]. In addition to functioning as the coreceptors for BRI1 to activate the ligand-bound BRI1, BAK1/SERKs were found to be versatile coreceptors that heterodimerize with many ligand-bound RLKs for their activation [17] and intracellular signal transduction.

3.1. Phosphorylation of BAK1

As discussed above, BAK1 activation requires its heterodimerization with BRI1 when assayed with *E. coli* expressed kinase domains of BRI1 and BAK1 [45] or with yeast expressed full-length proteins [18]. An earlier transgenic experiment, which expressed a BAK1-GFP fusion protein in a strong *bri1-1* mutant background, showed that exogenous BR application resulted in no detectable change in BAK1 phosphorylation [21], suggesting that BAK1/SERKs activation in vivo also requires its heterodimerization with BRI1.

MS analyses of *E. coli*-expressed recombinant BAK1 kinase, in vitro phosphorylated forms, and immunoprecipitated BAK1 fusion proteins from transgenic Arabidopsis plants identified a total of 23 phosphorylation sites in both N-lobe and C-lobe (Figure 4). The phosphorylated residues in the N-lobe include Ser^{286} of the αB-helix, Ser^{290} of the αB-β1 loop, Thr^{312} of the β2-β3 loop, Ser^{324} within the β3-αC loop (that is missing in the BAK1 kinase structure), Thr^{333} and Ser^{339} of the αC-helix, and Thr^{355} and Thr^{357} of the β4-β5 loop. The phosphorylated residues of the C-lobe include Ser^{370} and Ser^{373} on the αD-helix, Ser^{381} of the αD-αE loop, Tyr^{443}, Thr^{446}, Thr^{449}, Thr^{450}, and Thr^{455} on the activation segment, Ser^{465} and Thr^{466} located on a short α-helix that links the P+1 loop with the αF helix, Ser^{557} of the αI-helix, and 4 residues (Ser^{602}, Thr^{603}, Ser^{604}, and Ser^{612}) in the CT [21,59,65,130,132,133] (Figure 4). Among these phosphorylation sites, a total of 9 residues, including Ser^{290}, Thr^{312}, Thr^{446}, Thr^{449}, Thr^{455}, and the 4 Ser/Thr residues in the CT, were discovered from immunoprecipitated BAK1 fusion proteins of transgenic Arabidopsis plants [21,134]. Whereas no single pTyr residue of BRI1 was detected by MS, a pTyr residue, Tyr^{443} located at the tip of the activation loop (Figure 4), was identified from MS analysis of an *E. coli* expressed recombinant BAK1 kinase [65]. Two additional Tyr residues, Tyr^{463} (located at a short α-helix that links the P+1 loop with the αF-helix) (Figure 4) and Tyr^{610} (on the CT that is not shown on the BAK1 kinase crystal structure), were previously shown to be phosphorylated based on immunoblotting analysis with anti-pTyr antibodies and loss-of-phosphorylation (Tyr-Phe) mutagenesis [73,135].

Structural analysis and molecular dynamics simulations suggested that the Thr^{450} phosphorylation is likely the most critical for BAK1 activation as its attached phosphate group likely binds to a positively-charged phosphate-binding pocket, which consists of Arg^{415} (within the HRD motif), Lys^{439} (on the β9-strand), and Arg^{453} (on the activation segment), to stabilize the activation segment (Figure 4). It is interesting to note that the third Arg residue of the BAK1's phosphate binding pocket is from the activation segment itself instead the αC-helix that contributes the third Arg residue for the phosphate-binding pocket of the BRI1 kinase (Figure 2). The interaction between Arg^{922} with the phosphorylated Ser^{1044} is important to stabilize the N-lobe-C-lobe interaction in BRI1 to assemble the R-spine. It remains to be determined whether this is the likely cause for the apparent failure to incorporate a hydrophobic residue (likely Ile^{338}) from the αC-helix to the R-spine of BAK1 and an apparent gap between the Lys^{317} (from the β3-strand) and Glu^{334} (of the αC-helix) that should form the conserved salt bridge important for the kinase activity. Consistent with the structural analysis, Thr^{450} was found to be essential for its autophosphorylation and transphosphorylation activities when assayed in vitro [59,132,136]. However, these results were contradictory to an earlier study demonstrating that the Thr^{450}-Ala mutation only marginally affected the in vitro phosphorylation and in vivo signaling activities of BAK1 [21]. Further studies are needed to fully establish if the transphosphorylation of

BAK1 by BRI1 at the Thr[450] residue is absolutely required to activate the BAK1 kinase activity in vivo. It is also interesting to investigate if the gatekeeper residue Tyr[363] might contribute to the assembly of the R-spine crucial for BAK1 activation (Figure 4).

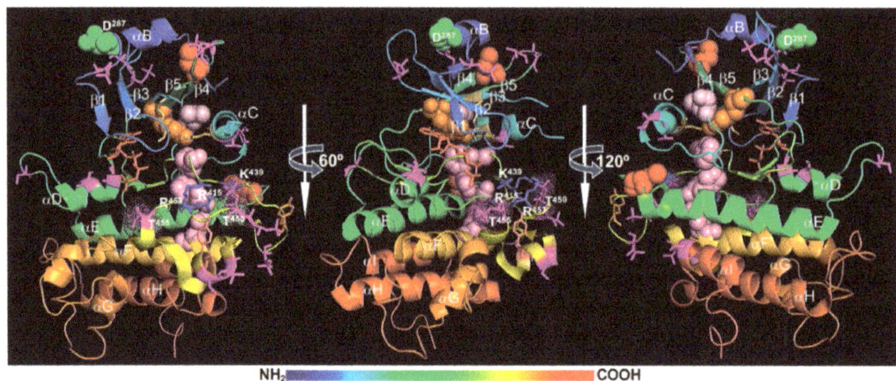

Figure 4. A crystal structure of the BAK1 kinase domain. Shown here are three rotation ($0°$, $60°$, and $180°$) views of a rainbow-colored ribbon model of the crystal structure of the BAK1 kinase domain (Protein Data Bank No: 3uim). Individual α-helices (αB-αI) and β-strands (β1-β5) were labelled. The purple sticks indicate phosphorylated Ser/Thr residues with the Thr[450] and Thr[455] residues surrounded with purple dots. The red sticks show the bound adenylyl-imidodiphosphate (a non-hydrolysable ATP analog), the light-orange sticks denote the two residues that form the conserved salt bridge between the β3-strand and the αC-helix, the blue sticks represent the three positively charged residues that make up the phosphate-binding pocket, and the dark-orange sticks indicate phosphorylated Tyr residues. The pink spheres mark the R-spine residues, the red spheres show the two Cys residues that were S-glutathionylated in vitro, the orange spheres denote the gatekeeper Tyr residue, and the green spheres designate the Asp[287] residue important for Ca^{2+}-dependent BAK1 cleavage. The rainbow bar indicates the order of AAs from the N-terminus (blue) to the C-terminus (red).

Similar to what was discovered with the BRI1 kinase domain, the Thr residue in the P+1 loop, Thr[455], is essential for the BAK1 activity. Mutating Thr[455] to Ala or Asp/Glu resulted in a strong loss of BAK1's phosphorylation activity in vitro and its BR-signaling function in transgenic Arabidopsis plants [21,59,132], suggesting that autophosphorylation of this highly conserved Ser/Thr residue might serve to inhibit rather activate BAK1 activity. Similar to what was discussed for the BRI1's Thr[1049] residue, the phosphorylation of Thr[455] could interfere with ATP binding or substrate binding due to its strategic position at the P+1 loop. However, it remains a possibility that the phosphorylation of Thr[455] is required for BAK1 activation but the Thr-Asp/Glu mutation might not be able to mimic the pThr[455] residue at this strategic position. Site-directed mutagenesis coupled with in vitro phosphorylation assays and Arabidopsis transgenic experiments revealed another Ser/Thr residue whose phosphorylation serves to attenuate the BAK1 activity. Mutating Ser[286] to Ala, which is located at the αB-helix capping the N-lobe, had little impact on the BAK1 activity; however, mutating Ser[286] to Asp resulted in almost complete loss of in vitro phosphorylation activity of BAK1 and caused a dominant negative effect in transgenic Arabidopsis plants [21]. The impact of pSer[286] on BAK1 activity is very similar to that of pThr[872] on BRI1, which is also localized on the αB-helix that caps the BRI1's N-lobe [22]. Further studies are needed to fully appreciate the negative impact of phosphorylating the Ser/Thr residues of the αB-helix. It is equally important to determine whether these Ser/Thr residues are intramolecularly autophosphorylated or intermolecularly transphosphorylated in order to fully understand the activation or autoinhibitory mechanisms of BAK1/SERKs and other LRR-RLKs.

In addition to the "*cis*" attenuation mechanism via auto/trans-phosphorylation of Ser^{286}/Thr^{455}, the BAK1 signaling activity could also be attenuated by a "*trans*" regulatory mechanism involving a phosphatase. A recent study suggested a role of a PP2A (consisting of subunits A1, C4, and B'η/ζ) in regulating the phosphorylation status of BAK1 in plant immunity response [137]. Given the demonstrated constitutive BAK1-PP2A association [137] and implication of PP2A in regulating BRI1 phosphorylation status [66], it is quite possible that this PP2A or its close homolog(s) could negatively influence BR signaling by controlling the phosphorylation level of the BRI1-associated BAK1/SERKs.

3.2. Regulating BAK1 Availability for BRI1 Interaction

Given the importance of BAK1/SERKs in activating the BR-bound BRI1 [49], it is no surprise that plant cells could regulate the availability of BAK1/SERKs to control BR signaling. The ligand-independent/dependent heterodimerization of BAK1 with BRI1 was previously known to induce endocytosis of both LRR-RLKs [15]. An Arabidopsis protein, MSBP1 (membrane steroid binding protein1) was previously found to inhibit BR signaling by stimulating BAK1 endocytosis to limit the amount of BAK1 on the PM [138]. Recently, several members of a small subfamily of LRR-RLKs known as BIRs were found to constitutively interact with BAK1/SERKs, thus interfering the signal initiation processes of many BAK1/SERKs-required LRR-RLKs [50,60,139–143]. However, only the BIR3-BAK1 interaction affects the BR-induced BRI1-BAK1/SERK heterodimerization and inhibits BR signaling [50]. A previously known gain-of-function mutant of BAK1, *elg1-D* (*elongated1-Dominant*, carrying an Asp^{122}-Asn mutation) that was previously identified as a suppressor of an Arabidopsis gibberellin-deficient dwarf mutant [144,145], exhibits a much weaker binding affinity to the extracellular domain of BIR3 [143], thus increasing availability of BAK1 for BRI1 interaction to enhance BR signaling. However, it remains to be studied why overexpression of BIR3 led to a strong *bri1*-like dwarf phenotype with complete BR insensitivity but its loss-of-function *bir3* mutation enhanced the dwarf phenotype of a weak *bri1* mutant and exhibited no morphological similarity to the *elg1-D* mutant or weak phenotype of BRI1-overexpression. Further studies are also needed to fully understand why BIR3, but not its two other homologs, BIR1 and BIR2, can inhibit BR signaling. One possible explanation is its ability to interact with BRI1 and form ligand-independent BRI1-BAK1/SERKs-BIR3 receptor nanoclusters [60,95]. In addition to BIRs, members of another small subfamily of LRR-RLK [LRR-RLK IX subfamily, named BAK1-Associated Receptor-like Kinase1 (BARK1) and BARK1-Like Kinase 1-3 (BLK1-3)], could also bind BAK1/SERKs to influence BR signaling [146]. However, overexpression of BARK1 resulted in a hypersensitive phenotype, suggesting a potentially positive role of BARK1 in BR signaling. Unlike BIR3 that likely carries a cytoplasmic pseudokinase domain [50,147], the kinase domain of BARK1 is predicted to be an active kinase that could potentially help to phosphorylate and activate BAK1/SERKs or downstream signaling components to enhance BR signaling. A recent rice study also implicated OsREM4.1, a member of the remorin family thought to be associated with micro/nanodomains on the PM [97], as an interactor of OsSERK1 to interfere with the OsSERK1-OsBRI1 heterodimerization [96], suggesting that competitive BAK1/SERK-binding is likely a conserved mechanism to control BR signal initiation on the cell surface.

3.3. BAK1 Regulation by Other Mechanisms

Increasing evidence suggested an important role of S-glutathionylation, a post-translational modification of a Cys residue via its disulfide linkage with the Cys residue of glutathione (a γ-Glu-Cys-Gly tripeptide) [148], in regulating protein stability and activity in response to cellular oxidative stress. A recent in vitro study showed that BAK1 could be S-glutathionylated at Cys^{353} and Cys^{408} (shown in red spheres in Figure 4) by AtGRXC2 (*Arabidopsis thaliana* glutaredoxin C2) via a thiol-dependent reaction with glutathione disulfide, leading to a reduction of the in vitro BAK1's kinase activity [149]. Molecular dynamics simulations suggested that S-glutathionylation of Cys^{408} promotes an inactive kinase conformation state [150]. The S-glutathionylation of Cys^{408} might directly affect the positioning of the αC-helix by steric hinderance or interfere with the interaction of the αC-β4 loop with the αE-helix, which is a crucial interaction that stabilizes the αC-helix to maintain an active kinase conformation [61]. It is interesting to note that Cys^{408} was mutated to Tyr in *bak1-5*, which compromises some plant innate immunity responses but has little effect on BR signaling [151]. It remains to be tested if BAK1/SERKs are S-glutathionylated in vivo and whether S-glutathionylation interferes their heterodimerization and transphosphorylation with BRI1 or other BAK1/SERKs-required LRR-RLKs and whether such a modification exhibits different impacts on BR signaling, plant immunity, and other BAK1/SERK-mediated processes.

A recent study suggested potential involvement of a Ca^{2+}-dependent BAK1 proteolytic cleavage process in BR-mediated plant development and growth [152]. Biochemical studies suggested that the Asp^{287} residue (shown in green spheres in Figure 4), which is conserved among SERK family members and located right after the important regulatory Ser^{286} residue [17], is critical for its proteolytic cleavage. However, functional verification of the cleavage process in BR signaling was complicated by retention of a mutated BAK1 (carrying Asp^{287}-Ala mutation) in the ER, most likely caused by ERQC-associated processes. This study suggested that considerable caution is needed in interpreting results with mutated transgenic constructs because some introduced mutations could potentially cause structural defects that could be detected by stringent quality control systems in plant cells, thus altering their protein abundance and/or subcellular locations.

One question that remains to be answered is whether or not BAK1/SERKs interact with downstream BR signaling components to directly influence the intracellular transmission of the extracellular BR signals in addition to their essential role of activating BRI1. Previous studies indicated that a C-terminally-tagged BAK1 or a *bak1* allele (known as *sobir7-1* for *suppressor of bir1 7-1*, carrying a non-sense mutation at Trp^{597} and lacking the CT) affected plant defense signaling but had little impact on BR signaling [131,153], leading to a conclusion that BAK1/SERKs require their CTs to interact with downstream signaling components to influence the plant immunity. It is interesting to note that the Arabidopsis BIK1 (Botrytis-Induced Kinase1), a receptor-like cytoplasmic kinase that is rapidly phosphorylated by FLS2-associated BAK1 and transduces the plant immunity signal to downstream targets [154], inhibits BR signaling by interacting with BRI1 but not the BRI1-interacting BAK1 [26]. Given the versatile roles of BAK1/SERKs in plant growth/development and plant defense [17], understanding the molecular mechanism(s) that determine the biochemical functions of BAK1/SERKs in BR signaling and FLS2/EFR-mediated plant defense responses will have a huge impact on plant biology.

4. BIN2, the Negative Regulator

The GSK3-like kinase BIN2 was originally discovered in a forward genetic screen for mutants similar to loss-of-function *bri1* mutants [30] and was subsequently shown to phosphorylate and negatively regulate BES1 and BZR1 [29,155,156] to block the intracellular transduction of the extracellular BR signals. The Arabidopsis genome encodes a total 10 GSK3-like kinases that are divided into four different subgroups [157], but genetic screens for BR-related dwarf mutants (*bin2* or *dwarf12*) [30,158] or leaf development mutants (*ucu* for *ultracurvata*) [159] discovered gain-of-function mutations only in BIN2 but not in any of the remaining 9 GSK3-like kinases. Interestingly, at least 7 of the 8 known *bin2/dwarf12/ucu1* mutants carry single AA changes in a 4-AA "TREE" (Thr^{261}Arg^{262}Glu^{263}Glu^{264}) motif that is a part of a surface-exposed α-helix (Figure 5) on the long connection fragment that links the αG and αH helices. However, subsequent reverse genetic studies showed that BIN2 functions redundantly with at least 6 other members of the Arabidopsis GSK3-like kinase family in regulating BR signaling [27,160–163].

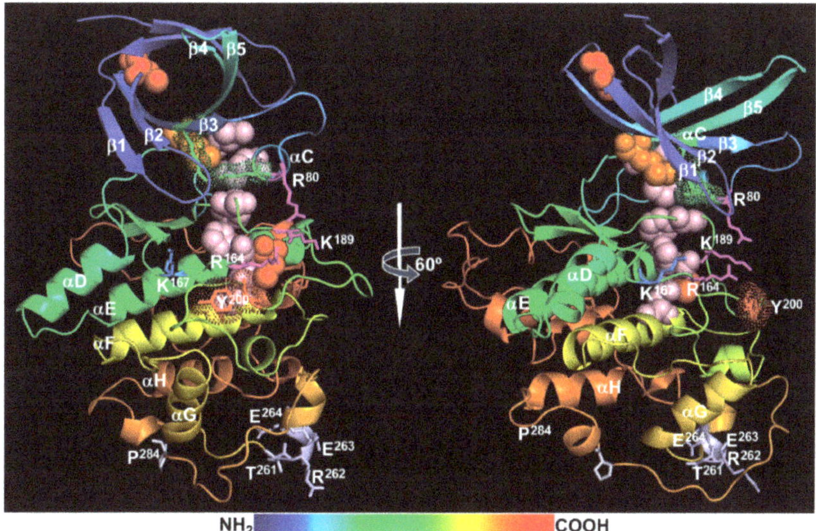

Figure 5. A structural model of BIN2. Shown here are two rotational (0° and 60°) views of a modeled BIN2 structure obtained at SWISS-MODEL (https://swissmodel.expasy.org/). Individual β-strands (β1-β5) and α-helices (αC-αH) were labeled. The two extra antiparallel-β-strands that cap the N-lobe are not labelled. The TREE motif and the Pro284 residue mutated in known *bin2/dwarf12/ucu1* alleles are indicated with grey-colored sticks. The magenta sticks show the three positively charged residues [Arg80, Arg164, and Lys189 (also acetylated)] of the conserved phosphate-binding pocket, the red sticks with red dots denote the Tyr200 residue, the green sticks with green dots mark the Lys69-Glu81 salt bridge between the β3-strand and the αC-helix, and the blue sticks represent the Lys167 residue that corresponds to the acetylated Lys183 residue of the human GSK3β. The red spheres show the Cys residues that were shown to be S-nitrosylated or S-glutathionylated in vitro, the orange spheres indicate the gatekeeper Met115 residue, and the pink spheres mark the 5 R-spine residues (from lower to top: Asp223, His163, Phe185, Met85, and Leu96). Important residues are also labelled with single letter codes with positions. The rainbow bar indicates the order of AAs from the N-terminus (blue) to the C-terminus (red).

4.1. BIN2 Regulation by Dephosphorylation

The signaling activity of BIN2 is likely regulated by post-translational modifications that include phosphorylation, acetylation, S-glutathionylation, S-nitrosylation, and ubiquitination. Like its animal homologs, BIN2 is a constitutively active kinase but is inactivated in response to BRI1/BAK1 activation. Molecular modeling revealed that BIN2 likely adopts an active kinase conformation with a well assembled R-spine consisting of Asp223 (the anchorage residue on the αF-helix), His163 (of the HRD motif), Phe185 (of the DFG motif), Met85 (of the αC-helix), and Leu96 (of the β4-strand), the Lys69-Glu81 salt bridge between the β3-strand and the αC-helix, and an opened activation segment (Figure 5). In animal cells, GSK3 phosphorylates its substrates via two different mechanisms: one requiring a priming phosphorylation of the substrate by a different kinase and the other requiring an adapter protein that binds both GSK3 and its substrate [164]. Accordingly, an animal GSK3 kinase can be inactivated by two general mechanisms: phosphorylation of a key Ser/Thr residue in its autoinhibitory N-terminal fragment and interaction of a GSK3-binding protein [164]. The phosphorylated N-terminal fragment competes with a "primed" substrate for the phosphate-binding pocket consisting of three positively-charged residues, Arg96, Arg180, and Lys205 in the human GSK3β (corresponding to Arg80, Arg164, and Lys189 in BIN2, respectively) (Figure 5), while a GSK3-binding protein competitively prevents the GSK3-substrate binding necessary to phosphorylate a non-primed GSK3 substrate.

There has been no evidence so far for a similar phosphorylation-mediated autoinhibitory mechanism in plant cells to inactivate BIN2 in response to BR. An earlier biochemical study showed that the BIN2-catalyzed BES1/BZR1 phosphorylation does not need a priming phosphorylation but instead requires a direct binding between BIN2 and BES1/BZR1 that carry a 12-AA docking motif [165]. However, phosphorylation is involved in BIN2 regulation, which is likely mediated by several protein Ser/Thr phosphatases (PSPs) instead of protein kinases [27,166]. It is widely believed that BIN2 is inhibited through dephosphorylating a phosphorylated Tyr residue (pTyr^{200}) in the activation segment by BSU1/BSLs [27]. The phosphorylation of the corresponding Tyr residue in the human GSK3 kinases was known to occur intramolecularly during HSP90-facilitated folding process [167,168], and peptides of BIN2 or its closest homologs containing the pTyr^{200} residue were identified previously by phosphoproteomic studies [169,170]. However, the requirement of the pTyr residue (pTyr^{216} in human GSK3β) for the GSK3 kinase activity remains controversial for animal GSK3 kinases [171]. An early study reported that the mammalian GSK3 kinases required the pTyr residue for their maximum enzymatic activities [172], but later studies showed that non-phosphorylated GSK3β adopted an active conformation [173] and mutating the Tyr residue to Phe had a marginal impact on the GSK3β kinase activity [174,175]. As expected from published GSK3 structures, structure modeling of BIN2 suggests that Tyr^{200} adopts an *anti*-conformation (away from the substrate binding site) to avoid steric clash with a substrate (Figure 5). However, mutating Tyr^{200} to Phe was shown to completely inactivate BIN2 in transgenic Arabidopsis plants [27], but the inhibitory impact of the Tyr-Phe mutation could be caused by lacking of a hydroxyl group or a phosphate group on the aromatic ring as the Tyr residue is absolutely conserved among GSK3 and GSK3-like kinases [158]. Thus, further investigation is needed to confirm whether dephosphorylation of the pTyr^{200} residue is capable of complete inactivating BIN2.

The current BR signaling model suggests that the phosphate group of pTyr^{200} is removed by BSU1/BSLs [27], members of a plant-specific PPKL family [176]. BSU1, which was originally discovered as an activation-tagged *bri1* suppressor [177] but later found to be exclusively present in the Brassicaceae family and exclusively expressed in pollen with yet unknown physiological function [178], was predicted and shown to be a functional PSP [176,177]. Consistently, molecular modeling revealed that BSU1 has the conserved PSP structure of an α/β fold with a β-sandwich surrounded by a large and a small α-helical domains (Figure 6) and contains three catalytic signature motifs, GDXHG (GlyAsp^{510}IleHis^{511}Gly in BSU1), GDXVDRG (GlyAsp^{544}TyrValAspArgGly in BSU1), and GNHE (GlyAsn^{576}HisGlu in BSU1), plus two conserved His residues (His^{629} and His^{707} in BSU1). It was well established that these conserved residues coordinate two metal ions that bind and activate a water molecule for its nucleophilic attack on the phosphate ester linkage with a Ser/Thr residue [179]. These signature motifs are quite different from the HCX$_5$R motif (Figure 6), the catalytic signature of PTP/DSPP with the Cys residue being the enzymatic nucleophile and Arg directly binding the phosphate group of pTyr/pSer/pThr [180]. Furthermore, a previous genetic study coupled with phylogeny/evolution analysis of BSU1/BSLs questioned a major BR signaling role of the pollen-expressed BSU1 and its three homologs whose loss of function mutations had a marginal impact on the in vivo BIN2 activity [178]. It is important to note that a role of BSU1/BSLs in BR signaling was supported by a recent study in rice that investigated genetic and biochemical interactions between a rice homolog of BSL1, qGL3 (quantitative trait locus regulating grain length3, also known as OsPPKL1), and OsGSK3 (a rice homolog of BIN2) [181]. However, the rice study did not test whether qGL3 was capable of dephosphorylating the conserved Tyr residue of OsGSK3. Given the crucial role of BIN2 and its homologs in regulating the intracellular transduction of the extracellular BR signals into the nucleus, further genetic, biochemistry, and structural studies are needed to fully understand the biochemical roles of pTyr^{200} and BSU1/BSLs in BR signaling and other relevant physiological processes. The results of these future studies will not only increase our knowledge of BR signaling, but will also significantly enhance our understating of GSK3 regulation in plants and catalytic mechanism of the plant specific PPKLs.

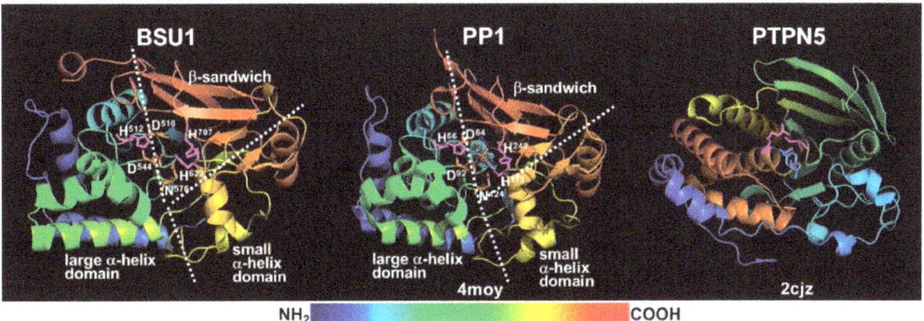

Figure 6. Comparison between a structure model of BSU1's phosphatase domain with crystal structures of PSP (protein Ser/Thr phosphatase) and PTP (protein tyrosine phosphatase). Shown here are a predicted structure of the BSU1's C-terminal phosphatase domain [obtained at SWISS-MODEL at https://swissmodel.expasy.org/ using the protein sequence of BSU1 (accession No: NP_171844)], a crystal structure of a human PSP (HsPP1; Protein Data Bank No. 4moy), and a crystal structure of a human PTP (HsPTPN5 for the human PTP non-receptor type 5; Protein Data Bank No. 2cjz). The conserved metal-coordinating amino acid residues are indicated by colored sticks: orange colored Asp(D)[510], Asp(D)[544], and Asn(N)[576], and three magenta-colored His(H) residues. The three dotted white lines separate the β-sandwich of two β-sheets from the two flanking α-helix domains in the structure models of BSU1 and PP1. The dotted spheres and the red sticks in PP1 represent two metal ions and the phosphate, respectively. The two conserved residues (Cys and Arg of the HCX_5R catalytic signature motif) are colored with magenta in the human PTPN5 structure and a phosphorylated tyrosine is shown with blue sticks. The rainbow bar indicates the order of AAs from the N-terminus (blue) to the C-terminus (red) of each peptide.

BIN2 could also be regulated through dephosphorylation by ABI1 (ABA-Insensitive1) and ABI2 [166], two protein phosphatase 2C-type PSPs known to play inhibitory roles in transducing the signal of the plant stress hormone abscisic acid (ABA) [182]. While this discovery seems to support earlier findings that the ABA-BR antagonism is mediated by a biochemical event located between BRI1 activation and BIN2 inhibition of the BR signaling cascade [183], it remains to be determined which phosphorylated residue(s) are dephosphorylated by ABA1/2. Quantitative analysis of the in vivo BIN2 phosphorylation dynamics in response to ABA and/or BR treatment coupled with structural analysis of BIN2 will not only lead to a better understanding of the well-known ABA-BR antagonism but will also provide biochemical insights into the regulatory mechanism(s) of plant GSK3-like kinases.

4.2. BIN2 Regulation by Other Post-Translational Modifications

In additional to phosphorylation/dephosphorylation, BIN2 can be regulated by other post-translational modifications. A recent study implicated a role of an Arabidopsis histone deacetylase HDA6 (histone deacetylase6) in BIN2 inactivation [184]. HDA6 interacts with BIN2 and deacetylates BIN2 at Lys[189] in vitro. Because Lys[189] is one of the positively charged residues that make up the highly conserved phosphate-binding pocket (Figure 5), its acetylation should neutralize its positive charge and likely reduce the BIN2 activity whereas deacetylation of the acetylated Lys[189] is expected to increase the BIN2 activity via enhanced BIN2 binding with BES1/BZR1 that has multiple conserved (Ser/Thr)X₃(Ser/Thr) GSK3 phosphorylation repeats. Consistent with the published GSK3 structures and the role of the conserved phosphate binding pocket [173,185], acetylation of Lys[205] in the human GSK3β (corresponding to Lys[189] of BIN2) was found to reduce its phosphorylation activity while a histone deacetylase sirtuin1 (Sirt1) and the Lys[205]-Arg mutation increased the GSK3β activity [186]. By contrast, the Lys[189]-Arg mutation, which should eliminate its presumed acetylation, actually reduced the in vitro kinase activity and in vivo signaling function of BIN2 [184], suggesting that acetylation of Lys[189] stimulates BIN2 activity via a yet to be defined biochemical mechanism in plants. However, the mutagenesis results should be interpreted with caution as previous sequence analyses showed that Lys[189] is absolutely conserved between BIN2 and its homologs from other plants or other eukaryotic organisms [30,158]. By contrast, both Arg and Lys were found at the corresponding position in plant LRR-RLKs [9,187]. It is noteworthy that acetylation at another conserved Lys residue, Lys[183] (corresponding to Lys[167] in BIN2) known to be important for ATP binding was recently shown to inhibit the mammalian GSK3β activity, whereas its deacetylation by another histone deacetylase Sirt7 served to promote ATP binding and to stimulate GSK3 activity [188]. Further studies, such as mapping the in vivo BIN2 acetylation sites and identification of BIN2 acetylation enzyme(s), are needed to fully understand the role of acetylation and deacetylation in BIN2 regulation.

Nitric oxide (NO), an important signaling molecular, was shown to inhibit the kinase activity of BIN2 by S-nitrosylation at the conserved Cys[162] site in an in vitro assay [189]. It is important to note that Cys[162] sits right before the conserved HRD motif (Figure 5) and its S-nitrosylation could thus interfere with the assembly of the R-spine (involving the His[163] residue) or the interaction between a phosphate group of BES1/BZR1 with the Arg[164] residue, one of the three positively charged residues of the BIN2's phosphate binding pocket (Figure 5). S-nitrosylation was recently shown to regulate the human GSK3β activity in a very interesting way that inhibits its phosphorylation of cytosolic substrates containing the (Ser/Thr)X$_3$(Ser/Thr) phosphorylation motif but promotes its nuclear localization to phosphorylate its nuclear substrates containing the (Ser/Thr)-Pro motif [190]. In addition to the NO-triggered S-nitrosylation of Cys residues, a recent study suggested that certain cysteine residues of BIN2, such as Cys[59] (sits at the end of the β2-strand) and Cys[162], could also be S-glutathionylated in vitro and are likely involved in the reactive oxygen species (ROS)-induced activation of BIN2 in vivo [191]. Further studies are needed to confirm these S-nitrosylation and S-glutathionylation sites in vivo and to investigate their biochemical/cellular impacts on the catalytic activity and subcellular localization of BIN2 and other plant GSK3-like kinases. It will also be interesting to determine if additional post-translational modifications, such as ribosylation, SUMOylation, and methylation that were implicated in regulating the mammalian GSK3 kinases [171], are also involved in regulating the in vivo BIN2 activity.

4.3. BIN2 Regulation by Protein–Protein Interactions

Given the earlier discovery of a direct BIN2-BES1/BZR1 binding [165] that was recently confirmed by a single-molecular analysis [191], it is no surprise that BIN2 is also regulated by its competitive binding with other cellular proteins in plant cells. Recent studies showed that BIN2 interacted in vivo with several well-studied light signaling components, including CRY1 (Cryptochrome1, a blue-light photoreceptor) [192], the COP1 (Constitutive Photomorphogenesis1)/SPA (Suppressor of *phyA-105*) complex (a light regulated E3 ubiquitin ligase) [193], and HY5 (long hypocotyl5, a bZIP-type transcriptional factor) [194]. CRY1 exhibits a blue light-dependent binding with both BIN2 and BZR1 to enhance the BIN2-catalyzed BZR1 phosphorylation [192], thus inhibiting its nuclear translocation (likely by increased interaction with 14-3-3 proteins [195]) and its DNA binding activity through CRY1's competitive binding to the BZR1's DNA binding domain. Thus, the blue light-activated CRY1 functions as an adapter to further promote the recruitment of BZR1, which carries a BIN2-binding motif near its C-terminus [165], to the BIN2 kinase. While the blue light-dependent CRY1-BIN2 interaction stimulates the BIN2-catalyzed BZR1 phosphorylation, COP1/SPA was recently found to inhibit the BIN2-catalyzed phosphorylation of PIF3 (phytochrome interacting factor3) [193]. COP1/SPA1 is a well-studied crucial photomorphogenesis repressor complex [196] that was previously shown to interact with CRY1 [197,198], while PIF3 functions together with its homologs to repress photomorphogenesis in the dark [199]. Despite being an E3 ubiquitin ligase [196], COP1/SPA did not promote BIN2 degradation to inhibit its PIF3 phosphorylation activity. Instead, the inhibitory effect of the COP1/SPA complex on the BIN2-catalyzed PIF3 phosphorylation is mediated by preventing the BIN2-PIF3 binding via competitive bindings of COP1/SPA with the kinase and the substrate [193]. Further experiments are needed to see whether these two studies actually revealed two general mechanisms to regulate the phosphorylation activities of BIN2 and other plant GSK3-like kinases: one using an adapter protein that facilitates the BIN2-substrate phosphorylation while the other relying on a disrupter that interferes the BIN2-substrate binding.

A more interesting discovery on the impact of protein–protein interaction on the BIN2 phosphorylation activity came from a recent study [194] that investigated the genetic and biochemical interactions between BIN2 and HY5, a bZIP-type transcription factor with versatile roles in plant growth and development [200] and a known substrate of the COP1/SPA E3 ligase [201]. Biochemical experiments coupled with molecular modeling and computational simulations suggested that the HY5-BIN2 binding stimulated the BIN2 catalytic activity assayed by a HY5-stimulated increase in pTyr^{200} signal [194], suggesting that HY5 might function similarly as the mammalian HSP90 known to facilitate GSK3β folding during which its Tyr^{206} (the equivalent of BIN2's Tyr^{200} residue) was intramolecularly autophosphorylated [167]. Further experimentation is needed to confirm its hypothetic "chaperone-like" function to assist BIN2 folding into an active kinase in vivo.

4.4. BIN2 Regulation by Subcellular Localization

Given that many of the known BIN2 substrates are transcriptional factors [157], it is no surprise that differential subcellular localization is an important regulatory mechanism to control the BIN2 phosphorylation activity in vivo. An earlier study with the GFP-tagged BIN2 or the mutant bin2-1 [carrying the Glu^{264}-Lys mutation in the TREE motif (Figure 5)] revealed that the bin2-1 protein mainly accumulated in the nucleus while the wild-type BIN2 seemed to exhibit more or less equal distributions at the PM, in the cytosol and the nucleus [163]. More importantly, adding a nuclear localization signal (NLS) to the wild-type BIN2 resulted in a stronger BIN2 activity to block BR signaling, whereas fusing a nuclear export signal (NES) greatly reduced its BR signaling-inhibition activity. Consistent with these findings, Arabidopsis OPS (OCTOPUS), a PM-associated protein crucial for phloem differentiation in Arabidopsis roots [202], binds and recruits BIN2 to the PM, thus reducing the BIN2 phosphorylation activity towards BES1/BZR1 and enhancing BR signaling to promote phloem differentiation [203]. Similarly, a stomata lineage cell-expressed POLAR (polar localization during asymmetric division and redistribution) protein [204], interacts with BIN2 and several other GSK3-like kinases to prevent their nuclear localization and transiently polarizes their subcellular distributions, thus driving asymmetric cell division essential for stomata development [205]. It is noteworthy that the rice PM-associated protein GW5 (grain width and weight5) interacts with OsGSK2, a rice homolog of BIN2, to inhibit the OsGSK2's phosphorylation activity, presumably by recruiting OsGSK2 to the PM, and to enhance BR signaling that regulates grain width and weight in rice [206]. The regulation of BIN2/GSKs via differential subcellular localization was also discovered in *Sorghum bicolor*, which uses DW1 (Dwarfing1), a PM/cytosol-localized protein, to interact with SbBIN2 (a *Sorghum bicolor* BIN2 homolog), thus preventing the nuclear localization of SbBIN2 and increasing BR signaling [207]. It will be interesting to know if their Arabidopsis homologs can also bind BIN2/GSK3s to enhance BR signaling. More importantly, further studies are needed to understand how these four distinct proteins with little sequence homology interact with BIN2/GSK3s to regulate their subcellular distributions. Careful biochemical analyses of their association with BIN2/GSK3s could define the minimum binding motifs while structural biology studies might reveal similar docking mechanisms to allow their BIN2/GSK3-binding.

In contrast to these four proteins that enhance BR signaling by preventing nuclear localization of BIN2/GSK3s, at least two members of the Arabidopsis HSP90 family whose mammalian homologs were known to assist the folding and activation of GSK3β [167], were shown to keep BIN2 inside the nucleus in an ATP-dependent manner to inhibit BR signaling [208]. Treatment with geldanamycin, a widely used inhibitor of HSP90-dependent ATPase, or active BR, resulted in nuclear export of the HSP90-BIN2 complexes [208]. Further studies are needed to determine 1) whether or not the Arabidopsis HSP90s are needed for the cytosolic folding and autoactivation of BIN2/GSK3s, 2) if the HSP90-bound BIN2 is still capable of phosphorylating its many substrates, and 3) how the extracellular BR signal is rapidly sensed by the nuclear-localized HSP90-BIN2 complexes that translocate into the cytosol [208].

4.5. BIN2 Inhibition by Degradation

Proteasome-mediated BIN2 degradation is another important mechanism to keep this negative regulator of BR signaling at low levels. An earlier study suggested that the *bin2-1* (Glu^{264}-Lys) mutation greatly stabilized the mutant kinase, which was partially responsible for its increased BES1/BZR1 phosphorylation activity [209]. A recent study demonstrated that the Arabidopsis KIB1, a Kelch-repeat-containing F-box E3 ubiquitin ligase that was identified as a genetic suppressor of a constitutively active form of BZR1, bzr1-1D, ubiquitinated BIN2 in vitro and stimulated the proteasome-mediated BIN2 degradation in vivo [28]. In addition to its role in BIN2 degradation, the Kelch repeat-mediated KIB1-BIN2 interaction blocked the BIN2-BZR1 binding, thus further reducing the BIN2 phosphorylation activity [28]. It remains to be determined whether proteasome-mediated degradation is an evolutionally conserved mechanism in other plant species to regulate the phosphorylation activity of BIN2 homologs given a recent discovery of restricted distribution of KIB1 and its homologs in the Brassicaceae family and their extremely low expression in vegetative tissues [210]. It is interesting to note that the proteasome-mediated GSK3 degradation was previously shown to be induced by glucocorticoid in mammalian cells although the identity of a GSK3-interacting E3 ubiquitin ligase remains unknown [211].

5. Conclusions and Remarks

The basic scheme of the BR signaling pathway is quite simple, involving a PM-localized BR receptor complex, a cytosolic/nuclear-localized inhibitor BIN2, and two key transcription factors, BES1 and BZR1. In the absence of BR, BIN2 is constitutively active to phosphorylate and inhibit BES1/BZR1. BR binding to BRI1 triggers its conformational changes to allow stable heterodimerization, transphosphorylation, and activation of BRI1 and BAK1, leading to inhibition of BIN2 and nuclear accumulation of non/de-phosphorylated BES1/BZR1. These BES1/BZR1 proteins bind to their target promoters to control expression of thousands of genes important for plant growth, development, and stress tolerance. Studies in the last twenty years have provided molecular understanding of key BR signaling events and discovered a wide range of biochemical and cellular mechanisms that regulate the abundance, subcellular locations, and biochemical activities of these key signaling components, which have dramatically enhanced our understanding of BR signaling processes. Despite these great achievements, there are still outstanding questions and unresolved controversies (discussed throughout this review) that require further investigation. Emerging technologies and novel experimental approaches will not only help to answer questions and resolve controversies, but will also lead to new discoveries that will provide a high-resolution atomic portrait of the BR signaling events to further our understanding of the BR signaling pathway and its interactions with other plant signaling processes.

Author Contributions: The authors confirm their contributions to this work: J.M. in conceptualization, original draft preparation, and writing; and J.L. in writing, review, and editing. All authors have read and agreed to the published version of the manuscript.

Funding: This work was partially supported by grants from the National Natural Science Foundation of China (No. 31900176 to Juan Mao and No. 31870253 to Jianming Li).

Acknowledgments: We thank Erik Nielsen (University of Michigan) for careful editing of the revised manuscript.

Conflicts of Interest: The authors declare no conflict of interest.

Abbreviations

AA	amino acids
ABA	abscisic acid
ABI1/2	ABA-Insensitive1/2
ALIX	apoptosis-linked gene 2-interacting protein X
AP2	adapter protein 2
ARF-GEF	ADP-ribosylation factor-guanine nucleotide exchange factor
AtFlot1	*Arabidopsis thaliana* Flot1
BAK1	BRI1-Associated receptor Kinase1
BARK1	BAK1-Associated Receptor-like Kinase1
BES1	bri1-EMS suppressor1
BIG3/5	brefeldin A-inhibited guanine nucleotide-exchange protein3/5
BIK1	Botrytis-Induced Kinase1
BIL4	Brassinazole-Insensitive-Long hypocotyl4
BIN2	Brassinosteroid-Insensitive2
BIR1-3	BAK1-Interacting Receptor-like kinase1-3
BKI1	BRI1 Kinase Inhibitor1
BLK1-3	BARK1-Like Kinase1-3
BR	Brassinosteroid
BRI1	Brassinosteroid-Insensitive1
BRL1-3	BRI1-Like1-3
BRRE	BR-response element
BSK1/3	BR-Signaling Kinase1/3
BSL1-3	BSU1-Like1-3
BSU1	bri1 suppressor1
BZR1	Brassinazole-Resistant1
CDG1	Constitutive Differential Growth1
CIE	clathrin-independent endocytosis
CLASP	cytoplasmic linker-associated protein
CME	clathrin-mediated endocytosis
COP1	Constitutive Photomorphogenesis1
CRY1	Cryptochrome1
CT	C-terminal tail
det2/DET3	*de-etiolated2/3*
DW1	Dwarfing1
EFR	Elongation Factor-Tu Receptor
elg1-D	*elongated 1-Dominant*
ER	endoplasmic reticulum
ERAD	ER-associated degradation
ERQC	ER-quality control
ESCRT	endosomal sorting complex required for transport
EXO70A1	exocyst subunit 70A1
FLS2	Flagellin Sensing2
G-loop	glycine-rich loop
AtGRXC2	*Arabidopsis thaliana* glutaredoxin C2
GSK3	Glycogen Synthase Kinase3
GW5	grain width and weight5
HDA6	histone deacetylase6
HSP90	heat shock protein 90
HsPTPN5	human PTP non-receptor type 5
HY5	long hypocotyl5
ID	island domain
ILVs	intraluminal vesicles
JM	juxtamembrane
KIB1	Kink suppressed in *bzr1-1D*1
LE/MVBs	late endosomes/multivesicular bodies

LRR-RLK	leucine-rich repeat receptor-like kinase
MS	mass spectrometry
MSBP1	membrane steroid binding protein1
NES	nuclear export signal
NLS	nuclear localization signal
NO	nitric oxide
OPS	OCTOPUS
OsGSK3	*Oryza sativa* GSK3-like kinase3
OsREM4.1	*Oryza sativa* Remorin 4.1
PIF3	Phytochrome-Interacting Factor3
PM	plasma membrane
POLAR	polar localization during asymmetric division and redistribution
PP2A	protein phosphatase 2A
PPKL	protein phosphatase with Kelch-like domains
PSP	protein Ser/Thr phosphatase
PTP/DSPP	protein tyrosine phosphatase/dual specificity protein phosphatase
pTyr	phosphorylated tyrosine residue
PUB12/13	plant U-box protein12/13
QC	quality control
qGL3	quantitative trait locus regulating grain length3
ROS	reactive oxygen species
R-spine	regulatory-spine
SbBIN2	*Sorghum bicolor* BIN2 homolog
SBI1	suppressor of bri1 1
SERK	Somatic Embryogenesis Receptor Kinase
Sirt1/7	Sirtuin1/7
SNARE	soluble N-ethylmaleimide-sensitive factor attachment proteins receptor
SNX1	sorting nexin1
sobir7-1	*suppressor of bir1 7-1*
SPA	Suppressor of *phyA-105*
TGN/EE	trans-Golgi network/early endosome
TPC	TPLATE complex
TTL	Tetratricopeptide Thioredoxin-Like
TWD1	Twisted Dwarf1
ucu	*ultracurvata*
V-ATPase	Vacuolar ATPase
VH1	Vascular Highway1

References

1. Clouse, S.D.; Sasse, J.M. Brassinosteroids: Essential regulators of plant growth and development. *Annu. Rev. Plant Physiol. Plant Mol. Biol.* **1998**, *49*, 427–451. [CrossRef] [PubMed]
2. Wang, Z.Y.; Bai, M.Y.; Oh, E.; Zhu, J.Y. Brassinosteroid signaling network and regulation of photomorphogenesis. *Annu. Rev. Genet.* **2012**, *46*, 701–724. [CrossRef] [PubMed]
3. Guo, H.; Li, L.; Aluru, M.; Aluru, S.; Yin, Y. Mechanisms and networks for brassinosteroid regulated gene expression. *Curr. Opin. Plant Biol.* **2013**, *16*, 545–553. [CrossRef] [PubMed]
4. Yang, C.J.; Zhang, C.; Lu, Y.N.; Jin, J.Q.; Wang, X.L. The mechanisms of brassinosteroids' action: From signal transduction to plant development. *Mol. Plant* **2011**, *4*, 588–600. [CrossRef]
5. Nawaz, F.; Naeem, M.; Zulfiqar, B.; Akram, A.; Ashraf, M.Y.; Raheel, M.; Shabbir, R.N.; Hussain, R.A.; Anwar, I.; Aurangzaib, M. Understanding brassinosteroid-regulated mechanisms to improve stress tolerance in plants: A critical review. *Environ. Sci. Pollut. Res. Int.* **2017**, *24*, 15959–15975. [CrossRef]
6. Saini, S.; Sharma, I.; Pati, P.K. Versatile roles of brassinosteroid in plants in the context of its homoeostasis, signaling and crosstalks. *Front Plant Sci.* **2015**, *6*, 950. [CrossRef]

7.   Li, Q.-F.; Lu, J.; Yu, J.-W.; Zhang, C.-Q.; He, J.-X.; Liu, Q.-Q. The brassinosteroid-regulated transcription factors BZR1/BES1 function as a coordinator in multisignal-regulated plant growth. *Biochim. Biophys. Acta Gene Regul. Mech.* **2018**, *1861*, 561–571. [CrossRef]

8.   Li, J.; Nagpal, P.; Vitart, V.; McMorris, T.C.; Chory, J. A role for brassinosteroids in light-dependent development of Arabidopsis. *Science* **1996**, *272*, 398–401. [CrossRef]

9.   Li, J.; Chory, J. A putative leucine-rich repeat receptor kinase involved in brassinosteroid signal transduction. *Cell* **1997**, *90*, 929–938. [CrossRef]

10.   Nolan, T.M.; Vukasinovic, N.; Liu, D.; Russinova, E.; Yin, Y. Brassinosteroids: Multidimensional regulators of plant growth, development, and stress responses. *Plant Cell* **2020**, *32*, 295–318. [CrossRef]

11.   Planas-Riverola, A.; Gupta, A.; Betegon-Putze, I.; Bosch, N.; Ibanes, M.; Cano-Delgado, A.I. Brassinosteroid signaling in plant development and adaptation to stress. *Development* **2019**, *146*, dev151894. [CrossRef] [PubMed]

12.   Belkhadir, Y.; Jaillais, Y. The molecular circuitry of brassinosteroid signaling. *New Phytol.* **2015**, *206*, 522–540. [CrossRef] [PubMed]

13.   Wang, Z.Y.; Seto, H.; Fujioka, S.; Yoshida, S.; Chory, J. BRI1 is a critical component of a plasma-membrane receptor for plant steroids. *Nature* **2001**, *410*, 380–383. [CrossRef] [PubMed]

14.   Li, J.; Wen, J.; Lease, K.A.; Doke, J.T.; Tax, F.E.; Walker, J.C. BAK1, an Arabidopsis LRR receptor-like protein kinase, interacts with BRI1 and modulates brassinosteroid signaling. *Cell* **2002**, *110*, 213–222. [CrossRef]

15.   Russinova, E.; Borst, J.-W.; Kwaaitaal, M.; Caño-Delgado, A.; Yin, Y.; Chory, J.; de Vries, S.C. Heterodimerization and endocytosis of Arabidopsis brassinosteroid receptors BRI1 and AtSERK3 (BAK1). *Plant Cell* **2004**, *16*, 3216–3229. [CrossRef]

16.   Santiago, J.; Henzler, C.; Hothorn, M. Molecular mechanism for plant steroid receptor activation by somatic embryogenesis co-receptor kinases. *Science* **2013**, *341*, 889–892. [CrossRef]

17.   Ma, X.; Xu, G.; He, P.; Shan, L. SERKing coreceptors for receptors. *Trends. Plant Sci.* **2016**, *21*, 1017–1033. [CrossRef]

18.   Nam, K.H.; Li, J. BRI1/BAK1, a receptor kinase pair mediating brassinosteroid signaling. *Cell* **2002**, *110*, 203–212. [CrossRef]

19.   Sun, Y.; Han, Z.; Tang, J.; Hu, Z.; Chai, C.; Zhou, B.; Chai, J. Structure reveals that BAK1 as a co-receptor recognizes the BRI1-bound brassinolide. *Cell Res.* **2013**, *23*, 1326–1329. [CrossRef]

20.   Wang, X.; Chory, J. Brassinosteroids regulate dissociation of BKI1, a negative regulator of BRI1 signaling, from the plasma membrane. *Science* **2006**, *313*, 1118–1122. [CrossRef]

21.   Wang, X.; Kota, U.; He, K.; Blackburn, K.; Li, J.; Goshe, M.B.; Huber, S.C.; Clouse, S.D. Sequential transphosphorylation of the BRI1/BAK1 receptor kinase complex impacts early events in brassinosteroid signaling. *Dev. Cell* **2008**, *15*, 220–235. [CrossRef] [PubMed]

22.   Wang, X.; Goshe, M.B.; Soderblom, E.J.; Phinney, B.S.; Kuchar, J.A.; Li, J.; Asami, T.; Yoshida, S.; Huber, S.C.; Clouse, S.D. Identification and functional analysis of in vivo phosphorylation sites of the Arabidopsis Brassinosteroid-Insensitive1 receptor kinase. *Plant Cell* **2005**, *17*, 1685–1703. [CrossRef] [PubMed]

23.   Wang, X.; Li, X.; Meisenhelder, J.; Hunter, T.; Yoshida, S.; Asami, T.; Chory, J. Autoregulation and homodimerization are involved in the activation of the plant steroid receptor BRI1. *Dev. Cell* **2005**, *8*, 855–865. [CrossRef] [PubMed]

24.   Tang, W.; Kim, T.W.; Oses-Prieto, J.A.; Sun, Y.; Deng, Z.; Zhu, S.; Wang, R.; Burlingame, A.L.; Wang, Z.Y. BSKs mediate signal transduction from the receptor kinase BRI1 in *Arabidopsis*. *Science* **2008**, *321*, 557–560. [CrossRef] [PubMed]

25.   Kim, T.W.; Guan, S.; Burlingame, A.L.; Wang, Z.Y. The CDG1 kinase mediates brassinosteroid signal transduction from BRI1 receptor kinase to BSU1 phosphatase and GSK3-like kinase BIN2. *Mol. Cell* **2011**, *43*, 561–571. [CrossRef]

26.   Lin, W.; Lu, D.; Gao, X.; Jiang, S.; Ma, X.; Wang, Z.; Mengiste, T.; He, P.; Shan, L. Inverse modulation of plant immune and brassinosteroid signaling pathways by the receptor-like cytoplasmic kinase BIK1. *Proc. Natl. Acad. Sci. USA* **2013**, *110*, 12114–12119. [CrossRef]

27.   Kim, T.W.; Guan, S.; Sun, Y.; Deng, Z.; Tang, W.; Shang, J.X.; Sun, Y.; Burlingame, A.L.; Wang, Z.Y. Brassinosteroid signal transduction from cell-surface receptor kinases to nuclear transcription factors. *Nat. Cell Biol.* **2009**, *11*, 1254–1260. [CrossRef]

28. Zhu, J.Y.; Li, Y.; Cao, D.M.; Yang, H.; Oh, E.; Bi, Y.; Zhu, S.; Wang, Z.Y. The F-box protein KIB1 mediates brassinosteroid-induced inactivation and degradation of GSK3-like kinases in *Arabidopsis*. *Mol. Cell* **2017**, *66*, 648–657. [CrossRef]

29. Yin, Y.; Wang, Z.Y.; Mora-Garcia, S.; Li, J.; Yoshida, S.; Asami, T.; Chory, J. BES1 accumulates in the nucleus in response to brassinosteroids to regulate gene expression and promote stem elongation. *Cell* **2002**, *109*, 181–191. [CrossRef]

30. Li, J.; Nam, K.H. Regulation of brassinosteroid signaling by a GSK3/SHAGGY-like kinase. *Science* **2002**, *295*, 1299–1301.

31. Tang, W.; Yuan, M.; Wang, R.; Yang, Y.; Wang, C.; Oses-Prieto, J.A.; Kim, T.W.; Zhou, H.W.; Deng, Z.; Gampala, S.S.; et al. PP2A activates brassinosteroid-responsive gene expression and plant growth by dephosphorylating BZR1. *Nat. Cell Biol.* **2011**, *13*, 124–131. [CrossRef] [PubMed]

32. He, J.X.; Gendron, J.M.; Sun, Y.; Gampala, S.S.; Gendron, N.; Sun, C.Q.; Wang, Z.Y. BZR1 is a transcriptional repressor with dual roles in brassinosteroid homeostasis and growth responses. *Science* **2005**, *307*, 1634–1638. [CrossRef] [PubMed]

33. Yin, Y.; Vafeados, D.; Tao, Y.; Yoshida, S.; Asami, T.; Chory, J. A new class of transcription factors mediates brassinosteroid-regulated gene expression in *Arabidopsis*. *Cell* **2005**, *120*, 249–259. [CrossRef] [PubMed]

34. Peres, A.; Soares, J.S.; Tavares, R.G.; Righetto, G.; Zullo, M.A.T.; Mandava, N.B.; Menossi, M. Brassinosteroids, the sixth class of phytohormones: A molecular view from the discovery to hormonal interactions in plant development and stress adaptation. *Int. J. Mol. Sci.* **2019**, *20*, 331. [CrossRef] [PubMed]

35. Shiu, S.H.; Bleecker, A.B. Plant receptor-like kinase gene family: Diversity, function, and signaling. *Sci. STKE* **2001**, *113*, re22. [CrossRef] [PubMed]

36. Kinoshita, T.; Cano-Delgado, A.; Seto, H.; Hiranuma, S.; Fujioka, S.; Yoshida, S.; Chory, J. Binding of brassinosteroids to the extracellular domain of plant receptor kinase BRI1. *Nature* **2005**, *433*, 167–171. [CrossRef] [PubMed]

37. Clay, N.K.; Nelson, T. VH1, a provascular cell-specific receptor kinase that influences leaf cell patterns in *Arabidopsis*. *Plant Cell* **2002**, *14*, 2707–2722. [CrossRef]

38. Cano-Delgado, A.; Yin, Y.; Yu, C.; Vafeados, D.; Mora-Garcia, S.; Cheng, J.C.; Nam, K.H.; Li, J.; Chory, J. BRL1 and BRL3 are novel brassinosteroid receptors that function in vascular differentiation in *Arabidopsis*. *Development* **2004**, *131*, 5341–5351. [CrossRef]

39. Zhou, A.; Wang, H.; Walker, J.C.; Li, J. BRL1, a leucine-rich repeat receptor-like protein kinase, is functionally redundant with BRI1 in regulating Arabidopsis brassinosteroid signaling. *Plant J.* **2004**, *40*, 399–409. [CrossRef]

40. Ceserani, T.; Trofka, A.; Gandotra, N.; Nelson, T. VH1/BRL2 receptor-like kinase interacts with vascular-specific adaptor proteins VIT and VIK to influence leaf venation. *Plant J.* **2009**, *57*, 1000–1014. [CrossRef]

41. Lozano-Elena, F.; Caño-Delgado, A.I. Emerging roles of vascular brassinosteroid receptors of the BRI1-like family. *Curr. Opin. Plant Biol.* **2019**, *51*, 105–113. [CrossRef]

42. Fàbregas, N.; Lozano-Elena, F.; Blasco-Escámez, D.; Tohge, T.; Martínez-Andújar, C.; Albacete, A.; Osorio, S.; Bustamante, M.; Riechmann, J.L.; Nomura, T.; et al. Overexpression of the vascular brassinosteroid receptor BRL3 confers drought resistance without penalizing plant growth. *Nat. Commun.* **2018**, *9*, 4680. [CrossRef]

43. Klok, E.J.; Wilson, I.W.; Wilson, D.; Chapman, S.C.; Ewing, R.M.; Somerville, S.C.; Peacock, W.J.; Dolferus, R.; Dennis, E.S. Expression profile analysis of the low-oxygen response in Arabidopsis root cultures. *Plant Cell* **2002**, *14*, 2481–2494. [CrossRef] [PubMed]

44. Ohashi-Ito, K.; Kubo, M.; Demura, T.; Fukuda, H. Class III homeodomain leucine-zipper proteins regulate xylem cell differentiation. *Plant Cell Physiol.* **2005**, *46*, 1646–1656. [CrossRef] [PubMed]

45. Wang, J.; Jiang, J.; Wang, J.; Chen, L.; Fan, S.L.; Wu, J.W.; Wang, X.; Wang, Z.X. Structural insights into the negative regulation of BRI1 signaling by BRI1-interacting protein BKI1. *Cell Res.* **2014**, *24*, 1328–1341. [CrossRef]

46. Bojar, D.; Martinez, J.; Santiago, J.; Rybin, V.; Bayliss, R.; Hothorn, M. Crystal structures of the phosphorylated BRI1 kinase domain and implications for brassinosteroid signal initiation. *Plant J.* **2014**, *78*, 31–43. [CrossRef]

47. Bucherl, C.A.; van Esse, G.W.; Kruis, A.; Luchtenberg, J.; Westphal, A.H.; Aker, J.; van Hoek, A.; Albrecht, C.; Borst, J.W.; de Vries, S.C. Visualization of BRI1 and BAK1(SERK3) membrane receptor heterooligomers during brassinosteroid signaling. *Plant Physiol.* **2013**, *162*, 1911–1925. [CrossRef]

48. Hutten, S.J.; Hamers, D.S.; den Toorn, M.A.; van Esse, W.; Nolles, A.; Bucherl, C.A.; de Vries, S.C.; Hohlbein, J.; Borst, J.W. Visualization of BRI1 and SERK3/BAK1 nanoclusters in arabidopsis roots. *PLoS ONE* **2017**, *12*, e0169905. [CrossRef] [PubMed]

49. Gou, X.; Yin, H.; He, K.; Du, J.; Yi, J.; Xu, S.; Lin, H.; Clouse, S.D.; Li, J. Genetic evidence for an indispensable role of somatic embryogenesis receptor kinases in brassinosteroid signaling. *PLoS Genet.* **2012**, *8*, e1002452. [CrossRef]

50. Imkampe, J.; Halter, T.; Huang, S.; Schulze, S.; Mazzotta, S.; Schmidt, N.; Manstretta, R.; Postel, S.; Wierzba, M.; Yang, Y.; et al. The Arabidopsis leucine-rich repeat receptor kinase BIR3 negatively regulates BAK1 receptor complex formation and stabilizes BAK1. *Plant Cell* **2017**, *29*, 2285–2303. [CrossRef]

51. Johnson, L.N.; Noble, M.E.; Owen, D.J. Active and inactive protein kinases: Structural basis for regulation. *Cell* **1996**, *85*, 149–158. [CrossRef]

52. Schlessinger, J. Cell signaling by receptor tyrosine kinases. *Cell* **2000**, *103*, 211–225. [CrossRef]

53. Heldin, C.H.; Moustakas, A. Signaling receptors for TGF-beta family members. *Cold Spring Harb. Perspect. Biol.* **2016**. [CrossRef]

54. Hink, M.A.; Shah, K.; Russinova, E.; de Vries, S.C.; Visser, A.J. Fluorescence fluctuation analysis of Arabidopsis thaliana somatic embryogenesis receptor-like kinase and BRASSINOSTEROID INSENSITIVE 1 receptor oligomerization. *Biophys. J.* **2008**, *94*, 1052–1062. [CrossRef] [PubMed]

55. Wang, L.; Li, H.; Lv, X.; Chen, T.; Li, R.; Xue, Y.; Jiang, J.; Jin, B.; Baluska, F.; Samaj, J.; et al. Spatiotemporal dynamics of the BRI1 receptor and its regulation by membrane microdomains in living arabidopsis cells. *Mol. Plant* **2015**, *8*, 1334–1349. [CrossRef]

56. Hohmann, U.; Ramakrishna, P.; Wang, K.; Lorenzo-Orts, L.; Nicolet, J.; Henschen, A.; Barberon, M.; Bayer, M.; Hothorn, M. Constitutive activation of leucine-rich repeat receptor kinase signaling pathways by BAK1-interacting receptor-like kinase 3 chimera. *bioRxiv* **2020**. [CrossRef]

57. She, J.; Han, Z.; Kim, T.W.; Wang, J.; Cheng, W.; Chang, J.; Shi, S.; Wang, J.; Yang, M.; Wang, Z.Y.; et al. Structural insight into brassinosteroid perception by BRI1. *Nature* **2011**, *474*, 472–476. [CrossRef]

58. Hothorn, M.; Belkhadir, Y.; Dreux, M.; Dabi, T.; Noel, J.P.; Wilson, I.A.; Chory, J. Structural basis of steroid hormone perception by the receptor kinase BRI1. *Nature* **2011**, *474*, 467–471. [CrossRef]

59. Yan, L.; Ma, Y.; Liu, D.; Wei, X.; Sun, Y.; Chen, X.; Zhao, H.; Zhou, J.; Wang, Z.; Shui, W.; et al. Structural basis for the impact of phosphorylation on the activation of plant receptor-like kinase BAK1. *Cell Res.* **2012**, *22*, 1304–1308. [CrossRef]

60. Grosseholz, R.; Feldman-Salit, A.; Wanke, F.; Schulze, S.; Glockner, N.; Kemmerling, B.; Harter, K.; Kummer, U. Specifying the role of BAK1-interacting receptor-like kinase 3 in brassinosteroid signaling. *J. Integr. Plant Biol.* **2020**, *62*, 456–469. [CrossRef]

61. Kornev, A.P.; Taylor, S.S. Dynamics-driven allostery in protein kinases. *Trends Biochem. Sci.* **2015**, *40*, 628–647. [CrossRef] [PubMed]

62. Kornev, A.P.; Haste, N.M.; Taylor, S.S.; Eyck, L.F. Surface comparison of active and inactive protein kinases identifies a conserved activation mechanism. *Proc. Natl. Acad. Sci. USA* **2006**, *103*, 17783–17788. [CrossRef]

63. Mitra, S.K.; Chen, R.; Dhandaydham, M.; Wang, X.; Blackburn, R.K.; Kota, U.; Goshe, M.B.; Schwartz, D.; Huber, S.C.; Clouse, S.D. An autophosphorylation site database for leucine-rich repeat receptor-like kinases in Arabidopsis thaliana. *Plant J.* **2015**, *82*, 1042–1060. [CrossRef] [PubMed]

64. Oh, M.H.; Ray, W.K.; Huber, S.C.; Asara, J.M.; Gage, D.A.; Clouse, S.D. Recombinant BRASSINOSTEROID INSENSITIVE 1 receptor-like kinase autophosphorylates on serine and threonine residues and phosphorylates a conserved peptide motif in vitro. *Plant Physiol.* **2000**, *124*, 751–766. [CrossRef] [PubMed]

65. Wu, X.; Oh, M.H.; Kim, H.S.; Schwartz, D.; Imai, B.S.; Yau, P.M.; Clouse, S.D.; Huber, S.C. Transphosphorylation of E. coli proteins during production of recombinant protein kinases provides a robust system to characterize kinase specificity. *Front. Plant Sci.* **2012**, *3*, 262. [CrossRef] [PubMed]

66. Wang, R.; Liu, M.; Yuan, M.; Oses-Prieto, J.A.; Cai, X.; Sun, Y.; Burlingame, A.L.; Wang, Z.Y.; Tang, W. The brassinosteroid-activated BRI1 receptor kinase is switched off by dephosphorylation mediated by cytoplasm-localized PP2A B' subunits. *Mol. Plant* **2016**, *9*, 148–157. [CrossRef]

67. Oh, M.H.; Wang, X.; Kota, U.; Goshe, M.B.; Clouse, S.D.; Huber, S.C. Tyrosine phosphorylation of the BRI1 receptor kinase emerges as a component of brassinosteroid signaling in *Arabidopsis*. *Proc. Natl. Acad. Sci. USA* **2009**, *106*, 658–663. [CrossRef]

68. Oh, M.H.; Clouse, S.D.; Huber, S.C. Tyrosine phosphorylation of the BRI1 receptor kinase occurs via a post-translational modification and is activated by the juxtamembrane domain. *Front. Plant Sci.* **2012**, *3*, 175. [CrossRef]

69. Amorim-Silva, V.; Garcia-Moreno, A.; Castillo, A.G.; Lakhssassi, N.; Del Valle, A.E.; Perez-Sancho, J.; Li, Y.; Pose, D.; Perez-Rodriguez, J.; Lin, J.; et al. TTL proteins scaffold brassinosteroid signaling components at the plasma membrane to optimize signal transduction in arabidopsis. *Plant Cell* **2019**, *31*, 1807–1828. [CrossRef]

70. Ren, H.; Willige, B.C.; Jaillais, Y.; Geng, S.; Park, M.Y.; Gray, W.M.; Chory, J. Brassinosteroid-signaling kinase 3, a plasma membrane-associated scaffold protein involved in early brassinosteroid signaling. *PLoS Genet.* **2019**, *15*, e1007904. [CrossRef]

71. Jaillais, Y.; Hothorn, M.; Belkhadir, Y.; Dabi, T.; Nimchuk, Z.L.; Meyerowitz, E.M.; Chory, J. Tyrosine phosphorylation controls brassinosteroid receptor activation by triggering membrane release of its kinase inhibitor. *Genes Dev.* **2011**, *25*, 232–237. [CrossRef] [PubMed]

72. Wang, H.; Yang, C.; Zhang, C.; Wang, N.; Lu, D.; Wang, J.; Zhang, S.; Wang, Z.X.; Ma, H.; Wang, X. Dual role of BKI1 and 14-3-3 s in brassinosteroid signaling to link receptor with transcription factors. *Dev. Cell* **2011**, *21*, 825–834. [CrossRef] [PubMed]

73. Oh, M.H.; Wang, X.; Clouse, S.D.; Huber, S.C. Deactivation of the Arabidopsis BRASSINOSTEROID INSENSITIVE 1 (BRI1) receptor kinase by autophosphorylation within the glycine-rich loop. *Proc. Natl. Acad. Sci. USA* **2012**, *109*, 327–332. [CrossRef] [PubMed]

74. Wu, G.; Wang, X.; Li, X.; Kamiya, Y.; Otegui, M.S.; Chory, J. Methylation of a phosphatase specifies dephosphorylation and degradation of activated brassinosteroid receptors. *Sci. Signal.* **2011**, *4*, ra29. [CrossRef] [PubMed]

75. Janssens, V.; Goris, J. Protein phosphatase 2A: A highly regulated family of serine/threonine phosphatases implicated in cell growth and signalling. *Biochem. J.* **2001**, *353*, 417–439. [CrossRef]

76. Macho, A.P.; Schwessinger, B.; Ntoukakis, V.; Brutus, A.; Segonzac, C.; Roy, S.; Kadota, Y.; Oh, M.H.; Sklenar, J.; Derbyshire, P.; et al. A bacterial tyrosine phosphatase inhibits plant pattern recognition receptor activation. *Science* **2014**, *343*, 1509–1512. [CrossRef]

77. Kerk, D.; Templeton, G.; Moorhead, G.B. Evolutionary radiation pattern of novel protein phosphatases revealed by analysis of protein data from the completely sequenced genomes of humans, green algae, and higher plants. *Plant Physiol.* **2008**, *146*, 351–367. [CrossRef]

78. Araki, K.; Nagata, K. Protein folding and quality control in the ER. *Cold Spring Harb. Perspect. Biol.* **2011**, *3*, a007526. [CrossRef]

79. Strasser, R. Protein quality control in the endoplasmic reticulum of plants. *Annu. Rev. Plant Biol.* **2018**, *69*, 147–172. [CrossRef]

80. Jin, H.; Hong, Z.; Su, W.; Li, J. A plant-specific calreticulin is a key retention factor for a defective brassinosteroid receptor in the endoplasmic reticulum. *Proc. Natl. Acad. Sci. USA* **2009**, *106*, 13612–13617. [CrossRef]

81. Jin, H.; Yan, Z.; Nam, K.H.; Li, J. Allele-specific suppression of a defective brassinosteroid receptor reveals a physiological role of UGGT in ER quality control. *Mol. Cell* **2007**, *26*, 821–830. [CrossRef] [PubMed]

82. Hong, Z.; Jin, H.; Tzfira, T.; Li, J. Multiple mechanism-mediated retention of a defective brassinosteroid receptor in the endoplasmic reticulum of *Arabidopsis*. *Plant Cell* **2008**, *20*, 3418–3429. [CrossRef]

83. Su, W.; Liu, Y.; Xia, Y.; Hong, Z.; Li, J. Conserved endoplasmic reticulum-associated degradation system to eliminate mutated receptor-like kinases in Arabidopsis. *Proc. Natl. Acad. Sci. USA* **2011**, *108*, 870–875. [CrossRef] [PubMed]

84. Su, W.; Liu, Y.; Xia, Y.; Hong, Z.; Li, J. The Arabidopsis homolog of the mammalian OS-9 protein plays a key role in the endoplasmic reticulum-associated degradation of misfolded receptor-like kinases. *Mol. Plant* **2012**, *5*, 929–940. [CrossRef]

85. Lin, L.; Zhang, C.; Chen, Y.; Wang, Y.; Wang, D.; Liu, X.; Wang, M.; Mao, J.; Zhang, J.; Xing, W.; et al. PAWH1 and PAWH2 are plant-specific components of an Arabidopsis endoplasmic reticulum-associated degradation complex. *Nat. Commun.* **2019**, *10*, 3492. [CrossRef] [PubMed]

86. Liu, Y.; Zhang, C.; Wang, D.; Su, W.; Liu, L.; Wang, M.; Li, J. EBS7 is a plant-specific component of a highly conserved endoplasmic reticulum-associated degradation system in Arabidopsis. *Proc. Natl. Acad. Sci. USA* **2015**, *112*, 12205–12210. [CrossRef]

87. Zhang, X.; Zhou, L.; Qin, Y.; Chen, Y.; Liu, X.; Wang, M.; Mao, J.; Zhang, J.; He, Z.; Liu, L.; et al. A temperature-sensitive misfolded bri1-301 receptor requires its kinase activity to promote growth. *Plant Physiol.* **2018**, *178*, 1704–1719. [CrossRef]

88. Lv, M.; Li, M.; Chen, W.; Wang, Y.; Sun, C.; Yin, H.; He, K.; Li, J. Thermal-enhanced bri1-301 instability reveals a plasma membrane protein quality control system in plants. *Front. Plant Sci.* **2018**, *9*, 1620. [CrossRef]

89. Wu, G.; Otegui, M.S.; Spalding, E.P. The ER-localized TWD1 immunophilin is necessary for localization of multidrug resistance-like proteins required for polar auxin transport in Arabidopsis roots. *Plant Cell* **2010**, *22*, 3295–3304. [CrossRef]

90. Zhao, B.; Lv, M.; Feng, Z.; Campbell, T.; Liscum, E.; Li, J. TWISTED DWARF 1 associates with BRASSINOSTEROID-INSENSITIVE 1 to regulate early events of the brassinosteroid signaling pathway. *Mol. Plant* **2016**, *9*, 582–592. [CrossRef]

91. Chaiwanon, J.; Garcia, V.J.; Cartwright, H.; Sun, Y.; Wang, Z.Y. Immunophilin-like FKBP42/TWISTED DWARF1 interacts with the receptor kinase BRI1 to regulate brassinosteroid signaling in *Arabidopsis*. *Mol. Plant* **2016**, *9*, 593–600. [CrossRef] [PubMed]

92. Geisler, M.; Kolukisaoglu, H.U.; Bouchard, R.; Billion, K.; Berger, J.; Saal, B.; Frangne, N.; Koncz-Kalman, Z.; Koncz, C.; Dudler, R.; et al. TWISTED DWARF1, a unique plasma membrane-anchored immunophilin-like protein, interacts with Arabidopsis multidrug resistance-like transporters AtPGP1 and AtPGP19. *Mol. Biol. Cell* **2003**, *14*, 4238–4249. [CrossRef] [PubMed]

93. Zhu, J.; Bailly, A.; Zwiewka, M.; Sovero, V.; Di Donato, M.; Ge, P.; Oehri, J.; Aryal, B.; Hao, P.; Linnert, M.; et al. TWISTED DWARF1 mediates the action of auxin transport inhibitors on actin cytoskeleton dynamics. *Plant Cell* **2016**, *28*, 930–948. [CrossRef]

94. Ott, T. Membrane nanodomains and microdomains in plant-microbe interactions. *Curr. Opin. Plant Biol.* **2017**, *40*, 82–88. [CrossRef] [PubMed]

95. Bucherl, C.A.; Jarsch, I.K.; Schudoma, C.; Segonzac, C.; Mbengue, M.; Robatzek, S.; MacLean, D.; Ott, T.; Zipfel, C. Plant immune and growth receptors share common signalling components but localise to distinct plasma membrane nanodomains. *Elife* **2017**, *6*, e25114. [CrossRef] [PubMed]

96. Gui, J.; Zheng, S.; Liu, C.; Shen, J.; Li, J.; Li, L. OsREM4.1 interacts with OsSERK1 to coordinate the interlinking between abscisic acid and brassinosteroid signaling in rice. *Dev. Cell* **2016**, *38*, 201–213. [CrossRef]

97. Jaillais, Y.; Ott, T. The nanoscale organization of the plasma membrane and its importance in signaling: A proteolipid perspective. *Plant Physiol.* **2020**, *182*, 1682–1696. [CrossRef]

98. Shiu, S.H.; Karlowski, W.M.; Pan, R.; Tzeng, Y.H.; Mayer, K.F.; Li, W.H. Comparative analysis of the receptor-like kinase family in Arabidopsis and rice. *Plant Cell* **2004**, *16*, 1220–1234. [CrossRef]

99. Geldner, N.; Hyman, D.L.; Wang, X.; Schumacher, K.; Chory, J. Endosomal signaling of plant steroid receptor kinase BRI1. *Genes Dev.* **2007**, *21*, 1598–1602. [CrossRef]

100. Claus, L.A.N.; Savatin, D.V.; Russinova, E. The crossroads of receptor-mediated signaling and endocytosis in plants. *J. Integr. Plant Biol.* **2018**, *60*, 827–840. [CrossRef]

101. Dettmer, J.; Hong-Hermesdorf, A.; Stierhof, Y.D.; Schumacher, K. Vacuolar H+-ATPase activity is required for endocytic and secretory trafficking in Arabidopsis. *Plant Cell* **2006**, *18*, 715–730. [CrossRef]

102. Viotti, C.; Bubeck, J.; Stierhof, Y.D.; Krebs, M.; Langhans, M.; van den Berg, W.; van Dongen, W.; Richter, S.; Geldner, N.; Takano, J.; et al. Endocytic and secretory traffic in Arabidopsis merge in the trans-Golgi network/early endosome, an independent and highly dynamic organelle. *Plant Cell* **2010**, *22*, 1344–1357. [CrossRef] [PubMed]

103. Jaillais, Y.; Fobis-Loisy, I.; Miege, C.; Gaude, T. Evidence for a sorting endosome in *Arabidopsis* root cells. *Plant J.* **2008**, *53*, 237–247. [CrossRef] [PubMed]

104. Irani, N.G.; Di Rubbo, S.; Mylle, E.; van den Begin, J.; Schneider-Pizon, J.; Hnilikova, J.; Sisa, M.; Buyst, D.; Vilarrasa-Blasi, J.; Szatmari, A.M.; et al. Fluorescent castasterone reveals BRI1 signaling from the plasma membrane. *Nat. Chem. Biol.* **2012**, *8*, 583–589. [CrossRef] [PubMed]

105. Geldner, N.; Robatzek, S. Plant receptors go endosomal: A moving view on signal transduction. *Plant Physiol.* **2008**, *147*, 1565–1574. [CrossRef]

106. Reynolds, G.D.; Wang, C.; Pan, J.; Bednarek, S.Y. Inroads into internalization: Five years of endocytic exploration. *Plant Physiol.* **2018**, *176*, 208–218. [CrossRef] [PubMed]

107. Banbury, D.N.; Oakley, J.D.; Sessions, R.B.; Banting, G. Tyrphostin A23 inhibits internalization of the transferrin receptor by perturbing the interaction between tyrosine motifs and the medium chain subunit of the AP-2 adaptor complex. *J. Biol. Chem.* **2003**, *278*, 12022–12028. [CrossRef]

108. Gadeyne, A.; Sánchez-Rodríguez, C.; Vanneste, S.; Di Rubbo, S.; Zauber, H.; Vanneste, K.; van Leene, J.; de Winne, N.; Eeckhout, D.; Persiau, G.; et al. The TPLATE adaptor complex drives clathrin-mediated endocytosis in plants. *Cell* **2014**, *156*, 691–704. [CrossRef]

109. Di Rubbo, S.; Irani, N.G.; Kim, S.Y.; Xu, Z.Y.; Gadeyne, A.; Dejonghe, W.; Vanhoutte, I.; Persiau, G.; Eeckhout, D.; Simon, S.; et al. The clathrin adaptor complex AP-2 mediates endocytosis of brassinosteroid insensitive1 in *Arabidopsis*. *Plant Cell* **2013**, *25*, 2986–2997. [CrossRef]

110. Li, R.; Liu, P.; Wan, Y.; Chen, T.; Wang, Q.; Mettbach, U.; Baluska, F.; Samaj, J.; Fang, X.; Lucas, W.J.; et al. A membrane microdomain-associated protein, Arabidopsis Flot1, is involved in a clathrin-independent endocytic pathway and is required for seedling development. *Plant Cell* **2012**, *24*, 2105–2122. [CrossRef]

111. Rodriguez-Furlan, C.; Minina, E.A.; Hicks, G.R. Remove, recycle, degrade: Regulating plasma membrane protein accumulation. *Plant Cell* **2019**, *31*, 2833–2854. [CrossRef] [PubMed]

112. Zhou, J.; Liu, D.; Wang, P.; Ma, X.; Lin, W.; Chen, S.; Mishev, K.; Lu, D.; Kumar, R.; Vanhoutte, I.; et al. Regulation of *Arabidopsis* brassinosteroid receptor BRI1 endocytosis and degradation by plant U-box PUB12/PUB13-mediated ubiquitination. *Proc. Natl. Acad. Sci. USA* **2018**, *115*, E1906–E1915. [CrossRef] [PubMed]

113. Cardona-Lopez, X.; Cuyas, L.; Marin, E.; Rajulu, C.; Irigoyen, M.L.; Gil, E.; Puga, M.I.; Bligny, R.; Nussaume, L.; Geldner, N.; et al. ESCRT-III-associated protein ALIX mediates high-affinity phosphate transporter trafficking to maintain phosphate homeostasis in arabidopsis. *Plant Cell* **2015**, *27*, 2560–2581. [CrossRef] [PubMed]

114. Mattern, M.; Sutherland, J.; Kadimisetty, K.; Barrio, R.; Rodriguez, M.S. Using ubiquitin binders to decipher the ubiquitin code. *Trends. Biochem. Sci.* **2019**, *44*, 599–615. [CrossRef] [PubMed]

115. Martins, S.; Dohmann, E.M.; Cayrel, A.; Johnson, A.; Fischer, W.; Pojer, F.; Satiat-Jeunemaitre, B.; Jaillais, Y.; Chory, J.; Geldner, N.; et al. Internalization and vacuolar targeting of the brassinosteroid hormone receptor BRI1 are regulated by ubiquitination. *Nat. Commun.* **2015**, *6*, 6151. [CrossRef]

116. Gomez-Gomez, L.; Boller, T. FLS2: An LRR receptor-like kinase involved in the perception of the bacterial elicitor flagellin in Arabidopsis. *Mol. Cell* **2000**, *5*, 1003–1011. [CrossRef]

117. Lu, D.; Lin, W.; Gao, X.; Wu, S.; Cheng, C.; Avila, J.; Heese, A.; Devarenne, T.P.; He, P.; Shan, L. Direct ubiquitination of pattern recognition receptor FLS2 attenuates plant innate immunity. *Science* **2011**, *332*, 1439–1442. [CrossRef]

118. Kalinowska, K.; Nagel, M.K.; Goodman, K.; Cuyas, L.; Anzenberger, F.; Alkofer, A.; Paz-Ares, J.; Braun, P.; Rubio, V.; Otegui, M.S.; et al. Arabidopsis ALIX is required for the endosomal localization of the deubiquitinating enzyme AMSH3. *Proc. Natl. Acad. Sci. USA* **2015**, *112*, E5543–E5551. [CrossRef]

119. Yamagami, A.; Saito, C.; Nakazawa, M.; Fujioka, S.; Uemura, T.; Matsui, M.; Sakuta, M.; Shinozaki, K.; Osada, H.; Nakano, A.; et al. Evolutionarily conserved BIL4 suppresses the degradation of brassinosteroid receptor BRI1 and regulates cell elongation. *Sci. Rep.* **2017**, *7*, 5739. [CrossRef] [PubMed]

120. Luo, Y.; Scholl, S.; Doering, A.; Zhang, Y.; Irani, N.G.; Rubbo, S.D.; Neumetzler, L.; Krishnamoorthy, P.; Van Houtte, I.; Mylle, E.; et al. V-ATPase activity in the TGN/EE is required for exocytosis and recycling in *Arabidopsis*. *Nat. Plants* **2015**, *1*, 15094. [CrossRef] [PubMed]

121. Schumacher, K.; Vafeados, D.; McCarthy, M.; Sze, H.; Wilkins, T.; Chory, J. The Arabidopsis det3 mutant reveals a central role for the vacuolar H(+)-ATPase in plant growth and development. *Genes Dev.* **1999**, *13*, 3259–3270. [CrossRef] [PubMed]

122. Ambrose, C.; Ruan, Y.; Gardiner, J.; Tamblyn, L.M.; Catching, A.; Kirik, V.; Marc, J.; Overall, R.; Wasteneys, G.O. CLASP interacts with sorting nexin 1 to link microtubules and auxin transport via PIN2 recycling in Arabidopsis thaliana. *Dev. Cell* **2013**, *24*, 649–659. [CrossRef] [PubMed]

123. Ruan, Y.; Halat, L.S.; Khan, D.; Jancowski, S.; Ambrose, C.; Belmonte, M.F.; Wasteneys, G.O. The microtubule-associated protein CLASP sustains cell proliferation through a brassinosteroid signaling negative feedback loop. *Curr. Biol.* **2018**, *28*, 2718–2729. [CrossRef] [PubMed]

124. Zhang, L.; Xing, J.; Lin, J. At the intersection of exocytosis and endocytosis in plants. *New Phytol.* **2019**, *224*, 1479–1489. [CrossRef] [PubMed]

125. Zhang, L.; Liu, Y.; Zhu, X.F.; Jung, J.H.; Sun, Q.; Li, T.Y.; Chen, L.J.; Duan, Y.X.; Xuan, Y.H. SYP22 and VAMP727 regulate BRI1 plasma membrane targeting to control plant growth in Arabidopsis. *New Phytol.* **2019**, *223*, 1059–1065. [CrossRef]

126. Drdova, E.J.; Synek, L.; Pecenkova, T.; Hala, M.; Kulich, I.; Fowler, J.E.; Murphy, A.S.; Zarsky, V. The exocyst complex contributes to PIN auxin efflux carrier recycling and polar auxin transport in Arabidopsis. *Plant J.* **2013**, *73*, 709–719. [CrossRef]

127. Xue, S.; Zou, J.; Liu, Y.; Wang, M.; Zhang, C.; Le, J. Involvement of BIG5 and BIG3 in BRI1 trafficking reveals diverse functions of BIG-subfamily ARF-GEFs in plant growth and gravitropism. *Int. J. Mol. Sci.* **2019**, *20*, 2339. [CrossRef]

128. Wu, W.; Wu, Y.; Gao, Y.; Li, M.; Yin, H.; Lv, M.; Zhao, J.; Li, J.; He, K. Somatic embryogenesis receptor-like kinase 5 in the ecotype Landsberg erecta of Arabidopsis is a functional RD LRR-RLK in regulating brassinosteroid signaling and cell death control. *Front Plant Sci.* **2015**, *6*, 852. [CrossRef]

129. Chinchilla, D.; Zipfel, C.; Robatzek, S.; Kemmerling, B.; Nürnberger, T.; Jones, J.D.G.; Felix, G.; Boller, T. A flagellin-induced complex of the receptor FLS2 and BAK1 initiates plant defence. *Nature* **2007**, *448*, 497–500. [CrossRef]

130. Karlova, R.; Boeren, S.; Russinova, E.; Aker, J.; Vervoort, J.; de Vries, S. The arabidopsis SOMATIC EMBRYOGENESIS RECEPTOR-LIKE KINASE1 protein complex includes BRASSINOSTEROID-INSENSITIVE1. *Plant Cell* **2006**, *18*, 626–638. [CrossRef]

131. Wu, D.; Liu, Y.; Xu, F.; Zhang, Y. Differential requirement of BAK1 C-terminal tail in development and immunity. *J. Integr. Plant Biol.* **2018**, *60*, 270–275. [CrossRef] [PubMed]

132. Wang, Y.; Li, Z.; Liu, D.; Xu, J.; Wei, X.; Yan, L.; Yang, C.; Lou, Z.; Shui, W. Assessment of BAK1 activity in different plant receptor-like kinase complexes by quantitative profiling of phosphorylation patterns. *J. Proteomics* **2014**, *108*, 484–493. [CrossRef] [PubMed]

133. Shah, K.; Vervoort, J.; de Vries, S.C. Role of threonines in the Arabidopsis thaliana somatic embryogenesis receptor kinase 1 activation loop in phosphorylation. *J. Biol. Chem.* **2001**, *276*, 41263–41269. [CrossRef]

134. Perraki, A.; DeFalco, T.A.; Derbyshire, P.; Avila, J.; Sere, D.; Sklenar, J.; Qi, X.; Stransfeld, L.; Schwessinger, B.; Kadota, Y.; et al. Phosphocode-dependent functional dichotomy of a common co-receptor in plant signalling. *Nature* **2018**, *561*, 248–252. [CrossRef]

135. Singh, V.; Perraki, A.; Kim, S.Y.; Shrivastava, S.; Lee, J.H.; Zhao, Y.; Schwessinger, B.; Oh, M.H.; Marshall-Colon, A.; Zipfel, C.; et al. Tyrosine-610 in the receptor kinase BAK1 does not play a major role in brassinosteroid signaling or innate immunity. *Front Plant Sci.* **2017**, *8*, 1273. [CrossRef] [PubMed]

136. Lin, W.; Li, B.; Lu, D.; Chen, S.; Zhu, N.; He, P.; Shan, L. Tyrosine phosphorylation of protein kinase complex BAK1/BIK1 mediates Arabidopsis innate immunity. *Proc. Natl. Acad. Sci. USA* **2014**, *111*, 3632–3637. [CrossRef] [PubMed]

137. Segonzac, C.; Macho, A.P.; Sanmartin, M.; Ntoukakis, V.; Sanchez-Serrano, J.J.; Zipfel, C. Negative control of BAK1 by protein phosphatase 2A during plant innate immunity. *EMBO J.* **2014**, *33*, 2069–2079. [CrossRef]

138. Song, L.; Shi, Q.M.; Yang, X.H.; Xu, Z.H.; Xue, H.W. Membrane steroid-binding protein 1 (MSBP1) negatively regulates brassinosteroid signaling by enhancing the endocytosis of BAK1. *Cell Res.* **2009**, *19*, 864–876. [CrossRef]

139. Gao, M.; Wang, X.; Wang, D.; Xu, F.; Ding, X.; Zhang, Z.; Bi, D.; Cheng, Y.T.; Chen, S.; Li, X.; et al. Regulation of cell death and innate immunity by two receptor-like kinases in Arabidopsis. *Cell Host Microbe* **2009**, *6*, 34–44. [CrossRef]

140. Halter, T.; Imkampe, J.; Mazzotta, S.; Wierzba, M.; Postel, S.; Bucherl, C.; Kiefer, C.; Stahl, M.; Chinchilla, D.; Wang, X.; et al. The leucine-rich repeat receptor kinase BIR2 is a negative regulator of BAK1 in plant immunity. *Curr. Biol.* **2014**, *24*, 134–143. [CrossRef]

141. Liu, Y.; Huang, X.; Li, M.; He, P.; Zhang, Y. Loss-of-function of Arabidopsis receptor-like kinase BIR1 activates cell death and defense responses mediated by BAK1 and SOBIR1. *New Phytol.* **2016**, *212*, 637–645. [CrossRef] [PubMed]

142. Halter, T.; Imkampe, J.; Blaum, B.S.; Stehle, T.; Kemmerling, B. BIR2 affects complex formation of BAK1 with ligand binding receptors in plant defense. *Plant Signal. Behav.* **2014**, *9*, e28944. [CrossRef] [PubMed]

143. Hohmann, U.; Nicolet, J.; Moretti, A.; Hothorn, L.A.; Hothorn, M. The SERK3 elongated allele defines a role for BIR ectodomains in brassinosteroid signalling. *Nat. Plants* **2018**, *4*, 345–351. [CrossRef] [PubMed]

144. Chung, Y.; Choe, V.; Fujioka, S.; Takatsuto, S.; Han, M.; Jeon, J.S.; Park, Y.I.; Lee, K.O.; Choe, S. Constitutive activation of brassinosteroid signaling in the Arabidopsis elongated-D/bak1 mutant. *Plant Mol. Biol.* **2012**, *80*, 489–501. [CrossRef]

145. Halliday, K.; Devlin, P.F.; Whitelam, G.C.; Hanhart, C.; Koornneef, M. The ELONGATED gene of Arabidopsis acts independently of light and gibberellins in the control of elongation growth. *Plant J.* **1996**, *9*, 305–312. [CrossRef]

146. Kim, M.H.; Kim, Y.; Kim, J.W.; Lee, H.S.; Lee, W.S.; Kim, S.K.; Wang, Z.Y.; Kim, S.H. Identification of Arabidopsis BAK1-associating receptor-like kinase 1 (BARK1) and characterization of its gene expression and brassinosteroid-regulated root phenotypes. *Plant Cell Physiol.* **2013**, *54*, 1620–1634. [CrossRef]

147. Blaum, B.S.; Mazzotta, S.; Noldeke, E.R.; Halter, T.; Madlung, J.; Kemmerling, B.; Stehle, T. Structure of the pseudokinase domain of BIR2, a regulator of BAK1-mediated immune signaling in Arabidopsis. *J. Struct. Biol.* **2014**, *186*, 112–121. [CrossRef]

148. Grek, C.L.; Zhang, J.; Manevich, Y.; Townsend, D.M.; Tew, K.D. Causes and consequences of cysteine S-glutathionylation. *J. Biol. Chem.* **2013**, *288*, 26497–26504. [CrossRef] [PubMed]

149. Bender, K.W.; Wang, X.; Cheng, G.B.; Kim, H.S.; Zielinski, R.E.; Huber, S.C. Glutaredoxin AtGRXC2 catalyses inhibitory glutathionylation of Arabidopsis BRI1-associated receptor-like kinase 1 (BAK1) in vitro. *Biochem. J.* **2015**, *467*, 399–413. [CrossRef] [PubMed]

150. Moffett, A.S.; Bender, K.W.; Huber, S.C.; Shukla, D. Allosteric control of a plant receptor kinase through s-glutathionylation. *Biophys. J.* **2017**, *113*, 2354–2363. [CrossRef] [PubMed]

151. Schwessinger, B.; Roux, M.; Kadota, Y.; Ntoukakis, V.; Sklenar, J.; Jones, A.; Zipfel, C. Phosphorylation-dependent differential regulation of plant growth, cell death, and innate immunity by the regulatory receptor-like kinase BAK1. *PLoS Genet.* **2011**, *7*, e1002046. [CrossRef] [PubMed]

152. Zhou, J.; Wang, P.; Claus, L.A.N.; Savatin, D.V.; Xu, G.; Wu, S.; Meng, X.; Russinova, E.; He, P.; Shan, L. Proteolytic processing of SERK3/BAK1 regulates plant immunity, development, and cell death. *Plant Physiol.* **2019**, *180*, 543–558. [CrossRef]

153. Ntoukakis, V.; Schwessinger, B.; Segonzac, C.; Zipfel, C. Cautionary notes on the use of C-terminal BAK1 fusion proteins for functional studies. *Plant Cell* **2011**, *23*, 3871–3878. [CrossRef] [PubMed]

154. Lu, D.; Wu, S.; Gao, X.; Zhang, Y.; Shan, L.; He, P. A receptor-like cytoplasmic kinase, BIK1, associates with a flagellin receptor complex to initiate plant innate immunity. *Proc. Natl. Acad. Sci. USA* **2010**, *107*, 496–501. [CrossRef]

155. Zhao, J.; Peng, P.; Schmitz, R.J.; Decker, A.D.; Tax, F.E.; Li, J. Two putative BIN2 substrates are nuclear components of brassinosteroid signaling. *Plant Physiol.* **2002**, *130*, 1221–1229. [CrossRef] [PubMed]

156. Wang, Z.Y.; Nakano, T.; Gendron, J.; He, J.; Chen, M.; Vafeados, D.; Yang, Y.; Fujioka, S.; Yoshida, S.; Asami, T.; et al. Nuclear-localized BZR1 mediates brassinosteroid-induced growth and feedback suppression of brassinosteroid biosynthesis. *Dev. Cell* **2002**, *2*, 505–513. [CrossRef]

157. Youn, J.H.; Kim, T.W. Functional insights of plant GSK3-like kinases: Multi-taskers in diverse cellular signal transduction pathways. *Mol. Plant* **2015**, *8*, 552–565. [CrossRef]

158. Choe, S.; Schmitz, R.J.; Fujioka, S.; Takatsuto, S.; Lee, M.O.; Yoshida, S.; Feldmann, K.A.; Tax, F.E. Arabidopsis brassinosteroid-insensitive dwarf12 mutants are semidominant and defective in a glycogen synthase kinase 3beta-like kinase. *Plant Physiol.* **2002**, *130*, 1506–1515. [CrossRef]

159. Perez-Perez, J.M.; Ponce, M.R.; Micol, J.L. The UCU1 Arabidopsis gene encodes a SHAGGY/GSK3-like kinase required for cell expansion along the proximodistal axis. *Dev. Biol.* **2002**, *242*, 161–173. [CrossRef]

160. Yan, Z.; Zhao, J.; Peng, P.; Chihara, R.K.; Li, J. BIN2 functions redundantly with other *Arabidopsis* GSK3-like kinases to regulate brassinosteroid signaling. *Plant Physiol.* **2009**, *150*, 710–721. [CrossRef]

161. Youn, J.H.; Kim, T.W.; Kim, E.J.; Bu, S.; Kim, S.K.; Wang, Z.Y.; Kim, T.W. Structural and functional characterization of Arabidopsis GSK3-like kinase AtSK12. *Mol. Cells* **2013**, *36*, 564–570. [CrossRef] [PubMed]

162. De Rybel, B.; Audenaert, D.; Vert, G.; Rozhon, W.; Mayerhofer, J.; Peelman, F.; Coutuer, S.; Denayer, T.; Jansen, L.; Nguyen, L.; et al. Chemical inhibition of a subset of Arabidopsis thaliana GSK3-like kinases activates brassinosteroid signaling. *Chem. Biol.* **2009**, *16*, 594–604. [CrossRef] [PubMed]

163. Vert, G.; Chory, J. Downstream nuclear events in brassinosteroid signalling. *Nature* **2006**, *441*, 96–100. [CrossRef] [PubMed]

164. Beurel, E.; Grieco, S.F.; Jope, R.S. Glycogen synthase kinase-3 (GSK3): Regulation, actions, and diseases. *Pharmacol. Ther.* **2015**, *148*, 114–131. [CrossRef] [PubMed]

165. Peng, P.; Zhao, J.; Zhu, Y.; Asami, T.; Li, J. A direct docking mechanism for a plant GSK3-like kinase to phosphorylate its substrates. *J. Biol. Chem.* **2010**, *285*, 24646–24653. [CrossRef] [PubMed]

166. Wang, H.; Tang, J.; Liu, J.; Hu, J.; Liu, J.; Chen, Y.; Cai, Z.; Wang, X. Abscisic acid signaling inhibits brassinosteroid signaling through dampening the dephosphorylation of BIN2 by ABI1 and ABI2. *Mol. Plant* **2018**, *11*, 315–325. [CrossRef]

167. Lochhead, P.A.; Kinstrie, R.; Sibbet, G.; Rawjee, T.; Morrice, N.; Cleghon, V. A chaperone-dependent GSK3beta transitional intermediate mediates activation-loop autophosphorylation. *Mol. Cell* **2006**, *24*, 627–633. [CrossRef]

168. Cole, A.; Frame, S.; Cohen, P. Further evidence that the tyrosine phosphorylation of glycogen synthase kinase-3 (GSK3) in mammalian cells is an autophosphorylation event. *Biochem. J.* **2004**, *377*, 249–255. [CrossRef]

169. Sugiyama, N.; Nakagami, H.; Mochida, K.; Daudi, A.; Tomita, M.; Shirasu, K.; Ishihama, Y. Large-scale phosphorylation mapping reveals the extent of tyrosine phosphorylation in Arabidopsis. *Mol. Syst. Biol.* **2008**, *4*, 193. [CrossRef]

170. Lin, L.L.; Hsu, C.L.; Hu, C.W.; Ko, S.Y.; Hsieh, H.L.; Huang, H.C.; Juan, H.F. Integrating phosphoproteomics and bioinformatics to study brassinosteroid-regulated phosphorylation dynamics in arabidopsis. *BMC Genomics* **2015**, *16*, 533. [CrossRef]

171. Hoffmeister, L.; Diekmann, M.; Brand, K.; Huber, R. GSK3: A kinase balancing promotion and resolution of inflammation. *Cells* **2020**, *9*, 820. [CrossRef] [PubMed]

172. Hughes, K.; Nikolakaki, E.; Plyte, S.E.; Totty, N.F.; Woodgett, J.R. Modulation of the glycogen synthase kinase-3 family by tyrosine phosphorylation. *EMBO J.* **1993**, *12*, 803–808. [CrossRef]

173. Dajani, R.; Fraser, E.; Roe, S.M.; Young, N.; Good, V.; Dale, T.C.; Pearl, L.H. Crystal structure of glycogen synthase kinase 3 beta: Structural basis for phosphate-primed substrate specificity and autoinhibition. *Cell* **2001**, *105*, 721–732. [CrossRef]

174. Buescher, J.L.; Phiel, C.J. A noncatalytic domain of glycogen synthase kinase-3 (GSK-3) is essential for activity. *J. Biol. Chem.* **2010**, *285*, 7957–7963. [CrossRef] [PubMed]

175. Itoh, K.; Tang, T.L.; Neel, B.G.; Sokol, S.Y. Specific modulation of ectodermal cell fates in Xenopus embryos by glycogen synthase kinase. *Development* **1995**, *121*, 3979–3988.

176. Moorhead, G.B.; de Wever, V.; Templeton, G.; Kerk, D. Evolution of protein phosphatases in plants and animals. *Biochem. J.* **2009**, *417*, 401–409. [CrossRef]

177. Mora-Garcia, S.; Vert, G.; Yin, Y.; Cano-Delgado, A.; Cheong, H.; Chory, J. Nuclear protein phosphatases with Kelch-repeat domains modulate the response to brassinosteroids in *Arabidopsis*. *Genes Dev.* **2004**, *18*, 448–460. [CrossRef]

178. Maselli, G.A.; Slamovits, C.H.; Bianchi, J.I.; Vilarrasa-Blasi, J.; Cano-Delgado, A.I.; Mora-Garcia, S. Revisiting the evolutionary history and roles of protein phosphatases with Kelch-like domains in plants. *Plant Physiol.* **2014**, *164*, 1527–1541. [CrossRef]

179. Shi, Y. Serine/threonine phosphatases: Mechanism through structure. *Cell* **2009**, *139*, 468–484. [CrossRef]

180. Tonks, N.K. Protein tyrosine phosphatases: From genes, to function, to disease. *Nat. Rev. Mol. Cell Biol.* **2006**, *7*, 833–846. [CrossRef]

181. Gao, X.; Zhang, J.Q.; Zhang, X.; Zhou, J.; Jiang, Z.; Huang, P.; Tang, Z.; Bao, Y.; Cheng, J.; Tang, H.; et al. Rice qGL3/OsPPKL1 functions with the GSK3/SHAGGY-like kinase OsGSK3 to modulate brassinosteroid signaling. *Plant Cell* **2019**, *31*, 1077–1093. [CrossRef] [PubMed]

182. Merlot, S.; Gosti, F.; Guerrier, D.; Vavasseur, A.; Giraudat, J. The ABI1 and ABI2 protein phosphatases 2C act in a negative feedback regulatory loop of the abscisic acid signalling pathway. *Plant J.* **2001**, *25*, 295–303. [CrossRef] [PubMed]

183. Zhang, S.; Cai, Z.; Wang, X. The primary signaling outputs of brassinosteroids are regulated by abscisic acid signaling. *Proc. Natl. Acad. Sci. USA* **2009**, *106*, 4543–4548. [CrossRef] [PubMed]

184. Hao, Y.; Wang, H.; Qiao, S.; Leng, L.; Wang, X. Histone deacetylase HDA6 enhances brassinosteroid signaling by inhibiting the BIN2 kinase. *Proc. Natl. Acad. Sci. USA* **2016**, *113*, 10418–10423. [CrossRef] [PubMed]

185. ter Haar, E.; Coll, J.T.; Austen, D.A.; Hsiao, H.M.; Swenson, L.; Jain, J. Structure of GSK3beta reveals a primed phosphorylation mechanism. *Nat. Struct. Biol.* **2001**, *8*, 593–596. [CrossRef]

186. Monteserin-Garcia, J.; Al-Massadi, O.; Seoane, L.M.; Alvarez, C.V.; Shan, B.; Stalla, J.; Paez-Pereda, M.; Casanueva, F.F.; Stalla, G.K.; Theodoropoulou, M. Sirt1 inhibits the transcription factor CREB to regulate pituitary growth hormone synthesis. *FASEB J.* **2013**, *27*, 1561–1571. [CrossRef]

187. Taylor, I.; Wang, Y.; Seitz, K.; Baer, J.; Bennewitz, S.; Mooney, B.P.; Walker, J.C. Analysis of phosphorylation of the receptor-like protein kinase HAESA during arabidopsis floral abscission. *PLoS ONE* **2016**, *11*, e0147203. [CrossRef]

188. Sarikhani, M.; Mishra, S.; Maity, S.; Kotyada, C.; Wolfgeher, D.; Gupta, M.P.; Singh, M.; Sundaresan, N.R. SIRT2 deacetylase regulates the activity of GSK3 isoforms independent of inhibitory phosphorylation. *Elife* **2018**, *7*, e32925. [CrossRef]

189. Wang, P.; Du, Y.; Hou, Y.J.; Zhao, Y.; Hsu, C.C.; Yuan, F.; Zhu, X.; Tao, W.A.; Song, C.P.; Zhu, J.K. Nitric oxide negatively regulates abscisic acid signaling in guard cells by S-nitrosylation of OST1. *Proc. Natl. Acad. Sci. USA* **2015**, *112*, 613–618. [CrossRef]

190. Wang, S.B.; Venkatraman, V.; Crowgey, E.L.; Liu, T.; Fu, Z.; Holewinski, R.; Ranek, M.; Kass, D.A.; O'Rourke, B.; van Eyk, J.E. Protein s-nitrosylation controls glycogen synthase kinase 3beta function independent of its phosphorylation state. *Circ. Res.* **2018**, *122*, 1517–1531. [CrossRef]

191. Song, S.; Wang, H.; Sun, M.; Tang, J.; Zheng, B.; Wang, X.; Tan, Y.W. Reactive oxygen species-mediated BIN2 activity revealed by single-molecule analysis. *New Phytol.* **2019**, *223*, 692–704. [CrossRef] [PubMed]

192. He, G.; Liu, J.; Dong, H.; Sun, J. The blue-light receptor CRY1 interacts with BZR1 and BIN2 to modulate the phosphorylation and nuclear function of BZR1 in repressing BR signaling in *Arabidopsis*. *Mol. Plant* **2019**, *12*, 689–703. [CrossRef] [PubMed]

193. Ling, J.J.; Li, J.; Zhu, D.; Deng, X.W. Noncanonical role of Arabidopsis COP1/SPA complex in repressing BIN2-mediated PIF3 phosphorylation and degradation in darkness. *Proc. Natl. Acad. Sci. USA* **2017**, *114*, 3539–3544. [CrossRef] [PubMed]

194. Li, J.; Terzaghi, W.; Gong, Y.; Li, C.; Ling, J.J.; Fan, Y.; Qin, N.; Gong, X.; Zhu, D.; Deng, X.W. Modulation of BIN2 kinase activity by HY5 controls hypocotyl elongation in the light. *Nat. Commun.* **2020**, *11*, 1592. [CrossRef] [PubMed]

195. Gampala, S.S.; Kim, T.W.; He, J.X.; Tang, W.; Deng, Z.; Bai, M.Y.; Guan, S.; Lalonde, S.; Sun, Y.; Gendron, J.M.; et al. An essential role for 14-3-3 proteins in brassinosteroid signal transduction in Arabidopsis. *Dev. Cell* **2007**, *13*, 177–189. [CrossRef]

196. Hoecker, U. The activities of the E3 ubiquitin ligase COP1/SPA, a key repressor in light signaling. *Curr. Opin. Plant Biol.* **2017**, *37*, 63–69. [CrossRef]

197. Liu, B.; Zuo, Z.; Liu, H.; Liu, X.; Lin, C. Arabidopsis cryptochrome 1 interacts with SPA1 to suppress COP1 activity in response to blue light. *Genes Dev.* **2011**, *25*, 1029–1034. [CrossRef]

198. Lian, H.L.; He, S.B.; Zhang, Y.C.; Zhu, D.M.; Zhang, J.Y.; Jia, K.P.; Sun, S.X.; Li, L.; Yang, H.Q. Blue-light-dependent interaction of cryptochrome 1 with SPA1 defines a dynamic signaling mechanism. *Genes Dev.* **2011**, *25*, 1023–1028. [CrossRef]

199. Pham, V.N.; Kathare, P.K.; Huq, E. Phytochromes and phytochrome interacting factors. *Plant Physiol.* **2018**, *176*, 1025–1038. [CrossRef]

200. Gangappa, S.N.; Botto, J.F. The multifaceted roles of HY5 in plant growth and development. *Mol. Plant* **2016**, *9*, 1353–1365. [CrossRef]

201. Osterlund, M.T.; Hardtke, C.S.; Wei, N.; Deng, X.W. Targeted destabilization of HY5 during light-regulated development of Arabidopsis. *Nature* **2000**, *405*, 462–466. [CrossRef] [PubMed]

202. Truernit, E.; Bauby, H.; Belcram, K.; Barthelemy, J.; Palauqui, J.C. OCTOPUS, a polarly localised membrane-associated protein, regulates phloem differentiation entry in Arabidopsis thaliana. *Development* **2012**, *139*, 1306–1315. [CrossRef] [PubMed]

203. Anne, P.; Azzopardi, M.; Gissot, L.; Beaubiat, S.; Hematy, K.; Palauqui, J.C. OCTOPUS negatively regulates BIN2 to control phloem differentiation in *Arabidopsis thaliana*. *Curr. Biol.* **2015**, *25*, 2584–2590. [CrossRef] [PubMed]

204. Pillitteri, L.J.; Peterson, K.M.; Horst, R.J.; Torii, K.U. Molecular profiling of stomatal meristemoids reveals new component of asymmetric cell division and commonalities among stem cell populations in Arabidopsis. *Plant Cell* **2011**, *23*, 3260–3275. [CrossRef]

205. Houbaert, A.; Zhang, C.; Tiwari, M.; Wang, K.; de Marcos Serrano, A.; Savatin, D.V.; Urs, M.J.; Zhiponova, M.K.; Gudesblat, G.E.; Vanhoutte, I.; et al. POLAR-guided signalling complex assembly and localization drive asymmetric cell division. *Nature* **2018**, *563*, 574–578. [CrossRef]

206. Liu, J.; Chen, J.; Zheng, X.; Wu, F.; Lin, Q.; Heng, Y.; Tian, P.; Cheng, Z.; Yu, X.; Zhou, K.; et al. GW5 acts in the brassinosteroid signalling pathway to regulate grain width and weight in rice. *Nat. Plants* **2017**, *3*, 17043. [CrossRef]

207. Hirano, K.; Kawamura, M.; Araki-Nakamura, S.; Fujimoto, H.; Ohmae-Shinohara, K.; Yamaguchi, M.; Fujii, A.; Sasaki, H.; Kasuga, S.; Sazuka, T. Sorghum DW1 positively regulates brassinosteroid signaling by inhibiting the nuclear localization of BRASSINOSTEROID INSENSITIVE 2. *Sci. Rep.* **2017**, *7*, 126. [CrossRef]

208. Samakovli, D.; Margaritopoulou, T.; Prassinos, C.; Milioni, D.; Hatzopoulos, P. Brassinosteroid nuclear signaling recruits HSP90 activity. *New Phytol.* **2014**, *203*, 743–757. [CrossRef]

209. Peng, P.; Yan, Z.; Zhu, Y.; Li, J. Regulation of the *Arabidopsis* GSK3-like kinase BRASSINOSTEROID-INSENSITIVE 2 through proteasome-mediated protein degradation. *Mol. Plant* **2008**, *1*, 338–346. [CrossRef]

210. Lama, S.; Broda, M.; Abbas, Z.; Vaneechoutte, D.; Belt, K.; Sall, T.; Vandepoele, K.; van Aken, O. Neofunctionalization of mitochondrial proteins and incorporation into signaling networks in plants. *Mol. Biol. Evol.* **2019**, *36*, 974–989. [CrossRef]

211. Failor, K.L.; Desyatnikov, Y.; Finger, L.A.; Firestone, G.L. Glucocorticoid-induced degradation of glycogen synthase kinase-3 protein is triggered by serum- and glucocorticoid-induced protein kinase and Akt signaling and controls beta-catenin dynamics and tight junction formation in mammary epithelial tumor cells. *Mol. Endocrinol.* **2007**, *21*, 2403–2415. [CrossRef] [PubMed]

International Journal of
Molecular Sciences

Article

Early Cellular Responses Induced by Sedimentary Calcite-Processed Particles in Bright Yellow 2 Tobacco Cultured Cells

Daniel Tran [1,2,*], Tingting Zhao [2], Delphine Arbelet-Bonnin [2,3], Takashi Kadono [2,4], Patrice Meimoun [2], Sylvie Cangémi [2], Tomonori Kawano [4,5,6,7], Rafik Errakhi [8] and François Bouteau [2,3,5,6]

1   Agroscope, Institute for Plant Production Systems, 1964 Conthey, Switzerland
2   Laboratoire Interdisciplinaire des Energies de Demain, Université de Paris, 75013 Paris, France; zhaotingting0325@gmail.com (T.Z.); delphine.bonnin@univ-paris-diderot.fr (D.A.-B.); kadono.takashi@gmail.com (T.K.); patrice.meimoun@upmc.fr (P.M.); sylvie.cangemi@univ-paris-diderot.fr (S.C.); francois.bouteau@univ-paris-diderot.fr (F.B.)
3   Cogitamus Laboratory, 75013 Paris, France
4   Graduate School of Environmental Engineering, University of Kitakyushu, 1-1 Hibikino, Wakamatsu-ku, Kitakyushu 808-0135, Japan; kawanotom@gmail.com
5   LINV Kitakyushu Research Center (LINV@Kitakyushu), Kitakyushu 808-0135, Japan
6   International Photosynthesis Industrialization Research Center, The University of Kitakyushu, Kitakyushu 808-0135, Japan
7   Paris Interdisciplinary Energy Research Institute (PIERI), Université de Paris, 75013 Paris, France
8   Eurofins Agriscience Service, Casablanca 20000, Morocco; r.errakhi@gmail.com
*   Correspondence: qnoctnandaniel.tran@agroscope.admin.ch

Received: 24 April 2020; Accepted: 12 June 2020; Published: 16 June 2020

Abstract: Calcite processed particles (CaPPs, Megagreen®) elaborated from sedimentary limestone rock, and finned by tribomecanic process were found to increase photosynthetic CO_2 fixation grapevines and stimulate growth of various cultured plants. Due to their processing, the CaPPs present a jagged shape with some invaginations below the micrometer size. We hypothesised that CaPPs could have a nanoparticle (NP)-like effects on plants. Our data show that CaPPs spontaneously induced reactive oxygen species (ROS) in liquid medium. These ROS could in turn induce well-known cellular events such as increase in cytosolic Ca^{2+}, biotic ROS generation and activation of anion channels indicating that these CaPPs could activate various signalling pathways in a NP-like manner.

Keywords: tobacco; calcium; calcite; reactive oxygen species; ion channels; cellular signalization

---

1. Introduction

Several minerals have been used in agriculture [1], among which sedimentary rock that emerges from calcareous seaweed. Megagreen® is a preparation from calcite processed particles (CaPPs), elaborated from sedimentary limestone rock, which is finned and activated by a tribomecanic process [2]. These processed calcite particles are supposedly small enough to enter the leaf and have a beneficial effect on plants. The application of CaPPs on grapevines submitted to water stress was shown to increase photosynthetic CO_2 fixation [3]. The benefits of CaPPs once inside the plant were supposed to be due the decomposition products, CO_2 and CaO, that could feed the plant. However, the cellular responses induced by the CaPPs are poorly understood. Due to the tribomecanic processing, the CaPPs present jagged shape with some invaginations below the micrometer size (Megagreen® data sheet: https://dokumen.tips/documents/megagreen-study.html, accessed on 06/04/2020). Nanoparticles (NPs) possess a large specific surface area allowing a greater reactivity compared to macrosized particles.

Since the high surface reactivity of NPs is important for their biological effects, we hypothesised that CaPPs could have NP-like effects on plants.

Recent reviews focused on beneficial applications of nanomaterials in agricultural production [4–8]. NPs notably could induce enhancement in growth and seed yield [9], and participate in crop protection [4,10]. Although some cerium oxide nanoparticles were shown to augment reactive oxygen species (ROS) scavenging in *Arabidopsis thaliana* plants [11], a part of the biological effects of various NPs is proposed to be due to their ability to produce ROS, possibly due to molecular size, shape, oxidation status, increased specific surface area, bonded surface species, surface coating, solubility, and degree of aggregation and agglomeration [12–14]. We effectively showed by using *Nicotiana tabacum* L. cv. Bright Yellow 2 (BY-2) cultured cells that TiO_2 NPs spontaneously generate ROS in the culture medium, but also induced a rapid biological ROS production and a ROS-dependent increase in cytosolic calcium ($[Ca^{2+}]_{cyt}$) [15]. Variations of $[Ca^{2+}]_{cyt}$ serve as secondary messenger involved in many adaptation and developmental processes in plants [16,17]. Reactive oxygen species also play a key signal transduction role in plant cells, such as growth regulation, development, responses to environmental stimuli and cell death [18,19]. However, the response of plants to NPs varies with the growth stages, type of plant species and the nature of NPs. Thus, they could have positive and negative effects on plants [20]. In this study, we tested the impact of CaPPs on cell viability and further checked if CaPPs as NPs could induce ROS generation due to their increased specific surface area and carried out an experimental layout on plant cultured cells to study the impact of CaPPs on variations of $[Ca^{2+}]_{cyt}$, biological ROS generation and ion fluxes variations, early cellular responses frequently involved in signalling processes [21].

2. Results

2.1. Non-Biological ROS Production by CaPPs

We made the hypothesis that CaPPs could have a NP-like effects and could thus generate ROS independently of living cells. According to this hypothesis, we checked if CaPPs could induce ROS generation independently of any living cells. We showed by using the Murashige and Skoog (MS) culture medium that CaPPs spontaneously generate in a dose- and time-dependent manner ROS production evidenced by chemiluminescence of *Cypridina* luciferin analogue (CLA) (Figure 1A,B, Supplemenary Figure S1A). It is noteworthy that, on the contrary to CaPPs, the addition of dissolved $CaCO_3$ (the main component of CaPPs) at 100 μg.mL^{-1} in free MS medium did not induce ROS generation (Supplementary Figure S1B), reducing the likelihood of a chemical effect for CaPPs and providing a NP-like effect. The CaPP-induced ROS production continues to increase for about 5 h and decreases slowly after 24 h (Figure 1B). The chemiluminescence of CLA indicates the generation of superoxide anion ($O_2^{\bullet-}$), and of singlet oxygen ($^1O_2$) to a lesser extent [22]. We then checked the effect of DABCO (1,4-diazabicyclo(2,2,2)octane, a scavenger of $^1O_2$) and tiron (sodium 4,5-dihydroxybenzene-1,3-disulfonate, a scavenger of $O_2^{\bullet-}$) on CaPP-induced ROS generation (Figure 1C,D). Only tiron allowed for a significant decrease of ROS generation. This suggests that CaPPs induced mainly $O_2^{\bullet-}$ generation in culture medium. Since hydroxyl radical ($HO^\bullet$) could be chemically generated from $O_2^{\bullet-}$ through Haber–Weiss or Fenton reactions, we further search for $HO^\bullet$ generation by using the specific probe hydroxyphenyl fluorescein (HPF) [23]. A time-dependent increase in HPF fluorescence could be detected upon treatment with 100 μg.mL^{-1} CaPPs (Figure 1E). This increase in HPF fluorescence was decreased by a pretreatment with 100 mM DMTU (Dimethylthiourea), a scavenger of $HO^\bullet$ (Figure 1F) supporting the hypothesis of $HO^\bullet$ generation.

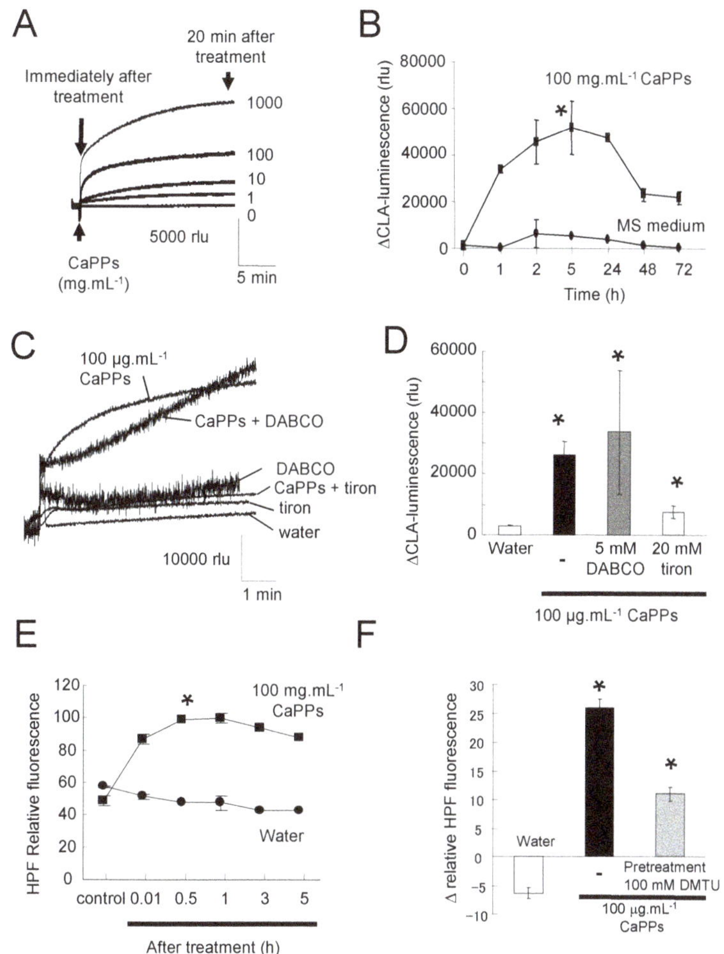

Figure 1. Calcite processed particles (CaPPs)-induced ROS generation in free Murashige and Skoog (MS) medium. (**A**). Typical time and dose *Cypridina* Luminescent Analog (CLA) luminescence recorded in MS medium free of cells after addition of CaPPs. (**B**). Mean values of CaPP-induced CLA luminescence. (**C,D**). Effect of singlet oxygen scavenger DABCO (5 mM), and superoxide anion scavenger tiron (20 mM) on CaPP-induced CLA luminescence. The histogram represents the mean values after 20 min. (**E**). Time-dependent hydroxyphenyl fluorescein (HPF) fluorescence in response to 100 μg.mL^{-1} CaPPs. (**F**). Effect of hydroxyl radical scavenger DMTU (100 mM) on CaPP-induced HPF fluorescence after 30 min. Data corresponded to mean values $\pm$ standard error (SE) of at least 4 independent experiments. * Significantly different from the water treatment. Data were analyzed by variance analysis (ANOVA) and when ANOVA gave a statistically significant result, the Newman–Keuls multiple range test was used to identify which specific pairs of means were different. All numeric differences in the data were considered significantly different for a *p*-value $\leq$ 0.05.

2.2. CaPP Particles Induced Cytosolic Calcium Variation in Tobacco BY-2 Cells

We showed that TiO$_2$ NPs induced a ROS-dependent increase in cytosolic calcium ([Ca^{2+}]$_{cyt}$) in BY-2 tobacco cells [15]. ROS were also shown to activate plasma membrane Ca^{2+} channels in plant cells [24]. We thus investigated the effect of CaPPs on cytosolic calcium level in BY-2 tobacco cultured cells expressing the Ca^{2+}-sensitive luminescent protein aequorin in their cytosol [25]. CaPPs induced a rapid dose-dependent and transient increase in $[Ca^{2+}]_{cyt}$ (Figure 2A,B). Influx of Ca^{2+} from the apoplast through plasma membrane was confirmed by using 500 µM La^{3+}, a blocker of Ca^{2+} channels, and 3 mM EGTA, a calcium chelator (Figure 2C,D). This Ca^{2+} influx was dependent on the early CaPP-induced ROS production since tiron, and DMTU could also reduce the $[Ca^{2+}]_{cyt}$ increase (Figure 2C,D).

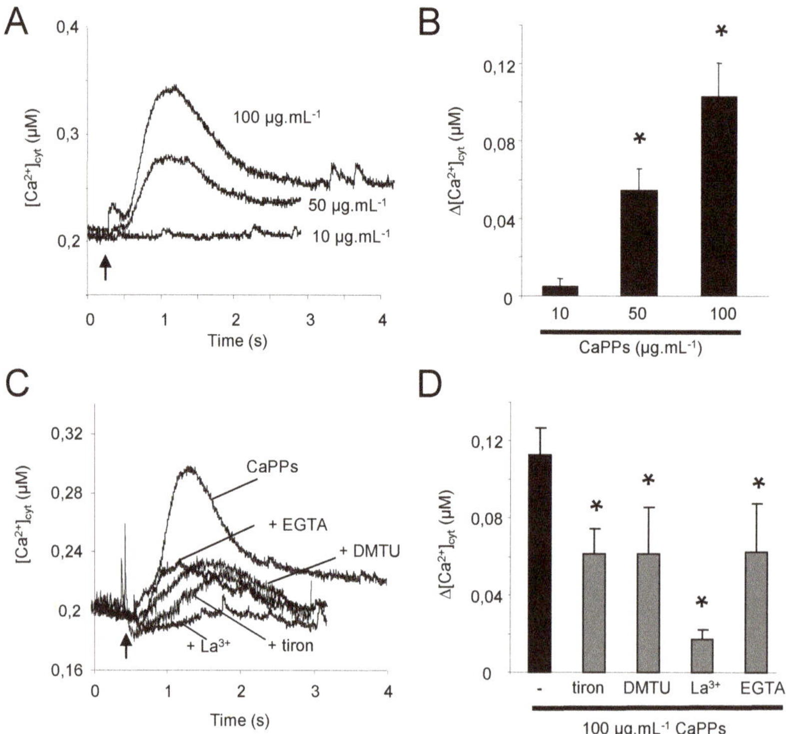

Figure 2. CaPP-induced variations of cytosolic Ca^{2+} in BY-2 cells. (**A**). A typical $[Ca^{2+}]_{cyt}$ variations of aequorin expressing *BY-2* cells in response to various concentrations of CaPPs. (**B**). Mean values of maximal $[Ca^{2+}]_{cyt}$ increase in response to various concentrations of CaPPs. * Significantly different from the treatment at 10 µg.mL^{-1} CaPPs. (**C**). Effect of calcium (La^{3+}, EGTA) and ROS (tiron and DMTU) pharmacology on 100 mg.ml^{-1} CaPPs induced $[Ca^{2+}]_{cyt}$ variations. (**D**). Mean values of maximal $[Ca^{2+}]_{cyt}$ increase in response to 100 µg.mL^{-1} of CaPPs in the presence of calcium and ROS pharmacology. Controls with pharmacology alone did not affect significantly the basal $[Ca^{2+}]_{cyt}$ (not shown). Data corresponded to mean values ± SD of at least six independent experiments. * Significantly different from the treatment at 100 µg.mL^{-1}. Data were analyzed by variance analysis (ANOVA) and when ANOVA gave a statistically significant result, the Newman–Keuls multiple range test was used to identify which specific pairs of means were different. All numeric differences in the data were considered significantly different for a *p*-value ≤ 0.05.

Variations in $[Ca^{2+}]_{cyt}$ and ROS generation are known to regulate different early events involved in signal transduction pathways such as ion channel activities and NADPH-oxidase activities induced in response to various biotic and abiotic stressors [21,26]. We then further checked if such events could be regulated by CaPPs.

2.3. CaPPs Induced a NADPH Oxidase-Dependent ROS Production

As expected from the spontaneously CaPP-induced ROS production in MS medium (Figure 1A), the chemiluminescence of CLA also rapidly increased after addition of 100 µg.mL^{-1} CaPPs in BY-2 cell cultures (Figure 3A). From analysis of luminol-chemilumiscence, we further showed that CaPP-induced ROS generation reached a maximum at about 8 h in BY-2 cultured cells when untreated cells presented no significant increase in chemilumiscence level during the time of experiments (Figure 3B). This effect was dose-dependent (Figure 3C). The addition of 50 µM diphenyleneiodonium (DPI), an inhibitor of NADPH-oxidase [27,28], into BY-2 cell medium diminished the chemilumiscence (Figure 3C). These data suggest the involvement of plant enzymes such NADPH-oxidase in this ROS production induced by CaPPs.

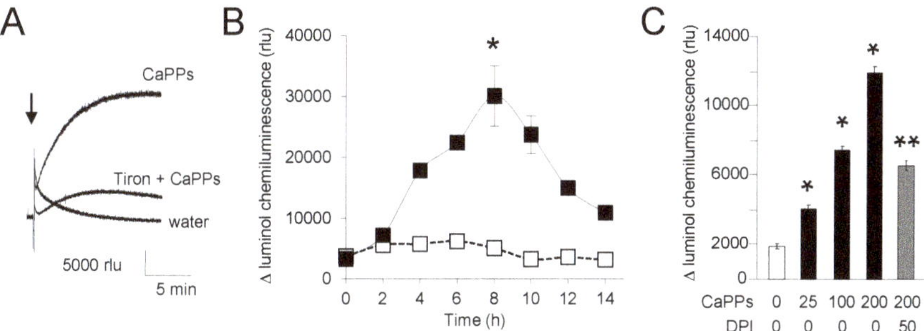

Figure 3. Biological CaPP-induced ROS generation by BY-2 cells. (**A**). Typical time CLA luminescence recorded with BY-2 cells after addition of 100 µg.mL^{-1} CaPPs with or without 20 mM tiron. (**B**). Kinetic of biological ROS generation detected with luminol during 14 h after addition of 100 µg.mL^{-1} CaPPs. (**C**). Mean values of maximal ROS increase (at 8h) in response to various CaPPs concentrations (in mg.mL^{-1}) and in the presence 50 µM diphenyleneiodonium (DPI), an inhibitor of NADPH-oxidase. Data corresponded to mean values ± SD of at least six independent experiments. * significantly different from the control. ** Significantly different from the treatment at 200 µg.mL^{-1} CaPPs. Data were analyzed by variance analysis (ANOVA) and when ANOVA gave a statistically significant result, the Newman–Keuls multiple range test was used to identify which specific pairs of means were different. All numeric differences in the data were considered significantly different for a *p*-value ≤ 0.05.

2.4. CaPPs Induce a Depolarization of Plasma Membrane Due to Anion Channel Activation

We used an electrophysiological approach to test the effect of CaPPs on membrane potentials and ion currents of cultured cells. Upon direct addition of CaPPs, we recorded a rapid dose-dependent depolarization of BY-2 cells (Figure 4A). The depolarization was correlated with a large increase in ion currents (Figure 4B). Because impalement of a single cells could not be maintained for a long time, we further analysed the mean plasma membrane potentials and ion currents of BY-2 cell populations exposed to CaPPs for different amounts of time (Figure 4C,D). The value of the resting membrane potential (V_m) of control cells (without treatment) was around -25 mV (Figure 4C), in the same range of previous studies [26,29]. As expected from the direct addition of CaPPs (Figure 4A), cells pretreated 15 min with CaPPs were drastically depolarized (Figure 4C), but these depolarizations were transient and the cell polarizations were partly recovered for cells pretreated during 45 min (Figure 4C). These membrane potential variations were correlated with a transient increase in ion currents (Figure 4B,D) presenting the main hallmarks of anion current as previously characterized [26,29–31]. This type of current was shown to be sensitive to structurally unrelated anion channel inhibitors [26,29]. Accordingly, the increases in ion currents and the depolarizations after 15 min CaPPs pretreatment were effectively partly avoided upon pretreatment with 200 µM of glibenclamide (gli) or 9-anthracen carboxylic acid (9AC), two structurally unrelated anion channel blockers (Figure 4D), confirming the anionic nature of these currents. These currents present the features of slow anion channels [32], but a part of the instantaneous current could be carried out by fast-activating anion channels [33]. However, these data show that increase in anion currents could be part of the early CaPP-induced signaling events.

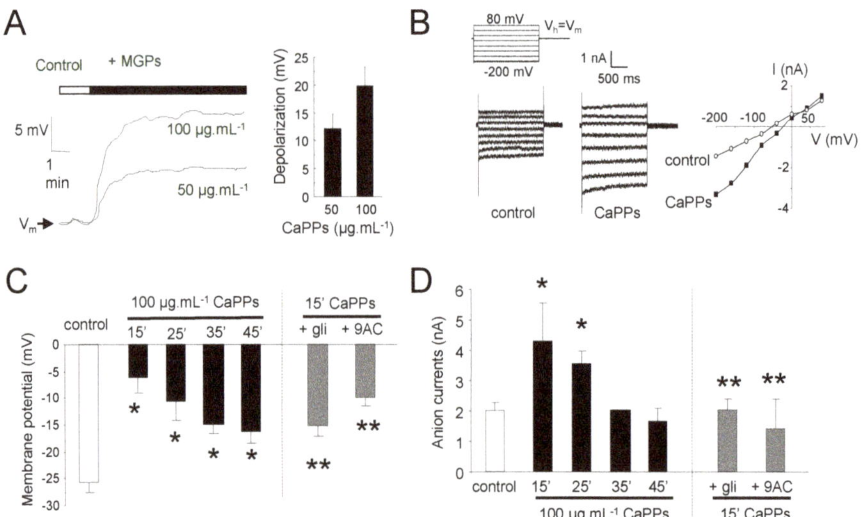

Figure 4. CaPP-induced depolarization and anion current increase in BY-2 cells. (**A**). Typical depolarizations of BY-2 cell observed in response to CaPPs at 50 or 100 $\mu g.mL^{-1}$ and mean values of depolarizations. (**B**). Whole currents measured under control conditions and 5 min after addition of 100 $\mu g.mL^{-1}$ CaPPs. The protocol was as illustrated, holding potential (V_h) was V_m. Corresponding current-voltage relationships at 1.8 s. (**C**). Mean values of polarizations for BY-2 cells treated during different times with 100 $\mu g.mL^{-1}$ CaPPs and mean values of polarizations for BY-2 cells treated 15 min with 100 $\mu g.mL^{-1}$ CaPPs in the presence of 200 μM glibenclamide (gli) or 200 μM 9-antharcen carboxylic acid (9AC), two unrelated anion channel inhibitors. (**D**). Mean values of anion currents for BY-2 cells treated during different times with 100 $\mu g.mL^{-1}$ CaPPs and mean values of anion currents for BY-2 cells treated 15 min with 100 $\mu g.mL^{-1}$ CaPPs in the presence of 200 μM gli or 200 μM 9AC. Currents were recorded at −200 mV and 1.8 s of voltage clamp. Control values corresponded to the value before CaPPs addition. Data corresponded to mean values ± SD of at least six independent experiments. * Significantly different from the control. ** Significantly different from the treatment at 15 min. Data were analyzed by variance analysis (ANOVA) and when ANOVA gave a statistically significant result, the Newman–Keuls multiple range test was used to identify which specific pairs of means were different. All numeric differences in the data were considered significantly different for a *p*-value ≤ 0.05.

2.5. CaPPs Toxicity?

Nanoparticles were shown to induce cell death in various models [6,13,34]. We thus checked if CaPPs could induce death of BY-2 cells. No increase in cell death was observed in BY-2 cultured cells, even after 24 h of treatment (Figure 5A). We further checked if these CaPPs could have an impact on BY-2 cell culture growth. As expected from the data of cell death, addition of CaPPs in the culture medium of BY-2 cells for 7 days has no impact on the culture cell growth (Figure 5B).

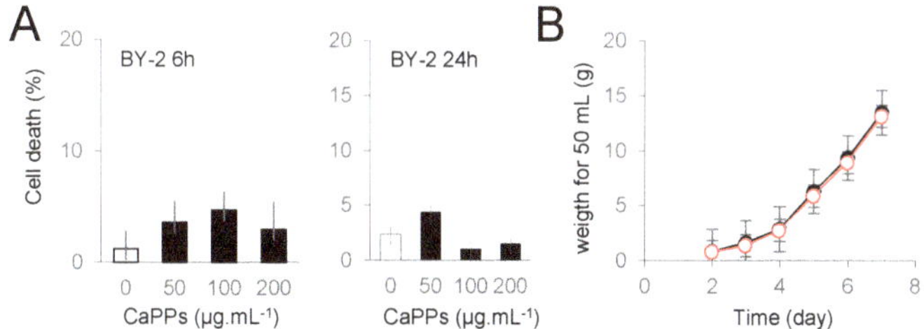

Figure 5. CaPPs cytotoxicity in *BY-2* cultured cells. (**A**). Cell death extent in BY-2 cultured cells detected by the Evans Blue staining after 6 or 24h of treatment with various concentrations of CaPPs. (**B**). BY-2 cultured cell growth during 7 days in the presence or not of 200 µg.mL^{-1} CaPPs. The data corresponded to means of at least 4 independent replicates and error bars corresponded to SE.

3. Discussion

The CaPPs application has been shown to be beneficial on several crops such as olive trees, maize, strawberry and lettuce, especially under drought conditions (technical data sheet for Megagreen®: https://dokumen.tips/documents/megagreen-study.html, accessed on 06/04/2020). The benefits of CaPPs once inside the leaves was attributed to the decomposition products CO_2 and CaO that could feed the plant. CaPPs application on grapevines subjected to water stress was shown to increase photosynthetic CO_2 fixation [3]. The CaPPs penetrating directly into the leaves are supposed to increase CO_2 saturation in the leave leading to stomatal closure and therefore a reduction in evapotranspiration a reduction of photorespiration in favor of photosynthesis [3]. Spray of CaO were also shown to correct Ca^{2+} deficiency in groundnut [35] although the mean levels of Ca^{2+} were not statistically different between CaPP-treated and untreated vines [36]. However, due to the size distribution of these CaPPs ranging from the nano- to the microparticle (0.1 to 20 µm), we hypothesized NP-like effects of CaPPs at the cellular level. By using nonphotosynthetic BY-2 cultured plant cells, we could discriminate the effects of NPs from already-reported effects on photosynthetic activity.

Our data showed that CaPPs induce ROS generation independently of any living cells. This ROS production is dose- and time-dependent and seemed to be mainly due to $O_2^{\bullet-}$ (detected by CLA and scavenged by tiron) and subsequently $HO^\bullet$ (detected by HPF and scavenged by DMTU) through Haber–Weiss or Fenton reactions after the dismutation of $O_2^{\bullet-}$ into H_2O_2. These data correlate with previous one indicating that NPs from different nature can produce ROS due to their increased specific surface area [12–14].

Our pharmacological data with ROS scavengers show that these CaPP-induced ROS could be responsible in BY-2 cells for the induction of well-known cellular events involved in the signalling process, such as calcium influx through plasma membrane Ca^{2+} channels, subsequent NADPH-oxidase stimulation and anion channel activation. The NADPH-oxidase stimulation and anion channel activations could also be recorded in response to CaPPs in *A. thaliana* cultured cells (Supplementary Figure S2). ROS generation and the cytosolic calcium increase are reminiscent with what was observed in response to TiO_2 NPs in BY-2 cells [15], or in response to ZnO NPs in *Salicornia* [37], but also in responses to O_3, another oxidative stress [38], on tobacco cells [39] and *A. thaliana* cultured cells [21,40–42]. Less data are available on the effect of NPs on ion channel regulation especially in plants, but it is noticeable that polystyrene NPs could activate CFTR-Cl$^-$ channels in hamster kidney cells [43] and O_3 anion currents in *A. thaliana* cells [21].

Although CaPPs do not seem to be toxic for BY-2 cells, such signalling events are frequently related to the induction of programmed cell death (PCD) [21,26,29]. Effectively we could observe in *A. thaliana* cells after addition of CaPPs an increase in cell death slowing the whole suspension growth (Supplementary Figure S3). Toxic effects of nanoparticles were already observed in response to various NPs such as ZnONPs or AgNPs in algae [44,45] or CuONPs, SiNPs and single-wall carbon nanotubes on a terrestrial model [34,46,47], sometimes due to the PCD process [34]. In *A. thaliana,* cell death was dependent on transcription and translation (Figure S3), effectively suggesting an active process, thus a PCD. The discrepancy observed in terms of cell death between the two cultured cell lines, since there was no record of cell death or the slowing of suspension cell growth for BY-2 cells, which could be explained by a difference in sensitivity. Effectively, carbon nanotubes were shown to induce the growth enhancement of tobacco cells [48] when they induce PCD in *A. thaliana* and rice [14,34]. However, the CaPP-induced PCD in *A. thaliana* cells could be reduced by the ROS scavengers DMTU and tiron, the blockers of Ca^{2+} influx, BAPTA and La^{3+} and the anion channel blockers 9AC and glibenclamide (Supplementary Figure S3). These data support the hypothesis that the CaPP-induced ROS generation induces the signaling pathways leading to the PCD process. It is also noteworthy that these cellular events are also involved in stomatal aperture regulation [49]. We could further confirm the decrease of stomatal aperture 30 min after application of 50 μM CaPPs on the epidermis *A. thaliana* leaves (Supplementary Figure S4). Thus, the CaPP-induced stomatal closure could be due to not only an increase in CO_2 saturation of the leaves [3], but also to the CaPP-induced ROS generation.

In summary, our study shows that CaPPs could have, in addition to its known effects on photosynthesis [3], NP-like effects due to their size distribution. The abiotic ROS generation induced by these CaPPs could induce cellular events that could be involved in various signaling pathways. More studies, particularly with different species, will be needed to clarify the possible outputs of these signaling pathways.

4. Materials and Methods

4.1. CaPP Particles

Megagreen® is composed of calcite processed particles (CaPPs) elaborated from sedimentary limestones rock, which is finned and activated by tribomecanic process (European Patent N° WO/2000/064586). These CaPPs present a distribution ranging from the nano- to the microparticle (0.1 to 20 μm). The chemical composition of CaPPs is: total calcium carbonate 823.0 $g{\cdot}kg^{-1}$; SiO_2 85.2 g kg^{-1}; MgO 30.2 $g{\cdot}kg^{-1}$; Fe 8.78 $g{\cdot}kg^{-1}$, and other trace elements. CaPPs were diluted in distilled water and pH adjusted to 5.8 with HCl.

4.2. Plant Cell Culture Conditions

Nicotiana tabacum BY-2 cultured cells were grown in Murashige and Skoog medium (MS medium) [50] complemented with 30 $g.L^{-1}$ sucrose, 0.1 $mg.L^{-1}$ 2,4 D (pH 5.8) and maintained by weekly dilution (2/80). The cell culture was agitated on a rotary shaker at 120 rpm at 22 ± 2 °C in the dark. Such cells are white and nonphotosynthetic. All experiments were performed at 22 ± 2 °C using the cells in log-phase (6 days after subculturing).

Cell growth was estimated for by recording each day after subculture the fresh weight of cells contained in 50 mL of culture for BY-2 cell cultures.

4.3. Monitoring of ROS Production

The production of ROS was monitored using different techniques and probes. The chemiluminescence of the *Cypridina* luciferin analog (CLA) react mainly with $O_2^{\bullet-}$ and $^1O_2$ with light emission [22]. Chemiluminescence from CLA was monitored using a FB12-Berthold luminometer (with a signal integrating time of 0.2 s). For data analysis, the luminescence ratio (L/Lbasal) was calculated by dividing the luminescence intensities of CLA-luminescence (L) with the luminescence intensity before treatment (Lbasal). Hydroxy radicals (HO•) formation was also checked using the specific probe hydroxyphenyl fluorescein (HPF) [23]. Briefly, HPF was added to 1 mL of MS medium to a final concentration of 10 μM at different times after the addition of 100 $mg.mL^{-1}$ of CaPPs. The fluorescence increase was monitored at 515 nm after an excitation at 490 nm using a F-2000 spectrofluorimeter (Hitachi, Tokyo, Japan).

For biological production of ROS, we used the chemiluminescence of luminol [51], which is dependent on the activity of cell-derived peroxidase. Briefly, 6 mL of the cultured cells were inoculated with CaPPs. Before each measurement, 200 µL of the cell culture was added prior to the addition of 5 µL luminol (1.1 mM). Chemiluminescence measurements were carried out at 30 min intervals using a FB12-Berthold luminometer (signal integrating time 0.2 s).

4.4. Aequorin Luminescence Measurements

Cytoplasmic Ca^{2+} variations were recorded from BY-2 cultured cells expressing the apoaequorin gene [25]. For Ca^{2+} measurement, aequorin was reconstituted by an overnight incubation of the cell cultures in MS medium supplemented with 2.5 µM native coelenterazine. Cell culture aliquots (450 µL in MS medium) were transferred carefully to a luminometer glass tube and luminescence was recorded continuously at 0.2 s intervals using a FB12-Berthold luminometer (Berthold Technologies, Bad Wildbad, Germany). Treatments were performed by 50 µL injections containing the CaPPs. At the end of each experiment, residual aequorin was discharged by addition of 500 µL of a 1M $CaCl_2$ solution dissolved in 100% methanol. The resulting luminescence was used to estimate the total amount of aequorin in each experiment. Calibration of the calcium measurement was performed using the equation: pCa = 0.332588(−logk) +5.5593, where k is a rate constant equal to luminescence counts per second divided by total remaining counts [25]. To test the effects of each different pharmacological treatment, BY-2 cells were pretreated for 15 min before the application of CaPPs.

4.5. Electrophysiology

Experiments were conducted on BY-2 cells maintained in their culture medium to limit stress (main ions in MS medium 28 mM NO_3^- and 16 mM K^+) [26]. Individual cells were immobilized by a microfunnel (approximately 30 to 80 µm outer diameter and controlled by a micromanipulator (WR6-1, Narishige, Tokyo, Japan). Impalement were carried out with a piezoelectric micromanipulator (PCS-5000, Burleigh Inst., New York, NY, USA) in a chamber (500 µL) made of Perspex. Voltage-clamp measurements of whole-cell currents from intact BY-2 cells presenting stable running membrane potential were carried out at room temperature (20–22 °C) using the technique of the discontinuous single voltage-clamp microelectrode [52] adapted to plant cells [40,53]. Microelectrodes were made from borosilicate capillary glass (Clark GC 150F, Clark Electromedical, Pangbourne Reading, UK) pulled on a vertical puller (Narishige PEII, Tokyo, Japan). Their tips were less than 1 µm diameter; they were filled with 600 mM KCl, and had electrical resistances between 20 and 50 MΩ with the culture medium. Specific software (pCLAMP 8) drives the voltage clamp amplifier (Axoclamp 2A, Molecular Devices, Sunnyvale, CAL, USA). Voltage and current were digitalised with a Digidata 1322A (Molecular Devices, Sunnyvale, CAL, USA). In whole-cell current measurements the membrane potential was held to the value of the resting membrane potential. Current recordings were obtained by hyperpolarizing pulses from −200 to +80 mV (20 mV, 2 s steps of current injection, 6s of settling time). We systematically checked that cells were correctly clamped by comparing the protocol voltage values with those really imposed. Only microelectrodes presenting a linear relationship were used.

4.6. Cell Viability Assays

Cell viability was checked using the vital dye, Evans Blue. Cells (50 µL) were incubated for 5 min in 1 mL phosphate buffer pH 7 supplemented with Evans blue to a final concentration of 0.005%.[21] Cells that accumulated Evans blue were considered dead. At least 1000 cells were counted for each independent treatment. The experiment was repeated at least 4 times for each condition.

4.7. Statistical Analysis

Data were analyzed by variance analysis (ANOVA) and when ANOVA gave a statistically significant result, the Newman–Keuls multiple range test was used to identify which specific pairs of means were different. All numeric differences in the data were considered significantly different for a *p*-value ≤ 0.05.

Supplementary Materials: Supplementary Materials can be found at http://www.mdpi.com/1422-0067/21/12/4279/s1. Figure S1. A. Mean values of CaPP-induced CLA luminescence B. Comparison of CaPP- and $CaCO_3$-induced ROS generation in free MS medium. Figure S2: **A.** Kinetic of biological ROS generation detected with luminol during 7 h after addition of 100 μg.mL^{-1} CaPPs. **B.** Mean values of polarizations for A. thaliana cells treated during different times with 100 μg.mL^{-1} CaPPs and mean values of polarizations for A. thaliana cells treated 15 min with 100 μg.mL^{-1} CaPPs in presence of 200 μM glibenclamide (gli) or 200 μM 9-antharcen carboxylic acid (9AC), two unrelated anion channel inhibitors. Figure S3: A. Dose-dependent cell death reaching about 50% of the *Arabidopsis thaliana* cell population was observed after 6 h after treatment with 200 μg. mL^{-1} CaPPs. B. Decrease of the culture growth induced by 200 μg. mL^{-1} CaPPs. C. Decrease of cell death extent by pretreatments with actinomycin D (AD, 20 μg/mL), cycloheximide (Chx, 20 μg/mL), inhibitors of traduction and translation, ROS scavengers Tiron (5 mM) and DMTU (100 mM), Ca^{2+} channel blocker La^{3+} (500 μM), Ca^{2+} chelator, BAPTA (3 mM), and anion channel blockers, glibenclamide (gli 200 μM) and 9AC (200 μM). For each pretreatment, cells were incubated for 15 min before CaPPs treatment. Figure S4: Applications of 100 μg mL^{-1} CaPPs reduce the stomatal aperture of *A. thaliana* leaves. Figure S4: Applications of 100 μg.mL^{-1} CaPPs reduce the stomatal aperture of *A. thaliana* leaves. In presence of 3 mM EGTA, the CaPPs-induced stomatal closure was reduced.

Author Contributions: D.T., T.Z., D.A.-B., T.K. (Takashi Kadono), R.E. and P.M. carried out the experiments. S.C. helped maintaining the cultures. F.B. supervised the project with the help of R.E. and T.K. (Tomonori Kawano), F.B. wrote the manuscript with support from D.T. The manuscript was written through contributions of all authors. All authors have read and agreed to the published version of the manuscript.

Funding: This study was supported by funds from Ministère de l'Enseignement supérieur, de la Recherche et de l'Innovation to LIED.

Acknowledgments: The authors thank Faouzi Attia, Joël Briand and Bernadette Biligui for their implication at early step of the project, and Agronutrition for providing us Megagreen®. This study was supported by funds from Ministère de l'Enseignement supérieur, de la Recherche et de l'Innovation to LIED.

Conflicts of Interest: The authors declare no conflict of interest

References

1. Van Straaten, P. Farming with Rocks and Minerals: Challenges and Opportunities. *An. Acad. Bras. Ciênc.* **2006**, *78*. [CrossRef]
2. Device for Micronizing Materials. EU Patent WO/2000/064586, 2000.
3. Attia, F.; Martinez, L.; Lamaze, T. Foliar Application of Processed Calcite Particles Improved Leaf Photosynthsis of Potted *Vitis vinifera* Grown Under Water Deficit. *OENO ONE* **2014**, *48*, 237–245. [CrossRef]
4. Khot, L.R.; Sankaran, S.; Maja, J.M.; Eshani, R.; Schuster, E.W. Applications of Nanomaterials in Agricultural Production and Crop Protection. *Crop Prot.* **2012**, *35*, 64–70. [CrossRef]
5. Parisi, C.; Vigani, M.; Rodríguez-Cerezo, E. Agricultural Nanotechnologies: What Are the Current Possibilities? *Nano Today* **2015**, *10*, 124–127. [CrossRef]
6. Wang, P.; Lombi, E.; Zhao, F.J.; Kopittke, P.M. Nanotechnology: A New Opportunity in Plant Sciences. *Trends Plant Sci.* **2016**, *21*, 699–712. [CrossRef]
7. Tripathi, D.K.; Singh, S.; Singh, S.; Pandey, R.; Singh, V.P.; Sharma, N.C.; Prasad, S.M.; Dubey, N.K.; Chauhan, D.K. An Overview on Manufactured Nanoparticles in Plants: Uptake, Translocation, Accumulation and Phytotoxicity. *Plant Physiol. Biochem.* **2017**, *110*, 2–12. [CrossRef]
8. Iavicoli, I.; Leso, V.; Beezhold, D.H.; Shvedova, A.A. Nanotechnology in Agriculture: Opportunities, Toxicological Implications, and Occupational risks. *Toxicol. Appli. Pharmacol.* **2017**, *329*, 96–111. [CrossRef]
9. Arora, S.; Sharma, P.; Kumar, S.; Nayan, R.; Khanna, P.K.; Zaidi, M.G.H. Gold-Nanoparticles Induced Enhancement in Growth and Seed Yield of *Brassica juncea*. *Plant Growth Regul.* **2012**, *66*, 303–310. [CrossRef]
10. Rai, M.; Ingle, A. Role of Nanotechnology in Agriculture with Special Reference to Management of Insect Pests. *Appl. Microbiol. Biotechnol.* **2012**, *94*, 287–293. [CrossRef]
11. Wu, H.; Tito, N.; Juan, P.; Giraldo, J.P. Anionic Cerium Oxide Nanoparticles Protect Plant Photosynthesis from Abiotic Stress by Scavenging Reactive Oxygen Species. *ACS Nano* **2017**, *11*, 11283–11297. [CrossRef]
12. Oberdöster, G.; Oberdöster, E.; Oberdöster, J. Nanotoxicology: An Emerging Discipline Evolving from Studies of Ultrafine Particles. *Environ. Health Respect* **2005**, *113*, 823–839. [CrossRef] [PubMed]

13. Voinov, M.A.; Sosa, J.A.; Morrison, P.E.; Smirnova, T.I.; Smirnov, A.I. Surface-Mediated Production of Hydroxyl Radicals as a Mechanism of Iron Oxide Nanoparticle Biotoxicity. *J. Am. Chem. Soc.* **2011**, *133*, 35–41. [CrossRef] [PubMed]

14. Fu, P.P.; Xia, Q.; Hwang, H.M.; Ray, C.P.; Yu, H. Mechanisms of Nanotoxicity: Generation of Reactive Oxygen Species. *J. Food Drug Anal.* **2014**, *22*, 64–75. [CrossRef] [PubMed]

15. Tran, D.; Kadono, T.; Meimoun, P.; Kawano, T.; Bouteau, F. TiO$_2$ Nanoparticles Induce ROS Generation and Cytosolic Ca^{2+} Increases on BY-2 Tobacco Cells: A Chemiluminescence Study. *Luminescence* **2010**, *25*, 140–142.

16. Lecourieux, D.; Ranjeva, R.; Pugin, A. Calcium in Plant Defence-Signalling Pathways. *New Phytol.* **2006**, *171*, 249–269. [CrossRef] [PubMed]

17. Batistič, O.; Kudla, J. Analysis of Calcium Signaling Pathways in Plants. *Biochim. Biophys. Acta* **2012**, *1820*, 1283–1293. [CrossRef]

18. Foyer, C.H.; Noctor, G. Redox Regulation in Photosynthetic Organisms: Signaling, Acclimation, and Practical Implications. *Antioxid. Redox Signal.* **2009**, *11*, 861–905. [CrossRef]

19. Suzuki, N.; Miller, G.; Morales, J.; Shulaev, V.; Torres, M.A.; Mittler, R. Respiratory Burst Oxidases: The Engines of ROS Signaling. *Curr. Opin. Plant Biol.* **2011**, *14*, 691–699. [CrossRef]

20. Liu, Q.; Zhao, Y.; Wan, Y.; Zheng, J.; Zhang, X.; Wang, C.; Fang, X.; Lin, J. Study of the Inhibitory Effect of Water-Soluble Fullerenes on Plant Growth at the Cellular Level. *ACS Nano* **2011**, *4*, 5743–5748. [CrossRef]

21. Kadono, T.; Tran, D.; Errakhi, R.; Hiramatsu, T.; Meimoun, P.; Briand, J.; Iwaya-Inoue, M.; Kawano, T.; Bouteau, F. Increased Anion Channel Activity Is an Unavoidable Event in Ozone-Induced Programmed Cell Death. *PLoS ONE* **2010**, *5*, e13373. [CrossRef] [PubMed]

22. Nakano, M.; Sugioka, K.; Ushijima, Y.; Goto, T. Chemiluminescence Probe with *Cypridina* Luciferin Analog, 2-methyl-6-phenyl-3,7-dihydroimidazo[1,2-*a*]pyrazin-3-one, for Estimating the Ability of Human Granulocytes to Generate O$_2{}^-$. *Anal. Biochem.* **1986**, *159*, 363–369. [CrossRef]

23. Setsukinai, K.; Urano, Y.; Kakinuma, K.; Majima, H.J.; Nagano, T. Development of Novel Fluorescence Probes That Can Reliably Detect Reactive Oxygen Species and Distinguish Specific Species. *J. Biol. Chem.* **2003**, *278*, 170–175. [CrossRef] [PubMed]

24. Foreman, J.; Demidchik, V.; Bothwell, J.H.; Mylona, P.; Miedema, H.; Torres, M.A.; Linstead, P.; Costa, S.; Brownlee, C.; Jones, J.D.; et al. Reactive Oxygen Species Produced by NADPH Oxidase Regulate Plant Cell Growth. *Nature* **2003**, *422*, 442–446. [CrossRef]

25. Knight, M.R.; Campbell, A.K.; Smith, S.M.; Trewavas, A.J. Transgenic Plant Aequorin Reports the Effects of Touch and Cold-Shock and Elicitors on Cytoplasmic Calcium. *Nature* **1991**, *352*, 524–526. [CrossRef]

26. Monetti, E.; Kadono, T.; Tran, D.; Azzarello, E.; Arbelet-Bonnin, D.; Biligui, B.; Briand, J.; Kawano, T.; Mancuso, S.; Bouteau, F. Deciphering in Early Events Involved in Hyperosmotic Stress-Induced Programmed Cell Death in Tobacco BY-2 Cells. *J. Exp. Bot.* **2014**, *65*, 1361–1375. [CrossRef] [PubMed]

27. Murphy, T.M.; Auh, C.K. The Superoxide Synthases of Plasma Membrane Preparations from Cultured Rose Cells. *Plant Physiol.* **1996**, *110*, 621–629. [CrossRef] [PubMed]

28. Van Gestelen, P.; Asard, H.; Caubergs, R.J. Solubilization and Separation of a Plant Plasma Membrane NADPH–O$_2$- Synthase from Other NADPH Oxidoreductases. *Plant Physiol.* **1997**, *115*, 543–550. [CrossRef]

29. Gauthier, A.; Lamotte, O.; Reboutier, D.; Bouteau, F.; Pugin, A.; Wendehenne, D. Cryptogein-Induced Anion Effluxes: Electrophysiological Properties and Analysis of the Mechanisms Through Which They Contribute to the Elicitor-Triggered Cell Death. *Plant Signal Behav.* **2007**, *2*, 86–95. [CrossRef]

30. Brault, M.; Amiar, Z.; Pennarun, A.M.; Monestiez, M.; Zhang, Z.; Cornel, D.; Dellis, O.; Knight, H.; Bouteau, F.; Rona, J.P. Plasma Membrane Depolarization Induced by Abscisic Acid in Arabidopsis Suspension Cells Involves Reduction of Proton Pumping in Addition to Anion Channel Activation, Which Are Both Ca^{2+} Dependent. *Plant Physiol.* **2004**, *135*, 231–243. [CrossRef] [PubMed]

31. Reboutier, D.; Bianchi, M.; Brault, M.; Roux, C.; Dauphin, A.; Rona, J.P.; Legue, V.; Lapeyrie, F.; Bouteau, F. The Indolic Compound Hypaphorine Produced by Ectomycorrhizal Fungus Interferes with Auxin Action and Evokes Early Responses in Non-Host *Arabidopsis Thaliana*. *Mol. Plant Microbe Interac.* **2002**, *15*, 932–938. [CrossRef] [PubMed]

32. Schroeder, J.I.; Keller, B.U. Two Types of Anion Channel Currents in Guard Cells with Distinct Voltage Regulation. *Proc. Natl. Acad. Sci. USA* **1992**, *89*, 5025–5029. [CrossRef] [PubMed]

33. Hedrich, R.; Busch, H.; Raschke, K. Ca^{2+} and Nucleotide Dependent Regulation of Voltage Dependent Anion Channels in the Plasma Membrane of Guard Cells. *EMBO J.* **1990**, *9*, 3889–3892. [CrossRef]

34. Shen, C.X.; Zhang, Q.F.; Li, J.; Bi, F.C.; Yao, N. Induction of Programmed Cell Death in Arabidopsis and Rice by Single-Wall Carbon Nanotubes. *Am. J. Bot.* **2010**, *97*, 1602–1609. [CrossRef] [PubMed]

35. Deepa, M.; Sudhakar, P.; Venkata, K.; Kota, N.; Reddy, B.; Krishna, T.G.; Krishna, N.V.; Prasad, V. First Evidence on Phloem Transport of Nanoscale Calcium Oxide in Groundnut Using Solution Culture Technique. *Appl Nanosci.* **2015**, *5*, 545–551. [CrossRef]

36. Sofo, A.; Scopa, A.; Manfra, M.; De Nisco, M.; Tenore, G.C.; Nuzzo, V. Different Water and Light Regimes Affect Ionome Composition in Grapevine (*Vitis Vinifera* L.). *Vitis* **2013**, *52*, 13–20.

37. Balážová, Ľ.; Babula, P.; Baláž, M.; Bačkorová, M.; Bujňáková, Z.; Briančin, J.; Kurmanbayeva, A.; Sagi, M. Zinc Oxide Nanoparticles Phytotoxity on Halophyte from Genus *Salicornia*. *Plant Physiol Biochem.* **2018**, *130*, 30–42. [CrossRef] [PubMed]

38. Vainonen, J.P.; Kangasjärvi, J. Plant Signalling in Acute Ozone Exposure. *Plant Cell Environ.* **2015**, *38*, 240–252. [CrossRef] [PubMed]

39. Kadono, T.; Yamaguchi, Y.; Furuichi, T.; Hirono, M.; Garrec, J.P. Ozone-Induced Cell Death Mediated with Oxidative and Calcium Signaling Pathways in Tobacco Bel-w3 and Bel-B Cell Suspension Cultures. *Plant Signal. Behav.* **2006**, *1*, 312–322. [CrossRef] [PubMed]

40. Tran, D.; El-Maarouf-Bouteau, H.; Rossi, M.; Biligui, B.; Briand, J.; Kawano, T.; Mancuso, S.; Bouteau, F. Post-Transcriptional Regulation of GORK Channels by Superoxide Anion Contributes Towards Increases in Outward Rectifying K$^+$ Currents. *New Phytol.* **2013**, *198*, 1039–1048. [CrossRef]

41. Tran, D.; Molas, M.L.; Kadono, T.; Errakhi, R.; Briand, J.; Biligui, B.; Kawano, T.; Bouteau, F. A Role for Oxalic Acid Generation in Ozone-Induced Programmed Cell Death. *Plant Cell Environ.* **2013**, *36*, 569–578. [CrossRef] [PubMed]

42. Tran, D.; Rossi, M.; Biligui, B.; Kawano, T.; Mancuso, S.; Bouteau, F. Ozone-Induced Caspase-Like Activities Are Dependent on Early Ion Channel Regulations and ROS Generation in *Arabidopsis Thaliana* Cells. *Plant Signal. Behav.* **2013**, *8*, e25170. [CrossRef] [PubMed]

43. McCarthy, J.; Gong, X.; Nahirney, D.; Duszyk, M.; Radomski, M. Polystyrene Nanoparticles Activate Ion Transport in Human Airway Epithelial Cells. *Int. J. Nanomed.* **2011**, *6*, 1343–1356. [CrossRef] [PubMed]

44. Oukarroum, A.; Bras, S.; Perreault, F.; Popovic, R. Inhibitory Effects of Silver Nanoparticles in Two Green Algae, *Chlorella Vulgaris* and *Dunaliella Tertiolecta*. *Ecotoxicol. Environ. Safety* **2012**, *78*, 80–85. [CrossRef] [PubMed]

45. Chen, P.; Powell, B.A.; Mortimer, M.; Ke, P.C. Adaptive Interactions between Zinc Oxide Nanoparticles and *Chlorella* sp. *Environ. Sci. Technol.* **2012**, *46*, 12178–12185. [CrossRef] [PubMed]

46. Atha, D.H.; Wang, H.; Petersen, E.J.; Cleveland, D.; Holbrook, R.D.; Jaruga, P.; Dizdaroglu, M.; Xing, B.; Nelson, B.C. Copper Oxide Nanoparticle Mediated DNA Damage in Terrestrial Plant Models. *Environ. Sci. Technol.* **2012**, *46*, 1819–1827. [CrossRef] [PubMed]

47. Slomberg, D.L.; Schoenfisch, M.H. Silica Nanoparticle Phytotoxicity to *Arabidopsis Thaliana*. *Environ. Sci. Technol.* **2012**, *46*, 10247–10254. [PubMed]

48. Khodakovskaya, M.V.; de Silva, K.; Biris, A.S.; Dervishi, E.; Villagarcia, H. Carbon Nanotubes Induce Growth Enhancement of Tobacco Cells. *ACS Nano* **2012**, *6*, 2128–2135. [CrossRef]

49. Schroeder, J.I.; Kwak, J.M.; Allen, G.J. Guard Cell Abscisic Acid Signalling and Engineering of Drought Hardiness in Plants. *Nature* **2001**, *410*, 327–330. [CrossRef]

50. Murashige, T.; Skoog, F. A Revised Medium for Rapid Growth and Bioassays with Tobacco Tissue Cultures. *Physiol. Plant.* **1962**, *15*, 473–497. [CrossRef]

51. Bouizgarne, B.; El-Maarouf-Bouteau, H.; Frankart, C.; Reboutier, D.; Madiona, K.; Pennarun, A.M.; Monestiez, M.; Trouvier, J.; Amiar, Z.; Briand, J.; et al. Early Physiological Responses of *Arabidopsis Thaliana* Cells to Fusaric Acid: Toxic and Signaling Effects. *New Phytol.* **2006**, *169*, 209–218. [CrossRef]

52. Finkel, A.S.; Redman, S. Theory and Operation of a Single Microelectrode Voltage Clamp. *J. Neurosci. Methods* **1984**, *11*, 101–127. [CrossRef]

53. Bouteau, F.; Tran, D. Plant response to stress: Microelectrode voltage clamp studies. In *Plant Electrophysiology*; Volkov, A.G., Ed.; Springer: Berlin/Heidelberg, Germany, 2012; Volume 3, pp. 69–90.

International Journal of
Molecular Sciences

MDPI

Erratum

Erratum: Tran et al. Early Cellular Responses Induced by Sedimentary Calcite-Processed Particles in Bright Yellow 2 Tobacco Cultured Cells. *Int. J. Mol. Sci.* 2020, 21, 4279

Daniel Tran [1,2,*], Tingting Zhao [2], Delphine Arbelet-Bonnin [2,3], Takashi Kadono [2,4], Patrice Meimoun [2], Sylvie Cangémi [2], Tomonori Kawano [4,5,6,7], Rafik Errakhi [8] and François Bouteau [2,3,5,6]

[1] Agroscope, Institute for Plant Production Systems, 1964 Conthey, Switzerland
[2] Laboratoire Interdisciplinaire des Energies de Demain, Université de Paris, 75013 Paris, France; zhaotingting0325@gmail.com (T.Z.); delphine.bonnin@univ-paris-diderot.fr (D.A.-B.); kadono.takashi@gmail.com (T.K.); patrice.meimoun@upmc.fr (P.M.); sylvie.cangemi@univ-paris-diderot.fr (S.C.); francois.bouteau@univ-paris-diderot.fr (F.B.)
[3] Cogitamus Laboratory, 75013 Paris, France
[4] Graduate School of Environmental Engineering, University of Kitakyushu, 1-1 Hibikino, Wakamatsu-ku, Kitakyushu 808-0135, Japan; kawanotom@gmail.com
[5] LINV Kitakyushu Research Center (LINV@Kitakyushu), Kitakyushu 808-0135, Japan
[6] International Photosynthesis Industrialization Research Center, The University of Kitakyushu, Kitakyushu 808-0135, Japan
[7] Paris Interdisciplinary Energy Research Institute (PIERI), Université de Paris, 75013 Paris, France
[8] Eurofins Agriscience Service, Casablanca 20000, Morocco; r.errakhi@gmail.com
* Correspondence: qnoctnandaniel.tran@agroscope.admin.ch

Citation: Tran, D.; Zhao, T.; Arbelet-Bonnin, D.; Kadono, T.; Meimoun, P.; Cangémi, S.; Kawano, T.; Errakhi, R.; Bouteau, F. Erratum: Tran et al. Early Cellular Responses Induced by Sedimentary Calcite-Processed Particles in Bright Yellow 2 Tobacco Cultured Cells. *Int. J. Mol. Sci.* **2020**, *21*, 4279. *Int. J. Mol. Sci.* **2021**, *22*, 6863. https://doi.org/10.3390/ijms22136863

Received: 7 June 2021
Accepted: 10 June 2021
Published: 25 June 2021

Publisher's Note: MDPI stays neutral with regard to jurisdictional claims in published maps and institutional affiliations.

The authors would like to remove the scientific consortium 'Camille Nous' from the author list and the Author Contributions section in the published paper [1], as suggested by the Editorial Office. To recognize the consortium's contribution, the authors claimed a double affiliation of two of the authors to the Cogitamus laboratory [2]. The original article has been updated.

References

1. Tran, D.; Zhao, T.; Arbelet-Bonnin, D.; Kadono, T.; Meimoun, P.; Cangémi, S.; Kawano, T.; Errakhi, R.; Bouteau, F. Early Cellular Responses Induced by Sedimentary Calcite-Processed Particles in Bright Yellow 2 Tobacco Cultured Cells. *Int. J. Mol. Sci.* **2020**, *21*, 4279. [CrossRef] [PubMed]
2. Cogitamus Laboratory. Available online: https://www.cogitamus.fr/faqen.

International Journal of
Molecular Sciences

Review

Arabidopsis Transmembrane Receptor-Like Kinases (RLKs): A Bridge between Extracellular Signal and Intracellular Regulatory Machinery

Jismon Jose [†], Swathi Ghantasala [†] and Swarup Roy Choudhury [*]

Department of Biology, Indian Institute of Science Education and Research (IISER) Tirupati, Tirupati, Andhra Pradesh 517507, India; jismon.jose@students.iisertirupati.ac.in (J.J.); swathighantasala@gmail.com (S.G.)
* Correspondence: srchoudhury@iisertirupati.ac.in
† These authors contributed equally to this work.

Received: 16 May 2020; Accepted: 1 June 2020; Published: 3 June 2020

Abstract: Receptors form the crux for any biochemical signaling. Receptor-like kinases (RLKs) are conserved protein kinases in eukaryotes that establish signaling circuits to transduce information from outer plant cell membrane to the nucleus of plant cells, eventually activating processes directing growth, development, stress responses, and disease resistance. Plant RLKs share considerable homology with the receptor tyrosine kinases (RTKs) of the animal system, differing at the site of phosphorylation. Typically, RLKs have a membrane-localization signal in the amino-terminal, followed by an extracellular ligand-binding domain, a solitary membrane-spanning domain, and a cytoplasmic kinase domain. The functional characterization of ligand-binding domains of the various RLKs has demonstrated their essential role in the perception of extracellular stimuli, while its cytosolic kinase domain is usually confined to the phosphorylation of their substrates to control downstream regulatory machinery. Identification of the several ligands of RLKs, as well as a few of its immediate substrates have predominantly contributed to a better understanding of the fundamental signaling mechanisms. In the model plant Arabidopsis, several studies have indicated that multiple RLKs are involved in modulating various types of physiological roles via diverse signaling routes. Here, we summarize recent advances and provide an updated overview of transmembrane RLKs in Arabidopsis.

Keywords: Arabidopsis; development; kinase; receptor; stress

1. Introduction

Responsiveness to extracellular or intracellular changes is the nub for the survival of any organism, and receptors act as trump cards. Receptors predominantly tweak their downstream gene expression, in accordance with the stimuli perceived and yield a suitable response that enables survival of the organism. Eukaryotic protein kinases (ePKs) are a superfamily of proteins that facilitate this signal transduction by catalyzing the transfer of γ-phosphate from ATP to the free hydroxyl groups of serine/threonine/tyrosine residues of the substrate protein. This post-translational modification or phosphorylation of the substrate alters its reactivity, which results in the activation or inactivation of the signaling circuit [1]. The ePKs are represented by several families of kinases like receptor-like kinases (RLKs), mitogen-activated protein kinases (MAPKs), calcium-dependent protein kinases (CDPKs), NIMA-related kinases (NEKs), glycogen synthase kinases (GSKs) etc., each with their unique structural and functional attributes [2].

Receptor-like kinases (RLKs), a multi-gene family, is the largest class of ePKs that is crucial for mediating growth, development and stress-responsive signals in plants. Their domain organization resembles the receptor tyrosine kinases (RTKs) and receptor serine/threonine kinases (RSKs) of the animal system, and their closest animal homologs are the Drosophila Pelle kinase family and human interleukin-1 receptor-associated kinases (IRAKs) [3,4]. RLKs include transmembrane receptor kinases as well as non-receptor or cytoplasmic kinases. The former consists of a signal peptide, an extracellular ectodomain, single membrane-spanning domain, intracellular juxta membrane domain, and the cytoplasmic kinase domain; while the latter has only the cytoplasmic kinase domain, and are, therefore, called receptor-like cytoplasmic kinases (RLCKs) [5]. In addition, another group of proteins called receptor-like proteins (RLPs) are similar to the RLKs, except that they do not possess the kinase domain [6]. RLKs and RLPs are the major cell-surface receptors observed in plants [7]. Throughout this review, 'RLKs' refer only to the transmembrane receptor kinases.

RLKs are known to exist in animals as well as plants, but are not yet reported in fungi, despite the presence of other soluble protein kinases in them [8,9]. Unlike plants, RLKs are represented by smaller gene numbers in the animal system. Except for transforming growth factor-β (TGF-β) receptors, all animal receptor kinases are tyrosine kinases, whereas the majority of plant RLKs possess serine/threonine kinase domain [10]. Some of the plant RLKs (nod factor receptor 1 (NFR1), brassinosteroid insensitive 1 (BRI1), BRI1-associated kinase 1 (BAK1), pollen-expressed receptor kinase 1 (PRK1), somatic embryogenesis receptor kinase 1 (SERK1), BAK1-like kinase 1 (HAESA)) have been found to behave as dual-specificity kinases, possessing conserved motifs of both types of kinases and, thus, efficiently phosphorylating at serine/threonine as well as tyrosine residues [11,12]. The structural configuration of animal receptor kinases is similar to plant RLKs. The three conserved motifs in their cytoplasmic domains, such as Valine–Alanine–Isoleucine–Lysine (VAIK), Histidine–Arginine–Aspartate (HRD), and Aspartate–Phenylalanine–Glycine (DFG), assign them to the kinase family, while a few (human epidermal growth factor receptor 3 (HER-3), serine threonine tyrosine kinase 1 (STYK1)) that have a variant residue in at least one of these motifs are called pseudokinases [13]. Intriguingly, both plant and animal RLKs have similar downstream targets like MAPKs and reactive oxygen species (ROS) and also undergo similar desensitization pathways, such as ubiquitination and endocytosis [14].

Despite the similarity of plant RLKs to their animal counterparts, it can be noted that these families belong to distinct monophyletic groups within the protein kinases, implying the independent evolution of these classes among plant and animal systems, whereas, the analogy in their biochemical events indicates convergent evolution [3,10]. The enormous representative members in RLKs are confined to the angiosperms only, whereas the numbers are fewer in the lower plant groups. Though the kinase domains (KD) and the conserved motifs of the ectodomain (ED) are encountered as discrete entities in algae, the receptor conformation, which is characterized by the presence of both ED and KD, has not yet been reported, except in the charophytes (*Nitella axillaris* and *Closterium ehrenbergii*), suggesting that the receptor conformation had been established just before the divergence of land plants from the charophytes [3,15]. Furthermore, exploration of the sequenced genomes of different groups of plants revealed that the RLKs in angiosperms range from 0.67%–1.39% of their protein-encoding genes, while that of bryophytes (*Physcomitrella patens*) and pteridophytes (*Selaginella moellendorffii*) account for only 0.36% and 0.30% respectively. These indicate greater expansion of this family in the flowering plant lineage within Viridiplantae, which might probably account for the acquisition of new roles that are essential for their survival. *Arabidopsis,* rice, and poplar possess 1.9, 3.3 and 3.6 times the number of RLKs detected in moss, validating that this expansion is not concomitant with an increase in genome size but with genome complexity [15,16]. Within the RLK family, the expansion is not uniform in the different taxa. Those subfamilies, which have a critical role in plant growth and development, tend to remain more conserved within the taxa, while those specific to plant defense tend to expand more, in order to co-evolve with their biotic counterparts [15].

This review focusses on RLKs in the model plant *Arabidopsis thaliana* providing insights into its domain organization, classification, signaling mechanism, their roles in plant growth and development, and in conferring resistance to biotic and abiotic stresses.

2. Classification of Arabidopsis RLKs

In Arabidopsis, RLKs represent the largest protein family with more than 600 members, constituting about 2.5% of its euchromatin; thus, eliciting the significance of this class of plant receptors. It is noteworthy that the phylogenetic analysis of RLKs with other protein kinases of Arabidopsis validates the monophyletic origin of RLKs. Out of the 610 genes encoding for RLKs, 417 encode for receptor kinases while the other 193 lack the signature signal sequence and/or transmembrane sequence indicating that these might be cytoplasmic kinases (RLCKs) [10]. Based on the signature motifs in the ectodomains of receptor kinases, Arabidopsis transmembrane RLKs can be classified into 14 types, viz., leucine-rich repeat (LRR), lectin (C-Lectin and L-Lectin), wall-associated kinase (WAK), extensin like, proline-rich extensin like (PERK), *Catharanthus roseus* like (CrRLK), self-incompatibility domain (S-domain), CRINKLY-like (CR-like), the domain of unknown function 26 (DUF26), lysin motif (LysM), thaumatin, leaf rust kinase-like (LRK), receptor-like kinase in flowers (RKF), unknown receptor kinase (URK), of which the biological role of only a few have been studied in detail [17–35] (Table 1). Some of these RLK types are placed under different subfamilies due to the phylogenetic distinctness of their kinase domains [5]. This suggests probable functional diversification such that single isoforms may comply with different specificities. The structural features of different types of RLKs are explained here (Figure 1).

Table 1. List of few representative members of each receptor-like kinase (RLK) type.

S. No.	RLK Type	Gene (s)	Function	Ref
1	LRR	CLAVATA1	Meristem and organ development	[17]
		SERK	Microsporogenesis, embryogenesis, and embryonic competence in tissue culture	[18]
		HAESA	Floral organ abscission	[19]
		FLS2	Senses bacterial flagellin	[20]
2	LecRLK	LecRK1	Oligosaccharide-mediated signal transduction	[33]
3	WAK-RLK	WAK1	Cell wall integrity, pathogen response	[21]
4	Extensin	LRX1	Root hair morphogenesis	[22]
5	PERK	PERK4	Cell wall integrity and drought response	[23]
6	CrRLK	HERK1	Determinants of pollen tube	[35]
		FER	Polar growth of root hair and pollen tube	[24,25]
7	S-domain	AtS1	Self-incompatibility	[34]
		ARK2, ARK3	Organ maturation	[34]
8	CR-like	ACR4	Epidermal patterning, integument development in ovules	[26]
			Plant defense	[27]
9	DUF26	CRK13	Biotic stress response	[28]
		CRK6, CRK7	Oxidative stress response	[29]
10	LysM-RLK	AtCERK1	Perception of MAMPs	[30]
11	Thaumatin	PR5K	Response to pathogenic signals	[31]
12	LRK 10-like	LRK10L1.2	Drought resistance	[32]

The functional significance of unknown receptor kinase (URK) and receptor-like kinase in flowers (RKF) in Arabidopsis has not yet been reported and is thus, not mentioned in this table.

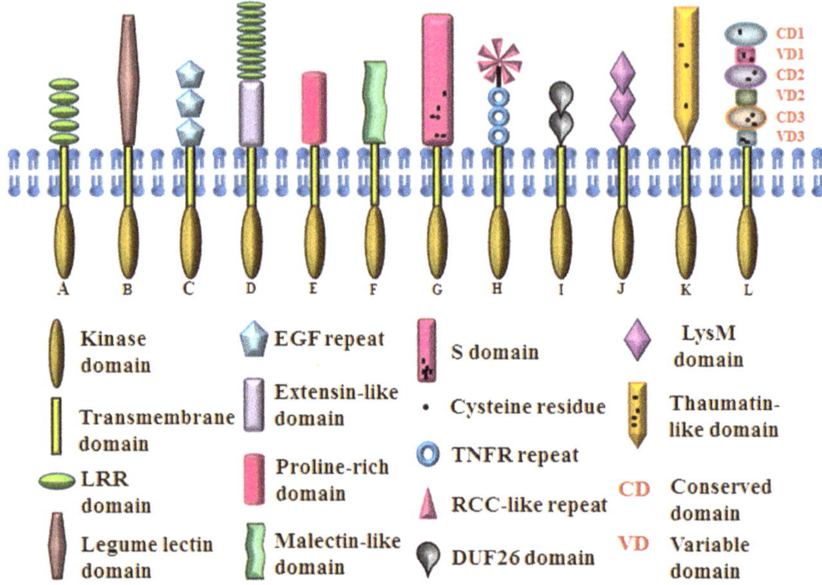

Figure 1. Domain architecture of Arabidopsis RLKs. A. SERK (LRR), B. LecRK1 (Lectin), C. WAK1 (WAK), D. LRX1 (Extensin + LRR), E. PERK4 (PERK), F. FER (CrRLK), G. AtS1 (S-domain), H. ACR4 (CR-like), I. CRK (DUF26), J. AtCERK1 (LysM), K. PR5K (Thaumatin), L. LRK10L1.2 (LRK). RCC, regulator of chromosome condensation.

Leucine-rich repeats (LRRs) are the largest represented class of RLKs, encoded by 239 genes and comprising 15 subfamilies in Arabidopsis [5]. LRRs are tandem repeats of about 24 amino acid residues, having conserved leucine residues and are homologous to the ectodomains of the toll-like receptor of the animal system [36,37]. The exact number, arrangement of residues, and the sequences interspersed between the leucine repeats determine the perception of diverse ligands by their ectodomain, which ultimately initiate various signaling events to modulate growth as well as stress responses [38,39]. Similarly, Lectin receptor-like kinases (LecRLKs), which are the second-largest group of RLKs, are known for their role in plant stress and developmental pathways. These Lectin RLKs are encoded by 47 genes belonging to two subfamilies in Arabidopsis [5]. They can bind to various homo and hetero-disaccharides, such as chitobiose, glucose-mannose, and galactose-GlcNAc, through the sugar-binding motifs in their ectodomains [33,40]. Broadly, LecRLKs are of three types: C, L, and G, while only C and L type have been known to exist in Arabidopsis. C-type lectin is encoded by a single gene in Arabidopsis and can be considered homologous to calcium-binding lectin motifs of the mammalian system [5]. The carbohydrate-binding domains of C-type lectin are calcium-dependent for ligand binding and maintenance of domain integrity [41]. The L-type lectins have carbohydrate-binding domains similar to the leguminous plant lectins and extracellular ATP is one of their chief ligands [42,43]. The lectin domain of L-type lectins is closely related to other RLKs like wall-associated kinase (WAK) and proline-rich extensin like kinase (PERK) [44].

Maintenance of cell wall integrity is crucial to cater efficient mechanical support during growth, development, injury, and exposure to abiotic/biotic stress. RLKs like lectin RLKs, wall-associated kinases (WAKs), extensin-like kinases, proline-rich extensin like kinases (PERKs), and *Catharanthus roseus* like kinases (CrRLKs), are the aides, which ensure it. WAKs are coupled with pectin to tether the cell wall to cytoplasm for providing structural integrity. Arabidopsis has 26 WAKs, all of which belong to the same subfamily. The ectodomain of WAKs possesses a cysteine-containing EGF motif, which is the only motif that is common in both plant and animal ectodomains. The kinase domains of

WAKs are known to facilitate protein-protein interactions and also respond to changes in cellulose biosynthesis during pathogen attacks [21]. On the other hand, extensin is a cell wall structural protein which consists of a repeating Ser-(Hyp)$_4$ motif and extensin-like kinases possess glycosylated Ser-(Hyp)$_{3-5}$ motifs to maintain the dynamicity of the cell wall [45–47]. The LRX1 of Arabidopsis is a chimeric RLK, possessing LRR, as well as extensin domains [22]. The ectodomains of PERKs share sequence similarity with extensins and are rich in proline. This type of RLKs perhaps interact with the positively charged pectin network and generate a repair response upon wall damage or injury, thus, maintaining wall integrity [48]. *Catharanthus roseus* like RLK possess a putative carbohydrate-binding malectin-like domain, essential for the supervision of cell wall tenacity [49]. This malectin-like domain is globular, membrane-anchored, and known to bind Glc2-N-glycans [50]. FERONIA (FER), ANXUR1 (ANX1), ANX2, THESEUS1 (THE1), HERCULES1 (HERK1) are important members of CrRLK1L family. Although FER, ANX1, and HERK1 have similar downstream targets, they are activated by diverse ligand interactions [35].

Accumulating evidence indicates that a few groups of RLKs participate in plant responses to a variety of biotic stresses, as well as during plant development, viz., S-RLK, CRINKLY-like RLK and domain of unknown function 26 (DUF26). The S-domain of S-RLK is homologous to the self-incompatibility-locus glycoproteins in wild cabbage [51]. In Arabidopsis, there are 40 different S-domain bearing RLKs, which belong to three different subfamilies. The S-domain has the signature WQSFDXPTDTFL, called the PTDT-box, where X and F represent any non-conserved and aliphatic amino acid residues, respectively. This S-domain also contains 12 conserved cysteine residues as well as agglutinin, EGF and PAN (plasminogen/apple/nematode) motifs [5,34]. On the other hand, Arabidopsis CRINKLY-like RLKs (ACR4) have tumor necrosis factor receptor (TNFR)-like repeats in their ligand-binding domain, i.e., seven tandem repeats of about 39 amino acid residues, followed by three cysteine-rich regions [26,27]. Another cysteine-rich domain-containing receptor-like kinase (CRK) is the domain of unknown function 26 (DUF26), which contains C-8X-C-2X-C motif in its ectodomain [52,53].

Few RLK types are known to play essential roles predominantly in plant defense and one of the major groups is LysM-RLK, which shows a critical role in chitin signaling and fungal resistance in Arabidopsis. For instance, chitin elicitor receptor kinase 1 (CERK1) is essential for perception of the fungal cell wall component, chitin and confers resistance against fungal pathogens. The ectodomain of LysM-RLK is comprised of three lysin motifs and each motif is a stretch of about 40 amino acid residues, discovered in most organisms, except Archaea [54–56]. This motif can interact with N-acetylglucosamine (GlcNAc) containing polymers; thus, mediating microbial interactions [55]. The other groups of kinases exhibiting anti-fungal and chitinase activity are the thaumatin and leaf rust kinase 10-like (LRK 10-like) RLK. The thaumatin group, also known as pathogenesis-related group 5 receptor kinase (PR5K), is encoded by three genes in Arabidopsis and their ectodomains possess 16 conserved cysteine residues [5,31]. The ectodomains of leaf rust kinase 10-like (LRK 10-like) RLKs are homologous to the LR10 protein, which belongs to the family of wheat leaf rust kinases (WLRKs). The 14 conserved cysteine residues are arranged in a specific manner in the ectodomain of these RLKs [32,57]. This diversity in the ectodomain architecture of RLKs facilitate the perception of distinct ligands and thus account for the diverse roles of RLKs throughout a plant's life.

3. Signaling Mechanism of RLKs

Ligand binding at ectodomain is essential for oligomerization and activation of the RLKs. The diverse ectodomains of RLKs help in the perception of lucrative and noxious stimuli; thus, enabling efficient survival of plants in the constantly changing environment. Ligands like plant growth regulators (brassinolide and phytosulfokine), peptides (PSY1-sulphated peptide, TPD1-cysteine-rich peptide, and CLV3-proline-rich peptide), and MAMPs (microbe-associated molecular patterns: Nod factors or other GlcNAc) stimulate plant developmental signaling, while PAMPs (pathogen-associated molecular patterns: chitin, lipopolysaccharides, ergosterol, transglutaminase, etc.) and DAMPs (damage-associated molecular patterns: cutin monomers, oligogalacturonic acid, cello oligomers, etc.) induce immune response via diverse signaling cascades and enable combat against the pathogen/injury for conferring tolerance or resistance to the plant cell [32,58]. An outline of the signal transduction mechanism, depicting only the conserved members involved in most of the signaling cascades, is illustrated in Figure 2.

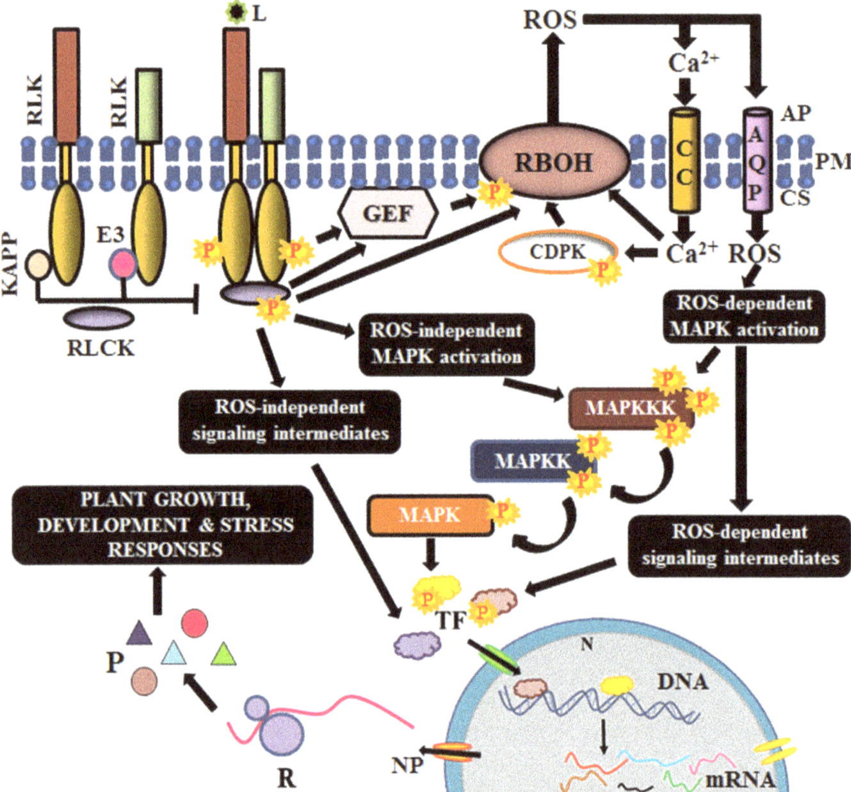

Figure 2. Schematic outline of signaling mechanism of Arabidopsis RLKs. Complex formation and interaction with receptor-like cytoplasmic kinases (RLCKs) with RLKs are prevented by kinase-associated protein phosphatases (KAPP) and E3 ubiquitin ligases. Upon perception of ligand (L), they dissociate to allow the stimulation of RLCK via phosphorylation. Activated RLCK has many possible routes of activation. The RLKs might also activate guanosine exchange factors (GEF) directly. RLCKs and G-proteins elicit gene expression via several intermediates like reactive oxygen species (ROS), calcium channels, calcium-dependent protein kinases, (CDPK), mitogen-activated protein (MAP) kinases (MAPKKK, MAPKK, MAPK), and transcription factors (TF). AP, apoplast; PM, plasma membrane; CS, cytosol; AQP, aquaporin; CC, calcium channel; N, nucleus; NP, nuclear pore; R, ribosome; P, protein.

Few RLKs require co-receptors (like BAK1) or scaffold proteins (like FERONIA) for the establishment of receptor complex [59,60]. Before ligand perception, the cytosolic kinase domains of RLKs are maintained inactive by intramolecular interactions or by phosphatases and other regulatory proteins like E3 ligases, calcium-dependent kinases, G-proteins etc. Binding to their cognate ligand causes a conformational change in the receptor, leading to the formation of homo or heterodimers. Homodimerization is observed in Arabidopsis CERK1, in which the two inactive LysM-RLK monomers interact and dimerize to activate immune signaling, in response to chitin oligomers [61,62]. On the other hand, an LRR-RLK, Flagellin sensitive 2 or FLS2 forms a complex with another LRR-RLK, BAK1 (co-receptor), upon the perception of bacterial flagellin, to form a heterodimer [63]. Heterodimerization is known to occur either between a pseudokinase (FLS2) and an RD (arginine-aspartate) kinase (BAK1) or between two RD kinases, like BRI1 and its co-receptor BAK1 [62,64]. Besides, RLKs are also known to form complexes with RLPs for establishing the signal response. For instance, CLAVATA1 (RLK) dimerizes with CLAVATA2 (RLP) upon the perception of a peptide ligand, CLV3 [6,65]. In all the above scenarios, complex formation negates the auto-inhibition effect on the kinase domains of the RLKs and makes it amenable for phosphorylation. The proximity of the kinase domains of the dimers induces auto and/or transphosphorylation, facilitating mutual activation [66].

Most often, the immediate substrates of the activated complex are the diverse families of RLCKs. On the other hand, guanine nucleotide exchange factors like GTPases and G-proteins have also been reported to be the immediate substrates of the activated complex [67]. Occasionally, RLKs are associated with their RLCKs in prior, in which the RLCKs are tethered to the membrane via palmitoylation or myristoylation, and their activation is prohibited by negative regulators. However, ligand binding induces dissociation of the regulators and thus, enable the stimulation of the RLCKs [62,68]. The specificity of different families of RLCKs, as well as their downstream targets, is regulated by the RLK complex and its configuration [69]. At times, the same RLCK interacts with different classes of RLKs and generates different responses as a result of differential phosphorylation of the RLCK [70,71]). For instance, BIK1 (RLCK) interacts with FLS2 (RLK) to positively regulate immune signaling, while it interacts with BRI1 (RLK) to negatively regulate brassinolide-mediated growth [70,72]. Eventually, RLCKs transduce the message from the apoplast to the interior of the cell via a phosphorelay [68].

One of the substrates of these RLCKs is the respiratory burst oxidase homologs (RBOHs), which are membrane-bound NADPH oxidases that cause accumulation of ROS in the apoplast [73]. RLCK-mediated phosphorylation of RBOHs is sensed by calcium channels, followed by an influx of calcium ions, which in turn, activates the RBOHs by feedback regulation. Calcium ions also activate calcium-dependent protein kinases (CDPKs), which are also essential for RBOH triggering [68,74]. Moreover, RBOH stimulation is also achieved via the Rac/Rho like guanine nucleotide exchange factors (Rac/ROP GEFs), which are GTPases, and also by G-proteins like XLG2 (extra-large G-protein 2) [75,76]. The subsequent accumulation of ROS in the apoplast stimulates the ROS-dependent signaling cascade via post-translational modification of its target proteins [77]. Although ROS outbursts can also occur in chloroplast, mitochondria, and peroxisomes, apoplastic burst has a rapid transduction rate [78]. Thus, ROS, calcium ions and Rac/ROP GEFs act as secondary messengers for the amplification of the signal.

Another class of targets for the RLCKs is the MAPKs, which are activated via phosphorylation of their regulatory domains. MAP kinases have known to be the core constituent of signal transduction cascade during the response to many extracellular stimuli [79]. It constitutes three members viz., MAP kinase kinase kinase (MAPKKK), MAP kinase kinase (MAPKK) and MAP kinase (MAPK). The MAPKKK acts on its substrate MAPKK, which in turn, activates MAPK by phosphorylation. MAPK subsequently, activates respective transcription factors to elicit a relevant response from the nucleus [80]. The MAPK activation by RLCKs might be ROS-dependent or independent [77,81]. Ultimately, these aid in the activation of respective transcription factors, which tweak the expression of their respective genes, culminating with appropriate cellular responses like growth, development, immunity, symbiosis and stress tolerance or resistance.

4. Functions of RLKs in the Regulation of Plant Growth and Development

Arabidopsis RLKs modulate growth and developmental responses by governing stem-cell maintenance, cell fate determination and patterning, male and female gametophyte development, pollen-pistil interactions, embryogenesis, hormone signaling, vascular patterning, organ development, and abscission. Some of these essential responses are discussed here.

4.1. Regulation in Anther and Ovule Development

The anther generally has four lobes and each lobe contains reproductive microsporocyte surrounded by various layers of somatic cells viz., tapetum, middle layer, endothecium, and epidermis. In Arabidopsis, multiple LRR-RLKs like excess microsporocytes1 (EMS1)/extra sporogenous cell (EXS), somatic embryogenesis receptor-like kinase 1/2 (SERK1/2), receptor-like protein kinase 2 (RPK2), barely any meristem 1/2 (BAM1/2), CLAVATA3 insensitive receptor kinase (CIK1/2/3/4), ERECTA (ER), and ERECTA-like 1/2 (ERL1/2) regulate anther development, especially, the differentiation and patterning of the somatic cell layers. EMS1/EXS was the first LRR-RLK to be identified that plays a crucial role in anther cell differentiation [82,83]. The anthers of *ems1/exs* mutants lack tapetum but produce large numbers of microsporocytes than the wild type. In addition, delayed expression of *EMS1* in the *ems1* mutant tapetal initials has been shown to aid in the generation of a functional tapetum and the diminution of microsporocyte numbers [84]. These results suggest that EMS1/EXS determines the fate of tapetal cells during early anther development. Tapetum determinant 1 (TPD1), a small secreted protein, is known to induce the phosphorylation of EMS1/EXS, thus, behaving as their ligand; and the signal is transduced downstream via phosphorylation of β-carbonic anhydrases (βCAs) [85,86]. Similarly, SERK1/2 has also been known to determine tapetal cell fate, as the anthers of *serk1serk2* double mutants are phenotypically similar to that of *ems1/exs* mutant [18,87]. Moreover, SERK1 interacts with and transphosphorylates EMS1 to enhance its activity for guiding a co-regulatory network (Figure 3A) [88]. Corroborated by the phenotype of *rpk2* mutants, it can be deduced that RPK2 is responsible for the differentiation of middle layers and tapetum during anther development. It essentially controls tapetal cell fate by triggering their degradation via modulation of the enzymes involved in cell wall metabolism and lignin biosynthesis [89] (Figure 3A). Both BAM1 and BAM2 are responsible for regulating early stages of anther differentiation, as confirmed by the lack of somatic cell layers, including endothecium, middle layer, and tapetum in *bam1bam2* double mutants [90]. CLAVATA3 insensitive receptor kinases (CIK1/2/3/4) are co-receptors of BAM1/2 and RPK2, which regulate the determination of parietal cell fate and archesporial cell division [91] (Figure 3A). ERECTA (ER), ERECTA-Like 1 (ERL1), and ERL2 are also known to play essential roles in healthy anther lobe formation and anther cell differentiation via mitogen-activated protein kinases like MPK3/MPK6 (Figure 3A). The sterility of *er-105 erl1-2 erl2-1* triple mutant and the phenotypic similarity of the anther lobes in single mutants of *er-105* or *erl1-2* or *erl2-1* with that of *mpk3* or *mpk6* mutants suggests the correlation of these genes in the regulation of anther cell division and differentiation [92]. Further, a Lectin RLK, small, glued together, collapsed (SGC) has also been validated as a regulator of pollen development as its knockout had led to the development of small, glued-together and collapsed pollen and resulted in male sterility [93] (Figure 3A).

Knowledge about the role of RLKs in ovule development is very scarce. In Arabidopsis ovules, *EMS1* is expressed in nucellar epidermis and chalaza, while *TPD1* is weakly restricted to the distal end of integuments. Altered expression of cell-cycle genes and auxin signaling genes during ovule development, concomitant with the ectopic expression of *TPD1,* indicates the regulation of ovule development by TPD1-EMS1 [94] (Figure 3A).

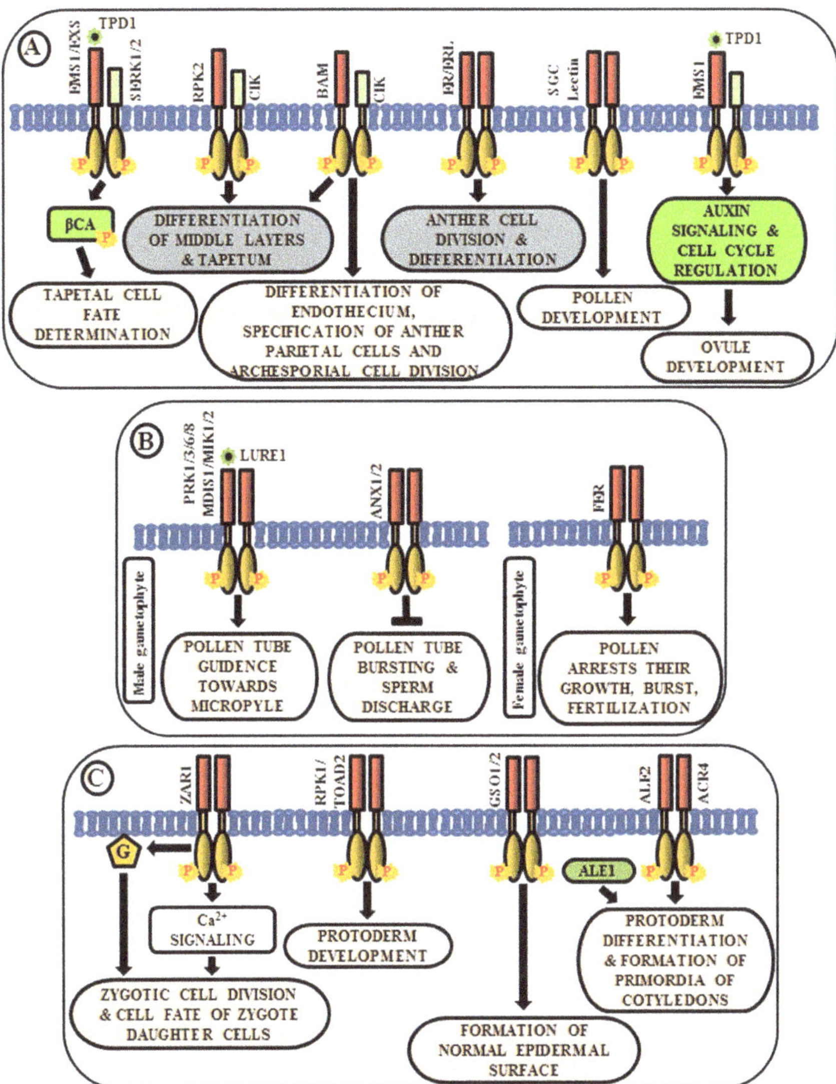

Figure 3. Arabidopsis RLKs in the regulation of growth and development. A few examples of RLKs that regulate (**A**) anther and ovule development, (**B**) pollen-pistil interaction, and (**C**) embryo development.

4.2. Pollen-Pistil Interactions

Reproduction in angiosperms involves the release of an immobile male gamete from the pollen tube onto the compatible pistil. A fruitful pollen-pistil interaction is a pre-requisite for successful fertilization and this requires an accurate perception of ovule-emitted guidance cues by the receptors in pollen tubes. LURE1, an ovule-secreted peptide is perceived by RLKs like pollen receptor kinase 1 (PRK1), PRK3, PRK6, PRK8 in the pollen tube [95]. Recent studies ascertain the presence of other LURE receptors like Male Discoverer 1 (MDIS1), MDIS1-interacting receptor-like kinase1 (MIK1), and MIK2 [96,97]. Once the pollen tube reaches the micropyle, its growth is ceased and the sperm cells are released by its rupture. These processes are regulated by the RLK FERONIA (FER), which is expressed in the synergids of female gametophyte [98] (Figure 3B). The phenotypic study of *fer* mutants exhibited overgrowth of pollen tube and loss of its rupturing ability [24]. ANXUR1 and ANXUR2 (ANX1, ANX2) are homologs of FER-RLK, expressed at the tip of the pollen tube. The *anx1anx2* double mutants have been found to arrest the growth of pollen tubes and promote bursting immediately after germination. These validate the clue that both FER-mediated and ANX-dependent signaling cascades act as a switch for accurate pollen tube growth and subsequent release of sperm cells for fertilization [99] (Figure 3B).

4.3. Role in Embryo Development

After successful fertilization, the zygote develops into embryo via repeated cell division and differentiation. Several genetic evidences suggest that multiple signaling cascades are essential for embryogenesis in Arabidopsis, and RLK is one amongst them. Predominantly, embryo development initiates from the asymmetric division of the zygote. Intriguingly, the transcript of *ZYGOTIC ARREST 1* (*ZAR1*), a LRR-RLK, has been detected in the embryo sac before and after fertilization. It has been noticed in an eight-nucleate stage of embryo sac to different cells of mature embryo sac including the central cell, egg cell, and synergids. Even after fertilization, it was observed in the endosperm. Phenotypic analysis of *zar1* mutants revealed the role of ZAR1 in the regulation of asymmetric division of zygote and determination of the cell fate of its daughter cells via the activation of calcium and G-protein signaling cascades [100] (Figure 3C). Besides ZAR1, receptor-like protein kinase 1 (RPK1) and Toadstool 2 (TOAD2) are considered indispensable for normal protoderm development, while GASSHO 1 (GSO1) and GSO2 are crucial for the formation of the proper epidermal surface during embryogenesis. The *gso1gso2* double mutants have shown abnormal bending of embryos, highly permeable epidermal structure, and irregular stomatal patterning [101,102] (Figure 3C). Further, molecular analysis has detected the interaction of ALE2 (Abnormal Leaf Shape 2) and ACR4 (CRINKLY 4) with a subtilisin-like serine protease ALE1, which is essential for the formation of primordia of cotyledons during embryogenesis [103] (Figure 3C).

4.4. Organ Development

Coordinated cell growth, differentiationand morphogenesis are the three fundamental aspects of development that cause an organism to procure its shape and an intricate cascade of gene regulatory networks comprising RLKs are known to be implicated in this. In higher plants, all the aerial organs develop from shoot apical meristem (SAM). The maintenance of undifferentiated cells of SAM and organ formation through differentiation from the progeny cells are two processes maintained in a balance during the common developmental process. Interestingly, different RLKs are known to suffice this role. In Arabidopsis, CLAVATA1 or CLV1 (RLK), CLV2 (RLP) and CLV3 (secreted polypeptide) perform a pivotal role in meristem and organ development [17,104,105]. The CLV3 polypeptide acts as a ligand for CLV1 and CLV2 complex. This ligand-receptor binding promotes the activation of cytosolic kinase domain of CLV1 and subsequently, it initiates a signal transduction cascade to control gene expression and stem cell fate in the SAM by elevation of cytosolic calcium as secondary messengers [17,106,107] (Figure 4A). Meristematic receptor-like kinase (MRLK), a LRR-RLK expressed in shoot and root apical meristems, interacts with and phosphorylates a MADS-box transcription factor, AGL24, to regulate floral transition [108] (Figure 4A). Another LRR-RLK, ERECTA, which is expressed in the entire shoot apical meristem and developing organs, monitors organ shape and inflorescence architecture, upon the perception of epidermal patterning factors (EPFs)/EPF-like proteins (EPFLs) [109] (Figure 4A). Moreover, mutants of ERECTA-family LRR-RLKs conferred extreme dwarfism and abnormal flower development, suggesting that ERECTA-family RLKs control cell proliferation as well as organ growth and patterning like stomata formation, the shoot apical meristem (SAM) and flower development [110]. ERECTA can form complexes with a range of co-receptors like SERKs and transmembrane receptor-like proteins like Too Many Mouths (TMM) to activate the signaling pathway [111,112]. Botrytis-induced kinase 1 or BIK1, an RLCK, interacts and phosphorylates ER-family proteins to modulate leaf morphogenesis and inflorescence architecture [113] (Figure 4A).

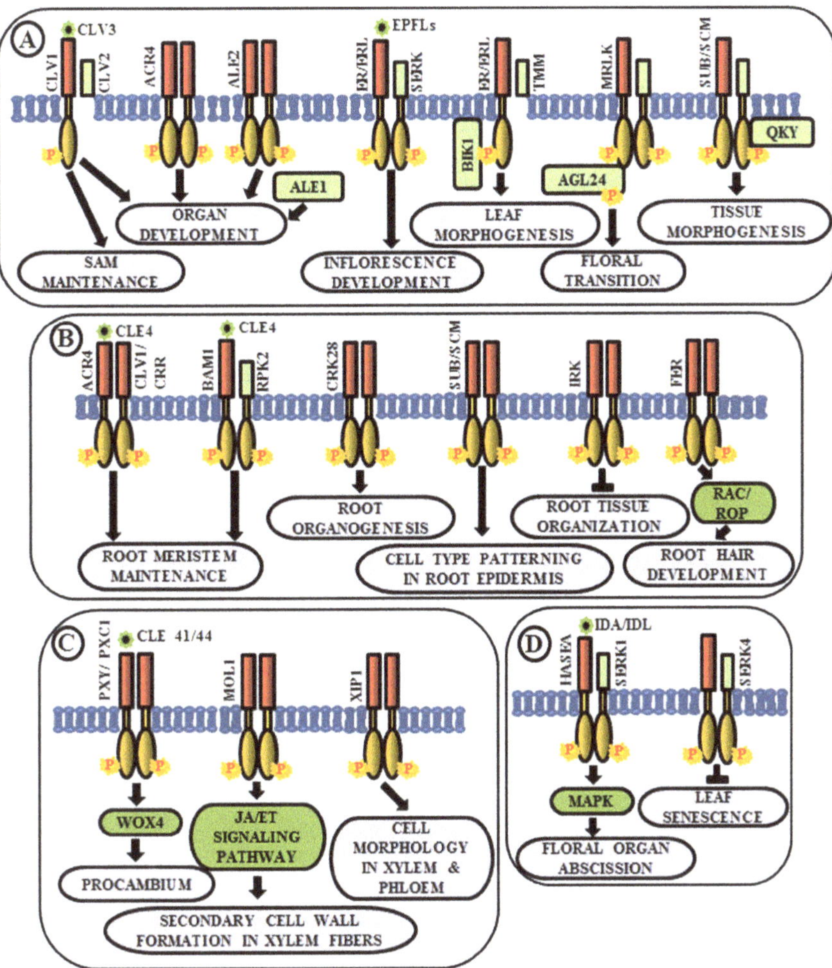

Figure 4. Arabidopsis RLKs in the regulation of growth and development. A few examples of RLKs that regulate (**A**) shoot development, (**B**) root development, (**C**) vascular tissue development, and (**D**) organ abscission.

Similar to aboveground organ development, several studies demonstrated the utmost importance of multiple RLKs in root development. Arabidopsis CRINKLY 4 (ACR4) is involved in the formation of proper lateral roots and columella stem cell differentiation in the root apical meristem [114,115]. ACR4 can regulate root meristem maintenance in response to the CLE4 peptide by forming heterodimers with CLV1 [116] (Figure 4B). Besides, ACR4, abnormal leaf shape 1 (ALE1) (a member of subtilisin-like serine protease family), and ALE2 (RLK) have been known to share partial overlapping roles in the formation of leafy organs [103] (Figure 4A). Similar to ACR4, cysteine-rich receptor-like kinases (CRKs), a member of one of the largest RLK families, is involved in root organogenesis. The *crk28* mutants have displayed longer and branched roots, while CRK28 overexpression lines have shown the contrasting phenotype, i.e., delayed root growth and reduced lateral root formation [117] (Figure 4B).

Plasmodesmata are microchannels between two cells, through which trafficking of molecules occur. STRUBBELIG (SUB) is a RLK involved in inter-cell layer signaling which is required for tissue morphogenesis. The *sub* mutants have shown defects in floral organ shape, integument initiation, and outgrowth, asymmetry in leaf shape and stem morphology, as well as a reduction in plant height. This indicates the functional role of SUB across several cells in the floral meristem, ovule, and shoot apex [118,119]. Further genetic screening has led to the identification of a putative membrane-anchored C2-domain protein, encoded by QUIRKY (QKY), which is known to act as a downstream component of SUB signaling [120]. SUB and QKY interact in plasmodesmata to promote tissue morphogenesis (Figure 4A). Apart from aerial organs, SUB or SCRAMBLED (SCM) also regulates cell-type patterning in the root epidermis [121] (Figure 4B). The BAM1 (barely any meristem 1), a member of CLV1 class LRR-RLKs, is expressed preferentially in the quiescent center and its surrounding stem cells at the root tip and known to bind to the CLE peptide. BAM1 is capable of forming heteromeric complexes with RPK2 and inhibit cell proliferation in the root meristem [122] (Figure 4B). Inflorescence and root apices receptor kinase (IRK), a typical meristematic LRR-RLK, is known to be expressed in the outer plasma membrane of root endodermal cells and negatively regulates cell division to maintain tissue organization [123] (Figure 4B). Further, FERONIA (FER) receptor-like kinase functions upstream of Rho-like small G-protein or RAC/ROP during reactive oxygen species (ROS)-mediated root hair development. The FER activates RAC/ROP by GDP-GTP exchange to stimulate NADPH oxidase for ROS formation [25] (Figure 4B).

4.5. Vascular Tissue Development

The development of xylem and phloem from the vascular meristem is a multifaceted process. The RLK, phloem intercalated with xylem (PXY), maintains cell polarity during vascular development, which is ascertained by the presence of partially interspersed xylem and phloem, and irregular vascular development in *pxy* mutants [124]. The ligand for PXY receptor is tracheary element differentiation factor (TDIF), a peptide, which is encoded by CLAVATA3/ESR 41/44 (CLE41/44) genes [125]. The PXY-TDIF interaction activates the WUSCHEL-related homeobox 4 (WOX4) signaling pathway to regulate cell division in the procambium. Another LRR-RLK, PXY/TDR-CORRELATED (PXC1), acts as a positive regulator of secondary cell wall formation in xylem fibers [126] (Figure 4C). The CLE41/PXY/WOX4 cascade is antagonistically directed by the LRR-RLK more lateral growth 1 (MOL1), via regulating the stem cell homeostasis within the cambium. This MOL1 also attenuates ethylene and jasmonic acid hormone signaling pathways that positively influence cambium activity [127] (Figure 4C). The maintenance of the cell morphology organization during vascular development is accomplished by a RLK, xylem intermixed with phloem 1 (XIP1). Genetic evidences also unveil that XIP1 prevents ectopic lignification in phloem cells [128] (Figure 4C).

4.6. Regulation of Organ Abscission

Arabidopsis LRR–RLK HAESA (formerly named RLK5) exhibits developmentally regulated expression in the abscission layers of floral organs. The antisense suppression of the HAESA is known to delay the abscission of floral organs such as sepals, petals, and stamens [19]. Inflorescence deficient in abscission (IDA) and IDA-Like (IDL) proteins are considered as the ligands of HAESA (HAE) and HAESA-Like RLKs [129] (Figure 4D). The phenotypic analysis of *ida* mutant and overexpression of *IDA* gene validates the role of HAE in floral organ abscission via IDA/IDL perception. A phosphorylation-based activation mechanism of HAE leads to the stimulation of a MAP kinase-signaling cascade and initiates cell wall hydrolysis at the base of the abscising organs. SERK1 acts as a co-receptor of HAE and allows the binding of IDA, eventually leading to floral abscission pathway [130,131]. In contrast, an early leaf senescence phenotype observed in *serk4-1* knockout mutant indicates that SERK4 acts as a co-receptor in negatively regulating leaf senescence, as well [132] (Figure 4D).

4.7. Modulation of Phytohormone Signaling

Brassinosteroids (BRs) are essential polyhydroxylated steroidal phytohormones crucial for plant development. The developmental defects of BR biosynthetic and signaling mutants are mostly similar, which include dwarfism, severely stunted and rounded leaf with a shorter petiole, delayed flowering, photomorphogenic malfunctions as well as senescence and reduced male fertility. The first BR signaling gene, whose mutation showed these phenotypes, has been named as brassinosteroid insensitive 1 (BRI1) [133]. BAK1 (BRI1-associated receptor kinase 1), a co-receptor of BRI1, is involved in BR perception and signaling via heterodimerization with BRI1 [59,134]. In addition, a close homologue of BRI1, BRI1-like receptor kinase (BRL1) is also responsible for BR perception [135] (Figure 5A). BAK1-associating receptor-like kinase 1 (BARK1), a LRR-RLK, specifically binds to BAK1 and its homologs. Overexpression of BARK1 enhances primary root growth and these roots are hypersensitive to BR-induced root growth inhibition, suggesting the role of BARK1 in BR-mediated lateral root development via auxin signaling [136] (Figure 5A). Apart from these, evidence achieved from *bir1* mutants helps us to comprehend how it modulates immune response pathways and plant architecture as an interacting partner of BAK1 [137]. A member of somatic embryogenesis receptor, SERK3 acts as a co-receptor, which directly interacts with BRI1 [64] (Figure 5A).

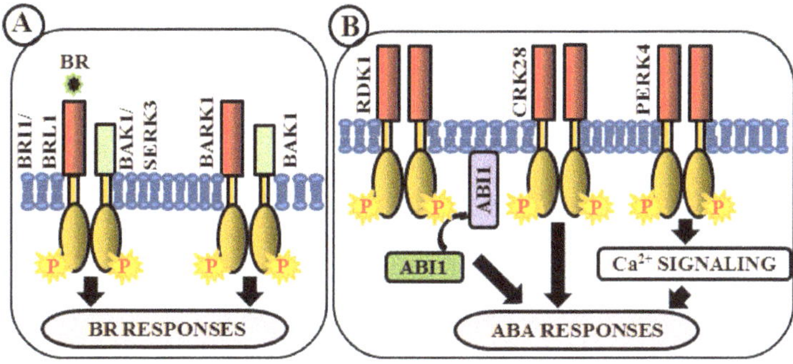

Figure 5. Arabidopsis RLKs in brassinosteroid (BR) and abscisic acid (ABA) signaling. RLK-mediated phosphorylation-based signaling circuits regulate BR (**A**) and ABA (**B**) responses.

Abscisic acid (ABA) is yet another vital phytohormone involved in the regulation of plant abiotic stress-related phenotype as well as developmental processes. Unlike BR, in Arabidopsis, RLKs are not accountable for direct ABA perception. A LRR-RLK, receptor dead kinase 1 (RDK1) is involved in ABA signal transduction via interaction with abscisic acid insensitive 1 (ABI1), a type 2C protein phosphatase, in the plasma membrane. Predominantly, this interaction is enhanced by exogenous application of ABA, underpinning the involvement of RDK1 to recruit ABI1 to the plasma membrane [138] (Figure 5B). Most recently, a cysteine-rich receptor-like kinase, CRK28, has shown an indirect relationship with ABA. The *CRK28* overexpression lines have displayed slow root growth, reduced lateral root formation, and also ABA hypersensitivity; thereby being an important modulator of ABA signaling [117] (Figure 5B). PERK4 is also known to play an important role in ABA response. The *perk4* mutants have shown reduced sensitivity to ABA concerning seed germination, seedling growth, and primary root tip growth. Moreover, *perk4* mutant cells have retained lower cytosolic calcium concentration and Ca²⁺ channel currents. These results suggest that PERK4 contributes to the early stage of ABA signaling and inhibits root cell elongation via intracellular calcium signaling [139] (Figure 5B). Other RLKs like CRK5, CRK36, LRK10L1.2, and RPK1 are also known to be involved in ABA signaling during response to drought and oxidative stresses.

5. RLKs in the Regulation of Plant Biotic Interactions

5.1. RLKs in Pathogen Triggered Immunity

Plants sense the invasion of pathogens through the perception of pathogen and host-derived elicitors, like MAMPs, PAMPs, DAMPs, and HAMPs (herbivore associated molecular patterns). To combat the attack of invading pathogens, RLK-mediated signaling boosts transcriptional activation of multiple defense and pathogenesis-related genes to eliminate the adversity caused by the pathogens. These kinds of RLKs are also termed as 'pattern recognition receptors' (PRRs) and the resulting immune response is called pathogen-triggered immunity (PTI). Predominantly, RLK-derived signals activate defense responses like hypersensitive response, stimulation of ion fluxes, ROS (reactive oxygen species) production, synthesis of phytoalexins, salicylic acid (SA) accumulation, and cell wall reinforcement [6,140,141]. Some important examples of Arabidopsis RLKs involved in defense responses are discussed here.

The flagellin sensitive 2 (FLS2) preferentially recognizes a PAMP, the flagellin epitope of bacteria (flg22), to trigger the recruitment of co-receptors or adaptor proteins and subsequent phosphorylation [20]. Usually, FLS2 heterodimerizes with BAK1 or its homolog BAK1-like kinase 1 (BKK1) and undergo transphosphorylation [72,142–144]. Subsequently, botrytis-induced kinase 1 (BIK1) (RLCK) is phosphorylated and released from the FLS2-BAK1 or FLS2-BKK1 complex. This is followed by rapid bursts of calcium and reactive oxygen species (ROS), activation of MAPKs and/or CDPKs, in order to regulate the PTI [145] (Figure 6). In contrast, BIR2 is an atypical LRR-RLK or pseudokinase, which competes with FLS2 for BAK1 and negatively regulates BAK1 mediated immune signaling and cell death responses [5,146,147] (Figure 6). The *bak1* mutants display enhanced susceptibility to the most commonly encountered necrotrophic pathogens *Alternaria brassicicola* or *Botrytis cinerea* and thus, BAK1 and its co-receptors are considered as important regulators of plant immunity [148]. Further, BAK1 is also involved in temporary desensitization of signaling as it promotes the ubiquitination and proteosomal degradation of FLS2 through phosphorylation of U-Box E3-ubiquitin ligases, PUB12 and PUB13 [149].

Another PAMP known as bacterial elongation factor Tu (EF-Tu) is perceived by an LRR-RLK, EF-Tu receptor (EFR), which activates plant defense responses, thereby reducing the efficiency of *Agrobacterium* transformation [150]. EFR physically interacts with BAK1 in a ligand-dependent manner and establishes the PTI signaling [151] (Figure 6). Another group of LRR-RLKs, PEPR1 (perception of the Arabidopsis danger signal peptide 1) and its close homolog PEPR2 stimulate the innate immune responses upon the perception of wound-induced or plant-derived peptides, PEP1 (perception of the damage-associated molecular pattern peptide 1) and PEP2 [152,153]. Unlike FLS2 and EFR, the signaling molecules of PEPR1 and PEPR2 are DAMPs, which are produced due to wounding, PAMP treatment, or microbial infection, at the early stage of invasion. Both PEPR1 and PEPR2 associate with BAK1 to activate downstream signaling for enhancing plant immunity [63,154] (Figure 6). RLK902 is also linked with plant immunity as it phosphorylates brassinosteroid-signaling kinase 1 (BSK1) and plays an essential role in conferring resistance to the bacterial pathogen *Pseudomonas syringae*. Enhanced disease resistance 4 (EDR4), a protein involved in endocytosis, regulates sub-cellular trafficking of RLK902 for proper modulation of plant immunity [155] (Figure 6).

Chitin, a fungal cell wall derivative, is recognized as a MAMP by a receptor complex comprising of chitin elicitor receptor kinase 1 (CERK1), LysM receptor-like kinase 1 (LYK1) and LYK5 [61,156]. CERK1 directly interacts and phosphorylates PBL27, an RLCK, to regulate chitin-induced defense gene expression and accumulation of callose [157]. Predominantly, PBL27 phosphorylates MAPKKK5, which activate MKK4/5 and MPK3/6 cascades for triggering defense responses (Figure 7) [158]. CERK1 is also involved in the perception of bacterial peptidoglycans (PGNs) and thereby, activate resistance against bacterial infections [30,159]. In addition to chitin, fungal 1,3-β-D-glucan oligosaccharides are perceived by LYK1 [160]. LYK4 augments chitin-induced signaling by acting as co-receptor or scaffold protein of LYK5 [161] (Figure 6). The homologues of LYKs in other angiosperms are involved in the maintenance of symbioses with beneficial mycorrhizal fungi and nitrogen-fixing bacteria [56,162,163]. In some instances, heterotrimeric G-protein components are known to participate immediately downstream to the PRRs. G-protein subunits Gα, Gγ1, and Gγ2 physically interact with the defense-related RD-type receptor-like kinases CERK1, BAK1, and BIR1 [67]. The Gβ, Gγ1, and Gγ2 are required for FLS2, EFR and CERK1-mediated PTI responses, because flg22, elf18 and chitin induced resistance is known to be compromised in *Gβ* single mutant (*agb1*) and *Gγ1* and *Gγ2* double mutant (*agg1agg2*) [164] (Figure 6).

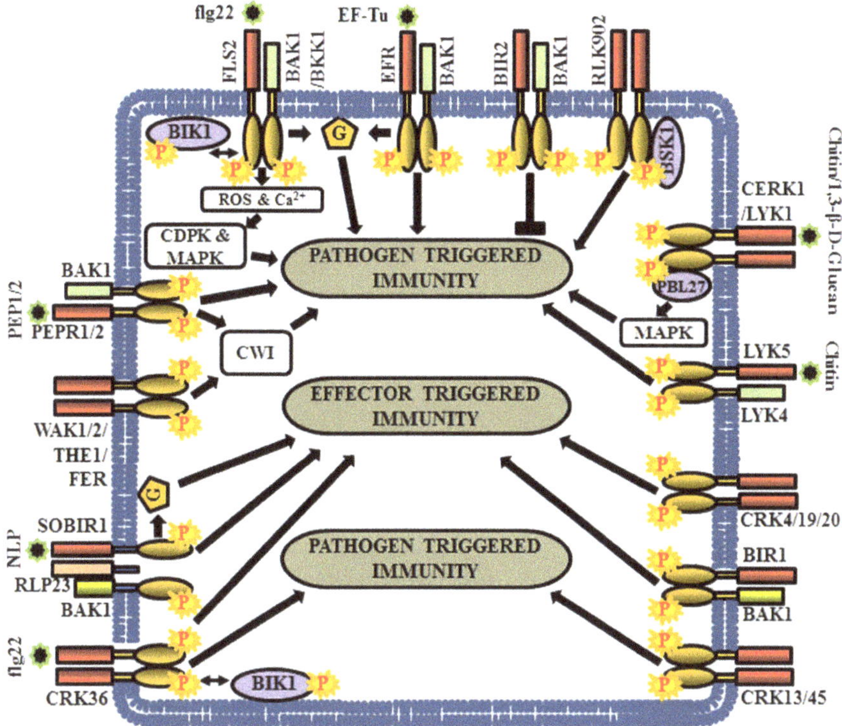

Figure 6. Role of RLKs in Arabidopsis biotic stress responses. This cartoon is representing a few examples of RLKs that regulate pathogen-triggered immunity (PTI), effector-triggered immunity (ETI) or both. G, heterotrimeric G-protein.

Cell wall damage (CWD) triggers cell wall integrity (CWI) maintenance and immune signaling systems to control stress responses. Multiple RLKs like FERONIA (FER), THESEUS 1 (THE1), Male discoverer 1 (MDIS1)-interacting receptor-like kinase 2 (MIK2), WAK1, and WAK2 are known to be involved in CWI maintenance [165–167]. Amongst them, FER, THE1, and MIK2 aid in conferring resistance to the plant against *Fusarium oxysporum*, a fungal pathogen [168,169] (Figure 6). In addition, BAK1, BIK1, BKK1, PEPR1, and PEPR2 modulate responses to CWD controlled by the CWI mechanism [23]. Both PEPR1and PEPR2 perceive DAMPs, like plant elicitor peptides (AtPeps). These AtPeps (AtPep1 and AtPep3) precursor peptides are encoded by the *PROPEP* (*PROPEP1* and *PROPEP3*) genes, which are induced by pathogen infection, wounding and CWD. Although the application of AtPep plant elicitor peptides enhances expression of their corresponding PROPEP genes, these peptides also inhibit CWD-induced Jasmonic acid (JA) and salicylic acid (SA) accumulation in a concentration-dependent manner. These results suggest that both PTI signaling and CWI maintenance mechanism contribute to biotic stress responses, coordinately [170].

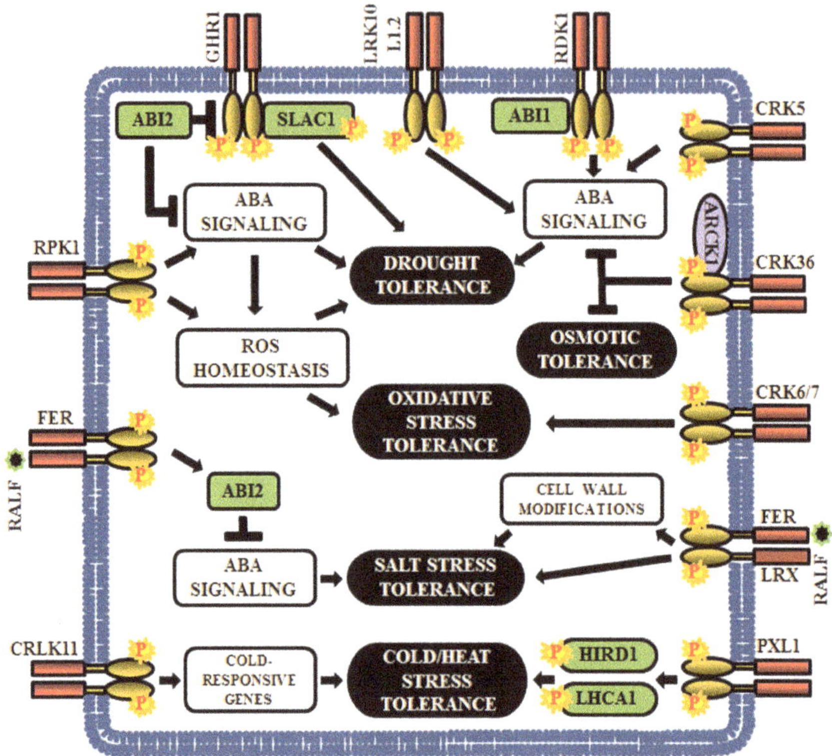

Figure 7. Role of RLKs in Arabidopsis abiotic stress responses. This cartoon is representing a few examples of RLKs that regulate various abiotic stresses in plants (drought, osmotic, oxidative, salt, cold, and heat).

5.2. RLKs in Effector Triggered Immunity

Effectors are the compounds secreted by bacterial and fungal pathogens, which translocate into the host cell for attenuation of the host's defense system (PTI). Impeding the formation of PRR complex is one of the key mechanisms of effectors to suppress immunity and in accordance with this, plants have evolved resistance (R) proteins to recognize pathogen effector proteins to establish effector-triggered immunity (ETI). AvrPto A and AvrPto B are the two types of effectors produced by *Pseudomonas syringae* to suppress the flagellin-induced PTI in Arabidopsis, by interacting with the cytosolic domain of BAK1 and thus, preventing FLS2-BAK1 heterodimerization [171,172]. BAK1-interacting RLK 1 (BIR1) is known to associate with BAK1 *in planta*. The *bir1-1* mutants display extensive cell death and activation of constitutive defense responses. Moreover, these *bir1-1* mutants show enhanced resistance to biotrophic oomycete, *Hyaloperonospora arabidopsidis*. These responses are similar to hypersensitive cell death (HR) observed during ETI, suggesting that BAK1 functions together with BIR1 to negatively regulate multiple plant resistance signaling pathways [173].

Genetic screening for suppressors of the *bir1-1* has led to the identification of *SOBIR1* gene, whose mutation showed impaired cell death in the *bir-1-1* mutant. However, in contrast, *SOBIR1* overexpression resulted in the activation of cell death, thereby indicating the role of SOBIR1 as a positive regulator of cell death [173]. The LRR-RLK, SOBIR1 also triggers defense responses by forming a complex with certain LRR-RLP like immune receptors. For example, RLP23 forms a complex with SOBIR1 and the perception of a necrosis and ethylene-inducing peptide-like 1 protein (NLP) initiates recruitment of BAK1 to the LRR-RLP/SOBIR1 complex, thereby activating LRR-RLP-mediated immunity [174] (Figure 6). A recent investigation has revealed that auto or transphosphorylation events between SOBIR1 and BAK1 are crucial for this ETI signaling [175]. Interestingly, G-protein β subunit mutant (*agb1-2*) has seemed to reduce the cell death and defense responses in *bir1-1* mutant as well as transgenic plants overexpressing *SOBIR1*. Furthermore, *agg1agg2* double mutant has shown suppression of cell death phenotype in the *bir1-1* mutant. These results exhibit the contribution of heterotrimeric G-protein subunits (AGB1, AGG1, and AGG2) in SOBIR1-mediated ETI signaling [164].

5.3. CRKs in Defense and Hypersensitive Responses

Cysteine-rich receptor-like kinases (CRKs) are one of the largest RLK groups, which are transcriptionally induced during pathogen attack, oxidative stress, and also by the application of salicylic acid (SA) [176]. Recent studies have demonstrated the implications of CRKs in the regulation of defense responses and programmed cell death by guiding both PTI and ETI [10,177,178]. For example, constitutive over-expression of *CRK5* and inducible expression of *CRK13* leads to enhanced defense against *Pseudomonas syringae* via up-regulation of defense-related genes, like *PR1* (*pathogenesis related protein 1*), *PR5*, and *ICS1* (*isochorismate synthase 1*). Similarly, overexpression of *CRK45* results in enhanced resistance to *P. syringae*, whereas *crk45* mutants display more sensitivity to *P. syringae* by attenuating the expression of defense-related genes [179]. In addition, the induced expression of *CRK4*, *CRK5*, *CRK19*, and *CRK20* triggered hypersensitive response-like cell death in transgenic plants [28,180,181]. Recently, a physical interaction study has established that CRK36 preferentially interacts with and phosphorylates BIK1 (RLCK) and boosts plant immunity in response to flg22 treatment by regulating stomatal defense against pathogens [182] (Figure 6).

6. RLKs in the Regulation of Plant Abiotic Stresses

Abiotic stresses, such as drought, cold, salinity, ozone, metals, and UV-B radiations, have adverse impact on plant growth and development. Plants have various tactics to survive in continuously changing environmental conditions and one such is the RLK-mediated signaling circuit [183–185].

Among the plant hormones, ABA is a crucial mediator of the abiotic stress response; it can regulate the expression of drought, salt and osmotic stress response genes [186–189]. Genetic screening in Arabidopsis has established the connection between several LRR-RLKs and ABA-mediated abiotic stress signal. The loss-of-function mutants of Arabidopsis *leaf rust 10 disease-resistance locus receptor-like protein kinase 1.2 (LRK10L1.2)* display ABA-insensitive and drought stress-sensitive phenotypes indicating that LRK10L1.2 acts as a positive regulator in response to drought tolerance, perhaps through ABA-mediated signaling [32] (Figure 7). The insensitivity to ABA and downregulation of various water stress-responsive genes are also observed in *RPK1* knockouts and further, overexpression of *RPK1* exhibits increased tolerance to both drought and oxidative stress as well as up-regulation of ROS related genes. These results indicate that RPK1 regulates water and oxidative stress response via ROS homeostasis and ABA signaling [190] (Figure 7). Another LRR-RLK, guard cell hydrogen peroxide resistant 1 (GHR1) is an early component in ABA signaling and is negatively regulated by ABI2. The *ghr1* mutants show impaired ABA and H_2O_2 regulated activation of S-type anion currents in guard cells. Predominantly, GHR1 physically interacts with and activates the slow anion channel-associated 1 (SLAC1) by phosphorylation, resulting in stomatal closure during drought stress [191] (Figure 7). In addition, Arabidopsis receptor dead kinase 1 (RDK1) plays an essential role in drought stress response in an ABA-dependent manner. The *rdk1* mutants are hypersensitive to drought stress as a result of down-regulation of ABA-responsive genes [138] (Figure 7).

Few CRKs are also involved in ABA-mediated drought resistance. Overexpression of *CRK5* promotes stomatal closure and inhibits stomatal opening, thereby acting as a positive regulator of drought response [192]. CRK36 physically interacts with and phosphorylates ARCK1 (RLCK) during abiotic stress. The *crk36* knockdown mutants exhibit osmotic stress response during post-seed germinative growth, increases ABA sensitivity, and upregulates ABA-responsive genes. Thus, CRK36 seems to function as a significant negative regulator of ABA and osmotic stress signal transduction [186]. Besides, CRK6 and CRK7 are essential for overaccumulation of ROS in the apoplast during exposure to O_3, and therefore, their mutants show increased sensitivity to O_3 [29] (Figure 7).

FERONIA (FER), a member of the CrRLK1L family, plays a crucial role in ABA and salt stress responses. FER promotes activation of ABI2, a PP2C member, and a negative regulator of ABA signaling, to attenuate the ABA signaling and it has been noticed that the *fer1* mutants show hypersensitivity to both ABA and salt. This confirms the clue that FER regulates salt stress response via ABI2-mediated ABA signaling [187,188] (Figure 7). Rapid alkalinization factor 22 (RALF22) peptides are considered as the ligands of FER, which are produced during salt stress, via S1P protease-dependent pathway. In addition, RALF22/23 physically associates with the cell-wall leucine-rich repeat extensins 3/4/5 - (LRX3/4/5), which are critical for salt tolerance. Strikingly, the *fer* mutant, *lrx3/4/5* triple mutant, and overexpressed RALF23/24 lines exhibit identical phenotypes, including increased sensitivity to salt stress and retarded growth. These results demonstrate that FER, LRXs and RALFs form a signaling network that regulate plant growth by conferring tolerance to salt stress [193] (Figure 7).

Phloem intercalated with xylem-like 1 (PXL1), a LRR-RLK, is induced by cold and heat stress. Moreover, Arabidopsis *pxl1* mutants display hypersensitive phenotypes when exposed to cold and heat during the germination stage, suggesting that PXL1 functions in the regulation of stress signaling pathways during temperature fluctuations. The downstream substrates for PXL1 are the histidine-rich dehydrin 1 (HIRD1) and light-harvesting protein complex 1 (LHCA1) [194] (Figure 7). Calcium/calmodulin-regulated RLK or CRLK1 is cold inducible and their expression is enhanced by cold and hydrogen peroxide treatments; thus, justifying the role of CRLK1 in cold-related oxidative stress signal transduction pathway. According to gene knockout studies, CRLK1 acts as a positive regulator of cold tolerance and establishes a link between calcium and cold signaling [195,196] (Figure 7).

7. Conclusions and Outlook

The cellular signaling pathway is a complex network. This review summarized how the different groups of RLK signaling pathways regulate developmental and stress responses in Arabidopsis. RLKs are evolutionarily conserved from algae to angiosperms and are known to monitor a wide variety of cellular processes. The abundance and diversity of RLKs provide insight into the significance of this receptor and its role in sustaining cellular homeostasis for the efficient survival of plants. It explains the reason for its continued expansion on par with the increasing complexity of the higher group of plants. As discussed above, RLKs perform a crucial role in almost every aspect in a plant cell, throughout its life, right from the embryonal stage to senescence. The involvement of RLKs in various developmental, as well as stress responses, can be attributed to the diversity in the architecture of their ectodomains, which aid in the recognition of a plethora of ligands. This is executed by recruiting transducers, which help in communicating the signal further downstream. One such important group of transducers belong to the RLCK family, which activate several other intermediates for establishing a successful response. Interestingly, some RLCKs are conserved between different RLK-mediated signaling pathways. Sporadically, the same RLCK interacts with one of the RLKs to elicit a particular response, while expressing a contrasting response upon interaction with another RLK, by activating a different downstream target. The RLKs can directly use guanosine exchange factors (GEFs) like G-proteins and ROP as transducers, or indirectly via RLCKs and other intermediates. Although differential phosphorylation might be one possible mechanism responsible for activating the transducers, the molecular insights of how this distinction is possible remain elusive.

Although a lot of research has been carried out on RLKs in the last few decades, the biochemical and molecular mechanisms of several RLKs modulating physiological responses are not well understood in detail. The most important challenge is to identify the range of signals for RLKs and to explain how plants integrate these signals downstream. In mechanistic concerns, the dependency of certain fully functional RLKs (like BRI1) upon another RLK (BAK1) for successful complex formation and activation is yet to be discovered. Furthermore, due to the presence of a lot of crosstalk in plants, the intermediate targets of many of the pathways tend to remain unidentified. However, irrespective of the transducer activated and the pathway used, the ultimate outcome is to express appropriate proteins and products that enable the plant to endure the environmental challenges, thus, prolonging its survival. More focus on these aspects might be beneficial for developing resistant/tolerant agronomic cultivars via plant breeding or transgenic approaches. Thus, RLKs can be considered as an inherent elixir for plants' life.

Author Contributions: J.J. and S.G. contributed equally to this manuscript. J.J., S.G., and S.R.C. were all involved in writing and editing the review. All authors have read and agreed to the published version of the manuscript. SRC secured the funding.

Funding: Research in the author's lab is supported by Indian Institute of Science Education and Research (IISER), Tirupati; DBT-Ramalingaswami Re-entry Fellowship of S.R.C. and SERB Start-up Research Grant (SRG/2019/000901) of S.R.C.

Acknowledgments: We apologize to all authors whose work was not cited due to the length limitations.

Conflicts of Interest: The authors declare no conflict of interest.

Abbreviations

BAK1	BRI1-associated receptor kinase 1
DAMP	Damage-associated molecular patterns
LRRs	Leucine-rich repeats
LysM	Lysin motif
MAMP	Microbe-associated molecular patterns
PAMP	Pathogen-associated molecular patterns
RLCK	Receptor-like cytoplasmic kinase
RLK	Receptor-like kinase
RLP	Receptor-like protein
SERK	Somatic embryogenesis receptor kinase

References

1. Lehti-Shiu, M.D.; Zou, C.; Shiu, S.-H. Origin, Diversity, Expansion history, and functional evolution of the plant receptor-like kinase/pelle family. In *Receptor-Like Kinases in Plants: From Development to Defense*; Tax, F., Kemmerling, B., Eds.; Springer: Berlin/Heidelberg, Germany, 2012; pp. 1–22.

2. Lehti-Shiu, M.D.; Shiu, S.H. Diversity, classification and function of the plant protein kinase superfamily. *Philos. Trans. R. Soc. B Biol. Sci.* **2012**, *367*, 2619–2639. [CrossRef] [PubMed]

3. Cock, J.M.; Vanoosthuyse, V.; Gaude, T. Receptor kinase signalling in plants and animals: Distinct molecular systems with mechanistic similarities. *Curr. Opin. Cell Biol.* **2002**, *14*, 230–236. [CrossRef]

4. Roux, M.; Zipfel, C. Receptor kinase interactions: Complexity of signalling. In *Receptor-Like Kinases in Plants: From Development to Defense*; Tax, F., Kemmerling, B., Eds.; Springer: Berlin/Heidelberg, Germany, 2012; pp. 145–172.

5. Shiu, S.H.; Bleecker, A.B. Plant receptor-like kinase gene family: Diversity, function, and signaling. *Sci. STKE* **2001**, *2001*, re22. [CrossRef] [PubMed]

6. He, Y.; Zhou, J.; Shan, L.; Meng, X. Plant cell surface receptor-mediated signaling—A common theme amid diversity. *J. Cell Sci.* **2018**, *131*, jcs209353. [CrossRef] [PubMed]

7. Tang, D.; Wang, G.; Zhou, J.M. Receptor kinases in plant-pathogen interactions: More than pattern recognition. *Plant Cell* **2017**, *29*, 618–637. [CrossRef] [PubMed]

8. Shiu, S.H.; Bleecker, A.B. Expansion of the receptor-like kinase/Pelle gene family and receptor-like proteins in Arabidopsis. *Plant Physiol.* **2003**, *132*, 530–543. [CrossRef]

9. Turra, D.; Segorbe, D.; Di Pietro, A. Protein kinases in plant-pathogenic fungi: Conserved regulators of infection. *Annu. Rev. Phytopathol.* **2014**, *52*, 267–288. [CrossRef]

10. Shiu, S.H.; Bleecker, A.B. Receptor-like kinases from Arabidopsis form a monophyletic gene family related to animal receptor kinases. *Proc. Natl. Acad. Sci. USA* **2001**, *98*, 10763–10768. [CrossRef]

11. Oh, M.H.; Wang, X.; Kota, U.; Goshe, M.B.; Clouse, S.D.; Huber, S.C. Tyrosine phosphorylation of the BRI1 receptor kinase emerges as a component of brassinosteroid signaling in Arabidopsis. *Proc. Natl. Acad. Sci. USA* **2009**, *106*, 658–663. [CrossRef]

12. Madsen, E.B.; Antolin-Llovera, M.; Grossmann, C.; Ye, J.; Vieweg, S.; Broghammer, A.; Krusell, L.; Radutoiu, S.; Jensen, O.N.; Stougaard, J.; et al. Autophosphorylation is essential for the in vivo function of the *Lotus japonicus* Nod factor receptor 1 and receptor-mediated signalling in cooperation with Nod factor receptor 5. *Plant J.* **2011**, *65*, 404–417. [CrossRef]

13. Boudeau, J.; Miranda-Saavedra, D.; Barton, G.J.; Alessi, D.R. Emerging roles of pseudokinases. *Trends Cell Biol.* **2006**, *16*, 443–452. [CrossRef] [PubMed]

14. Liang, X.; Zhou, J.M. Receptor-Like Cytoplasmic Kinases: Central Players in Plant Receptor Kinase-Mediated Signaling. *Annu. Rev. Plant Biol.* **2018**, *69*, 267–299. [CrossRef] [PubMed]

15. Lehti-Shiu, M.D.; Zou, C.; Hanada, K.; Shiu, S.H. Evolutionary history and stress regulation of plant receptor-like kinase/pelle genes. *Plant Physiol.* **2009**, *150*, 12–26. [CrossRef] [PubMed]

16. Liu, P.L.; Du, L.; Huang, Y.; Gao, S.M.; Yu, M. Origin and diversification of leucine-rich repeat receptor-like protein kinase (LRR-RLK) genes in plants. *BMC Evol. Biol.* **2017**, *17*, 47. [CrossRef]

17. Jeong, S.; Trotochaud, A.E.; Clark, S.E. The Arabidopsis CLAVATA2 gene encodes a receptor-like protein required for the stability of the CLAVATA1 receptor-like kinase. *Plant Cell* **1999**, *11*, 1925–1934. [CrossRef] [PubMed]

18. Albrecht, C.; Russinova, E.; Hecht, V.; Baaijens, E.; de Vries, S. The *Arabidopsis thaliana* SOMATIC EMBRYOGENESIS RECEPTOR-LIKE KINASES1 and 2 control male sporogenesis. *Plant Cell* **2005**, *17*, 3337–3349. [CrossRef]

19. Jinn, T.L.; Stone, J.M.; Walker, J.C. HAESA, an Arabidopsis leucine-rich repeat receptor kinase, controls floral organ abscission. *Genes Dev.* **2000**, *14*, 108–117.

20. Gomez-Gomez, L.; Boller, T. FLS2: An LRR receptor-like kinase involved in the perception of the bacterial elicitor flagellin in Arabidopsis. *Mol. Cell* **2000**, *5*, 1003–1011. [CrossRef]

21. Verica, J.A.; He, Z.H. The cell wall-associated kinase (WAK) and WAK-like kinase gene family. *Plant Physiol.* **2002**, *129*, 455–459. [CrossRef]

22. Baumberger, N.; Ringli, C.; Keller, B. The chimeric leucine-rich repeat/extensin cell wall protein LRX1 is required for root hair morphogenesis in *Arabidopsis thaliana*. *Genes Dev.* **2001**, *15*, 1128–1139. [CrossRef]

23. Engelsdorf, T.; Hamann, T. An update on receptor-like kinase involvement in the maintenance of plant cell wall integrity. *Ann. Bot.* **2014**, *114*, 1339–1347. [CrossRef]

24. Huck, N.; Moore, J.M.; Federer, M.; Grossniklaus, U. The Arabidopsis mutant feronia disrupts the female gametophytic control of pollen tube reception. *Development* **2003**, *130*, 2149–2159. [CrossRef]

25. Duan, Q.; Kita, D.; Li, C.; Cheung, A.Y.; Wu, H.M. FERONIA receptor-like kinase regulates RHO GTPase signaling of root hair development. *Proc. Natl. Acad. Sci. USA* **2010**, *107*, 17821–17826. [CrossRef]

26. Gifford, M.L.; Dean, S.; Ingram, G.C. The Arabidopsis ACR4 gene plays a role in cell layer organisation during ovule integument and sepal margin development. *Development* **2003**, *130*, 4249–4258. [CrossRef] [PubMed]

27. Czyzewicz, N.; Nikonorova, N.; Meyer, M.R.; Sandal, P.; Shah, S.; Vu, L.D.; Gevaert, K.; Rao, A.G.; De Smet, I. The growing story of (ARABIDOPSIS) CRINKLY 4. *J. Exp. Bot.* **2016**, *67*, 4835–4847. [CrossRef] [PubMed]

28. Acharya, B.R.; Raina, S.; Maqbool, S.B.; Jagadeeswaran, G.; Mosher, S.L.; Appel, H.M.; Schultz, J.C.; Klessig, D.F.; Raina, R. Overexpression of CRK13, an Arabidopsis cysteine-rich receptor-like kinase, results in enhanced resistance to Pseudomonas syringae. *Plant J.* **2007**, *50*, 488–499. [CrossRef]

29. Idanheimo, N.; Gauthier, A.; Salojarvi, J.; Siligato, R.; Brosche, M.; Kollist, H.; Mahonen, A.P.; Kangasjarvi, J.; Wrzaczek, M. The *Arabidopsis thaliana* cysteine-rich receptor-like kinases CRK6 and CRK7 protect against apoplastic oxidative stress. *Biochem. Biophys. Res. Commun.* **2014**, *445*, 457–462. [CrossRef] [PubMed]

30. Willmann, R.; Lajunen, H.M.; Erbs, G.; Newman, M.A.; Kolb, D.; Tsuda, K.; Katagiri, F.; Fliegmann, J.; Bono, J.J.; Cullimore, J.V.; et al. Arabidopsis lysin-motif proteins LYM1 LYM3 CERK1 mediate bacterial peptidoglycan sensing and immunity to bacterial infection. *Proc. Natl. Acad. Sci. USA* **2011**, *108*, 19824–19829. [CrossRef] [PubMed]

31. Wang, X.; Zafian, P.; Choudhary, M.; Lawton, M. The PR5K receptor protein kinase from *Arabidopsis thaliana* is structurally related to a family of plant defense proteins. *Proc. Natl. Acad. Sci. USA* **1996**, *93*, 2598–2602. [CrossRef] [PubMed]

32. Lim, C.W.; Yang, S.H.; Shin, K.H.; Lee, S.C.; Kim, S.H. The AtLRK10L1.2, *Arabidopsis* ortholog of wheat LRK10, is involved in ABA-mediated signaling and drought resistance. *Plant Cell Rep.* **2015**, *34*, 447–455. [CrossRef]

33. Bellande, K.; Bono, J.J.; Savelli, B.; Jamet, E.; Canut, H. Plant Lectins and lectin receptor-like kinases: How do they sense the outside? *Int. J. Mol. Sci.* **2017**, *18*, 1164. [CrossRef]

34. Dwyer, K.G.; Kandasamy, M.K.; Mahosky, D.I.; Acciai, J.; Kudish, B.I.; Miller, J.E.; Nasrallah, M.E.; Nasrallah, J.B. A superfamily of S locus-related sequences in *Arabidopsis*: Diverse structures and expression patterns. *Plant Cell* **1994**, *6*, 1829–1843. [CrossRef] [PubMed]

35. Kessler, S.A.; Lindner, H.; Jones, D.S.; Grossniklaus, U. Functional analysis of related CrRLK1L receptor-like kinases in pollen tube reception. *EMBO Rep.* **2015**, *16*, 107–115. [CrossRef] [PubMed]

36. Dunne, A.; O'Neill, L.A. The interleukin-1 receptor/Toll-like receptor superfamily: Signal transduction during inflammation and host defense. *Sci. STKE* **2003**, *2003*, re3. [CrossRef] [PubMed]

37. Ten Hove, C.A.; Bochdanovits, Z.; Jansweijer, V.M.; Koning, F.G.; Berke, L.; Sanchez-Perez, G.F.; Scheres, B.; Heidstra, R. Probing the roles of LRR RLK genes in *Arabidopsis thaliana* roots using a custom T-DNA insertion set. *Plant Mol. Biol.* **2011**, *76*, 69–83. [CrossRef]

38. Kobe, B.; Deisenhofer, J. The leucine-rich repeat: A versatile binding motif. *Trends Biochem. Sci.* **1994**, *19*, 415–421. [CrossRef]

39. Wang, H.; Chevalier, D.; Larue, C.; Ki Cho, S.; Walker, J.C. The protein phosphatases and protein kinases of *Arabidopsis thaliana*. *Arab. Book* **2007**, *5*, e0106. [CrossRef]

40. Sharon, N.; Lis, H. Legume lectins—A large family of homologous proteins. *FASEB J.* **1990**, *4*, 3198–3208. [CrossRef]

41. Cambi, A.; Koopman, M.; Figdor, C.G. How C-type lectins detect pathogens. *Cell Microbiol.* **2005**, *7*, 481–488. [CrossRef]

42. Herve, C.; Serres, J.; Dabos, P.; Canut, H.; Barre, A.; Rouge, P.; Lescure, B. Characterization of the *Arabidopsis* lecRK-a genes: Members of a superfamily encoding putative receptors with an extracellular domain homologous to legume lectins. *Plant Mol. Biol.* **1999**, *39*, 671–682. [CrossRef]

43. Choi, J.; Tanaka, K.; Cao, Y.; Qi, Y.; Qiu, J.; Liang, Y.; Lee, S.Y.; Stacey, G. Identification of a plant receptor for extracellular ATP. *Science* **2014**, *343*, 290–294. [CrossRef] [PubMed]

44. Vaid, N.; Macovei, A.; Tuteja, N. Knights in action: Lectin receptor-like kinases in plant development and stress responses. *Mol. Plant* **2013**, *6*, 1405–1418. [CrossRef]

45. Cassab, G.I. Plant cell wall proteins. *Annu. Rev. Plant Physiol. Plant Mol. Biol.* **1998**, *49*, 281–309. [CrossRef] [PubMed]

46. Showalter, A.M.; Keppler, B.; Lichtenberg, J.; Gu, D.; Welch, L.R. A bioinformatics approach to the identification, classification, and analysis of hydroxyproline-rich glycoproteins. *Plant Physiol.* **2010**, *153*, 485–513. [CrossRef] [PubMed]

47. Borassi, C.; Sede, A.R.; Mecchia, M.A.; Salgado Salter, J.D.; Marzol, E.; Muschietti, J.P.; Estevez, J.M. An update on cell surface proteins containing extensin-motifs. *J. Exp. Bot.* **2016**, *67*, 477–487. [CrossRef]

48. Silva, N.F.; Goring, D.R. The proline-rich, extensin-like receptor kinase-1 (PERK1) gene is rapidly induced by wounding. *Plant Mol. Biol.* **2002**, *50*, 667–685. [CrossRef] [PubMed]

49. Nguyen, Q.N.; Lee, Y.S.; Cho, L.H.; Jeong, H.J.; An, G.; Jung, K.H. Genome-wide identification and analysis of *Catharanthus roseus* RLK1-like kinases in rice. *Planta* **2015**, *241*, 603–613. [CrossRef]

50. Schallus, T.; Jaeckh, C.; Feher, K.; Palma, A.S.; Liu, Y.; Simpson, J.C.; Mackeen, M.; Stier, G.; Gibson, T.J.; Feizi, T.; et al. Malectin: A novel carbohydrate-binding protein of the endoplasmic reticulum and a candidate player in the early steps of protein N-glycosylation. *Mol. Biol. Cell* **2008**, *19*, 3404–3414. [CrossRef]

51. Nasrallah, J.B.; Yu, S.M.; Nasrallah, M.E. Self-incompatibility genes of *Brassica oleracea*: Expression, isolation, and structure. *Proc. Natl. Acad. Sci. USA* **1988**, *85*, 5551–5555. [CrossRef]

52. Wrzaczek, M.; Brosché, M.; Salojärvi, J.; Kangasjärvi, S.; Idänheimo, N.; Mersmann, S.; Robatzek, S.; Karpiński, S.; Karpińska, B.; Kangasjärvi, J. Transcriptional regulation of the CRK/DUF26 group of receptor-like protein kinases by ozone and plant hormones in *Arabidopsis*. *BMC Plant Biol.* **2010**, *10*, 95. [CrossRef]

53. Vaattovaara, A.; Brandt, B.; Rajaraman, S.; Safronov, O.; Veidenberg, A.; Luklová, M.; Kangasjärvi, J.; Löytynoja, A.; Hothorn, M.; Salojärvi, J.; et al. Mechanistic insights into the evolution of DUF26-containing proteins in land plants. *Commun. Biol.* **2019**, *2*, 56. [CrossRef]

54. Buist, G.; Steen, A.; Kok, J.; Kuipers, O.P. LysM, a widely distributed protein motif for binding to (peptido) glycans. *Mol. Microbiol.* **2008**, *68*, 838–847. [CrossRef] [PubMed]

55. Buendia, L.; Girardin, A.; Wang, T.; Cottret, L.; Lefebvre, B. LysM Receptor-like kinase and lysm receptor-like protein families: An update on phylogeny and functional characterization. *Front. Plant Sci.* **2018**, *9*, 1531. [CrossRef] [PubMed]

56. Choudhury, S.R.; Pandey, S. Phosphorylation-dependent regulation of G-protein cycle during nodule formation in soybean. *Plant Cell* **2015**, *27*, 3260–3276. [CrossRef] [PubMed]

57. Feuillet, C.; Travella, S.; Stein, N.; Albar, L.; Nublat, A.; Keller, B. Map-based isolation of the leaf rust disease resistance gene Lr10 from the hexaploid wheat (*Triticum aestivum* L.) genome. *Proc. Natl. Acad. Sci. USA* **2003**, *100*, 15253–15258. [CrossRef]

58. Hou, S.; Liu, Z.; Shen, H.; Wu, D. Damage-associated molecular pattern-triggered immunity in plants. *Front. Plant Sci.* **2019**, *10*, 646. [CrossRef]

59. Nam, K.H.; Li, J. BRI1/BAK1, a receptor kinase pair mediating brassinosteroid signaling. *Cell* **2002**, *110*, 203–212. [CrossRef]

60. Stegmann, M.; Monaghan, J.; Smakowska-Luzan, E.; Rovenich, H.; Lehner, A.; Holton, N.; Belkhadir, Y.; Zipfel, C. The receptor kinase FER is a RALF-regulated scaffold controlling plant immune signaling. *Science* **2017**, *355*, 287–289. [CrossRef]

61. Liu, T.; Liu, Z.; Song, C.; Hu, Y.; Han, Z.; She, J.; Fan, F.; Wang, J.; Jin, C.; Chang, J.; et al. Chitin-induced dimerization activates a plant immune receptor. *Science* **2012**, *336*, 1160–1164. [CrossRef]

62. Macho, A.P.; Zipfel, C. Plant PRRs and the activation of innate immune signaling. *Mol. Cell* **2014**, *54*, 263–272. [CrossRef]

63. Schulze, B.; Mentzel, T.; Jehle, A.K.; Mueller, K.; Beeler, S.; Boller, T.; Felix, G.; Chinchilla, D. Rapid heteromerization and phosphorylation of ligand-activated plant transmembrane receptors and their associated kinase BAK1. *J. Biol. Chem.* **2010**, *285*, 9444–9451. [CrossRef]

64. Gou, X.; Yin, H.; He, K.; Du, J.; Yi, J.; Xu, S.; Lin, H.; Clouse, S.D.; Li, J. Genetic evidence for an indispensable role of somatic embryogenesis receptor kinases in brassinosteroid signaling. *PLoS Genet.* **2012**, *8*, e1002452. [CrossRef] [PubMed]

65. Trotochaud, A.E.; Hao, T.; Wu, G.; Yang, Z.; Clark, S.E. The CLAVATA1 receptor-like kinase requires CLAVATA3 for its assembly into a signaling complex that includes KAPP and a Rho-related protein. *Plant Cell* **1999**, *11*, 393–406. [CrossRef]

66. Becraft, P.W. Receptor kinase signaling in plant development. *Annu. Rev. Cell Dev. Biol.* **2002**, *18*, 163–192. [CrossRef] [PubMed]

67. Aranda-Sicilia, M.N.; Trusov, Y.; Maruta, N.; Chakravorty, D.; Zhang, Y.; Botella, J.R. Heterotrimeric G proteins interact with defense-related receptor-like kinases in *Arabidopsis*. *J. Plant Physiol.* **2015**, *188*, 44–48. [CrossRef] [PubMed]

68. Zhou, Z.; Zhao, Y.; Bi, G.; Liang, X.; Zhou, J.M. Early signalling mechanisms underlying receptor kinase-mediated immunity in plants. *Philos. Trans. R. Soc. B Biol. Sci.* **2019**, *374*, 20180310. [CrossRef]

69. Brutus, A.; Sicilia, F.; Macone, A.; Cervone, F.; De Lorenzo, G. A domain swap approach reveals a role of the plant wall-associated kinase 1 (WAK1) as a receptor of oligogalacturonides. *Proc. Natl. Acad. Sci. USA* **2010**, *107*, 9452–9457. [CrossRef]

70. Lin, W.; Lu, D.; Gao, X.; Jiang, S.; Ma, X.; Wang, Z.; Mengiste, T.; He, P.; Shan, L. Inverse modulation of plant immune and brassinosteroid signaling pathways by the receptor-like cytoplasmic kinase BIK1. *Proc. Natl. Acad. Sci. USA* **2013**, *110*, 12114–12119. [CrossRef]

71. Boisson-Dernier, A.; Franck, C.M.; Lituiev, D.S.; Grossniklaus, U. Receptor-like cytoplasmic kinase MARIS functions downstream of CrRLK1L-dependent signaling during tip growth. *Proc. Natl. Acad. Sci. USA* **2015**, *112*, 12211–12216. [CrossRef]

72. Lu, D.; Wu, S.; Gao, X.; Zhang, Y.; Shan, L.; He, P. A receptor-like cytoplasmic kinase, BIK1, associates with a flagellin receptor complex to initiate plant innate immunity. *Proc. Natl. Acad. Sci. USA* **2010**, *107*, 496–501. [CrossRef]

73. Kärkönen, A.; Kuchitsu, K. Reactive oxygen species in cell wall metabolism and development in plants. *Phytochemistry* **2015**, *112*, 22–32. [CrossRef] [PubMed]

74. Dubiella, U.; Seybold, H.; Durian, G.; Komander, E.; Lassig, R.; Witte, C.P.; Schulze, W.X.; Romeis, T. Calcium-dependent protein kinase/NADPH oxidase activation circuit is required for rapid defense signal propagation. *Proc. Natl. Acad. Sci. USA* **2013**, *110*, 8744–8749. [CrossRef] [PubMed]

75. Wu, H.M.; Hazak, O.; Cheung, A.Y.; Yalovsky, S. RAC/ROP GTPases and auxin signaling. *Plant Cell* **2011**, *23*, 1208–1218. [CrossRef] [PubMed]

76. Liang, X.; Ding, P.; Lian, K.; Wang, J.; Ma, M.; Li, L.; Li, M.; Zhang, X.; Chen, S.; Zhang, Y.; et al. *Arabidopsis* heterotrimeric G proteins regulate immunity by directly coupling to the FLS2 receptor. *eLife* **2016**, *5*, e13568. [CrossRef]

77. Kimura, S.; Waszczak, C.; Hunter, K.; Wrzaczek, M. Bound by fate: The role of reactive oxygen species in receptor-like kinase signaling. *Plant Cell* **2017**, *29*, 638–654. [CrossRef]

78. Huang, H.; Ullah, F.; Zhou, D.X.; Yi, M.; Zhao, Y. Mechanisms of ROS Regulation of Plant Development and Stress Responses. *Front. Plant Sci.* **2019**, *10*, 800. [CrossRef]

79. Zulawski, M.; Schulze, G.; Braginets, R.; Hartmann, S.; Schulze, W.X. The *Arabidopsis* Kinome: Phylogeny and evolutionary insights into functional diversification. *BMC Genom.* **2014**, *15*, 548. [CrossRef]

80. Lee, Y.; Kim, Y.J.; Kim, M.H.; Kwak, J.M. MAPK cascades in guard cell signal transduction. *Front. Plant Sci.* **2016**, *7*, 80. [CrossRef]

81. Kawasaki, T.; Yamada, K.; Yoshimura, S.; Yamaguchi, K. Chitin receptor-mediated activation of MAP kinases and ROS production in rice and *Arabidopsis*. *Plant Signal. Behav.* **2017**, *12*, e1361076. [CrossRef]

82. Canales, C.; Bhatt, A.M.; Scott, R.; Dickinson, H. EXS, a putative LRR receptor kinase, regulates male germline cell number and tapetal identity and promotes seed development in *Arabidopsis*. *Curr. Biol.* **2002**, *12*, 1718–1727. [CrossRef]

83. Zhao, D.Z.; Wang, G.F.; Speal, B.; Ma, H. The excess microsporocytes1 gene encodes a putative leucine-rich repeat receptor protein kinase that controls somatic and reproductive cell fates in the *Arabidopsis* anther. *Genes Dev.* **2002**, *16*, 2021–2031. [CrossRef] [PubMed]

84. Feng, X.; Dickinson, H.G. Packaging the male germline in plants. *Trends Genet.* **2007**, *23*, 503–510. [CrossRef] [PubMed]

85. Jia, G.; Liu, X.; Owen, H.A.; Zhao, D. Signaling of cell fate determination by the TPD1 small protein and EMS1 receptor kinase. *Proc. Natl. Acad. Sci. USA* **2008**, *105*, 2220–2225. [CrossRef] [PubMed]

86. Huang, J.; Li, Z.; Biener, G.; Xiong, E.; Malik, S.; Eaton, N.; Zhao, C.Z.; Raicu, V.; Kong, H.; Zhao, D. Carbonic anhydrases function in anther cell differentiation downstream of the receptor-like kinase EMS1. *Plant Cell* **2017**, *29*, 1335–1356. [CrossRef]

87. Colcombet, J.; Boisson-Dernier, A.; Ros-Palau, R.; Vera, C.E.; Schroeder, J.I. *Arabidopsis* SOMATIC EMBRYOGENESIS RECEPTOR KINASES1 and 2 are essential for tapetum development and microspore maturation. *Plant Cell* **2005**, *17*, 3350–3361. [CrossRef] [PubMed]

88. Li, Z.; Wang, Y.; Huang, J.; Ahsan, N.; Biener, G.; Paprocki, J.; Thelen, J.J.; Raicu, V.; Zhao, D. Two SERK receptor-like kinases interact with ems1 to control anther cell fate determination. *Plant Physiol.* **2017**, *173*, 326–337. [CrossRef]

89. Mizuno, S.; Osakabe, Y.; Maruyama, K.; Ito, T.; Osakabe, K.; Sato, T.; Shinozaki, K.; Yamaguchi-Shinozaki, K. Receptor-like protein kinase 2 (RPK 2) is a novel factor controlling anther development in *Arabidopsis thaliana*. *Plant J.* **2007**, *50*, 751–766. [CrossRef]

90. Hord, C.L.; Chen, C.; Deyoung, B.J.; Clark, S.E.; Ma, H. The BAM1/BAM2 receptor-like kinases are important regulators of *Arabidopsis* early anther development. *Plant Cell* **2006**, *18*, 1667–1680. [CrossRef]

91. Cui, Y.; Hu, C.; Zhu, Y.; Cheng, K.; Li, X.; Wei, Z.; Xue, L.; Lin, F.; Shi, H.; Yi, J.; et al. CIK receptor kinases determine cell fate specification during early anther development in *Arabidopsis*. *Plant Cell* **2018**, *30*, 2383–2401. [CrossRef]

92. Hord, C.L.; Sun, Y.J.; Pillitteri, L.J.; Torii, K.U.; Wang, H.; Zhang, S.; Ma, H. Regulation of *Arabidopsis* early anther development by the mitogen-activated protein kinases, MPK3 and MPK6, and the ERECTA and related receptor-like kinases. *Mol. Plant* **2008**, *1*, 645–658. [CrossRef]

93. Wan, J.; Patel, A.; Mathieu, M.; Kim, S.Y.; Xu, D.; Stacey, G. A lectin receptor-like kinase is required for pollen development in *Arabidopsis*. *Plant Mol. Biol.* **2008**, *67*, 469–482. [CrossRef]

94. Huang, J.; Wijeratne, A.J.; Tang, C.; Zhang, T.; Fenelon, R.E.; Owen, H.A.; Zhao, D. Ectopic expression of TAPETUM DETERMINANT1 affects ovule development in *Arabidopsis*. *J. Exp. Bot.* **2016**, *67*, 1311–1326. [CrossRef] [PubMed]

95. Takeuchi, H.; Higashiyama, T. A species-specific cluster of defensin-like genes encodes diffusible pollen tube attractants in *Arabidopsis*. *PLoS Biol.* **2012**, *10*, e1001449. [CrossRef] [PubMed]

96. Takeuchi, H.; Higashiyama, T. Tip-localized receptors control pollen tube growth and LURE sensing in *Arabidopsis*. *Nature* **2016**, *531*, 245–248. [CrossRef] [PubMed]

97. Wang, T.; Liang, L.; Xue, Y.; Jia, P.F.; Chen, W.; Zhang, M.X.; Wang, Y.C.; Li, H.J.; Yang, W.C. Corrigendum: A receptor heteromer mediates the male perception of female attractants in plants. *Nature* **2016**, *536*, 360. [CrossRef]

98. Escobar-Restrepo, J.M.; Huck, N.; Kessler, S.; Gagliardini, V.; Gheyselinck, J.; Yang, W.C.; Grossniklaus, U. The FERONIA receptor-like kinase mediates male-female interactions during pollen tube reception. *Science* **2007**, *317*, 656–660. [CrossRef]

99. Boisson-Dernier, A.; Roy, S.; Kritsas, K.; Grobei, M.A.; Jaciubek, M.; Schroeder, J.I.; Grossniklaus, U. Disruption of the pollen-expressed FERONIA homologs ANXUR1 and ANXUR2 triggers pollen tube discharge. *Development* **2009**, *136*, 3279–3288. [CrossRef]

100. Yu, T.Y.; Shi, D.Q.; Jia, P.F.; Tang, J.; Li, H.J.; Liu, J.; Yang, W.C. The *Arabidopsis* receptor kinase ZAR1 is required for zygote asymmetric division and its daughter cell fate. *PLoS Genet.* **2016**, *12*, e1005933. [CrossRef]

101. Nodine, M.D.; Yadegari, R.; Tax, F.E. RPK1 and TOAD2 are two receptor-like kinases redundantly required for *Arabidopsis* embryonic pattern formation. *Dev. Cell* **2007**, *12*, 943–956. [CrossRef]

102. Tsuwamoto, R.; Fukuoka, H.; Takahata, Y. GASSHO1 and GASSHO2 encoding a putative leucine-rich repeat transmembrane-type receptor kinase are essential for the normal development of the epidermal surface in *Arabidopsis* embryos. *Plant J.* **2008**, *54*, 30–42. [CrossRef]

103. Tanaka, H.; Watanabe, M.; Sasabe, M.; Hiroe, T.; Tanaka, T.; Tsukaya, H.; Ikezaki, M.; Machida, C.; Machida, Y. Novel receptor-like kinase ALE2 controls shoot development by specifying epidermis in *Arabidopsis*. *Development* **2007**, *134*, 1643–1652. [CrossRef] [PubMed]

104. Clark, S.E. Organ formation at the vegetative shoot meristem. *Plant Cell* **1997**, *9*, 1067–1076. [CrossRef] [PubMed]

105. Fletcher, J.C.; Brand, U.; Running, M.P.; Simon, R.; Meyerowitz, E.M. Signaling of cell fate decisions by CLAVATA3 in *Arabidopsis* shoot meristems. *Science* **1999**, *283*, 1911–1914. [CrossRef]

106. Lenhard, M.; Laux, T. Stem cell homeostasis in the *Arabidopsis* shoot meristem is regulated by intercellular movement of CLAVATA3 and its sequestration by CLAVATA1. *Development* **2003**, *130*, 3163–3173. [CrossRef] [PubMed]

107. Chou, H.; Zhu, Y.; Ma, Y.; Berkowitz, G.A. The CLAVATA signaling pathway mediating stem cell fate in shoot meristems requires Ca(2+) as a secondary cytosolic messenger. *Plant J.* **2016**, *85*, 494–506. [CrossRef] [PubMed]

108. Fujita, H.; Takemura, M.; Tani, E.; Nemoto, K.; Yokota, A.; Kohchi, T. An *Arabidopsis* MADS-box protein, AGL24, is specifically bound to and phosphorylated by meristematic receptor-like kinase (MRLK). *Plant Cell Physiol.* **2003**, *44*, 735–742. [CrossRef]

109. Yokoyama, R.; Takahashi, T.; Kato, A.; Torii, K.U.; Komeda, Y. The *Arabidopsis* ERECTA gene is expressed in the shoot apical meristem and organ primordia. *Plant J.* **1998**, *15*, 301–310. [CrossRef]

110. Shpak, E.D.; Lakeman, M.B.; Torii, K.U. Dominant-negative receptor uncovers redundancy in the *Arabidopsis* ERECTA Leucine-rich repeat receptor-like kinase signaling pathway that regulates organ shape. *Plant Cell* **2003**, *15*, 1095–1110. [CrossRef]

111. Lee, J.S.; Kuroha, T.; Hnilova, M.; Khatayevich, D.; Kanaoka, M.M.; McAbee, J.M.; Sarikaya, M.; Tamerler, C.; Torii, K.U. Direct interaction of ligand-receptor pairs specifying stomatal patterning. *Genes Dev.* **2012**, *26*, 126–136. [CrossRef]

112. Meng, X.; Chen, X.; Mang, H.; Liu, C.; Yu, X.; Gao, X.; Torii, K.U.; He, P.; Shan, L. Differential function of *Arabidopsis* SERK family receptor-like kinases in stomatal patterning. *Curr. Biol.* **2015**, *25*, 2361–2372. [CrossRef]

113. Chen, S.; Liu, J.; Liu, Y.; Chen, L.; Sun, T.; Yao, N.; Wang, H.B.; Liu, B. BIK1 and ERECTA Play opposing roles in both leaf and inflorescence development in. *Front. Plant Sci.* **2019**, *10*, 1480. [CrossRef] [PubMed]

114. De Smet, I.; Vassileva, V.; De Rybel, B.; Levesque, M.P.; Grunewald, W.; Van Damme, D.; Van Noorden, G.; Naudts, M.; Van Isterdael, G.; De Clercq, R.; et al. Receptor-like kinase ACR4 restricts formative cell divisions in the *Arabidopsis* root. *Science* **2008**, *322*, 594–597. [CrossRef] [PubMed]

115. Stahl, Y.; Wink, R.H.; Ingram, G.C.; Simon, R. A signaling module controlling the stem cell niche in *Arabidopsis* root meristems. *Curr. Biol.* **2009**, *19*, 909–914. [CrossRef] [PubMed]

116. Stahl, Y.; Grabowski, S.; Bleckmann, A.; Kuhnemuth, R.; Weidtkamp-Peters, S.; Pinto, K.G.; Kirschner, G.K.; Schmid, J.B.; Wink, R.H.; Hulsewede, A.; et al. Moderation of *Arabidopsis* root stemness by CLAVATA1 and *ARABIDOPSIS* CRINKLY4 receptor kinase complexes. *Curr. Biol.* **2013**, *23*, 362–371. [CrossRef] [PubMed]

117. Pelagio-Flores, R.; Muñoz-Parra, E.; Barrera-Ortiz, S.; Ortiz-Castro, R.; Saenz-Mata, J.; Ortega-Amaro, M.A.; Jiménez-Bremont, J.F.; López-Bucio, J. The cysteine-rich receptor-like protein kinase CRK28 modulates *Arabidopsis* growth and development and influences abscisic acid responses. *Planta* **2019**, *251*, 2. [CrossRef] [PubMed]

118. Kwak, S.H.; Schiefelbein, J. A feedback mechanism controlling SCRAMBLED receptor accumulation and cell-type pattern in *Arabidopsis*. *Curr. Biol.* **2008**, *18*, 1949–1954. [CrossRef] [PubMed]

119. Yadav, R.K.; Fulton, L.; Batoux, M.; Schneitz, K. The *Arabidopsis* receptor-like kinase STRUBBELIG mediates inter-cell-layer signaling during floral development. *Dev. Biol.* **2008**, *323*, 261–270. [CrossRef]

120. Vaddepalli, P.; Herrmann, A.; Fulton, L.; Oelschner, M.; Hillmer, S.; Stratil, T.F.; Fastner, A.; Hammes, U.Z.; Ott, T.; Robinson, D.G.; et al. The C2-domain protein QUIRKY and the receptor-like kinase STRUBBELIG localize to plasmodesmata and mediate tissue morphogenesis in *Arabidopsis thaliana*. *Development* **2014**, *141*, 4139–4148. [CrossRef]

121. Kwak, S.H.; Schiefelbein, J. The role of the SCRAMBLED receptor-like kinase in patterning the *Arabidopsis* root epidermis. *Dev. Biol.* **2007**, *302*, 118–131. [CrossRef]

122. Shimizu, N.; Ishida, T.; Yamada, M.; Shigenobu, S.; Tabata, R.; Kinoshita, A.; Yamaguchi, K.; Hasebe, M.; Mitsumasu, K.; Sawa, S. BAM 1 and RECEPTOR-LIKE PROTEIN KINASE 2 constitute a signaling pathway and modulate CLE peptide-triggered growth inhibition in *Arabidopsis* root. *New Phytol.* **2015**, *208*, 1104–1113. [CrossRef]

123. Campos, R.; Goff, J.; Rodriguez-Furlan, C.; Van Norman, J.M. The *Arabidopsis* receptor kinase IRK is polarized and represses specific cell divisions in roots. *Dev. Cell* **2020**, *52*, 183–195. [CrossRef] [PubMed]

124. Fisher, K.; Turner, S. PXY, a receptor-like kinase essential for maintaining polarity during plant vascular-tissue development. *Curr. Biol.* **2007**, *17*, 1061–1066. [CrossRef] [PubMed]

125. Hirakawa, Y.; Shinohara, H.; Kondo, Y.; Inoue, A.; Nakanomyo, I.; Ogawa, M.; Sawa, S.; Ohashi-Ito, K.; Matsubayashi, Y.; Fukuda, H. Non-cell-autonomous control of vascular stem cell fate by a CLE peptide/receptor system. *Proc. Natl. Acad. Sci. USA* **2008**, *105*, 15208–15213. [CrossRef] [PubMed]

126. Wang, J.; Kucukoglu, M.; Zhang, L.; Chen, P.; Decker, D.; Nilsson, O.; Jones, B.; Sandberg, G.; Zheng, B. The *Arabidopsis* LRR-RLK, PXC1, is a regulator of secondary wall formation correlated with the TDIF-PXY/TDR-WOX4 signaling pathway. *BMC Plant Biol.* **2013**, *13*, 94. [CrossRef] [PubMed]

127. Gursanscky, N.R.; Jouannet, V.; Grunwald, K.; Sanchez, P.; Laaber-Schwarz, M.; Greb, T. MOL1 is required for cambium homeostasis in *Arabidopsis*. *Plant J.* **2016**, *86*, 210–220. [CrossRef]

128. Bryan, A.C.; Obaidi, A.; Wierzba, M.; Tax, F.E. XYLEM INTERMIXED WITH PHLOEM1, a leucine-rich repeat receptor-like kinase required for stem growth and vascular development in *Arabidopsis thaliana*. *Planta* **2012**, *235*, 111–122. [CrossRef]

129. Stenvik, G.E.; Butenko, M.A.; Aalen, R.B. Identification of a putative receptor-ligand pair controlling cell separation in plants. *Plant Signal. Behav.* **2008**, *3*, 1109–1110. [CrossRef]

130. Taylor, I.; Wang, Y.; Seitz, K.; Baer, J.; Bennewitz, S.; Mooney, B.P.; Walker, J.C. Analysis of phosphorylation of the receptor-like protein kinase HAESA during *Arabidopsis* floral abscission. *PLoS ONE* **2016**, *11*, e0147203. [CrossRef]

131. Santiago, J.; Brandt, B.; Wildhagen, M.; Hohmann, U.; Hothorn, L.A.; Butenko, M.A.; Hothorn, M. Mechanistic insight into a peptide hormone signaling complex mediating floral organ abscission. *eLife* **2016**, *5*, e15075. [CrossRef]

132. Li, X.; Ahmad, S.; Ali, A.; Guo, C.; Li, H.; Yu, J.; Zhang, Y.; Gao, X.; Guo, Y. Characterization of somatic embryogenesis receptor-like kinase 4 as a negative regulator of leaf senescence in *Arabidopsis*. *Cells* **2019**, *8*, 50. [CrossRef]

133. Li, J.; Chory, J. A putative leucine-rich repeat receptor kinase involved in brassinosteroid signal transduction. *Cell* **1997**, *90*, 929–938. [CrossRef]

134. Li, J.; Wen, J.; Lease, K.A.; Doke, J.T.; Tax, F.E.; Walker, J.C. BAK1, an *Arabidopsis* LRR receptor-like protein kinase, interacts with BRI1 and modulates brassinosteroid signaling. *Cell* **2002**, *110*, 213–222. [CrossRef]

135. Zhou, A.; Wang, H.; Walker, J.C.; Li, J. BRL1, a leucine-rich repeat receptor-like protein kinase, is functionally redundant with BRI1 in regulating *Arabidopsis* brassinosteroid signaling. *Plant J.* **2004**, *40*, 399–409. [CrossRef] [PubMed]

136. Kim, M.H.; Kim, Y.; Kim, J.W.; Lee, H.S.; Lee, W.S.; Kim, S.K.; Wang, Z.Y.; Kim, S.H. Identification of *Arabidopsis* BAK1-associating receptor-like kinase 1 (BARK1) and characterization of its gene expression and brassinosteroid-regulated root phenotypes. *Plant Cell Physiol.* **2013**, *54*, 1620–1634. [CrossRef]

137. Wierzba, M.P.; Tax, F.E. An Allelic Series of bak1 Mutations Differentially Alter bir1 Cell Death, Immune Response, Growth, and Root Development Phenotypes in *Arabidopsis thaliana*. *Genetics* **2016**, *202*, 689–702. [CrossRef]

138. Kumar, D.; Kumar, R.; Baek, D.; Hyun, T.K.; Chung, W.S.; Yun, D.J.; Kim, J.Y. *Arabidopsis thaliana* RECEPTOR DEAD KINASE1 functions as a positive regulator in plant responses to ABA. *Mol. Plant* **2017**, *10*, 223–243. [CrossRef]

139. Bai, L.; Zhang, G.; Zhou, Y.; Zhang, Z.; Wang, W.; Du, Y.; Wu, Z.; Song, C.P. Plasma membrane-associated proline-rich extensin-like receptor kinase 4, a novel regulator of Ca signalling, is required for abscisic acid responses in *Arabidopsis thaliana*. *Plant J.* **2009**, *60*, 314–327. [CrossRef]

140. Corwin, J.A.; Kliebenstein, D.J. Quantitative resistance: More Than just perception of a pathogen. *Plant Cell* **2017**, *29*, 655–665. [CrossRef]

141. Zhang, J.; Zhou, J.M. Plant immunity triggered by microbial molecular signatures. *Mol. Plant* **2010**, *3*, 783–793. [CrossRef]

142. Chinchilla, D.; Boller, T.; Robatzek, S. Flagellin signalling in plant immunity. *Adv. Exp. Med. Biol.* **2007**, *598*, 358–371. [CrossRef]

143. Heese, A.; Hann, D.R.; Gimenez-Ibanez, S.; Jones, A.M.; He, K.; Li, J.; Schroeder, J.I.; Peck, S.C.; Rathjen, J.P. The receptor-like kinase SERK3/BAK1 is a central regulator of innate immunity in plants. *Proc. Natl. Acad. Sci. USA* **2007**, *104*, 12217–12222. [CrossRef] [PubMed]

144. Roux, M.; Schwessinger, B.; Albrecht, C.; Chinchilla, D.; Jones, A.; Holton, N.; Malinovsky, F.G.; Tor, M.; de Vries, S.; Zipfel, C. The *Arabidopsis* leucine-rich repeat receptor-like kinases BAK1/SERK3 and BKK1/SERK4 are required for innate immunity to hemibiotrophic and biotrophic pathogens. *Plant Cell* **2011**, *23*, 2440–2455. [CrossRef]

145. Segonzac, C.; Zipfel, C. Activation of plant pattern-recognition receptors by bacteria. *Curr. Opin. Microbiol.* **2011**, *14*, 54–61. [CrossRef] [PubMed]

146. Halter, T.; Imkampe, J.; Blaum, B.S.; Stehle, T.; Kemmerling, B. BIR2 affects complex formation of BAK1 with ligand binding receptors in plant defense. *Plant Signal. Behav.* **2014**, *9*, e28944. [CrossRef] [PubMed]

147. Blaum, B.S.; Mazzotta, S.; Noldeke, E.R.; Halter, T.; Madlung, J.; Kemmerling, B.; Stehle, T. Structure of the pseudokinase domain of BIR2, a regulator of BAK1-mediated immune signaling in *Arabidopsis*. *J. Struct. Biol.* **2014**, *186*, 112–121. [CrossRef] [PubMed]

148. Kemmerling, B.; Schwedt, A.; Rodriguez, P.; Mazzotta, S.; Frank, M.; Qamar, S.A.; Mengiste, T.; Betsuyaku, S.; Parker, J.E.; Mussig, C.; et al. The BRI1-associated kinase 1, BAK1, has a brassinolide-independent role in plant cell-death control. *Curr. Biol.* **2007**, *17*, 1116–1122. [CrossRef] [PubMed]

149. Lu, D.; Lin, W.; Gao, X.; Wu, S.; Cheng, C.; Avila, J.; Heese, A.; Devarenne, T.P.; He, P.; Shan, L. Direct ubiquitination of pattern recognition receptor FLS2 attenuates plant innate immunity. *Science* **2011**, *332*, 1439–1442. [CrossRef] [PubMed]

150. Zipfel, C.; Kunze, G.; Chinchilla, D.; Caniard, A.; Jones, J.D.; Boller, T.; Felix, G. Perception of the bacterial PAMP EF-Tu by the receptor EFR restricts Agrobacterium-mediated transformation. *Cell* **2006**, *125*, 749–760. [CrossRef]

151. Schwessinger, B.; Roux, M.; Kadota, Y.; Ntoukakis, V.; Sklenar, J.; Jones, A.; Zipfel, C. Phosphorylation-dependent differential regulation of plant growth, cell death, and innate immunity by the regulatory receptor-like kinase BAK1. *PLoS Genet.* **2011**, *7*, e1002046. [CrossRef]

152. Huffaker, A.; Pearce, G.; Ryan, C.A. An endogenous peptide signal in *Arabidopsis* activates components of the innate immune response. *Proc. Natl. Acad. Sci. USA* **2006**, *103*, 10098–10103. [CrossRef]

153. Yamaguchi, Y.; Pearce, G.; Ryan, C.A. The cell surface leucine-rich repeat receptor for AtPep1, an endogenous peptide elicitor in *Arabidopsis*, is functional in transgenic tobacco cells. *Proc. Natl. Acad. Sci. USA* **2006**, *103*, 10104–10109. [CrossRef] [PubMed]

154. Postel, S.; Kufner, I.; Beuter, C.; Mazzotta, S.; Schwedt, A.; Borlotti, A.; Halter, T.; Kemmerling, B.; Nurnberger, T. The multifunctional leucine-rich repeat receptor kinase BAK1 is implicated in *Arabidopsis* development and immunity. *Eur. J. Cell Biol.* **2010**, *89*, 169–174. [CrossRef] [PubMed]

155. Zhao, Y.; Wu, G.; Shi, H.; Tang, D. RECEPTOR-LIKE KINASE 902 associates with and phosphorylates BRASSINOSTEROID-SIGNALING KINASE1 to regulate plant immunity. *Mol. Plant* **2019**, *12*, 59–70. [CrossRef] [PubMed]

156. Cao, Y.; Liang, Y.; Tanaka, K.; Nguyen, C.T.; Jedrzejczak, R.P.; Joachimiak, A.; Stacey, G. The kinase LYK5 is a major chitin receptor in *Arabidopsis* and forms a chitin-induced complex with related kinase CERK1. *eLife* **2014**, *3*, e03766. [CrossRef]

157. Shinya, T.; Yamaguchi, K.; Desaki, Y.; Yamada, K.; Narisawa, T.; Kobayashi, Y.; Maeda, K.; Suzuki, M.; Tanimoto, T.; Takeda, J.; et al. Selective regulation of the chitin-induced defense response by the *Arabidopsis* receptor-like cytoplasmic kinase PBL27. *Plant J.* **2014**, *79*, 56–66. [CrossRef]

158. Yamada, K.; Yamaguchi, K.; Shirakawa, T.; Nakagami, H.; Mine, A.; Ishikawa, K.; Fujiwara, M.; Narusaka, M.; Narusaka, Y.; Ichimura, K.; et al. The *Arabidopsis* CERK1-associated kinase PBL27 connects chitin perception to MAPK activation. *EMBO J.* **2016**, *35*, 2468–2483. [CrossRef]

159. Gimenez-Ibanez, S.; Ntoukakis, V.; Rathjen, J.P. The LysM receptor kinase CERK1 mediates bacterial perception in *Arabidopsis*. *Plant Signal. Behav.* **2009**, *4*, 539–541. [CrossRef]

160. Mélida, H.; Sopeña-Torres, S.; Bacete, L.; Garrido-Arandia, M.; Jordá, L.; López, G.; Muñoz-Barrios, A.; Pacios, L.F.; Molina, A. Non-branched β-1,3-glucan oligosaccharides trigger immune responses in *Arabidopsis*. *Plant J.* **2018**, *93*, 34–49. [CrossRef]

161. Xue, D.X.; Li, C.L.; Xie, Z.P.; Staehelin, C. LYK4 is a component of a tripartite chitin receptor complex in *Arabidopsis thaliana*. *J. Exp. Bot.* **2019**, *70*, 5507–5516. [CrossRef]

162. Chiu, C.H.; Paszkowski, U. Receptor-like kinases sustain symbiotic scrutiny. *Plant Physiol.* **2020**, *182*, 1597–1612. [CrossRef]

163. Choudhury, S.R.; Pandey, S. Specific subunits of heterotrimeric G proteins play important roles during nodulation in soybean. *Plant Physiol.* **2013**, *162*, 522–533. [CrossRef] [PubMed]

164. Liu, J.; Ding, P.; Sun, T.; Nitta, Y.; Dong, O.; Huang, X.; Yang, W.; Li, X.; Botella, J.R.; Zhang, Y. Heterotrimeric G proteins serve as a converging point in plant defense signaling activated by multiple receptor-like kinases. *Plant Physiol.* **2013**, *161*, 2146–2158. [CrossRef] [PubMed]

165. Kohorn, B.D.; Hoon, D.; Minkoff, B.B.; Sussman, M.R.; Kohorn, S.L. Rapid oligo-galacturonide induced changes in protein phosphorylation in *Arabidopsis*. *Mol. Cell. Proteom.* **2016**, *15*, 1351–1359. [CrossRef] [PubMed]

166. Franck, C.M.; Westermann, J.; Boisson-Dernier, A. Plant malectin-like receptor kinases: From cell wall integrity to immunity and beyond. *Annu. Rev. Plant Biol.* **2018**, *69*, 301–328. [CrossRef]

167. Feng, W.; Kita, D.; Peaucelle, A.; Cartwright, H.N.; Doan, V.; Duan, Q.; Liu, M.C.; Maman, J.; Steinhorst, L.; Schmitz-Thom, I.; et al. The FERONIA receptor kinase maintains cell-wall integrity during salt stress through Ca. *Curr. Biol.* **2018**, *28*, 666–675. [CrossRef]

168. Masachis, S.; Segorbe, D.; Turrà, D.; Leon-Ruiz, M.; Fürst, U.; El Ghalid, M.; Leonard, G.; López-Berges, M.S.; Richards, T.A.; Felix, G.; et al. A fungal pathogen secretes plant alkalinizing peptides to increase infection. *Nat. Microbiol.* **2016**, *1*, 16043. [CrossRef]

169. Van der Does, D.; Boutrot, F.; Engelsdorf, T.; Rhodes, J.; McKenna, J.F.; Vernhettes, S.; Koevoets, I.; Tintor, N.; Veerabagu, M.; Miedes, E.; et al. The *Arabidopsis* leucine-rich repeat receptor kinase MIK2/LRR-KISS connects cell wall integrity sensing, root growth and response to abiotic and biotic stresses. *PLoS Genet.* **2017**, *13*, e1006832. [CrossRef]

170. Engelsdorf, T.; Gigli-Bisceglia, N.; Veerabagu, M.; McKenna, J.F.; Vaahtera, L.; Augstein, F.; Van der Does, D.; Zipfel, C.; Hamann, T. The plant cell wall integrity maintenance and immune signaling systems cooperate to control stress responses in Arabidopsis thaliana. *Sci. Signal.* **2018**, *11*, eaao3070. [CrossRef]

171. He, P.; Shan, L.; Lin, N.C.; Martin, G.B.; Kemmerling, B.; Nurnberger, T.; Sheen, J. Specific bacterial suppressors of MAMP signaling upstream of MAPKKK in *Arabidopsis* innate immunity. *Cell* **2006**, *125*, 563–575. [CrossRef]

172. Shan, L.; He, P.; Li, J.; Heese, A.; Peck, S.C.; Nürnberger, T.; Martin, G.B.; Sheen, J. Bacterial effectors target the common signaling partner BAK1 to disrupt multiple MAMP receptor-signaling complexes and impede plant immunity. *Cell Host Microbe* **2008**, *4*, 17–27. [CrossRef]

173. Gao, M.; Wang, X.; Wang, D.; Xu, F.; Ding, X.; Zhang, Z.; Bi, D.; Cheng, Y.T.; Chen, S.; Li, X.; et al. Regulation of cell death and innate immunity by two receptor-like kinases in *Arabidopsis*. *Cell Host Microbe* **2009**, *6*, 34–44. [CrossRef]

174. Albert, I.; Böhm, H.; Albert, M.; Feiler, C.E.; Imkampe, J.; Wallmeroth, N.; Brancato, C.; Raaymakers, T.M.; Oome, S.; Zhang, H.; et al. An RLP23-SOBIR1-BAK1 complex mediates NLP-triggered immunity. *Nat. Plants* **2015**, *1*, 15140. [CrossRef] [PubMed]

175. Van der Burgh, A.M.; Postma, J.; Robatzek, S.; Joosten, M.H.A.J. Kinase activity of SOBIR1 and BAK1 is required for immune signalling. *Mol. Plant Pathol.* **2019**, *20*, 410–422. [CrossRef] [PubMed]

176. Ohtake, Y.; Takahashi, T.; Komeda, Y. Salicylic acid induces the expression of a number of receptor-like kinase genes in *Arabidopsis thaliana*. *Plant Cell Physiol.* **2000**, *41*, 1038–1044. [CrossRef] [PubMed]

177. Du, L.; Chen, Z. Identification of genes encoding receptor-like protein kinases as possible targets of pathogen- and salicylic acid-induced WRKY DNA-binding proteins in *Arabidopsis*. *Plant J.* **2000**, *24*, 837–847. [CrossRef] [PubMed]

178. Chen, Z. A superfamily of proteins with novel cysteine-rich repeats. *Plant Physiol.* **2001**, *126*, 473–476. [CrossRef]

179. Zhang, X.; Han, X.; Shi, R.; Yang, G.; Qi, L.; Wang, R.; Li, G. *Arabidopsis* cysteine-rich receptor-like kinase 45 positively regulates disease resistance to *Pseudomonas syringae*. *Plant Physiol. Biochem.* **2013**, *73*, 383–391. [CrossRef]

180. Chen, K.; Du, L.; Chen, Z. Sensitization of defense responses and activation of programmed cell death by a pathogen-induced receptor-like protein kinase in *Arabidopsis*. *Plant Mol. Biol.* **2003**, *53*, 61–74. [CrossRef]

181. Chen, K.; Fan, B.; Du, L.; Chen, Z. Activation of hypersensitive cell death by pathogen-induced receptor-like protein kinases from *Arabidopsis*. *Plant Mol. Biol.* **2004**, *56*, 271–283. [CrossRef]

182. Lee, D.S.; Kim, Y.C.; Kwon, S.J.; Ryu, C.M.; Park, O.K. The *Arabidopsis* cysteine-rich receptor-like kinase CRK36 Regulates immunity through interaction with the cytoplasmic kinase BIK1. *Front. Plant Sci.* **2017**, *8*, 1856. [CrossRef]

183. Osakabe, Y.; Yamaguchi-Shinozaki, K.; Shinozaki, K.; Tran, L.S. Sensing the environment: Key roles of membrane-localized kinases in plant perception and response to abiotic stress. *J. Exp. Bot.* **2013**, *64*, 445–458. [CrossRef] [PubMed]

184. Ye, Y.; Ding, Y.; Jiang, Q.; Wang, F.; Sun, J.; Zhu, C. The role of receptor-like protein kinases (RLKs) in abiotic stress response in plants. *Plant Cell Rep.* **2017**, *36*, 235–242. [CrossRef]

185. Manabe, Y.; Bressan, R.A.; Wang, T.; Li, F.; Koiwa, H.; Sokolchik, I.; Li, X.; Maggio, A. The *Arabidopsis* kinase-associated protein phosphatase regulates adaptation to Na$^+$ stress. *Plant Physiol.* **2008**, *146*, 612–622. [CrossRef] [PubMed]

186. Tanaka, H.; Osakabe, Y.; Katsura, S.; Mizuno, S.; Maruyama, K.; Kusakabe, K.; Mizoi, J.; Shinozaki, K.; Yamaguchi-Shinozaki, K. Abiotic stress-inducible receptor-like kinases negatively control ABA signaling in *Arabidopsis*. *Plant J.* **2012**, *70*, 599–613. [CrossRef] [PubMed]

187. Chen, J.; Yu, F.; Liu, Y.; Du, C.; Li, X.; Zhu, S.; Wang, X.; Lan, W.; Rodriguez, P.L.; Liu, X.; et al. FERONIA interacts with ABI2-type phosphatases to facilitate signaling cross-talk between abscisic acid and RALF peptide in *Arabidopsis*. *Proc. Natl. Acad. Sci. USA* **2016**, *113*, E5519–E5527. [CrossRef] [PubMed]

188. Yu, F.; Qian, L.; Nibau, C.; Duan, Q.; Kita, D.; Levasseur, K.; Li, X.; Lu, C.; Li, H.; Hou, C.; et al. FERONIA receptor kinase pathway suppresses abscisic acid signaling in *Arabidopsis* by activating ABI2 phosphatase. *Proc. Natl. Acad. Sci. USA* **2012**, *109*, 14693–14698. [CrossRef] [PubMed]

189. Kim, H.; Hwang, H.; Hong, J.W.; Lee, Y.N.; Ahn, I.P.; Yoon, I.S.; Yoo, S.D.; Lee, S.; Lee, S.C.; Kim, B.G. A rice orthologue of the ABA receptor, OsPYL/RCAR5, is a positive regulator of the ABA signal transduction pathway in seed germination and early seedling growth. *J. Exp. Bot.* **2012**, *63*, 1013–1024. [CrossRef]

190. Osakabe, Y.; Mizuno, S.; Tanaka, H.; Maruyama, K.; Osakabe, K.; Todaka, D.; Fujita, Y.; Kobayashi, M.; Shinozaki, K.; Yamaguchi-Shinozaki, K. Overproduction of the membrane-bound receptor-like protein kinase 1, RPK1, enhances abiotic stress tolerance in *Arabidopsis*. *J. Biol. Chem.* **2010**, *285*, 9190–9201. [CrossRef]

191. Hua, D.; Wang, C.; He, J.; Liao, H.; Duan, Y.; Zhu, Z.; Guo, Y.; Chen, Z.; Gong, Z. A plasma membrane receptor kinase, GHR1, mediates abscisic acid- and hydrogen peroxide-regulated stomatal movement in *Arabidopsis*. *Plant Cell* **2012**, *24*, 2546–2561. [CrossRef]

192. Lu, K.; Liang, S.; Wu, Z.; Bi, C.; Yu, Y.T.; Wang, X.F.; Zhang, D.P. Overexpression of an *Arabidopsis* cysteine-rich receptor-like protein kinase, CRK5, enhances abscisic acid sensitivity and confers drought tolerance. *J. Exp. Bot.* **2016**, *67*, 5009–5027. [CrossRef]
193. Zhao, C.; Zayed, O.; Yu, Z.; Jiang, W.; Zhu, P.; Hsu, C.C.; Zhang, L.; Tao, W.A.; Lozano-Durán, R.; Zhu, J.K. Leucine-rich repeat extensin proteins regulate plant salt tolerance in. *Proc. Natl. Acad. Sci. USA* **2018**, *115*, 13123–13128. [CrossRef] [PubMed]
194. Jung, C.G.; Hwang, S.G.; Park, Y.C.; Park, H.M.; Kim, D.S.; Park, D.H.; Jang, C.S. Molecular characterization of the cold- and heat-induced *Arabidopsis* PXL1 gene and its potential role in transduction pathways under temperature fluctuations. *J. Plant Physiol.* **2015**, *176*, 138–146. [CrossRef] [PubMed]
195. Yang, T.; Chaudhuri, S.; Yang, L.; Du, L.; Poovaiah, B.W. A calcium/calmodulin-regulated member of the receptor-like kinase family confers cold tolerance in plants. *J. Biol. Chem.* **2010**, *285*, 7119–7126. [CrossRef] [PubMed]
196. Yang, T.; Shad Ali, G.; Yang, L.; Du, L.; Reddy, A.S.; Poovaiah, B.W. Calcium/calmodulin-regulated receptor-like kinase CRLK1 interacts with MEKK1 in plants. *Plant Signal. Behav.* **2010**, *5*, 991–994. [CrossRef] [PubMed]

International Journal of
Molecular Sciences

Article

Internalization of miPEP165a into *Arabidopsis* Roots Depends on both Passive Diffusion and Endocytosis-Associated Processes

Mélanie Ormancey [1,*], Aurélie Le Ru [2], Carine Duboé [1], Hailing Jin [3], Patrice Thuleau [1], Serge Plaza [1] and Jean-Philippe Combier [1,*]

[1] Laboratoire de Recherche en Sciences Végétales, UMR5546, Université de Toulouse, UPS, CNRS, 31320 Auzeville-Tolosan, France; carine.duboe@lrsv.ups-tlse.fr (C.D.); thuleau@lrsv.ups-tlse.fr (P.T.); serge.plaza@lrsv.ups-tlse.fr (S.P.)

[2] Fédération de Recherche FR3450, CNRS, Université de Toulouse, 31326 Castanet-Tolosan, France; leru@lrsv.ups-tlse.fr

[3] Department of Microbiology & Plant Pathology, Center for Plant Cell Biology, Institute for Integrative Genome Biology, University of California, Riverside, CA 92521, USA; hailing.jin@ucr.edu

* Correspondence: ormancey@lrsv.ups-tlse.fr (M.O.); combier@lrsv.ups-tlse.fr (J.-P.C.); Tel.: +33-(0)5-34-32-38-44 (M.O.); +33-(0)5-34-32-38-11 (J.-P.C.)

Received: 25 February 2020; Accepted: 22 March 2020; Published: 25 March 2020

Abstract: MiPEPs are short natural peptides encoded by microRNAs in plants. Exogenous application of miPEPs increases the expression of their corresponding miRNA and, consequently, induces consistent phenotypical changes. Therefore, miPEPs carry huge potential in agronomy as gene regulators that do not require genome manipulation. However, to this end, it is necessary to know their mode of action, including where they act and how they enter the plants. Here, after analyzing the effect of *Arabidopsis thaliana* miPEP165a on root and aerial part development, we followed the internalization of fluorescent-labelled miPEP165a into roots and compared its uptake into endocytosis-altered mutants to that observed in wild-type plants treated or not with endocytosis inhibitors. The results show that entry of miPEP165a involves both a passive diffusion at the root apex and endocytosis-associated internalization in the differentiation and mature zones. Moreover, miPEP165a is unable to enter the central cylinder and does not migrate from the roots to the aerial part of the plant, suggesting that miPEPs have no systemic effect.

Keywords: *Arabidopsis*; endocytosis; microRNAs; miPEPs; peptides

1. Introduction

Gene expression is the consequence of the transcription of an RNA molecule from a gene—modulated by transcription factors and modifications of the chromatin structure—and post-transcriptional mechanisms acting on the RNA stability of translation or on the protein it encodes. One of the best-known mechanisms of post-transcriptional regulation of gene expression is gene silencing induced by microRNAs (miRNAs). MiRNAs are small, regulatory RNA molecules (21–24 nucleotides) first discovered in the worm *Caenorhabditis elegans* and later in plants and humans [1–4]. Each miRNA regulates the expression of specific target gene(s) either by cleaving the mRNA transcribed from it or by inhibiting its translation. Target genes of miRNAs are often key regulatory genes encoding, for example, transcription factors or hormone receptors. MiRNAs are therefore required for the correct regulation of most developmental processes in plants and animals, and dysregulation of miRNA expression is a feature of many human pathologies.

MiRNAs are themselves encoded by genes and are transcribed in the form of long primary transcripts (pri-miRNAs). One of the first steps in the maturation of pri-miRNAs involves a nuclear protein complex containing an enzyme called dicer-like 1 (DCL1), which cleaves pri-miRNAs to form precursor miRNAs (pre-miRNAs). A second cleavage step then forms mature miRNAs. In the cytoplasm, the mature miRNA anneals by homology with the mRNA of its target gene(s). This heteroduplex molecule is recognized by a protein complex called RISC, containing the enzyme Argonaute (AGO1), which either cleaves the targeted mRNA or inhibits its translation. Because the main role of miRNAs is to act as regulatory small RNAs and not in the direct translation of proteins, miRNAs have always been thought to be non-coding RNAs.

Surprisingly, the characterization of plant pri-miRNAs revealed that they encode small regulatory peptides, which were called miPEPs for miRNA-encoded peptides [5]. MiPEPs are involved in a positive autoregulatory feedback loop. They specifically activate transcription of their primary transcript and consequently enhance the synthesis of the mature miRNA, thus turning down the expression of specific genes. Interestingly, the application of exogenous synthetic miPEPs to plants is sufficient to stimulate the synthesis of their corresponding miRNAs and to modify plant development accordingly [6,7]. Given their efficiency simply by an external application on plants, miPEPs are promising molecules for many agronomic applications. In particular, they offer a new way of modulating plant development, stimulating plant symbioses, or increasing plant fitness, to name a few potential uses. Moreover, as natural and endogenous peptides, they are likely to be much less harmful to the environment than chemical treatments and more acceptable to the general public than genetically modified organisms.

Endocytosis plays a crucial role in the internalization of extracellular molecules and plasma membrane proteins into eukaryotic cells [8]. Clathrin-mediated endocytosis (CME) remains the most extensively studied and characterized endocytosis and constitutes the major route of entry and pathway in eukaryotes [8,9]. Clathrin is a triskelion-shaped scaffold protein composed of three clathrin light chains (CLCs) and three clathrin heavy chains (CHCs). The formation of clathrin-coated vesicle at the plasma membrane requires adaptor proteins, including AP2 complex [10]. In plants, CME is involved in multiple important biological processes, including growth, development, nutrient uptake, and biotic and abiotic stress responses [8,10–15]. For instance, clathrin is required for plasma membrane-located receptor endocytosis upon peptide perception, leading to peptide-mediated responses and thus to plant immunity [15,16]. Moreover, recent studies have also reported the existence of sterol-sensitive clathrin-independent pathways in plants, although this alternative endocytosis pathway is far less understood [17,18]. The best-studied clathrin-independent pathway in plants corresponds to flotillin-1-mediated endocytosis, a membrane microdomain-associated protein involved in plant development and promoted by flg22, a flagellin-derived 22-amino acid peptide [19,20]. Alternatively, proteins can assemble into clusters in membrane microdomains [8]. For instance, remorins form clusters at the plasma membrane and interact with a symbiotic receptor that allows bacterial infection in *Medicago truncatula* [21]. Finally, both clathrin-dependent and -independent pathways can be constitutive or differentially regulated in response to stimuli [17,18,22]. In summary, different endocytosis pathways have been reported to be involved in many biological outcomes.

Due to their capacity to modulate plant development, miPEPs are of interest in agronomy as an alternative to chemicals to stimulate plant development. Nevertheless, to achieve this goal, a better understanding of their mode of action at the molecular level, including the mechanisms of their entry into plants, is required. In this study, we investigated how miPEPs enter into plants. We first reported in detail the phenotypes observed after treatment of *Arabidopsis thaliana* with miPEP165a, previously used to decipher the mode of action of miPEPs [5]. By using this miPEP labelled with a fluorescent dye, we followed the internalization of the peptide into plants. The peptide entered rapidly into the root cap and the meristematic zone and it took longer to penetrate the other parts of the root. Using mutants potentially altered in endocytic pathways or chemical inhibitors affecting endocytosis, we identified two mechanisms of miPEP165a entry into roots, passive diffusion followed by an endocytosis process.

2. Results

2.1. MiPEP165a Promotes Cell Division in the Meristematic Zone to Increase Primary Root Length and Acts on Flowering Time in Arabidopsis

It has been previously shown that *A. thaliana* miPEP165a, as well as miR165a, is expressed in endodermis cells [5,23]. Exogenous treatment of *A. thaliana* seedlings with synthetic miPEP165a is sufficient to increase the primary root length [5]. However, the precise mechanisms (spatial and temporal) involved in the peptide uptake remained unknown. To study the entry of miPEPs, especially miPEP165a, we first defined the best experimental conditions to obtain a significant effect of miPEP165a on plant development. We first observed that watering plants with 100 μM of peptide was much more efficient at increasing the primary root length than treatments performed with only 10 μM of peptide, the concentration used in the previous study [5] (Figure 1A). In addition, similar to the concentration of 10 μM previously used [5], applying miPEP165a at 100 μM also induced the activation of the pri-miRNA from which it originates (Figure S1A). In addition, during the initial stages of the study, when the effect of miPEP165a on primary root length was studied, whatever the control used, i.e., scrambled miPEP165a, irrelevant peptides, or their corresponding solvents (acetonitrile or water), no response was observed compared to miPEP165a treatments (Figure S1B). Similarly, water and scrambled miPEP165a had no effect on the expression of pri-miR165a compared to miPEP165a (Figure S1A). Finally, we observed that several freeze/thaw cycles of the peptide were detrimental to its activity on the length of primary roots (Figure S1C). For these reasons, we used aliquots of unfrozen peptides only once and kept water as a reference in all the following experiments.

The increase in root length upon treatment may be a consequence of higher cell elongation or increased cell proliferation. To address this point, we analyzed the effect of miPEP165a at the cellular level on the meristematic zone since root growth was often determined by meristematic activity [24]. We revealed that more cells were present in the meristematic zone when roots were treated with miPEP165a (Figure 1B–E). Therefore, these experiments suggest that the increase in root length induced by the miPEP165a treatment is likely due to the stimulation of cellular proliferation rather than an increase in cell length.

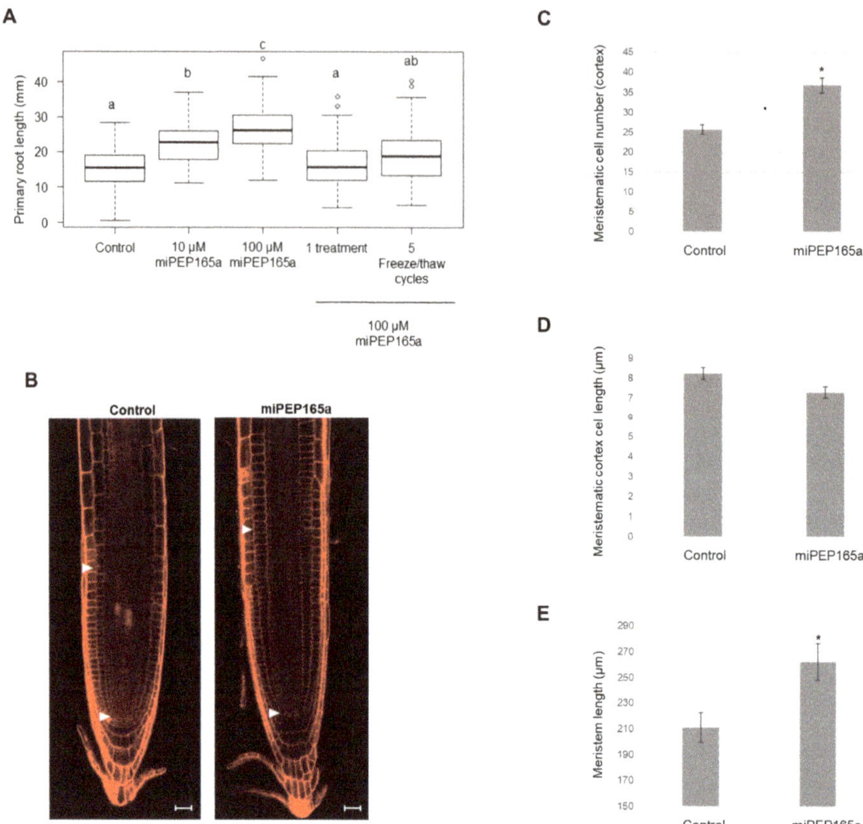

Figure 1. MiPEP165a promotes root growth by enhancing cell division. (**A**) Effect of miPEP165a on primary root length. *Arabidopsis thaliana* seedlings were treated with various concentrations of miPEP165a. Seedlings were treated daily with the peptide for 4 days, with the exception of those that received only one treatment. Peptides were thawed once, except those that underwent five freeze/thaw cycles. Two-way analysis of variance (ANOVA) significance levels were based on Tukey's post-test (1-way ANOVA), (a–c, $p < 0.05$, $n = 70$). At least three biological replicates were performed (**B–E**). Three-day-old seedlings were treated daily with water or 100 µM miPEP165a for a further 3 days and stained with 10 µg/mL propidium iodide for 20 min. (**B**) Confocal images showing the meristematic zone for the cortex cells, defined as the region between quiescent center cells and the first elongating cell that was twice the length compared to its distal neighbor (distance between white arrows) [25,26]. Meristematic cell number (**C**) and cell length (**D**) were determined with the software tool Cell-o-Tape, an open source ImageJ/Fiji macro [27–29]. (**E**) Quantification of root apical meristem length. (**B–E**) Four biological replicates were performed with at least 20 seedlings. Errors bars represent SEM. Asterisks indicate a significant difference at $p < 0.01$ (*) according to the t-test. Scale bar = 25 µm. Water was used as a control.

MiR165a and its target genes, *REVOLUTA* (*REV*), *PHABULOSA* (*PHB*), and *PHAVOLUTA* (*PHV*), are also known to be involved in flowering [30]. To investigate whether miPEP165a could have an effect on flowering, we treated the shoot apical meristem with a droplet of 100 µM miPEP165a three times a week during plant development. Treatments with miPEP165a accelerated plant development as illustrated by the decrease of the flowering day (Figure 2A,B) and the increase of the length of the inflorescence stem (Figure 2C,D). Interestingly, watering the roots with 10 µM peptide had no effect on the flowering, suggesting that peptides cannot migrate throughout the plant (Figure 2).

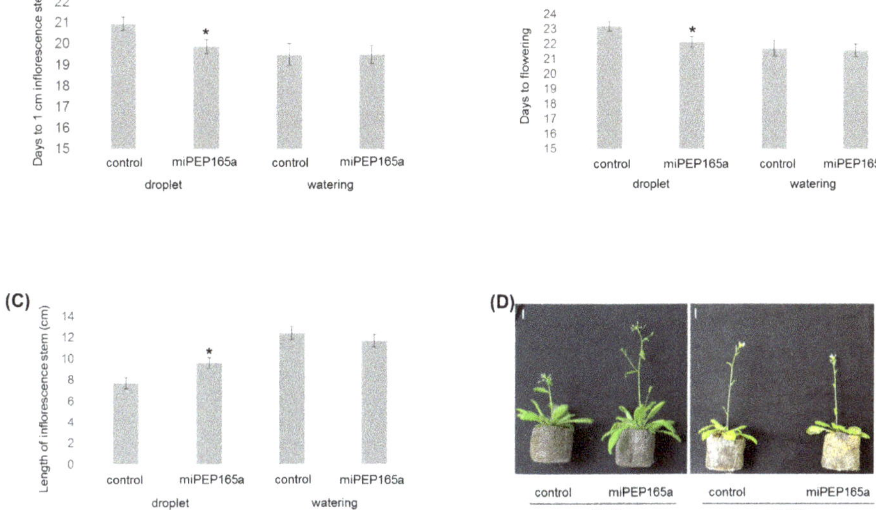

Figure 2. Flowering phenotypes of *Arabidopsis* plants in response to miPEP165a treatment. *Arabidopsis* plants were treated with either water (control) or a droplet of 100 µM miPEP165a placed on the shoot apical meristem or by watering with 10 µM miPEP165a three times a week until analyses. Flowering time measurements were determined using the number of days to obtain an inflorescence stem of 1 cm (**A**) and the number of days to obtain the first flowers (**B**). (**C**) The length of the *Arabidopsis* inflorescence stem was determined 24 days after sowing. Error bars indicate SEM. Statistical analysis was performed using a *t*-test ($p < 0.01$). (**D**) Representative pictures showing the flowering phenotype according to the miPEP165a treatment. Experiments were performed at least 4 independent times ($n > 78$ plants). Bar = 1 cm.

2.2. MiPEP165a Entry Involves both Passive Diffusion at the Root Apex and Endocytic Pathways in the Differentiation and Mature Zones

To document this observation, we used the miPEP165a labelled with FAM, a fluorescent dye derived from fluorescein. As illustrated in Figure S2, the physiochemical properties of the miPEP165a-FAM are similar to those of the non-modified peptide. Although slightly less active, the labelled peptide was still able to increase the primary root length (Figure 3).

Interestingly, while the labelled peptide penetrated rapidly (~2 h) into the root cap and the meristematic zone, it took longer to penetrate the other parts of the root (Figure 4). Twenty-four hours after the application of the labelled peptide, the latter was present in most external parts of the roots. The central cylinder was never labelled by the peptide, which seemed to be blocked by the pericycle (Figure 4).

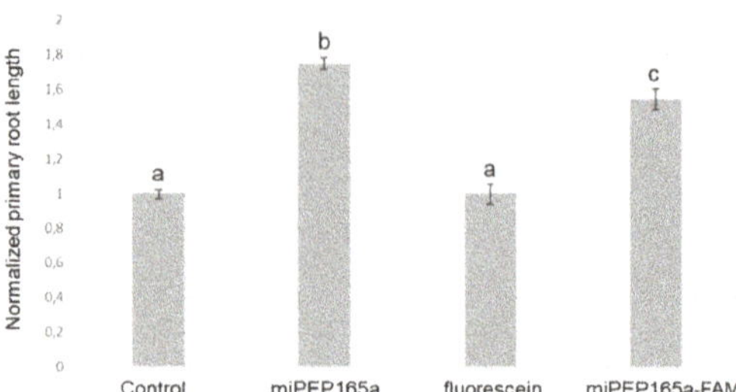

Figure 3. MiPEP165a-FAM is biologically active. Seedlings were treated with water (control), 100 μM miPEP165a, miPEP165a-FAM, or fluorescein. At least 70 seedlings were used to determine the normalized *Arabidopsis* root length. Data are given as ± SEM and statistical analysis was performed using a t-test (a–c, $p < 0.01$).

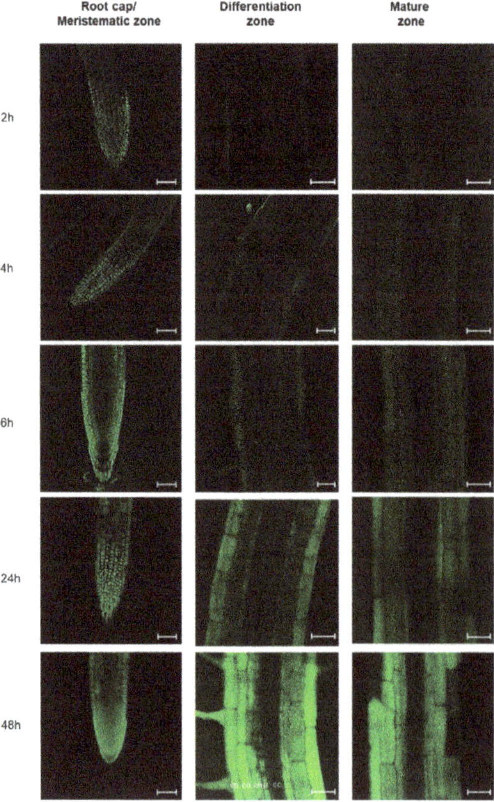

Figure 4. Kinetics of miPEP165a uptake into *Arabidopsis* roots. The mobility of miPEP165a-FAM was followed at the indicated time in different zones of *Arabidopsis* roots, as defined by [31]. Confocal images are representative of four independent experiments, with at least 6 seedlings for each condition. Bar = 50 μm (root cap/meristematic zone) or 25 μm (differentiation and mature zones). The different cell layers are indicated in the differentiation zone image at 48 h as follows: cc, central cylinder; p, pericycle; en, endodermis; co, cortex; ep, epidermis.

The entry of peptides into plants might occur passively, by diffusion, or actively, via specific transporters or by endocytosis. Because of the huge diversity of miPEPs in a plant and the lack of conservation between species [5], we hypothesized that specific transporters for each peptide are unlikely to exist and, more likely, the miPEPs might be internalized by generic internalization machinery or, more simply, by passive diffusion. To decipher the mechanisms involved in the entry of peptides into cells, we used *A. thaliana* mutants impaired on genes encoding proteins associated to the clathrin pathway (*chc1-1*, *chc2-1*, *ap2σ2*) [12,16,32] or to the membrane microdomain (*rem1-2*, *rem1-3*) [33–35]. Internalization of miPEP165a was not affected in most of the mutants tested, except in the root cap/meristematic zone of the *chc1-1* mutant and in the differentiation zone of the *chc1-1* and *rem1-2* mutants, suggesting that uptake in these parts was mainly passive (Figure 5, Figure S3). Conversely, the entry of the peptide into the mature zone of all mutants was strongly impaired (Figure 5, Figure S3). These data suggest that peptide entry in plants involves, in addition to passive diffusion, both clathrin and membrane microdomain-mediated pathways.

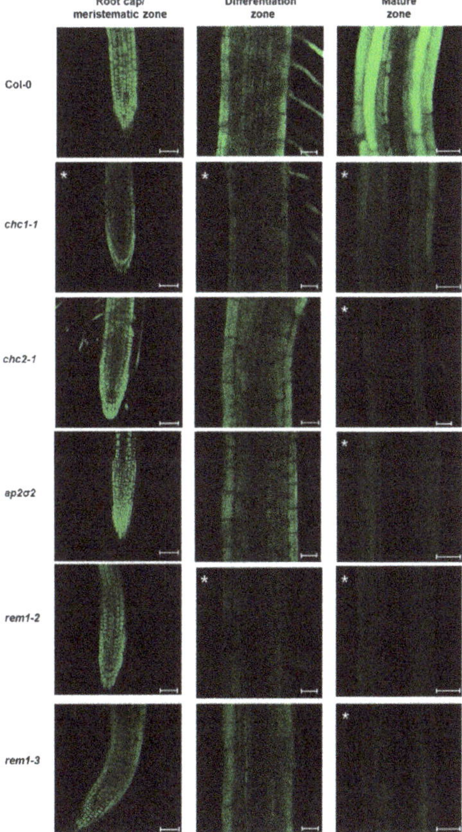

Figure 5. Internalization of miPEP165a is clathrin and remorin dependent. Representative confocal images showing the uptake of miPEP165-FAM 48 h after treatment in wild-type seedlings and *chc1-1*, *chc2-1*, *ap2σ2*, *rem1-2*, and *rem1-3* mutants. A significant fluorescence decrease for each condition is indicated in each panel by asterisks. Quantifications of the fluorescence intensity from more than 15 seedlings are shown in Figure S3. Bar = 50 μm (root cap/meristematic zone) or 25 μm (differentiation and mature zones).

In order to determine how and to what extent a defect in the peptide entry affects its biological effect on plant development, we treated the roots of *chc1-1*, *rem1-2*, and *rem1-3* mutants with the peptide in parallel with the wild-type roots. While the mutants showed a longer primary root in the control conditions compared to the wild type plants, they were unable to respond to the peptide by increasing their primary root length (Figure 6). Indeed, the *rem1-2* mutant, which was strongly affected in the peptide uptake, was unable to respond to miPEP165a.

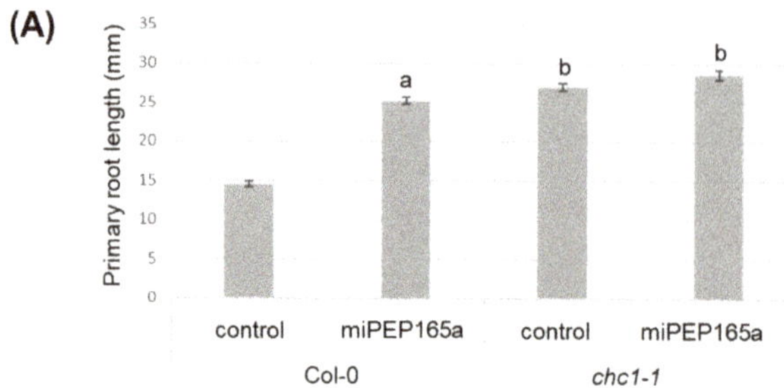

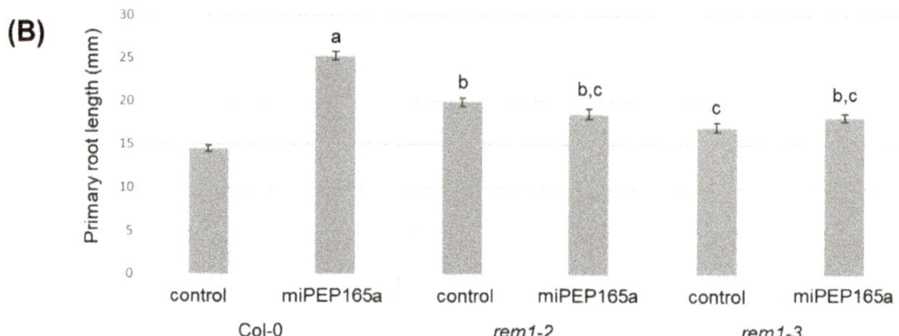

Figure 6. MiPEP165a-mediated root growth induction involves clathrin and remorin proteins. Measurement of the primary root length in *chc1-1* (**A**) and remorin (*rem1-2* and *rem1-3*) mutants (**B**) after water (control) or miPEP165a (100 μM) treatment. The error bars indicate SEM of at least three biological replicates (n > 110 seedlings) and statistical analyses were performed using a t-test (a–c, $p < 0.01$).

We next treated the aerial parts of the mutants with miPEP165a, and we observed similar results on the flowering time (Figure 7). These results suggest that the mechanisms of miPEP165a uptake into roots and aerial parts could be similar.

(A)

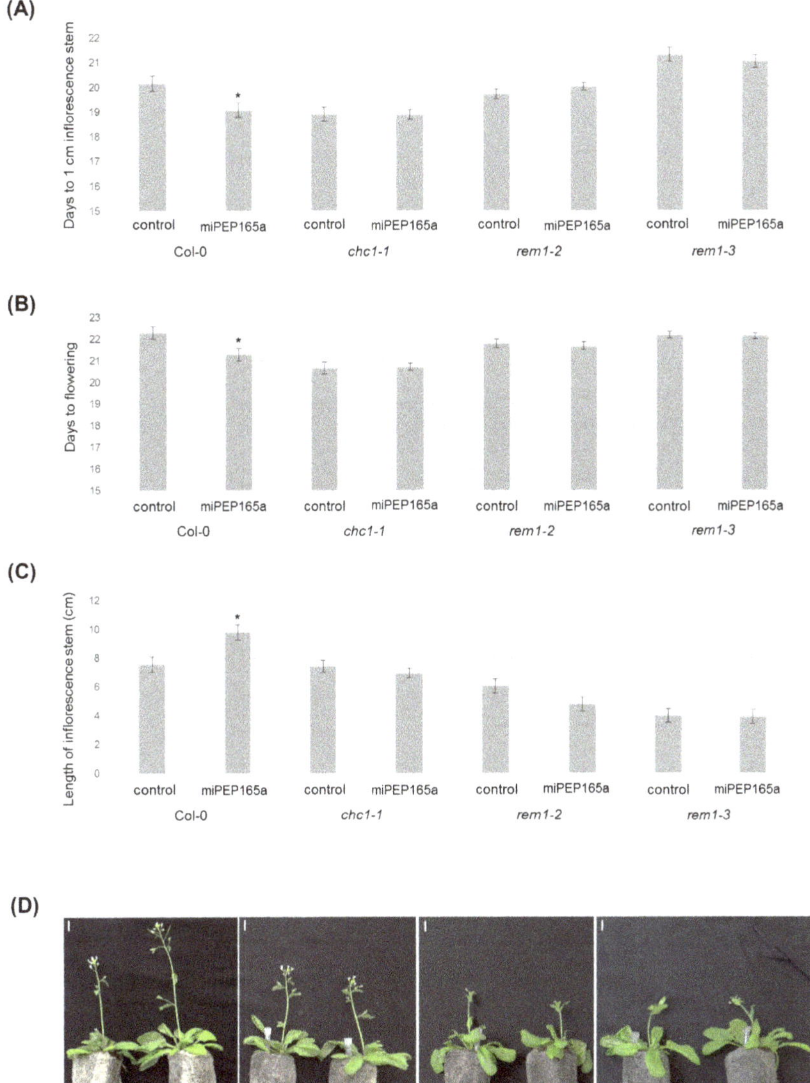

(B)

(C)

(D)

Figure 7. Flowering time depends on clathrin- and membrane microdomain-associated pathways. The number of days to obtain a 1-cm inflorescence stem (**A**) and the number of days to observe the first flowers (**B**) were determined for wild-type plants as well as for *chc1-1*, *rem1-2*, and *rem1-3* mutant plants. (**C**) Measurement of the inflorescence stem length was determined 24 days after sowing for wild-type and mutant plants. Data are representative of the average of at least four independent experiments with at least 10 plants per condition, for each experiment. Error bars represent SEM and statistical analyses were performed using a t-test (*, $p < 0.01$). (**D**) Representative images comparing wild-type and mutant plants treated with water (control) or miPEP165a. Bar = 1 cm.

Finally, we used TyrA23, a chemical inhibitor known to affect clathrin-mediated endocytosis [22,32,36], and MβCD, a cholesterol-depleting agent, which have been suggested to block microdomain-dependent endocytosis [17,18,22]. Interestingly, both molecules were able to inhibit the miPEP165a-activated root length phenotype, suggesting that peptide entry in plant involves clathrin-mediated endocytosis and membrane microdomain-dependent pathways (Figure 8).

(A)

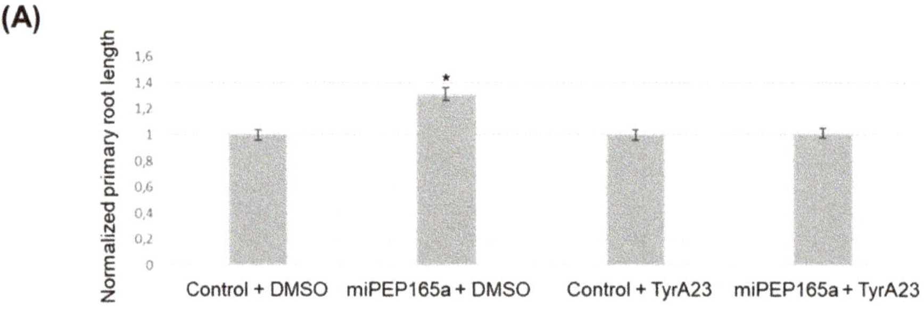

(B)

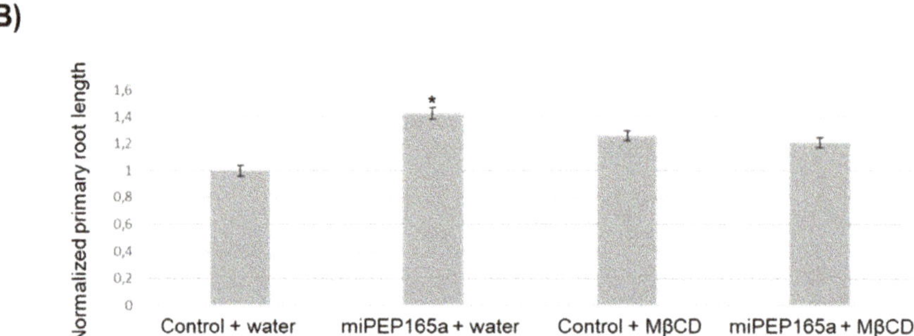

Figure 8. Disruption of endocytic pathways prevents miPEP165a-induced root growth. Normalized primary root growth analysis after treatment with miPEP165a and TyrA23 (**A**) or miPEP165a and MβCD (**B**). Three biological replicates were performed by using at least 100 seedlings for each condition and root lengths were statistically analyzed using a *t*-test ($p < 0.01$, *). The data represent the mean value ± SEM. Water was used as a control for the miPEP165a treatment.

Altogether, our results showed that miPEP165a entry used passive diffusion at the root apex followed by endocytosis in the differentiation or mature zone of plant roots. All pathways are required to mediate full peptide uptake (and activity).

3. Discussion

MiRNAs have been considered for a long time as non-coding RNAs. However, a few years ago, it was shown that pri-miRNAs can encode regulatory peptides, which were named miPEPs. These miPEPs activate the transcription of their associated miRNA and thus downregulate the expression of their target genes [5]. Among miPEPs, miPEP165a induces the accumulation of mature miR165a, known to repress the expression of all five class III homeodomain-leucine zipper (HD-ZIP III) transcription factors, i.e., *REV*, *PHB*, *PHV*, *CORONA* (*CAN/AtHB15*), and *AtHB8* [5,37]. In *Arabidopsis*, the overexpression of all HD-ZIP III results in plants with shorter roots whereas *phb*, *phv* double mutants and *phv-11* mutants display longer roots as well as an increase in the number of meristem cells compared to wild-type plants [38,39]. Moreover, the overexpression of miR166, differing by only one nucleotide from miR165 and targeting the expression of three HD-ZIP III genes, also promotes primary root growth in *Arabidopsis* [39]. These results can be correlated with those of the present study, since we showed that miPEP165a promotes primary root growth by increasing cell division in the root apical meristem (Figure 1). Moreover, misexpression of the HD-ZIP III genes by making them resistant to miR165/166 and a reduction in the expression of HD-ZIP IIIs by overexpression of miR165/166 induces prolonged activity of floral stem cells [30]. Here, we observed that miPEP165a accelerates the appearance of the inflorescence stem and the flowering time of *Arabidopsis* wild-type plants (Figure 2).

Since some small peptides were considered as long-distance signaling molecules, we wondered whether miPEP165a was involved in root/shoot communication [40–42]. By tracking the FAM-labelled miPEP165a across all layers of *Arabidopsis* roots, we showed that the labelled peptide entered into the epidermis and migrated up to the pericycle but did not reach the root vessels (Figure 4). Moreover, the acceleration of flowering observed in response to the miPEP165a treatment of the shoot apical meristem was not observed after watering *Arabidopsis* roots with miPEP165a (Figure 2). Taken together, these results indicate that miPEP165a is not a root-to-shoot mobile signal molecule.

Consequently, in order to have a better understanding of miPEP uptake into plants, we investigated the mobility of FAM-labelled miPEP165a in *Arabidopsis* roots. Clathrin-mediated endocytosis is the major and the most studied route of entry in plants [8]. A recent study showed that this endocytic pathway is necessary for the internalization of the elicitor peptide *At*pep1 and its receptor, leading to *At*pep1-induced responses [16]. Here, we showed that the entry of miPEP165a could also be dependent on clathrin since miPEP165a uptake was significantly decreased in the primary roots of *chc1-1* and strongly reduced in the mature zone in the three mutants *chc1-1*, *chc2-1*, and *ap2σ2* (Figure 5). These results were confirmed by the fact that the increase of the root length by miPEP165a was not observed in the *chc1-1* mutant or after treatment with TyrA23 (Figure 6A, Figure 8A), the most commonly used CME inhibitor [8,32,36]. Similarly, the acceleration of the flowering time induced by miPEP165a in wild-type plants was not observed in the *chc1-1* mutant (Figure 7).

Besides clathrin-mediated endocytosis, membrane microdomain-associated endocytosis has been described in plants as an alternative route of entry pathway [8]. This endocytosis pathway is sensitive to sterol depletion and consequently to the sterol-depleting agent MβCD [8,17,18]. In the present study, we showed that MβCD prevented miPEP165a-FAM entry and correlatively the increase of root length induced by miPEP165a (Figure 8B, Figure S4). Collectively, our results indicate that both clathrin-dependent pathways and microdomain-associated events may cooperate in peptide entry into *Arabidopsis* roots. Previous results have demonstrated that internalization of the aquaporin PIP2;1 and RbohD involved both dependent and independent clathrin-mediated endocytosis, the latter being stimulated in saline stress conditions [17,22]. Stimulation of the endocytic pathway under salt stress requires the simultaneous action of both clathrin-dependent and membrane microdomain-associated endocytosis [17,22]. In addition, Baral and his colleagues have shown that clathrin-mediated endocytosis allows the internalization of transmembrane proteins in all cell root layers whereas a sterol-sensitive clathrin-independent pathway internalizes lipid-anchored cargoes only in the epidermal cell layer [18]. Moreover, these authors showed that salt stress activates an additional clathrin-independent endocytosis pathway across all cell root layers that takes up both molecule types [18]. Considering membrane microdomain-associated endocytosis, it is known that proteins assemble into clusters in lipid rafts [8]. Among these proteins, remorins are considered as markers of membrane microdomains [35]. In *Medicago truncatula*, the symbiotic remorin 1 forms clusters and interacts with symbiotic receptors at the plasma membrane, playing a key role in bacterial signal perception [21]. Here, we showed that remorins 1-2 and 1-3, which are among the 10% of the most highly expressed genes in *Arabidopsis* [43], were also involved in miPEP165a entry into *Arabidopsis* roots (Figure 6B, Figure 7). Indeed, miPEP165a-FAM failed to enter the differentiation zone of *Arabidopsis* roots in *rem1-2* and *rem1-3* mutants. Moreover, root length and flowering acceleration induced by miPEP165a were perturbed in both remorin mutants (Figure 6B, Figure 7).

To conclude, we showed that endocytic pathways participate in miPEP uptake in plants. Thus, clathrin-mediated endocytosis as well as membrane microdomain-associated pathways seem to cooperate, allowing miPEPs to regulate their corresponding miRNAs and consequently modulate the plant phenotype, such as flowering and root development. Due to the simplicity of the mode of administration of miPEPs, a better understanding of miPEP uptake into plants is a first step towards the possible agronomic application of peptides.

4. Materials and methods

4.1. Peptide Synthesis

miPEP165a (MRVKLFQLRGMLSGSRIL), miPEP165a fused to fluorescein (miPEP165a-FAM), scrambled miPEP165a (LMGRQGLKISSLVFRMLR), PEP1 (KSNKTRVNFPS), PEP2 (MCFSFPDL), and PEP3 (MASAAKVYMA) were synthetized by Smart Bioscience (https://www.smart-bioscience. com/). They were dissolved in water (control) as a 10 mM stock solution (except for PEP2, which was dissolved in 50% acetonitrile as a 2 mM stock solution), aliquoted, and conserved at −80 °C until use.

4.2. Plant Materials

Different *Arabidopsis thaliana* plant lines (Columbia Col-0 ecotype) were used: the *chc1-1* (At3g11130), *chc2-1* (At3g08530), *ap2σ2* (At1g47830), *rem1-2* (At2g45820), and *rem1-3* (At3g61260) *Arabidopsis* mutants.

4.3. Peptide Treatment of Arabidopsis Roots

Surface-sterilized *Arabidopsis* seeds were sown on the surface of cellophane membrane placed on $\frac{1}{2}$ MS solid medium and stratified for one day at 4 °C in the dark. Seeds were vertically grown in controlled environmental chambers at 22/20 °C, with a photoperiod of 16h light/8h dark, an irradiance of ~ 97.5 μmol photons.m^{-2}.s^{-1}, and a relative humidity of 40%. Three days after sowing, seedlings were treated daily for 4 days either with water, 2.5% acetonitrile, 100 μM scrambled miPEP165a, 100 μM irrelevant peptides (PEP1, PEP2, PEP3), or fluorescein (control conditions) or with 100 μM miPEP165a or miPEP165a-FAM (treated conditions). Twenty-four hours after the last treatment, seedlings were scanned in order to measure primary root lengths using NeuronJ plugin of ImageJ.

4.4. Peptide Uptake in Arabidopsis Roots

Surface-sterilized wild-type and mutant *Arabidopsis* seeds were grown onto $\frac{1}{2}$ MS solid medium in the same conditions as those described in the previous section. Three days after germination, three seedlings were transferred to each well of a 48-well plate containing 200 μL of $\frac{1}{2}$ MS liquid medium. One day later, medium was replaced by 10 μM miPEP165a-FAM diluted in $\frac{1}{2}$ MS liquid medium until confocal microscopy observations. FAM fluorescence was analyzed with a confocal laser scanning microscope (Leica TCS SP2-AOBS using a 40 X water immersion objective lens (numerical aperture 0.80; HCX APO). FAM fluorescence was excited with the 488-nm ray line of the argon laser and recorded in the 511–551-nm emission range.

For quantification of miPEP165a-FAM entry into wild-type and mutant *Arabidopsis* roots, the fluorescence intensity was determined per surface unit in the different root zones using ImageJ software.

4.5. Inhibitor Treatment

TyrA23 was dissolved in dimethyl sulfoxide to yield a 50 mM stock solution and MβCD was prepared in deionized water at a final concentration of 38 mM. For each experiment, 3-day-old seedlings germinated on $\frac{1}{2}$ MS solid medium + 1% sucrose (wt/vol) were pre-treated with 50 μM TyrA23 or 10 mM MβCD for 30 min [17]. Seedlings were then treated with the inhibitors supplemented with

100 µM miPEP165a. Treatments were performed daily for an additional 3 days and plates were scanned for analysis of the primary root length with NeuronJ, an Image J plugin [29,44].

4.6. Flowering Phenotype

Arabidopsis seeds were grown on Jiffy® under a 16 h light/8 h dark cycle (22/20 °C), with a relative humidity of 80%. Fifteen days after seed sowing, either a 2-µL droplet of 100 µM miPEP165a was put on the shoot apical meristem or seedlings were watered with 500 µL of 10 µM miPEP165a three times a week. Analyses of the aerial parts were performed 24 days after sowing.

4.7. Propidium Iodide Staining

Wild-type seeds were grown for 3 days on $\frac{1}{2}$ MS solid medium + 1% sucrose (wt/vol) in the same growth conditions as described above. Seedlings were then treated with water or 100 µM miPEP165a daily for 3 additional days and placed in the growth chamber at the same settings. Seedlings were then stained with 10 µg/mL propidium iodide for 20 min and *Arabidopsis* cell roots were analyzed with a laser scanning confocal microscope (Leica TCS SP8-AOBS) with a ×25 water immersion objective lens (numeral aperture 0.95; Fluotar Visir). The excitation and emission wavelengths of propidium iodide were 561 and 570–640 nm, respectively.

The meristematic zone for the cortex cells was defined as the region between quiescent center cells and the first elongating cell that was twice the length compared to its distal neighbor [20,26]. The meristematic cell length and cell number were determined with the software tool Cell-o-Tape, an open source ImageJ/Fiji macro [27–29]. At least 20 roots were analyzed for each treatment.

4.8. Immunoblots and RT-qPCR

Seven-week-old *Arabidopsis* seedlings were treated with 100 µM miPEP165a or its corresponding control for 24 h, and then the expression of pri-miR165a was evaluated by RT-qPCR according to Lauressergues et al. [5].

To evaluate miPEP165a stability, 5 nanomoles of miPEP165a were subjected to several freeze/thaw cycles and its degradation was detected by immunoblotting with an anti-miPEP165a antibody as previously described [5].

Supplementary Materials: The following are available online at http://www.mdpi.com/1422-0067/21/7/2266/s1, Figure S1. Effect of miPEP165a and importance of its stability. Figure S2. Physiochemical properties of miPEP165a. Figure S3. Quantification of miPEP165a-FAM uptake in Arabidopsis roots. Figure S4. MβCD impairs the miPEP165a-FAM entry in the Arabidopsis root cap/meristematic zone.

Author Contributions: Conceptualization, M.O., S.P. and J.-P.C.; Methodology, M.O., A.L.R., S.P. and J.-P.C.; Validation, M.O., A.L.R. and J.-P.C; Formal Analysis, M.O., J.-P.C; Investigation, M.O., A.L.R., C.D. and J.-P.C; Resources, H.J.; Writing—Original Draft Preparation, M.O., S.P. and J.-P.C.; Writing—Review and Editing, M.O., H.J., P.T., S.P. and J.-P.C.; Project Administration, S.P. and J.-P.C.; Funding Acquisition, S.P. and J.-P.C. All authors have read and agreed to the published version of the manuscript.

Funding: This work was funded by the *Laboratoire d'Excellence* entitled TULIP (ANR-10-LABX-41), and National Institute of Health (R01 GM093008) to H.J.

Acknowledgments: We thank Martina Beck (Micropep Technologies, France), Jean-Malo Couzigou (LRSV, France) and Nathalie Leborgne-Castel (Université Bourgogne Franche-Comté, Dijon, France) for their helpful advices.

Abbreviations

AGO1	Argonaute 1
AP2	adaptor protein 2
CAN/AtHB15	CORONA
CHC	clathrin heavy chain
CLC	clathrin light chain
CME	clathrin-mediated endocytosis
DCL1	dicer-like1
FAM	5-carboxyfluorescein
HB8	homeobox gene 8
HD-ZIP III	class III homeodomain-leucine zipper
MβCD	methyl-β-cyclodextrin
miPEP	miRNA-encoded peptide
miRNA	micro-RNA
MS	Murashige and Skoog medium
PHB	PHABULOSA
PHV	PHAVOLUTA
PIP2;1	plasma membrane intrinsic protein 2
pri-miRNA	primary-microRNA
pre-miRNA	precursor-microRNA
RbohD	respiratory burst oxidase protein D
REM	remorin
REV	REVOLUTA
TyrA23	Tyrphostin A23.

References

1. Lagos-Quintana, M.; Rauhut, R.; Lendeckel, W.; Tuschl, T. Identification of novel genes coding for small expressed RNAs. *Science* **2001**, *294*, 853–858. [CrossRef]
2. Lau, N.C.; Lim, L.P.; Weinstein, E.G.; Bartel, D.P. An abundant class of tiny RNAs with probable regulatory roles in *Caenorhabditis elegans*. *Science* **2001**, *294*, 858–862. [CrossRef] [PubMed]
3. Lee, R.C.; Ambros, V. An extensive class of small RNAs in *Caenorhabditis elegans*. *Science* **2001**, *294*, 862–864. [CrossRef] [PubMed]
4. Reinhart, B.J.; Weinstein, E.G.; Rhoades, M.W.; Bartel, B.; Bartel, D.P. MicroRNAs in plants. *Genes Dev.* **2002**, *16*, 1616–1626. [CrossRef]
5. Lauressergues, D.; Couzigou, J.M.; San Clemente, H.; Martinez, Y.; Dunand, C.; Bécard, G.; Combier, J.P. Primary transcripts of microRNAs encode regulatory peptides. *Nature* **2015**, *520*, 90–93. [CrossRef] [PubMed]
6. Couzigou, J.M.; André, O.; Guillotin, B.; Alexandre, M.; Combier, J.P. Use of microRNA-encoded peptide miPEP172c to stimulate nodulation in soybean. *New Phytol.* **2016**, *211*, 379–381. [CrossRef] [PubMed]
7. Couzigou, J.M.; Lauressergues, D.; André, O.; Gutjahr, C.; Guillotin, B.; Bécard, G.; Combier, J.P. Positive Gene Regulation by a Natural Protective miRNA Enables Arbuscular Mycorrhizal Symbiosis. *Cell Host Microbe* **2017**, *21*, 106–112. [CrossRef]
8. Fan, L.; Li, R.; Pan, J.; Ding, Z.; Lin, J. Endocytosis and its regulation in plants. *Trends Plant Sci.* **2015**, *20*, 388–397. [CrossRef]
9. Dhonukshe, P.; Aniento, F.; Hwang, I.; Robinson, D.G.; Mravec, J.; Stierhof, Y.D.; Friml, J. Clathrin-mediated constitutive endocytosis of PIN auxin efflux carriers in *Arabidopsis*. *Curr. Biol.* **2007**, *17*, 520–527. [CrossRef]
10. Yamaoka, S.; Shimono, Y.; Shirakawa, M.; Fukao, Y.; Kawase, T.; Hatsugai, N.; Tamura, K.; Shimada, T.; Hara-Nishimura, I. Identification and dynamics of *Arabidopsis* adaptor protein-2 complex and its involvement in floral organ development. *Plant Cell* **2013**, *25*, 2958–2969. [CrossRef]
11. Barberon, M.; Zelazny, E.; Robert, S.; Conéjéro, G.; Curie, C.; Friml, J.; Vert, G. Monoubiquitin-dependent endocytosis of the iron-regulated transporter 1 (IRT1) transporter controls iron uptake in plants. *Proc. Natl. Acad. Sci. USA* **2011**, *108*, 450–458. [CrossRef] [PubMed]

12. Kitakura, S.; Vanneste, S.; Robert, S.; Löfke, C.; Teichmann, T.; Tanaka, H.; Friml, J. Clathrin mediates endocytosis and polar distribution of PIN auxin transporters in *Arabidopsis*. *Plant Cell* **2011**, *23*, 1920–1931. [CrossRef] [PubMed]

13. Gadeyne, A.; Sánchez-Rodríguez, C.; Vanneste, S.; Di Rubbo, S.; Zauber, H.; Vanneste, K.; Van Leene, J.; De Winne, N.; Eeckhout, D.; Persiau, G.; et al. The TPLATE adaptor complex drives clathrin-mediated endocytosis in plants. *Cell* **2014**, *156*, 691–704. [CrossRef] [PubMed]

14. Zwiewka, M.; Nodzyński, T.; Robert, S.; Vanneste, S.; Friml, J. Osmotic Stress Modulates the Balance between Exocytosis and Clathrin-Mediated Endocytosis in *Arabidopsis thaliana*. *Mol. Plant* **2015**, *8*, 117–1187. [CrossRef]

15. Mbengue, M.; Bourdais, G.; Gervasi, F.; Beck, M.; Zhou, J.; Spallek, T.; Bartels, S.; Boller, T.; Ueda, T.; Kuhn, H.; et al. Clathrin-dependent endocytosis is required for immunity mediated by pattern recognition receptor kinases. *Proc. Natl. Acad. Sci. USA* **2016**, *113*, 11034–11039. [CrossRef]

16. Ortiz-Morea, F.A.; Savatin, D.V.; Dejonghe, W.; Kumar, R.; Luo, Y.; Adamowski, M.; Van den Begin, J.; Dressano, K.; Pereira de Oliveira, G.; Zhao, X.; et al. Danger-associated peptide signaling in *Arabidopsis* requires clathrin. *Proc. Natl. Acad. Sci. USA* **2016**, *113*, 11028–11033. [CrossRef]

17. Li, X.; Wang, X.; Yang, Y.; Li, R.; He, Q.; Fang, X.; Luu, D.-T.; Maurel, C.; Lin, J. Single-molecule analysis of PIP2;1 dynamics and partitioning reveals multiple modes of *Arabidopsis* plasma membrane aquaporin regulation. *Plant Cell* **2011**, *23*, 3780–3797. [CrossRef]

18. Baral, A.; Irani, N.G.; Fujimoto, M.; Nakano, A.; Mayor, S.; Mathew, M.K. Salt-induced remodeling of spatially restricted clathrin-independent endocytic pathways in *Arabidopsis* root. *Plant Cell* **2015**, *27*, 1297–1315. [CrossRef]

19. Li, R.; Liu, P.; Wan, Y.; Chen, T.; Wang, Q.; Mettbach, U.; Baluska, F.; Samaj, J.; Fang, X.; Lucas, W.J.; et al. A membrane microdomain-associated protein, *Arabidopsis* Flot1, is involved in a clathrin-independent endocytic pathway and is required for seedling development. *Plant Cell* **2012**, *24*, 2105–2122. [CrossRef]

20. Yu, M.; Liu, H.; Dong, Z.; Xiao, J.; Su, B.; Fan, L.; Komis, G.; Samaj, J.; Lin, J.; Li, R. The dynamics and endocytosis of Flot1 protein in response to flg22 in *Arabidopsis*. *J. Plant Physiol.* **2017**, *215*, 73–84. [CrossRef]

21. Lefebvre, B.; Timmers, T.; Mbengue, M.; Moreau, S.; Hervé, C.; Tóth, K.; Bittencourt-Silvestre, J.; Klaus, D.; Deslandes, L.; Godiard, L.; et al. A remorin protein interacts with symbiotic receptors and regulates bacterial infection. *Proc. Natl. Acad. Sci. USA* **2010**, *107*, 2343–2348. [CrossRef] [PubMed]

22. Hao, H.; Fan, L.; Chen, T.; Li, R.; Li, X.; He, Q.; Botella, M.; Lin, J. Clathrin and membrane microdomains cooperatively regulate RbohD dynamics and activity in *Arabidopsis*. *Plant Cell* **2014**, *26*, 1729–1745. [CrossRef] [PubMed]

23. Carlsbecker, A.; Lee, J.Y.; Roberts, C.J.; Dettmer, J.; Lehesranta, S.; Zhou, J.; Lindgren, O.; Moreno-Risueno, M.A.; Vatén, A.; Thitamadee, S.; et al. Cell signalling by microRNA165/6 directs gene dose-dependent root cell fate. *Nature* **2010**, *465*, 316–321. [CrossRef] [PubMed]

24. Perilli, S.; Di Mambro, R.; Sabatini, S. Growth and development of the root apical meristem. *Curr. Opin. Plant Biol.* **2012**, *15*, 17–23. [CrossRef] [PubMed]

25. Dello Ioio, R.; Linhares, F.S.; Scacchi, E.; Casamitjana-Martinez, E.; Heidstra, R.; Costantino, P.; Sabatini, S. Cytokinins determine *Arabidopsis* root-meristem size by controlling cell differentiation. *Curr. Biol.* **2007**, *17*, 678–682. [CrossRef] [PubMed]

26. Liu, Y.; Lai, N.; Gao, K.; Chen, F.; Yuan, L.; Mi, G. Ammonium inhibits primary root growth by reducing the length of meristem and elongation zone and decreasing elemental expansion rate in the root apex in *Arabidopsis thaliana*. *PLoS ONE* **2013**, *8*, e61031. [CrossRef]

27. French, A.P.; Wilson, M.H.; Kenobi, K.; Dietrich, D.; Voβ, U.; Ubeda-Tomás, S.; Pridmore, T.P.; Wells, D.M. Identifying biological landmarks using a novel cell measuring image analysis tool: Cell-o-Tape. *Plant Methods* **2012**, *8*, 7. [CrossRef]

28. Schindelin, J.; Arganda-Carreras, I.; Frise, E.; Kaynig, V.; Longair, M.; Pietzsch, T.; Preibisch, B.; Rueden, C.; Saalfeld, S.; Schmid, B.; et al. Fiji: An open-source platform for biological-image analysis. *Nat. Methods* **2012**, *9*, 676–682. [CrossRef]

29. Schneider, C.A.; Rasband, W.S.; Eliceiri, K.W. NIH Image to ImageJ: 25 years of image analysis. *Nat. Methods* **2012**, *9*, 671–675. [CrossRef]

30. Ji, L.; Liu, X.; Yan, J.; Wang, W.; Yumul, R.E.; Kim, Y.J.; Dinh, T.T.; Liu, J.; Cui, X.; Zheng, B.; et al. ARGONAUTE10 and ARGONAUTE1 regulate the termination of floral stem cells through two microRNAs in *Arabidopsis*. *PLoS Genet.* **2011**, *7*, e1001358. [CrossRef]

31. Balcerowicz, D.; Schoenaers, S.; Vissenberg, K. Cell fate determination and the switch from diffuse growth to planar polarity in *Arabidopsis* root epidermal cells. *Front. Plant Sci.* **2015**, *6*, 1163. [CrossRef]

32. Fan, L.; Hao, H.; Xue, Y.; Zhang, L.; Song, K.; Ding, Z.; Botella, M.A.; Wang, H.; Lin, J. Dynamic analysis of *Arabidopsis* AP2 σ subunit reveals a key role in clathrin-mediated endocytosis and plant development. *Development* **2013**, *140*, 3826–3837. [CrossRef] [PubMed]

33. Mongrand, S.; Morel, J.; Laroche, J.; Claverol, S.; Carde, J.P.; Hartmann, M.A.; Bonneu, M.; Simon-Plas, F.; Lessire, R.; Bessoule, J.J. Lipid rafts in higher plant cells: Purification and characterization of Triton X-100-insoluble microdomains from tobacco plasma membrane. *J. Biol. Chem.* **2004**, *279*, 36277–36286. [CrossRef] [PubMed]

34. Kierszniowska, S.; Seiwert, B.; Schulze, W.X. Definition of *Arabidopsis* sterol-rich membrane microdomains by differential treatment with methyl-beta-cyclodextrin and quantitative proteomics. *Mol. Cell Proteom.* **2009**, *8*, 612–623. [CrossRef]

35. Jarsch, I.K.; Ott, T. Perspectives on remorin proteins, membrane rafts, and their role during plant–microbe interactions. *Mol. Plant Microbe Interact.* **2011**, *24*, 7–12. [CrossRef]

36. Banbury, D.N.; Oakley, J.D.; Sessions, R.B.; Banting, G. Tyrphostin A23 inhibits internalization of the transferrin receptor by perturbing the interaction between tyrosine motifs and the medium chain subunit of the AP-2 adaptor complex. *J. Biol. Chem.* **2003**, *278*, 12022–12028. [CrossRef] [PubMed]

37. Zhou, G.K.; Kubo, M.; Zhong, R.; Demura, T.; Ye, Z.H. Overexpression of miR165 affects apical meristem formation, organ polarity establishment and vascular development in *Arabidopsis*. *Plant Cell Physiol.* **2007**, *48*, 391–404. [CrossRef] [PubMed]

38. Dello Ioio, R.; Galinha, C.; Fletcher, A.G.; Grigg, S.P.; Molnar, A.; Willemsen, V.; Scheres, B.; Sabatini, S.; Baulcombe, D.; Maini, P.K.; et al. A PHABULOSA/cytokinin feedback loop controls root growth in *Arabidopsis*. *Curr. Biol.* **2012**, *22*, 1699–1704. [CrossRef]

39. Singh, A.; Singh, S.; Panigrahi, K.C.; Reski, R.; Sarkar, A.K. Balanced activity of microRNA166/165 and its target transcripts from the class III homeodomain-leucine zipper family regulates root growth in *Arabidopsis thaliana*. *Plant Cell Rep.* **2014**, *33*, 945–953. [CrossRef]

40. Chen, A.; Komives, E.A.; Schroeder, J.I. An improved grafting technique for mature *Arabidopsis* plants demonstrates long-distance shoot-to-root transport of phytochelatins in *Arabidopsis*. *Plant Physiol.* **2006**, *141*, 108–120. [CrossRef]

41. Delay, C.; Imin, N.; Djordjevic, M.A. Regulation of *Arabidopsis* root development by small signaling peptides. *Front. Plant Sci.* **2013**, *4*, 352. [CrossRef] [PubMed]

42. Ko, D.; Helariutta, Y.A. Shoot—Root communication in flowering plants. *Curr. Biol.* **2017**, *27*, 973–978. [CrossRef] [PubMed]

43. Raffaele, S.; Mongrand, S.; Gamas, P.; Niebel, A.; Ott, T. Genome-wide annotation of remorins, a plant-specific protein family: Evolutionary and functional perspectives. *Plant Physiol.* **2007**, *145*, 593–600. [CrossRef] [PubMed]

44. Meijering, E.; Jacob, M.; Sarria, J.C.; Steiner, P.; Hirling, H.; Unser, M. Design and validation of a tool for neurite tracing and analysis in fluorescence microscopy images. *Cytom. A* **2004**, *58*, 167–176. [CrossRef]

Communication

GAI Functions in the Plant Response to Dehydration Stress in *Arabidopsis thaliana*

Zhijuan Wang [1,†], **Liu Liu** [2,3,†], **Chunhong Cheng** [2,3,†], **Ziyin Ren** [1], **Shimin Xu** [1] and **Xia Li** [1,*]

[1] State Key Laboratory of Agricultural Microbiology, College of Plant Science and Technology, Huazhong Agricultural University, Wuhan 430070, China; wzhijuan@163.com (Z.W.); rzy1819345962@163.com (Z.R.); 15527847403@163.com (S.X.)

[2] The State Key Laboratory of Plant Cell & Chromosome Engineering, Center for Agricultural Research Resources, Institute of Genetics and Developmental Biology, Chinese Academy of Sciences, 286 Huaizhong Road, Shijiazhuang 050021, China; liuliu19890729@163.com (L.L.); xiaobei15109217512@163.com (C.C.)

[3] Graduate University of Chinese Academy of Sciences, No.19A Yuquan Road, Beijing 100049, China

* Correspondence: xli@mail.hzau.edu.cn; Tel.: +86-027-87856637

† These authors contributed equally to this work.

Received: 20 December 2019; Accepted: 23 January 2020; Published: 27 January 2020

Abstract: DELLA (GAI/RGA/RGL1/RGL2/RGL3) proteins are key negative regulators in GA (gibberellin) signaling and are involved in regulating plant growth as a response to environmental stresses. It has been shown that the DELLA protein PROCERA (PRO) in tomato promotes drought tolerance, but its molecular mechanism remains unknown. Here, we showed that the *gai-1* (gibberellin insensitive 1) mutant (generated from the *gai-1* (L*er*) allele (with a 17 amino acid deletion within the DELLA domain of GAI) by backcrossing *gai-1* (L*er*) with Col-0 three times), the gain-of-function mutant of GAI (GA INSENSITIVE) in *Arabidopsis*, increases drought tolerance. The stomatal density of the *gai-1* mutant was increased but its stomatal aperture was decreased under abscisic acid (ABA) treatment conditions, suggesting that the drought tolerance of the *gai-1* mutant is a complex trait. We further tested the interactions between DELLA proteins and ABF2 (abscisic acid (ABA)-responsive element (ABRE)-binding transcription factors) and found that there was a strong interaction between DELLA proteins and ABF2. Our results provide new insight into DELLA proteins and their role in drought stress tolerance.

Keywords: drought; GA; DELLA; ABF2; protein–protein interaction

---

1. Introduction

A water deficit is a restrictive factor for plant development, productivity, and geographical distribution. Plants have evolved varied strategies to cope with decreased water availability, including promoting stomatal closure and altered plant growth and development. The stress-induced hormone abscisic acid (ABA) plays an important role in a plant's response to drought tolerance [1–3]. Increasing evidence has proven that gibberellin (GA) plays a negative role in drought response. The over-accumulation of the GA mutant or increased GA activity shows an increased water deficit sensitivity, whereas a GA-deficient mutant or decreased GA activity shows an increased water deficit tolerance [4–8].

DELLA (GAI/RGA/RGL1/RGL2/RGL3) proteins are major negative regulators of GA signaling. In the absence of GA, DELLA proteins inhibit the GA-dependent processes, including germination, growth, and flowering. Under increased GA levels, GA binds to its nuclear receptor GID1 (GA insensitive dwarf1) and changes its conformation, leading to its interaction with the N-terminal end of DELLA proteins [9–11]. The interaction of DELLA proteins with GID1 causes its ubiquitination and subsequent degradation by the 26S proteasome, leading to the activation of GA responses [12,13]. DELLA proteins are involved in most GA-mediated plant growth and environmental stresses, including dehydration stress. Recently, it was reported that the DELLA protein PRO (PROCERA) in tomato functions positively in the plant response to drought stress. The loss-of-function of the *PRO* mutant shows a reduced tolerance to drought, whereas the overexpression of the constitutively active stable *PRO* increases drought tolerance [8]. However, the molecular mechanism of DELLA proteins remains unclear.

There are five DELLA members in Arabidopsis: GAI (GA INSENSITIVE), RGA (REPRESSOR OR GAI3), RGL1 (RGA-LIKE1), RGL2, and RGL3. To uncover the molecular mechanism that determines how DELLA proteins function in drought tolerance, we used GAI as a sample to analyze the function of DELLA proteins in response to drought stress. We made use of a gain-of-function mutant *gai-1* generated from the *gai-1* (L*er*) allele (with a 17 amino acid deletion within the DELLA domain of GAI) by backcrossing *gai-1* (L*er*) with Col-0 (Columbia-0) three times. We showed that this mutant has an increased drought tolerance phenotype. Further, we found that GAI and other DELLA proteins interacted with ABF2 (abscisic acid (ABA)-responsive element (ABRE)-binding transcription factors), the transcriptional factor that plays a pivotal role in ABA signaling for drought tolerance. Our results thus shed some light on the mechanism behind how DELLA proteins function in drought stress tolerance.

2. Results

To study the function of DELLA proteins in drought tolerance in Arabidopsis, we firstly analyzed the phenotype of the *gai-1* mutant under the condition of drought treatment. Three-week-old seedlings of wild type (Col-0) and mutant *gai-1* were withheld from water for 21 days. The wild type plants were severely wilted, whereas the *gai-1* mutant did not wilt and continued to grow. After rewatering, all of the *gai-1* plants recovered, whereas none of the wild type plants survived (Figure 1A), indicating that the *gai-1* mutant is more tolerant to drought and that GAI is a positive regulator in the plant response to drought tolerance. The function of GAI in drought tolerance is consistent with that of PRO in tomato, suggesting that this tolerance is a conserved function of DELLA proteins in the plant kingdom.

Changes in transpiration rate could account for the altered tolerance to drought. We then tested the water loss rate of the detached leaves. Leaves of 3 week old seedlings were cut and exposed to air and were weighted at regular time points. To our surprise, the *gai-1* mutant leaves lost their water at a much higher rate than the wild type leaves (Figure 1B). The water loss of the wild type sample was only 20%, whereas the water loss of the *gai-1* mutant was over 30% at 4 h after exposure to air, suggesting that the *gai-1* mutant is sensitive to dehydration when detached leaves are exposed to air.

The stomata are key channels that control gas exchange and water evaporation. We then tested the stomatal density and aperture from leaves of wild type and *gai-1* plants grown in soil. The stomatal density of the *gai-1* mutant was significantly higher than that of the wild type plant (2.6×) (Figure 1C,D). This may be the reason for the higher rate water loss in the *gai-1* mutant for the detached leaves. For the stomatal apertures, the wild type and *gai-1* mutant were comparable under KCl-treated control conditions. However, under ABA treatment, the stomatal aperture of the *gai-1* mutant was much smaller than that of the wild type (Figure 1E,F). The stomatal density and aperture of the *gai-1* mutant under stress conditions are consistent with those of the *PRO* gain-of-function mutant, suggesting that this is a conserved mechanism for DELLA proteins in regulating plant development and environmental adaption.

Generally, DELLA proteins function by interacting with other transcriptional factors. As ABF2 (abscisic acid (ABA)-responsive element (ABRE)-binding transcription factors) is a key regulator in drought tolerance, we hypothesized that GAI interacts with ABF2 to increase drought tolerance. To test this hypothesis, we tested the interaction between GAI and ABF2 via yeast two-hybrid and BiFC (bimolecular fluorescent complimentary) assays. In the yeast two-hybrid assay, GAI and ABF2 were recombined to the gateway-compatible destination vectors pGADT7-DEST (AD) and pGBKT7-DEST (BD), respectively. The AD and BD constructs were cotransformed to the yeast strain AH109, and their interaction was determined by the growth on the SD (synthetic dropout) medium lacking

Trp (tryptophan), Leu (leucine), His (histidine), and Ade (adenine). There was a strong interaction between GAI and ABF2 in the yeast two-hybrid assay (Figure 2A). For the BiFC assay, GAI and ABF2 were recombined to pEarleyGate201-YN (N-terminal YFP (yellow fluorescent protein)) and pEarleyGate202-YC (C-terminal YFP), respectively. Both constructs were coinfiltrated into *Nicotiana benthamiana* leaves. The YFP signal was observed in the nucleus of the plant cell coexpressing GAI-YFP^N and ABF2-YFP^C (Figure 2B), but no YFP signal was detected in the plant cell coexpressing GAI-YFP^N and empty YFP^C or ABF2-YFP^C and empty YFP^N (Figure S1), indicating that GAI and ABF2 interacted in the nucleus of the plant cell. We also tested the interactions between other DELLA proteins, including RGA, RGL1, RGL2, and RGL3, with ABF2. The yeast two-hybrid and BiFC assays both showed that all of the DELLA proteins interacted with ABF2 (Figure 2A,B).

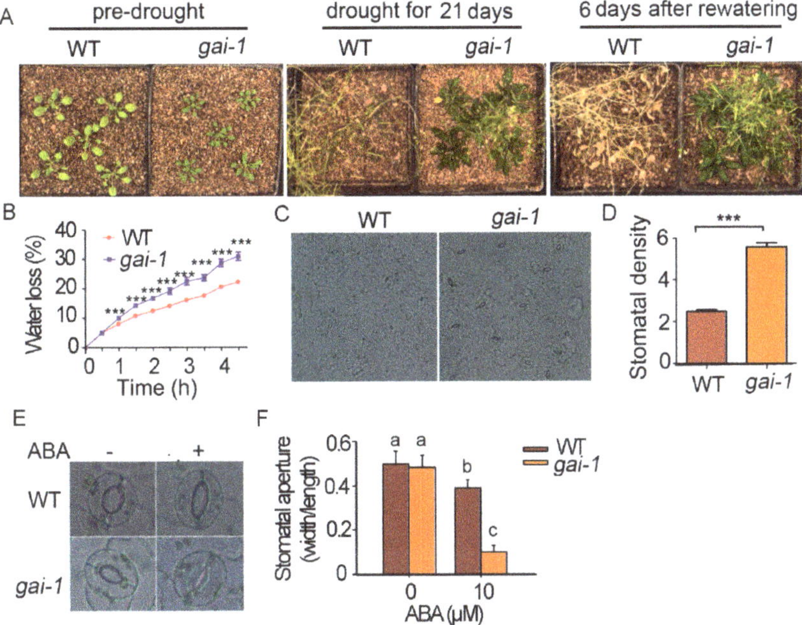

Figure 1. The *gai-1* (gibberellin insensitive 1) mutant is more tolerant to drought stress than WT (wild type). (**A**) *gai-1* mutant plants showed tolerance to dehydration stress. *gai-1* mutant plants showed the ability to withstand long drought conditions without negative effects whereas the wild type under the same conditions completely wilted. (**B**) *gai-1* plants showed increased water loss compared to WT. Data shown are the means ± SDs from three biological repeats ($n = 3$, eight leaves from eight plants were used for each repeat, $p < 0.001$). (**C,D**) Stomatal density of WT and *gai-1* mutant. Stomatal density was observed from comparable age leaves of 3 week old wild type and *gai-1* plants. The stomatal density was represent by number of stomata per millimeters squared. Data shown are the means ± SDs from three biological repeats ($n = 3$, five leaves from five plants were used for each repeat, $p < 0.001$). (**E**) Representative stomata of the WT and *gai-1* mutant under control and abscisic acid (ABA) treatment conditions. Leaves of the WT and *gai-1* mutant were treated with 10 µM ABA for 2 h (+), and (−) represents leaves without ABA treatment. (**F**) Stomatal apertures of the WT and *gai-1* mutant corresponding to (**E**). Values are mean ratios of width to length ± SDs of three independent experiments. Letters indicate significant differences from the WT (0 ABA treatment) according to the Student's Newman–Kuels test (*** $p < 0.05$).

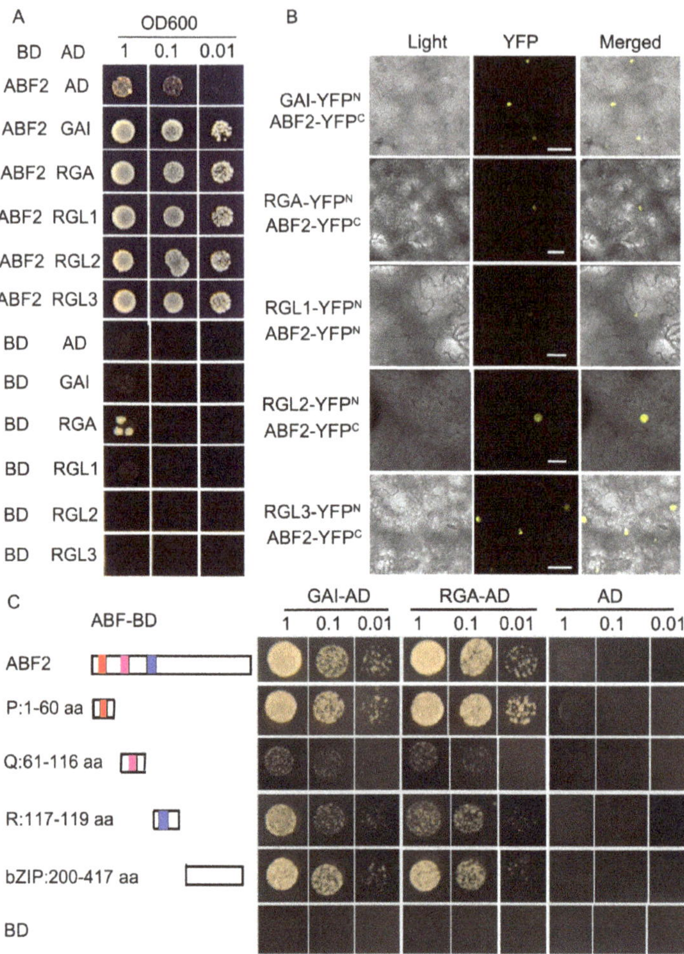

Int. J. Mol. Sci. **2020**, 21, 819

Figure 2. DELLA (GAI/RGA/RGL1/RGL2/RGL3) proteins interacted with ABF2 (abscisic acid (ABA)-responsive element (ABRE)-binding transcription factors). (**A**) DELLA proteins interacted with ABF2 in yeast two-hybrid assay. The yeast cells expressing the indicated constructs were spotted as a series of three dilutions. The yeast cells expressing the constructs of ABF2-pGBKT7-DEST (BD)/GAI-pGADT7-DEST (AD), ABF2-BD/RGA (REPRESSOR OR GAI3)-AD, ABF2-BD/RGL1 (RGA-LIKE1)-AD, ABF2-BD/RGL2-AD, and ABF2-BD/RGL3-AD grew better on the SD medium than that of yeast growth cells expressing the control's constructs. (**B**) BiFC (bimolecular fluorescent complimentary) assay between DELLA proteins and ABF2. *Nicotiana benthamiana* leaves were co-transformed with the constructs containing the indicated YFP (yellow fluorescent protein) N-terminal (YFP^N) and YFP C-terminal (YFP^C) fusions, and YFP was imaged 48 h after transformation. Bars = 50 μm. (**C**) Interaction assay between GAI and RGA with ABF2 fragments. P: 1–60 amino acid; Q: 61–116 amino acid; R: 117–199 amino acid; bZIP (basic region/leucine zipper): 200–417 amino acid. The yeast cells expressing the indicated constructs were spotted as a series of three dilution. The yeast cell expressing the constructs of ABF2-BD/GAI-AD, ABF2-BD/RGA-AD, ABF2P-BD/GAI-AD, ABF2P-BD/RGA-AD, ABF2bZIP-BD/GAI-AD, ABF2bZIP-BD/RGA-AD, grew more effectively on the SD medium than that of yeast cells expressing ABF2R-BD/RGA-AD, ABF2R-BD/GAI-AD, ABF2Q-BD/RGA-AD, and ABF2Q-BD/GAI-AD.

There are three conserved motifs in the N-terminal of ABF2: P, Q, and R. The P motif is responsible for transactivation activity and activates downstream gene expression [14]. To test which motif is responsible for interacting with GAI, we tested the interaction between GAI and the P, Q, R, and bZIP (basic region/leucine zipper) motifs of ABF2. Our yeast two-hybrid assay showed that there was a strong interaction between the GAI and the P and bZIP motifs, whereas the interaction between GAI and the Q and R motifs was much weaker (Figure 2C). We also tested the interaction between RGA and the P, Q, R, and bZIP motifs of ABF2. RGA also showed a strong interaction with the P and bZIP motifs but a weak interaction with the Q and R motifs (Figure 2C).

3. Discussion

GA is an important phytohormone that regulates plant growth. Increasing evidence has demonstrated that GA plays a role in the response to environmental stresses. DELLA proteins are key negative regulators in GA signaling, and our results showed that DELLA proteins increased drought tolerance by interacting with ABF2, a positive regulator in the ABA signaling pathway.

The cellular mechanism of the drought tolerance of the *gai-1* mutant seems complex and confusing. The stomatal density of the *gai-1* mutant is higher than that of the wild type, which makes water loss occur more quickly. Indeed, we found that the water loss rate of the detached leaves in the *gai-1* mutant was higher than that of the wild type (Figure 1B). The stomatal aperture in ABA treatment was found to be smaller in the *gai-1* mutant than that of the wild type sample, which may be responsible for the drought tolerance. The stomata's phenotype, density, and aperture of the *gai-1* mutant are similar to those of the gain-of-function of *PRO* in tomato [8], suggesting that it is conserved for DELLA proteins in regulating plant development and stress response.

As DELLA proteins do not have a DNA binding domain, it is common for them to interact with other transcription factors to regulate downstream target genes. For example, DELLA proteins interact with ABI3 and ABI5 to activate *SOM* (*SOMNUS*) expression at high temperatures [15]; RGA interacts with BZR1 (Brassinazole-resistant 1) to inhibit its transcriptional activities to downstream genes to regulate cell growth [16]; RGA interacts with WRKY6 (WRKYGQK) to block its transcriptional activities on its downstream genes, *SAG13* (Senescence-associated gene13) and *SGR* (Stay green), to regulate senescence [17]; and RGL2 interacts with the NF-YC (NUCLEAR FACTOR-Y C) homologues NF-YC3, NF-YC4, and NF-YC9 to activate the downstream gene ABI5 to regulate seed germination [18]. Here, we showed that DELLA proteins interact with ABF2 to regulate drought tolerance. AREB/ABF (abscisic acid-responsive element binding) proteins play pivotal roles in the regulation of plant responses to abiotic stresses. By binding to the ABRE element in the promoter region of stress-responsive genes, AREB/ABF factors regulate gene expression under drought stress [19,20]. In Arabidopsis, four AREB/ABF factors, namely, AREB1/ABF2, AREB2/ABF4, ABF1, and ABF3, are induced by ABA and osmotic stress [21,22]. Overexpressing *AREB1/ABF2*, *AREB2/ABF4*, or *ABF3* promotes drought tolerance, and a loss-of-function of these genes enhances drought sensitivity [14,16,23,24]. Many stress-inducible genes, including *RD29B* and *RAB18*, were downregulated in the *areb1 areb2 abf3 abf1-2* quadruple mutant [20]. Our results showed a strong interaction between DELLA proteins and the ABF2 protein (Figure 2A,B). It is reasonable that the interaction of DELLA proteins with ABF2 could activate ABF2 transcriptional activity to promote drought tolerance. A further expression assay of the downstream gene in the *gai-1* mutant and the binding assay of ABF2 to the promoter of *RD29B* or *RAB18* in the presence or absence of DELLA proteins would allow for the determination of the role of the interaction between DELLA proteins and ABF2. We also cannot exclude the possibility that DELLA proteins interact with other AREB/ABF factors, such as AREB2/ABF4, ABF1, or ABF3. A further interaction assay between the DELLA proteins and other AREB/ABF factors will deepen our understanding of the role of DELLA proteins in drought tolerance.

In summary, our results showed that GAI increased drought tolerance in Arabidopsis. GAI had conserved functions in increasing the stomatal density and decreasing the stomatal aperture under ABA treatment conditions. Further, we showed that GAI interacted with ABF2, especially the

N-terminal end P domain and the bZIP domain of ABF2. Our results provide new insight into DELLA protein functions in drought stress tolerance and the crosstalk between ABA and GA in response to drought tolerance.

4. Materials and Methods

4.1. Plant Materials and Growth Conditions

Arabidopsis thaliana ecotype Col-0 was used as the wild type in this study. The *gai-1* mutant was kindly gifted by Dr. Xiangdong Fu (Institute of Genetics and Developmental Biology, CAS). The seeds were germinated and grown on MS (Murashige & Skoog) medium and transplanted into soil at 10 days after germination. The plants were grown under a 16 h light/8 h dark photoperiod at 23 °C.

4.2. Drought Treatment, Water Loss Analysis, and Stomatal Aperture Measurement

For measurement of drought tolerance, water was withheld from 21 day old wild-type and *gai-1* mutant plants. After 21 days of drought treatment, the plants were rewatered; the plants were photographed 6 days after re-watering. For measurement of water loss, eight rosette leaves from eight plants were detached from 3 week old well-watered plants and weighted at the indicated times. For the stomatal function, rosette leaves from well-watered plant were incubated in a solution containing 50 mM KCl, 10 mM $CaCl_2$, and 10 mM MES (pH 6.15) for 2 h under light. ABA was then added to the solution to a final concentration of 10 µM. After ABA treatment for 2 h, stomatal apertures were measured as described previously [2].

4.3. Protein–Protein Analysis

The constructs were created in two pairs of Gateway-compatible destination vectors: pGBKT7-DEST (BD) with pGBAD7-DEST (AD) and pEarleyGate201-YN (N-terminal YFP) with pEarleyGate202-YC (C-terminal YFP) [25]. The coding sequences of *GAI*, *RGA*, *RGL1*, *RGL2*, *RGL3*, *ABF2*, and different deletion fragments of *ABF2* were amplified from Col-0 cDNA, inserted into pDONR207, and then recombined in the appropriate destination vector. Yeast two-hybrid and BiFC assays were performed as previously described [2]. For Y2H, *Saccharomyces cerevisiae* strain AH109 was used for co-transformation of the AD and BD constructs. A series of 5 µL aliquots of diluted co-transformed AH109 culture was spotted onto SD plates lacking Trp, Leu, His, and Ade, and incubated at 30 °C for 2–5 days. Plasmids pGBKT7 and pGADT7-Rec were used as negative controls. For the BiFC assay, *Agrobacterium tumefaciens* GV3101 carrying the YFP N-terminal and YFP C-terminal fusion constructs was infiltrated into *N. benthamiana* leaves, as described by Luo et al. [2]. The reconstituted YFP signals were observed using confocal imaging 48 h after infiltration. Empty vectors were used as negative controls.

Supplementary Materials: The following are available online at http://www.mdpi.com/1422-0067/21/3/819/s1.

Author Contributions: Conceptualization, X.L.; methodology, Z.W., C.C., L.L., Z.R., and S.X.; validation, Z.W., and Z.R.; formal analysis, Z.W. and L.L; investigation, X.L.; writing—original draft preparation, Z.W.; writing—review and editing, X.L.; funding acquisition, X.L. All authors have read and agreed to the published version of the manuscript.

Funding: This research was funded by the National Key Research and Development Program of China, grant number 2016YFA0500503; Fundamental Research Funds for Central Universities, grant number 2662018PY075; and Huazhong Agricultural University's Scientific and Technological Self-innovation Foundation, grant number 2015RC014.

Acknowledgments: We thank Xiangdong Fu (Institute of Genetics and Developmental Biology, CAS) for providing *gai-1* mutant seed.

Conflicts of Interest: The authors declare no conflict of interest. The funders had no role in the design of the study; in the collection, analyses, or interpretation of data; in the writing of the manuscript, or in the decision to publish the results.

Abbreviations

ABA	Abscisic acid
AREB/ABF	Abscisic acid-responsive element binding
GA	Gibberellin
GAI	GA INSENSITIVE
GID1	GA insensitive dwarf1
PRO	PROCERA

References

1. Munns, R. Comparative physiology of salt and water stress. *Plant Cell Environ.* **2002**, *25*, 239–250. [CrossRef] [PubMed]
2. Luo, Y.; Wang, Z.; Ji, H.; Fang, H.; Wang, S.; Tian, L.; Li, X. An Arabidopsis homolog of importin β1 is required for ABA response and drought tolerance. *Plant J.* **2013**, *75*, 377–389. [CrossRef] [PubMed]
3. Ku, Y.S.; Sintaha, M.; Cheung, M.Y.; Lam, H.M. Plant Hormone Signaling Crosstalks between Biotic and Abiotic Stress Responses. *Int. J. Mol. Sci.* **2018**, *19*, 3206. [CrossRef] [PubMed]
4. Achard, P.; Cheng, H.; De Grauwe, L.; Decat, J.; Schoutteten, H.; Moritz, T.; Van Der Straeten, D.; Peng, J.; Harberd, N.P. Integration of plant responses to environmentally activated phytohormonal signals. *Science* **2006**, *311*, 91–104. [CrossRef]
5. Li, J.; Sima, W.; Ouyang, B.; Wang, T.; Ziaf, K.; Luo, Z.; Liu, L.; Li, H.; Chen, M.; Huang, Y.; et al. Tomato SlDREB gene restricts leaf expansion and internode methylation and chromatin patterning elongation by downregulating key genes for gibberellin biosynthesis. *J. Exp. Bot.* **2012**, *63*, 695–709. [CrossRef]
6. Colebrook, E.H.; Thomas, S.G.; Phillips, A.L.; Hedden, P. The role of gibberellin signalling in plant responses to abiotic stress. *J. Exp. Biol.* **2014**, *217*, 67–75. [CrossRef]
7. Nir, I.; Moshelion, M.; Weiss, D. The Arabidopsis gibberellin methyl transferase 1 suppresses gibberellin activity, reduces whole-plant transpiration and promotes drought tolerance in transgenic tomato. *Plant Cell Environ.* **2014**, *37*, 113–123. [CrossRef]
8. Nir, I.; Shohat, H.; Panizel, I.; Olszewski, N.; Aharoni, A.; Weiss, D. The Tomato DELLA Protein PROCERA Acts in Guard Cells to Promote Stomatal Closure. *Plant Cell* **2017**, *29*, 3186–3197. [CrossRef]
9. Harberd, N.P.; Belfield, E.; Yasumura, Y. The angiosperm gibberellin-GID1-DELLA growth regulatory mechanism: How an "inhibitor of an inhibitor" enables flexible response to fluctuating environments. *Plant Cell* **2009**, *21*, 1328–1339. [CrossRef] [PubMed]
10. Schwechheimer, C. Gibberellin signaling in plants-the extended version. *Front. Plant Sci.* **2012**, *2*, 107. [CrossRef] [PubMed]
11. Sun, T.P. The molecular mechanism and evolution of the GA-GID1-DELLA signaling module in plants. *Curr. Biol.* **2011**, *21*, R338–R345. [CrossRef] [PubMed]
12. Gao, X.H.; Xiao, S.L.; Yao, Q.F.; Wang, Y.J.; Fu, X.D. An updated GA signaling 'relief of repression' regulatory model. *Mol. Plant* **2011**, *4*, 601–606. [CrossRef] [PubMed]
13. Daviere, J.M.; Achard, P. A pivotal role of DELLAs in regulating multiple hormone signals. *Mol. Plant* **2016**, *9*, 10–20. [CrossRef] [PubMed]
14. Fujita, Y.; Fujita, M.; Satoh, R.; Maruyama, K.; Parvez, M.M.; Seki, M.; Hiratsu, K.; Ohme-Takagi, M.; Shinozaki, K.; Yamaguchi-Shinozaki, K. AREB1 is a transcription activator of novel ABRE-dependent ABA signaling that enhances drought stress tolerance in Arabidopsis. *Plant Cell* **2005**, *17*, 3470–3488. [CrossRef] [PubMed]
15. Lim, S.; Park, J.; Lee, N.; Jeong, J.; Toh, S.; Watanabe, A.; Kim, J.; Kang, H.; Kim, D.H.; Kawakami, N.; et al. ABA-insensitive3, ABA-insensitive5, and DELLAs Interact to activate the expression of SOMNUS and other high-temperature-inducible genes in imbibed seeds in Arabidopsis. *Plant Cell* **2013**, *25*, 4863–4878. [CrossRef]
16. Li, Q.F.; Wang, C.; Jiang, L.; Li, S.; Sun, S.S.; He, J.X. An interaction between BZR1 and DELLAs mediates direct signaling crosstalk between brassinosteroids and gibberellins in Arabidopsis. *Sci. Signal.* **2012**, *5*, ra72. [CrossRef]
17. Zhang, Y.; Liu, Z.; Wang, X.; Wang, J.; Fan, K.; Li, Z.; Lin, W. DELLA proteins negatively regulate dark-induced senescence and chlorophyll degradation in Arabidopsis through interaction with the transcription factor WRKY6. *Plant Cell Rep.* **2018**, *37*, 981–992. [CrossRef]

18. Liu, X.; Hu, P.; Huang, M.; Tang, Y.; Li, Y.; Li, L.; Hou, X. The NF-YC-RGL2 module integrates GA and ABA signalling to regulate seed germination in Arabidopsis. *Nat. Commun.* **2016**, *7*, 12768. [CrossRef]

19. Nakashima, K.; Ito, Y.; Yamaguchi-Shinozaki, K. Transcriptional regulatory networks in response to abiotic stresses in Arabidopsis and grasses. *Plant Physiol.* **2009**, *149*, 88–95. [CrossRef]

20. Yoshida, T.; Fujita, Y.; Maruyama, K.; Mogami, J.; Todaka, D.; Shinozaki, K.; Yamaguchi-Shinozaki, K. Four Arabidopsis AREB/ABF transcription factors function predominantly in gene expression downstream of SnRK2 kinases in abscisic acid signaling in response to osmotic stress. *Plant Cell Environ.* **2015**, *38*, 35–49. [CrossRef]

21. Fujita, Y.; Fujita, M.; Shinozaki, K.; Yamaguchi-Shinozaki, K. ABA mediated transcriptional regulation in response to osmotic stress in plants. *J. Plant Res.* **2011**, *124*, 509–525. [CrossRef] [PubMed]

22. Chen, K.; Li, G.; Bressan, R.A.; Song, C.; Zhu, J.; Zhao, Y. Abscisic acid dynamics, signaling and functions in plants. *J. Integr. Plant Biol.* **2020**, *62*, 25–54. [CrossRef] [PubMed]

23. Yoshida, T.; Fujita, Y.; Sayama, H.; Kidokoro, S.; Maruyama, K.; Mizoi, J.; Shinozaki, K.; Yamaguchi-Shinozaki, K. AREB1, AREB2, and ABF3 are master transcription factors that cooperatively regulate ABRE-dependent ABA signaling involved in drought stress tolerance and require ABA for full activation. *Plant J.* **2010**, *61*, 672–685. [CrossRef] [PubMed]

24. Singh, D.; Laxmi, A. Transcriptional regulation of drought response: A tortuous network of transcriptional factors. *Front. Plant Sci.* **2015**, *6*, 895. [CrossRef]

25. Lu, Q.; Tang, X.; Tian, G.; Wang, F.; Liu, K.; Nguyen, V.; Kohalmi, S.E.; Keller, W.A.; Tsang, E.W.; Harada, J.J.; et al. Arabidopsis homolog of the yeast TREX2 mRNA export complex: Components and anchoring nucleoporin. *Plant J.* **2010**, *61*, 259–270. [CrossRef]

International Journal of
Molecular Sciences

Review

An Overview of Hazardous Impacts of Soil Salinity in Crops, Tolerance Mechanisms, and Amelioration through Selenium Supplementation

Muhammad Kamran [1,†]**, Aasma Parveen** [2,†]**, Sunny Ahmar** [3,*]**, Zaffar Malik** [2]**, Sajid Hussain** [4]**, Muhammad Sohaib Chattha** [3]**, Muhammad Hamzah Saleem** [3]**, Muhammad Adil** [5]**, Parviz Heidari** [6] **and Jen-Tsung Chen** [7,*]

[1]   Key Laboratory of Arable Land Conservation, Ministry of Agriculture, College of Resources and Environment, Huazhong Agricultural University, Wuhan 430070, China; kamiagrarian763@gmail.com

[2]   Department of Soil Science, University College of Agriculture and Environmental Sciences, The Islamia University of Bahawalpur, Bahawalpur 63100, Punjab, Pakistan; aasmaparveen452@gmail.com (A.P.); zafar_agrarian@yahoo.com (Z.M.)

[3]   College of Plant Science and Technology, Huazhong Agricultural University, Wuhan 430070, China; sohaibchattha08@gmail.com (M.S.C.); saleemhamza312@webmail.hzau.edu.cn (M.H.S.)

[4]   Stat Key Laboratory of Rice Biology, China National Rice Research Institute, Hangzhou 310006, China; hussainsajid@caas.cn

[5]   College of Urban and Environmental Sciences, Northwest University, Xi'an 710127, China; adilmuhammad@stumail.nwu.edu.cn

[6]   Department of Agronomy and Plant Breeding, Faculty of Agriculture, Shahrood University of Technology, Shahrood 3619995161, Iran; heidarip@shahroodut.ac.ir

[7]   Department of Life Sciences, National University of Kaohsiung, Kaohsiung 811, Taiwan

*   Correspondence: sunny.ahmar@yahoo.com (S.A.); jentsung@nuk.edu.tw (J.-T.C.)

†   These authors contributed equally to this work.

Received: 26 October 2019; Accepted: 20 December 2019; Published: 24 December 2019

Abstract: Soil salinization is one of the major environmental stressors hampering the growth and yield of crops all over the world. A wide spectrum of physiological and biochemical alterations of plants are induced by salinity, which causes lowered water potential in the soil solution, ionic disequilibrium, specific ion effects, and a higher accumulation of reactive oxygen species (ROS). For many years, numerous investigations have been made into salinity stresses and attempts to minimize the losses of plant productivity, including the effects of phytohormones, osmoprotectants, antioxidants, polyamines, and trace elements. One of the protectants, selenium (Se), has been found to be effective in improving growth and inducing tolerance against excessive soil salinity. However, the in-depth mechanisms of Se-induced salinity tolerance are still unclear. This review refines the knowledge involved in Se-mediated improvements of plant growth when subjected to salinity and suggests future perspectives as well as several research limitations in this field.

Keywords: salinity; selenium (Se); crops; reactive oxygen species (ROS); enzymatic anti-oxidative system

---

1. Introduction

Various abiotic stresses, such as drought, heat, heavy metals, soil salinity, flooding, and cold, are responsible for the reduction of the growth, development, and productivity of crops worldwide [1]. Soil salinity is an overwhelming environmental threat to world food production and agricultural sustainability [2]. A soil with an electrical conductivity (EC) of saturated soil paste extract (ECe) in the plant root zone more significant than 4 dSm^{-1} (about 40 mM NaCl), 0.2 MPa osmotic stress [3] and an exchangeable sodium percentage (ESP) of 15% at 25 °C is termed as salt-affected soil [4]. Some of the most discussed reasons for soil salinity are poor soil-sustainable practices, excessive saline water irrigation and a severe usage of mineral fertilizers in arid and semi-arid regions (characterized by high evapotranspiration, high temperature, and low rainfall) across the globe [5]. The area under soil salinity is further enhanced by the conversion of fertile agricultural land into urban area, placing the efforts of scientists to produce 70% more food to feed the population of the world in 2050 of 9.3 billion at risk [6]. In 2001, almost 7% of the soils of the entire world were salt-affected in nature [7]. Globally, salinity is a significant abiotic stress, affecting one-quarter to one-third of the crop productivity of agricultural soils [8]. It was estimated in 2003 that up to the middle of the 21st century, the salinity-induced loss of cultivated soil will reach up to 50% [9]. In 2008, it was reported that, due to high salinization, 77 million hectares of the world's total cultivated area (1.5 billion hectares) was adversely affected [10]. At present, about 10% of the global land area and 50% of irrigated areas are exposed to salinity, causing a loss of about 12 billion US$ in the agricultural sector [11].

Soil salinity is a complex mechanism that is responsible for adverse effects on the physiological and biochemical pathways of crop plants [12]. Excess accumulation of Na^+ induces efflux of cytosolic K^+ and Ca^{2+}, consequently, leading to imbalance in their cellular homeostasis, nutrient deficiency, oxidative stress, retarded growth, and cell death [13]. It has been reported in many previous studies that a high level of salinization drastically affects plant photosynthesis due to some stomatal restrictions; for example, stomatal closure [4] and/or non-stomatal restrictions comprising chlorophyll malfunctioning [14], deprivation of enzymatic proteins and membranes of photosynthetic apparatus [15], and chloroplast ultrastructure destruction [16]. Salt-affected soils have higher Na^+/K^+ and Na^+/Ca^{2+} ratios because of the higher amount of Na^+ in the soil solution. Hence, a reduction in K^+ and Ca^{+2} uptakes cause the inhibition of the proper functioning of the cell, instability of cell membranes, and hindrance of enzymatic activities [17]. Moreover, some other secondary stresses, such as oxidative stress followed by osmotic pressure and ionic toxicity, are involved in the production of excessive reactive oxygen species (ROS) in cytosol, chloroplast, and mitochondria [2,4] such as O_2^- (superoxide radicals), H_2O_2 (hydrogen peroxide), O_2 (singlet oxygen) and OH^- (hydroxyl ions). These reactive oxygen species with strong oxidation ability can cause injuries to plant tissue, DNA mutation, cell membrane disruption [18], and the degradation of lipids, proteins, and photosynthetic pigments [19] (Figure 1).

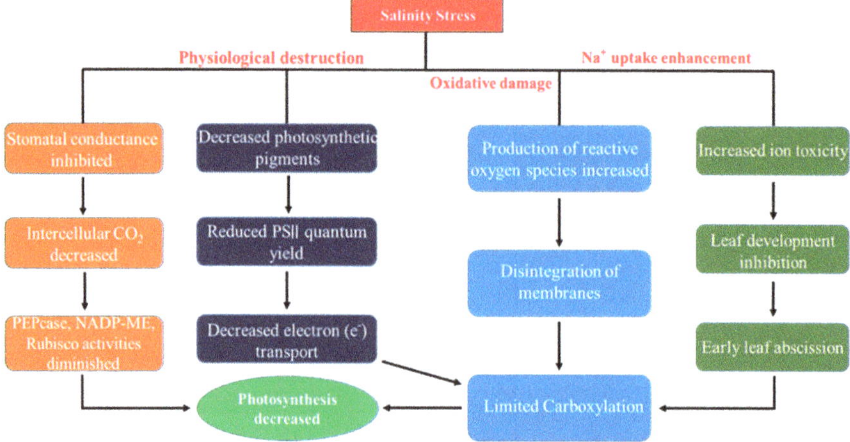

Figure 1. Schematic diagram interpreting the hazardous impacts of soil salinity stress in crop plants. The figure is briefly modified from the literature [20].

The application of macro- and micro-nutrients is one of the management approaches for coping with environmental stresses such as soil salinity [21]. Selenium (Se) has been considered as a beneficial element for crop production which plays an important role in physio-biochemical processes [22,23]. Although higher plants do not require Se for their growth and development [24,25], supplementation of Se at lower dosages not only protects plants from ROS induced oxidative damage by activating the antioxidative mechanisms [22], but also improves the Se content in the edible parts of the crop plants [26]. Some studies have shown that Se is an essential element for human and animal, which plays some beneficial roles in higher plants. Selenium application caused an increasing growth in rice (*Oryza sativa* L.) [27] and wheat (*Triticum aestivum* L.) [28], under both stressed and non-stressed conditions. Se has been demonstrated to regulate plant growth by strengthening the stress tolerance mechanisms such as antioxidant and secondary metabolite metabolism [29]. It has also been reported previously that Se reversed the negative impacts of soil salinity on the photochemical efficiency of photosystem II [30]. Moreover, Se could also protect the metabolism and cellular functioning by up-regulating the ROS neutralizing pathways and the osmoregulatory mechanisms [28].

Although several excellent investigations have been done on Se induced salinity tolerance mechanisms in various crops, there is no comprehensive review on Se-mediated improvements in crops. In this review article, the role of Se in the improvement of common morpho-physiological and molecular responses of various crop plants subjected to salt stress are briefly discussed and some practical options have been proposed on how Se could play its role to induce salinity tolerance in crops.

2. Hazardous Impacts of Soil Salinity in Crops

Salinity stress is exceptional among all the abiotic stresses limiting crop yield efficiency in arid and semi-arid zones where natural conditions favor salinization due to insufficient precipitation for the leaching of salts [31]. According to the biphasic model of growth reduction via salinization [32], the detrimental impacts of salt-affected soils are coupled with a reduction of osmosis (primary phase) and ion cytotoxicity (secondary phase), in addition to the production of reactive oxygen species (ROS) and nutrient imbalance [4]. A high osmotic stress is linked with the accumulation of soluble salts in soil solution, leading to water stress due to a reduction in the stomatal aperture, which eventually hampers plant growth [33]. Ion cytotoxicity is the effect of the substitution of K^+ and Ca^{+2} by Na^+ and Cl^- in different biochemical reactions due to a higher salt concentration in the root zone of crop plants [34,35] (Figure 1).

2.1. Impacts of Salinity on Plant Agronomic Traits

Soil salinity is known for its adverse effects on plant growth and development [36]. However, the inhibitory effect of salt stress depends on various factors such as salt concentration, time interval, plant species and varieties, photochemical quenching capacity, plant growth stages, stress type, gas exchange characteristics, photosynthetic pigments, and environmental conditions [21]. It was concluded in various studies on *Zea mays* L. [37], *Oryza sativa* L. seedlings [38], *Vigna unguiculata* L. [39], *Brassica campestris* L. [40], and *Vicia faba* L. [41] that a low level of salinization increased plant length. However, higher concentrations of sodium chloride salt reduced the plant height of *Vigna mungo* L., [42], *Helianthus annuus* L. [43], and *Tanacetum parthenium* L. [44]. The increment in plant height was might be an effect of an adjustment of osmotic activity due to fewer soluble salts in the growth medium, while plant height reduction was an indication of adverse effects of excessive salts on the photosynthetic rate, a decreased level of carbohydrates and growth hormones (causing growth inhibition) and a reduction in protein synthesis by changing antioxidant enzyme activities [45].

Various studies revealed that the plant biomass (fresh and dry biomass), number of leaves and leaf area were drastically affected by salinity levels up to 8 dSm^{-1} [44,46,47]. In the context of plant growth, it has been reported by many researchers that dry matter production and plant growth retardation under salt-affected soils could be subjected to the inhibition of cell elongation [21] through the direct impairment of the activities of transport proteins such as H^+-ATPase and H^+-PPase [48]. Another reason for plant growth reduction could be the detrimental effects of salinity stress on photosynthesis, ultimately limiting plant and leaf growth and chlorophyll contents [49]. Furthermore, the fresh and dry biomass of *Brassica napus* L. cv. Talaye was significantly decreased, while root growth was less affected compared to shoot growth under salinity stress [47]. It was hypothesized that, under salinization, a low water uptake efficiency leads to lesser leaf area development than root growth, due to which soil moisture is conserved to prevent the accumulation of the vast amount of soluble salts in the soil [4,47]. Several studies have revealed that a high accumulation of Na^+ and Cl^- ions in cell sap excites a low osmotic gradient in the nutrient medium, resulting in reduced water uptake, which in turn affects plant morphological characteristics [50]. It has been documented that high salt density is responsible for lower N accretion in plants due to the interaction between Cl^- and NO_3^- and between Na^+ and NH_4^+, which subsequently reduces plant growth and crop yield [51]. Another mechanism behind the reduction of plant growth under saline conditions might be the reduction in photosynthesis due to the plant stomatal closure and the resulting reduction of carbon uptake [21]. A significant reduction in the absorption of nutrient elements due to reduced osmotic pressure has also been reported as a secondary impact of salinity stress on reduced plant nourishment [52].

2.2. Impacts of Salinity on Physiological Traits

Soil salinization has been recognized as a severe threat to crop growth and yield, even in irrigated areas, worldwide [2]. It is estimated that salinity can reduce crop production in up to 20% of irrigated lands across the globe, and this loss will increase to about 50% of arable land up to mid-21st century [9]. Recently, various studies have reported that soil salinity stress causes reduction in the physiological attributes of cereal crops such as wheat (*Triticum aestivum* L.) [13,28] and mung bean (*Vigna radiata* L.) [53]. Plant growth and yield reduction induced by soil salinity might occur due to the changes in numerous physiological and biochemical attributes, i.e., the reduction of leaf chlorophyll content (Chl a, b, carotenoids) and photosynthesis capacity, as well as the alteration of energy in the mechanisms of ion exclusion, osmotic adjustment, and nutrient imbalance [54]. Mostly, salt-affected soils affect crops in three ways: osmotic stress, ion imbalance, and oxidative damage [55]. The main response of salt-affected soils is the toxic effects of sodium (Na^+) and chloride (Cl^-) ion accumulation in plant tissues [55,56]. It has been proven that plants under salinity stress accumulate more Na^+ ions, resulting in the agitation of ionic balance and plant metabolism and stimulation of oxidative damage, while the K^+ ion status in plant tissues helps plants develop tolerance towards soil salinity [9]. Rice (*Oryza sativa* L.) grown in salt-affected soil slightly impacted the K^+ ion contents; however, it enhanced the Na^+ contents in leaves and significantly lowered the K^+/Na^+ ratio [56,57]. Furthermore, a significant reduction was reported in the growth of strawberry plants [58]. These growth retardations could partially be attributed to reduced photosynthetic activity due to decreased *Chl a* and *Chl b* under various salinity levels [59]. The entrance of Na^+ and Cl^- ions into the plant cell causes ion imbalance in plant and soil, and this ion imbalance in the plant might cause crucial physiological problems [60]. A high concentration of salts in the soil profile may cause physiological drought due to the reduction in water uptake and salt accumulation in the plant's root zone [54], a decrease of plant osmotic potential, and thereby, the disturbance of cell metabolic functions due to ion toxicity [33,60]. Excess Na^+ in plants harms the cell membrane and organelles of the plant, resulting in a reduction in plant physiological mechanisms such as the net photosynthesis rate (*Pn*), stomatal conductance (*Gs*), transpiration rate (*Tr*), intracellular carbon dioxide (*Ci*), and soil plant analysis development (SPAD) value, which lead to plant cell death [56,61,62]. In addition, these physiological changes in the plant might include the disruption of the cell membrane, leading to an inability to detoxify the reactive oxygen species (ROS) in the cytoplasm, a reduced photosynthetic rate and transformations of the antioxidant enzymes [62]. These oxidative systems can interrupt the routine functions of various plant cellular components such as proteins, DNA, and lipids, interfering with dynamic cellular functions in plants under abiotic stress, especially soil salinity [63]. Furthermore, plants grown in a saline environment might inhibit chlorophyll formation and trigger various modifications in the functions and structure of the pigment protein complex [64]. The inhibition of chlorophyll pigment synthesis under salt stress might be attributed to the declined activity of various enzymes, i.e., porphyrinogen IX oxidase, porphobilinogen deaminase, coproporphyrinogen III oxidase, 5-aminolevulinic acid dehydratase, protochlorophyllide oxidoreductase, and Mg chelatase [65]. Theses enzymes in turn are responsible for the upgradation of chlorophyllase activity [66] or a reduction in leaf water potential, N uptake, and thereby, the reduced photosynthetic capacity of plants [53]. Chlorophyll degradation might also be carried out by salinity-induced superoxide radicals and H_2O_2, which degrade the membranes of thylakoids and chloroplast [27].

2.3. Impacts of Salinity on Enzymatic and Non-Enzymatic Antioxidants

Soil salinity stress is accompanied with a robust accumulation of ROS and hampers plant growth and development. Under stressful circumstances (biotic and abiotic), reactive oxygen species (ROS): (O_2, O^{2-}, H_2O_2, and OH^-) production is a stress indicator at a cellular level and is known as a secondary messenger which plays its role in the biological activities of plants, ranging from gene expression and translocation to enzymatic chemistry [67,68]. Ultimately, these ROS might cause alterations in the structures of lipids, proteins and nucleic acids, and thereby, cause an interruption of the normal plant metabolism [69]. It has been reported that soil salinity-stimulated oxidative stress due to the accretion of higher levels of H_2O_2 might induce apoptosis, cell shrinkage, chromatin condensation, and DNA fragmentation [70]. Under salinity stress, higher levels of ROS production might result in the production of malondialdehyde contents (MDA) in the thylakoid membranes. MDA concentration, which is known to be an effective indicator of lipid peroxidation, helps to calculate the lipid peroxidation of plant cells [71]. The balance between ROS production and their elimination by the antioxidative defense mechanism defines the degree of collateral damage to these molecules involved in plant metabolism [72]. Moreover, soil salinization causes acute oxidative damage in the plant tissues, and as a result, plants develop their own complex natural antioxidant defense system to combat with the salinity-induced oxidative stress [73]. The antioxidant enzymes inhibit the cell structural damages caused by salinity-induced ROS [74]. In the presence of an efficient antioxidant system in crop plants, it is believed that salt tolerance is better than for other types of plants. Previously, various researchers have reported the differential impacts of salinity stress on antioxidative enzymatic and non-enzymatic activities in *Tanacetum parthenium* L. [44], *Brassica napus* L. [47], *Oryza sativa* L. [75], and *Glycine max* L. [76]. The non-enzymatic antioxidative system mainly includes carotenoids, ascorbic acid (vitamin C), α-tocopherol, and flavonoids, while the enzymatic antioxidative system includes peroxidase (POD), superoxide dismutase (SOD), ascorbate peroxidase (APX), glutathione reductase (GR), polyphenol oxidase (PPO), etc. The major role of the enzymatic antioxidative system is to scavenge the injurious radicals produced during oxidative stress and thus help the crop plants to survive under abiotic stress such as soil salinity [67,77]. There are some natural antioxidants in almost all parts of the plant. These natural antioxidants are vitamins, carotenoids, phenols, dietary glutathione, flavonoids, and endogenous metabolites [78]. In salt-affected soils, the production and scavenging of these antioxidants makes up the first line of defense in plants to handle the oxidative stress.

3. Salinity Tolerance Mechanisms Adopted by Crop Plants

Plants have developed various adaptations at cellular, subcellular and organ levels for their nourishment under salt-affected soils. Some important salt resistance mechanisms are ion homeostasis, stomatal regulation, ion compartmentalization, osmoregulation/osmotic adjustment, hormonal balance changes, stimulation of the antioxidative defense mechanism, and the accumulation/exclusion of toxic ions from cells and tissues. However, all these salt-tolerant mechanisms are complex and vary from specie to specie [4]. According to biomass production under soil salinization, four plant groups are differentiated: (1) true halophytes (*Sued* sp. and *Atriplex* sp.), which can invigorate biomass production under salt stress; (2) optional halophytes (*Plantago maritima* and *Aster trripolium*), which show a minor increase in biomass at lower salt concentration; (3) nonresistant halophytes (*Hordeum* sp.), which can tolerate lower salt concentrations; and (4) glycophytes/halophytes (*Phaseolus vulgaris*), which are much more sensitive to salinization [79,80].

It has been suggested in many studies that salt tolerance is linked with the sequestration of Na^+ ions into vacuoles after their entry into leaf cells to maintain a low Na^+ concentration in the cytosol. This sodium and chloride ion compartmentalization phenomenon is carried out by proton gradient drove tonoplast Na^+/H^+ antiporters [46]. Once excess Na^+ and Cl^{-1} are vacuolated, this significantly lowers the osmotic potential without any change in the metabolic process rate and ultimately contributes to osmoregulation [57]. Many experiments have emphasized this strategy, where the overexpression of vacuolar Na^+/H^+ antiporter gene (*NHX1*) family has enhanced the salinity tolerance of tomato (*Solanum lycopersicum* L.) [81], rice (*Oryza sativa* L.) [82], and maize (*Zea mays* L.) [29]. More recently, a novel virus-induced gene silencing (VIGS) method has been applied to study the function of GhBI-1 gene in cotton regarding the salt-stress response [83].

Excessive Na^+ ion accumulation in plants is highly toxic because of its ability to interact with K^+ ions, causing disturbed stomatal regulation. Therefore, the maintenance of a higher K^+/Na^+ ratio is an essential strategy for salt resistance in plants [2,6]. Two essential findings support this strategy: 1) the presence of *CED-9* gene enhanced salinity tolerance in tobacco by accumulating K^+ ions [6] under salt-stressed conditions—more potassium is retained in the cell cytoplasm by caspase activity, i.e., proteases and endonucleases, [84]. Moreover, in salt-affected soils, the transfer of sodium and chloride ions in stem and leaf sheaths is another adaptation of crop plants to reduce the accumulation of these ions in more vulnerable leaf blades [85]. More precisely, it has been concluded that the K^+/Na^+ ratio in the cytosol can be retained by K^+ absorption maintenance, the reduction of K^+ efflux from cells, the prevention of Na^+ uptake, and the enhancement of Na^+ efflux from cells [86].

Generally, under stressful conditions, plant growth is also regulated by the synthesis of several phytohormones, such as jasmonic acid, salicylic acid, auxins, gibberellins and cytokinins (growth promoters) [87,88], ethylene, and abscisic acid (growth retardants). It has been reported that soil salinity enhanced the abscisic acid level in *Zea mays* L. at the expense of auxins (IAA) [89]. This modification may lead to the closing of stomata to reduce water loss as a consequence of osmotic stress under salinization. Methyl jasmonate, a natural plant growth regulator, can ameliorate the inhibitory effects of soil salinization on the photosynthetic rate to improve plant growth and development [90].

Another crucial physiological trait of salinity tolerance is the accumulation of organic compounds such as certain amino acids (proline, proline betaine, glycine betaine, and β-alanine betaine) and soluble sugars (fructose, glucose, fructans, raffinose, and trehalose). The accumulation of these compounds is positively correlated with salinity tolerance in *Zea mays* L. [91], *Pistacia vera* L. seedlings [46], and *Tanacetum parthenium* L. [44]. These compounds allow the maintenance of the turgor potential by decreasing the osmotic potential and minimizing the deleterious effects of Na^+ ions against ribosomes and proteins. Recently, the exogenous application of different amino acids, proline, and glycine betaine was also considered as an ameliorative strategy for soil salinity [75,92–94].

A variety of adaptive mechanisms at the molecular level are involved in overcoming the harmful effects of salinity-induced oxidative stress. Some of the most important are the up and downregulation of gene transcripts [29,95,96], changes of chemical composition and the rigidness of plant's cell wall [97]. It was reported that the expression of antioxidant defense genes is stimulated in *Zea mays* L. shoots [98], while Rodríguez-Kessler found that two genes, *Zmodc* and *Zmspds2A*, are responsible for salinity tolerance in maize roots through the accumulation of polyamine and spermidine [99].

4. Role of Selenium under Abiotic Stresses

Selenium (Se) has already been proven to be beneficial for humans and animals. However, Se is considered to be a double-edged sword due to its dual response to plants (beneficial or toxic) depending on its concentration and the nature of plant species [100]. Se is available in many forms to plants, such as selenate (Se, VI), selenite (IV), thioselenate, selenide, and elemental Se [101]. The optimum level of Se plays a crucial role in human and animal metabolism, e.g., a low concentration of Se in the diet is essential for antioxidant production and a healthy life and is recommended in many countries of the world. Thus, the effects of Se on humans and animals are linked with Se in the soil–plant system, because Se contents in edible parts of the plant come from the soil and are consumed by other organisms.

Under low Se levels, it acts as an important protectant in plants grown under different abiotic and abiotic stresses. Selenium causes the disputation of ROS and protects plants from toxic elements-induced oxidative stress. At high levels, Se acts as a pro-oxidant as with other heavy metals/metalloids and enhances the production of ROS, causes protein oxidation, lipid peroxidation, and genotoxicity [102]. Selenium shows a hermetic effect in plants, but the mechanisms, as well as the optimal, essential, and toxic values of Se in the soil, are not well-established for different plant species and soil types. The essentiality of Se in plants depends on the plant species and Se concentration. For example, a hyper-accumulator species of *Brassica* species (*Helianthus*, *Camelina*, and *Aster*) could accumulate Se up to 100–1000 mg·kg^{-1} DW without showing toxicity symptoms. On the other hand, non-hyper-accumulator species of food crops, grasses, and vegetables hardly accumulate 100 mg·kg^{-1} DW of Se in plant tissues [103]. However, the Se response to salinity stress is not very clear and needs to be explored (Figure 2).

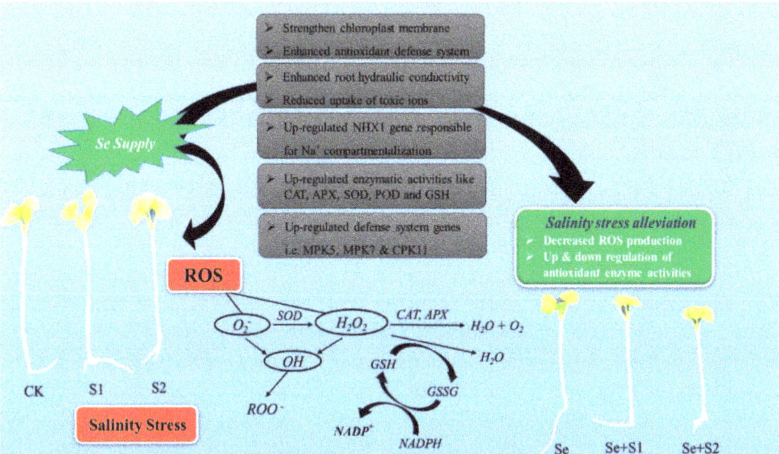

Figure 2. Schematic presentation showing the possible causes that overproduce reactive oxygen species (ROS), which might disturb the normal function of plant cells. The mechanism of antioxidants shown here scavenges the ROS effects as well as ameliorative effects of Se to induce salinity tolerance in crop plants. Se represents "selenium" (25 μM Na_2SeO_4) and S1 and S2 represent salinity stress (100 and 200 mM NaCl), respectively. The seedlings are representative of *Brassica napus* L. (Source: [24]). POD: peroxidase; SOD: superoxide dismutase; APX: ascorbate peroxidase; GSH: reduced glutathione; GSSG: oxidized glutathione; H_2O_2: hydrogen peroxide; NADP$^+$: nicotinamide adenine dinucleotide phosphate; MPK: mitogen activated protein kinase gene; CPK: calcium dependent protein kinase gene; NHX: sodium/hydrogen (Na$^+$/H$^+$) exchanger gene.

4.1. Selenium Speciation and Mobility in Soil

Selenium (Se) is present in excess in the Earth's crust and can be beneficial or toxic to plants depending on the concentration of Se, speciation, and nature of plant species. Se occurs in organic and inorganic forms in soil with different oxidation states (+6, +4, 0, and −2) for selenate, selenite, elemental Se, and selenides, respectively. The most mobile and water-soluble inorganic Se is selenate (SeO_4^{2-}), which is present abundantly under oxic soil conditions with low adsorption affinity to oxide surfaces [104]. Selenate could be reduced into selenite due to poor adsorption ability onto the oxide surface under poor redox potential [102]. It has been demonstrated that selenite (SeO_3^2, $HSeO_3^-$, H_2SeO_3) might be the most abundant inorganic Se speciation under an anaerobic soil environment (pH: 7.5–15) [105]. At low pH, selenite has a greater ability to be adsorbed on an oxide surface than selenate and thus has reduced bioavailability to crop plants [104]. Selenite could be reduced into elemental Se, Se^0, or selenides, Se^{2-} (unavailable to plants), under strong reducing conditions [102]. Various factors which are responsible for Se mobility and solubility in soil are soil pH, sorption, and desorption reactions, redox potential, organic/inorganic compounds, and dissolution processes in sediments and soils [106].

Soil Se is mainly inorganic but it can also be present in organic forms, such as complexes with organic matter, and incorporated into organic or organo-mineral colloids [107]. Se in organo-Se compounds (e.g., seleno-aminoacids) presents a valence state of −2 and is highly bioavailable. In addition, volatile organic forms of Se such as dimethyl selenide (DMSe) and dimethyl diselenide (DMDSe) may be present in soils. Se accumulation in plants is higher when seleno-amino acids are added to the hydroponic growth medium compared with inorganic forms of Se at the same concentration [108]. Organo-selenium compounds can either be released into the soil from biological decompositions of plant and soil microbial tissues or by Se-based fertilizer addition. Soil organic matter (OM) is shown to influence the retention of Se in [109]; however, the mechanisms of Se–OM interactions are poorly understood. Basically, three hypotheses explaining the OM-mediated retention of Se are generally discussed: (i) OM has increased sorption sites, which facilitates direct complexation with Se [109,110]; (ii) indirect complexation via OM–metal complexes [109]; (iii) microbial reduction and incorporation into amino acids, proteins, and natural organic matter [110]. Depending on the type of binding, Se may be easily mobilized (e.g., through pH adjustment) or immobilized (e.g., covalent incorporation to OM).

4.2. Selenium Uptake and Mobility within the Plants

The Se toxicity or deficiency margin is very small. This small gap between toxicity and essentiality is based on the nature of the organism and Se speciation [100]. It has been reported that a low-Se diet is important for antioxidant protection and a healthy life [111]. Therefore, threshold levels of Se have been added to the nutritional recommendation in various parts of the world such as China (essentiality: > 0.125 mg kg^{-1}; toxicity: > 3 mg kg^{-1}) [112]. Se deficiency or Se excess due to the intake of low or high-Se containing food may cause many health problems in living organisms [102]. Therefore, it is essential to understand and monitor the behavior of Se in the soil–plant system.

The majority of crop plants are able to uptake various inorganic forms such as selenite (+4), selenate (+6) [104], and/or various selenium based organic compounds such as SeCys (methylselenocysteine) and SeMet (selenomethionine) [105]. In contrast, plants are incapable to uptake elemental Se (0), selenide (−2) from the root zone. Even though Se is not an essential element for plants, it plays many significant roles in the plant, which depends on its applied concentration in the growth medium. Lower Se concentrations play a beneficial role and improve plant growth, whereas higher Se concentrations disturb the metabolic processes of the plant and reduce plant growth. The pathway of Se accumulation in plant roots is through specific and non-specific channels of essential nutrients (sulfur and phosphate), whereas the xylem channels and sinks transport *Se (VI)* into the shoot tissues within plants. Previously, it has been reported that phosphate transporter families (Pht1 and NIP2;1 transporter) are used to take up Se by root cells such as $HSeO_3^-$ and H_2SeO_3 (selenite) using aquaporins [113]. Afterwards, these Se speciations are translocated from root cells to the plant shoot as selenate via the root symplast and stele. During this whole process, selenite is persuaded into Se-based organic compounds, which stay behind in the plant roots [114,115]. Therefore, selenate and small amounts of SeMet and selenomethionine Se-oxide (SeOMet) have been considered important Se species in the plant xylem [116]. The family of aluminum-activated malate transporter (ALMT) genes are thought to be responsible for carrying selenate in the shoot xylem sap [117], whereas, following the delivery of selenate from root to shoot via the xylem, the members of the Sulfate transporters (SULTRs) family take it to leaf cells [118], where it is stored in the cell vacuoles [114].

In addition to inorganic Se, plant uptake of organic Se is known to occur and has been reported at much higher rates (20–100 fold greater) than the uptake of inorganic species [108]. Evidence suggests that amino acid transporters are important. To date, no Se-specific uptake mechanisms have been reported [119]. However, SeMet (selenomethionine), SeMeSeCys (Se-methyl selenocysteine) and SeCys (methylselenocysteine) forms of Se are taken up by the plant roots through transporters with the ability to catalyze the uptake of Met and Cys, respectively [120]. A synchrotron-based X-ray fluorescence microtomographic analysis was performed to demonstrate the transport mechanisms of organic species of Se. The authors observed that organic Se (SeMet and SeMeSeCys) was translocated in *Oryza sativa* L. exclusively via the phloem. The results indicated that, for SeMeSeCys- and SeMet-fed grain, Se was distributed throughout the external grain layers and into the endosperm, while SeMeSeCys Se was partitioned into the embryo. They demonstrated that organic Se species (SeMeSeCys and SeMet) are rapidly loaded into the phloem and transported to grain more efficiently than inorganic species [121].

5. Selenium-Mediated Alleviation of Salinity Stress in Plants

The findings to date have shown that Se is not ranked as an essential element for crop plants; however, a low Se concentration exerts beneficial effects on plant growth and development under biotic and abiotic stresses, especially soil salinization (Figure 2). Many studies have reported the effects of the application of Se to evoke tolerance against salt stress depending on the application method, dose of Se, salinity levels, and plant species [58]. For example, a foliar application of selenate (20 mg·L^{-1}) mitigated the adverse effects of salinity stress (12 dS m^{-1}) on the growth and development parameters of maize (*Zea mays* L.) [122]. Likewise, another study reported that Se application (20 µM) in the form of sodium selenite causes improvements in the growth and yield of eggplants under varying levels of soil salinity [123]. However, higher doses of selenite were found to have deleterious effects on the growth and development stages of maize under a salt stress of 100 mM NaCl [29]. Even though Se is an essential trace nutrient to humans and other animals as an antioxidant, Se toxicity might appear at higher concentrations due to the substitution of S with Se in the structure of amino acids, followed by the inaccurate folding of proteins and thus the creation of nonfunctional proteins and enzymes [102]. Conclusively, higher doses of Se hamper the growth and development of crop plants, while low doses cause improvements in growth and development mechanisms.

5.1. Improvement in Agronomic Traits

The maintenance of plant growth is directly associated with the survival of crop plants under salt-affected soils. The application of minute levels of Se under salinity stress significantly improved plant growth characteristics such as the shoot length, shoot diameter, and fresh and dry biomass of cucumber (*Cucumis sativus* L.), lemon balm (*Melissa officinalis* L.), cowpea (*Vigna unguiculata* L.), wheat (*Triticum aestivum* L.), and maize (*Zea mays* L.) as compared to salt stress alone [122,124–126] (Table 1). Likewise, Se showed a great potential to improve stem growth (diameter and biomass) in melon (*Cucumis melo* L.) and tomato (*Solanum lycopersicum* L.) when cultivated in salt-affected soils [30,106]. Recently, Astaneh suggested that growth parameters such as the bulb height, fresh and dry biomass of bulbs, bulb diameter, and the number of cloves in one bulb of Garlic (*Allium Sativum* L.) were significantly improved with the addition of Se under salinity stress [127]. Growth characteristics related to plant roots such as length, fresh, and dry weight were significantly improved with the supplementation of smaller amounts of Se alone and/or in combination with NaCl, compared to salinity stress alone [30,122]. Se applications significantly promoted root and shoot fresh weight and shoot dry weight as well as improving relative water contents in tomato (*Solanum lycopersicum* L.) and antioxidants activity and photosynthetic pigments in lettuce plants [100,128]. In addition, added Se also improved the growth parameters of ryegrass (*Lolium perenne* L.) and spinach (*Spinacia oleracea* L.) by improving nutritive values [106].

Table 1. Protective effects and mechanisms of Se supplementation on growth, physiological, and biochemical attributes of plants grown under salinity stress.

Salinity Stress	Plant Species	Se Dosages	Se Speciation	Experimental Details	Various Protective Effects and Mechanisms of Se in Salinity Stressed Plants	References
150 mM	*Oryza sativa* L.	2, 4, 6, 8, 10, 12 mg·L^{-1}	*Se (VI)*	Sand culture	Enhances plant biomass, K$^+$/Na$^+$ ratio, and Se accumulation; reduces malondialdehyde contents (MDA) and H$_2$O$_2$ contents; increases chlorophyll and water contents; causes upregulation of *OsNHX1* gene transcript levels	[27]
0, 30, 60, 90 mM	*Allium sativum* L.	0, 4, 8, 16 mg·L^{-1}	*Se (VI)*	Hydroponic culture	Increases root biomass, bulb diameter, bulb height, and photosynthetic pigments; reduces ion leakage and lipid peroxidation; improves K$^+$ and Na$^+$ contents, chlorophyll index, carotenoids, and water balance	[127,129]
100 mM	*Triticum aestivum* L.	5, 10 μM	*Se (VI)*	Reconstituted soil culture (Peat, compost, sand)	Improves wheat growth; promotes the synthesis of photosynthetic pigments, proline, and sugars; reduces H$_2$O$_2$ contents, Na$^+$ uptake, and Na$^+$/K$^+$ ratio	[28]
10, 30, 60, 90 mM	*Stevia rebaudiana* Bertoni	20 g/ha (2 ppt)	*Se (IV)*	Field experiment	Increases leaf and plant biological yields; enhances rebaudioside-A and stevioside of stevia leaves; improved the accumulation of sweet steviol glycosides contents	[21]
0.12, 0.30, 0.60 S m^{-1}	*Triticum aestivum* L.	0, 0.5, 1, 4 mg·kg^{-1}	*Se (IV)*	Pot soil culture	Dramatic decrease in shoot dry biomass; chlorophyll a, chlorophyll b, and carotenoid contents increase at lower Se, while they decrease at higher Se; enhances free proline and Se contents in shoots;	[130]
8 dS m^{-1}	*Allium cepa* L.	0, 0.5, 1 kg·ha^{-1}	*Se (IV)*	Field experiment	Increases bulb yield and dry matter; improves water and chlorophyll contents; causes bulb Se and K enrichment; causes a decrease in Na	[131]
0, 100 mM	*Phaseolus vulgaris* L.	0, 5, 10 μM	*Se (IV)*	Pot soil culture	Enhances plant growth and seed yield; promotes membrane stability index, photosynthetic capacity, and RuBPCase activity; reduces (MDA) and electrolyte leakage	[132]
0, 100 mM	*Zea mays* L.	0, 1, 5, 25 μM	*Se (IV)*	Pot vermiculite culture	Enhances growth and biomass; improves gas exchange attributes and the shape of thylakoids by alleviation of damage in the ultrastructure of chloroplasts; upregulates *ZmMPK5*, *ZmMPK7*, *ZmCPK11*, and *ZmNHX1* genes transcript levels in roots	[29]
0, 80 mM	*Petroselinum crispum* L.	1 mg·L^{-1}	*Se (VI)*	Hydroponic culture	Decreases root to shoot transport of Na$^+$; improves photochemical efficiency of photosystem II (PSII) and chlorophyll contents; protects photosynthetic apparatus by upregulation of non-photochemical quenching (*NPQ*); decreases cell sap Na$^+$	[133]

Table 1. *Cont.*

Salinity Stress	Plant Species	Se Dosages	Se Speciation	Experimental Details	Various Protective Effects and Mechanisms of Se in Salinity Stressed Plants	References
3.22 dS m^{-1}	*Lactuca sativa* L.	16, 32 μM	*Se (VI)*	Field experiment	Improves growth characteristics, yield, and relative water contents; decreases cell membrane permeability and malondialdehyde; enhances chlorophyll, carotenoids, K$^+$/Na$^+$, and total soluble sugars	[134]
0, 25, 50 mM	*Lycopersicon esculentum*-Mill.	0, 5, 10 μM	*Se (IV)*	Hydroponic culture	Enhances growth by improving water balance and cell membrane integrity; increases photosynthetic pigments; decreases proline and phenolics	[128]
0, 30, 60, 120 mM	*Solanum melongena* L. cv. Baladi	0, 5, 10, 20, 30 μM	*Se (IV)*	Bedding sand culture	Increases vegetative growth, yield, nitrogen, phosphorous and potassium NPK contents in leaves and fruits; improves chlorophyll contents (SPAD value) and proline contents; Enhances K$^+$/Na$^+$ ratio	[135]
0, 40 mM	*Lactuca sativa* L. var. capitate	0, 2, 6 μM	*Se (IV, VI)*	Hydroponic culture	Enhances fresh biomass, leaf area, chlorophyll, proline, and carotenoid contents; reduces H$_2$O$_2$ and *TBARS*; improves shoot ionic concentrations	[100]
0, 40 mM	*Melissa officinalis* L.	10 mM	-	Hydroponic culture	Improves growth rate; increases photosynthetic pigments, protein, and total amino acid contents; reduces lipid peroxidation to alleviate membrane damage	[125]
0, 2000, 4000, 6000 mg L^{-1}	*Brassica napus* L.	0, 2.5, 5, 10 mg·L^{-1}	*Se (VI)*	Pot clay soil culture	Enhances growth, photosynthetic pigments, canola oil quality; increases soluble sugar, polysaccharides, and total carbohydrates; significantly improves saturated and unsaturated fatty acids composition	[136]
0, 2000 ppm	*Cucumis sativus* L. cv Zena	0, 1 ppm	*Se (IV)*	Pot soil culture	Improves plant biomass; increases reduction of oxygen radicals and osmotic regulation by synthesis of osmoregulatory compound such as proline; reduces malondialdehyde concentration and electrolyte leakage	[137]
0, 50 mM	*Cucumis sativus* L.	0, 5, 10, 20 μM	*Se (VI)*	Hydroponic culture	Induces salt tolerance by protection of cell membranes against lipid peroxidation; improves growth rate, photosynthesis, and proline contents; reduces Cl$^-$ contents, while showing no effect on Na$^+$ ions and K$^+$/Na$^+$ ratio	[124]
100 mM	*Rumex patientia* × *R. tianshanicus*	0, 1, 3, 5, 10, 30 μM	*Se (IV)*	Sand culture	Increases seedling growth; lower Se supply improves total water-soluble sugars, K$^+$, and Na$^+$ concentrations; alleviates integrity of cytoplasmic organelles, plasma and nuclear membranes, root tip cells; makes more legible and increases mitochondrial cristae in leaf mesophyll	[138]

The abbreviations are explained in the list of abbreviations.

The accumulation of higher levels of Na$^+$ ions in plant roots under salinity stress causes a reduction in hydraulic conductivity and ultimately lowered relative water contents (RWC); however, the Se (Na$_2$SeO$_4$) supply reduced Na$^+$ ions and improved root growth, and thereby, might have enhanced the water supply to shoots and sustained plant growth [27,139]. Salt-affected soils cause hindrances in nitrogen assimilation, accumulation, and metabolism, and hence, disturb the proline (a molecular chaperone responsible for maintaining protein integrity) biosynthetic mechanism [140,141]. The improvement in the phenological parameters of crop plants could also be a consequence of Se-mediated increments in proline contents through the promotion of nitrogen (N) contents and nitrate reductase activity [53]. Furthermore, Se has been involved in the improvement of nutrient elements absorption and their transfer within the body of various crop plants, which ultimately improves growth and production [142]. It was stated that suitable Se supplementation might be involved in boosting the expression of tonoplast H$^+$ ATPase and Na$^+$/H$^+$ antiport at the root membranes, limiting Na$^+$ ion translocation to the upper plant tissues, thus, decreasing its toxic impacts [143]. Moreover, cations such as nitrogen, potassium, and calcium are required for growth regulation through their impact on the vital metabolic pathways such as antioxidant metabolism, nitrogen assimilation, and cellular stress signaling [72,91,144]. The Se supply has been reported to be beneficial to increasing the nitrogen, potassium, and calcium uptake from soils, thereby, leading to a larger production of amino acids, metabolites, and stress signaling for better induction of salinity tolerance in wheat (*Triticum aestivum* L.) [28]. Another important mechanism is Se-accelerated reduction in the Na$^+$/K$^+$ ratio in plants grown in salt-affected soils, which ultimately induces the protection of some essential processes and balanced osmotic potential [127]. Na$^+$ ions are responsible for inhibiting K$^+$ ion uptake at the membrane transport level, whereas Se might have the ability to influence the expression of Na$^+$ transporters and H$^+$ pumps [145].

5.2. Se-Mediated Improvement in Physiological Attributes

To situate the scientific context compiled in this review article, it should be taken into account that Se at low concentrations helped plants to alleviate exposed stress from its exterior environment, especially regarding soil salinity. Therefore, an exogenous application of Se has gained considerable interest in the scientific community around the world [22,24,94]. For instance, exogenously applied Se played a significant role in appraising the physiological and biochemical mechanisms (Table 1) involved in salinity tolerance in cucumber [124], canola [24], and parsley [133], which as a result helped plants to survive better in salt-stressed environments. Salinity stress in particular not only damages a plant's osmotic potential, but also accompanies various secondary stresses, such as cellular oxidative damage by the over-generation of reactive oxygen species (ROS) [122]. The maintenance of ROS homeostasis and other physiological functions such as photosynthesis are the chief priorities of plants exposed to salinity stress [29]. Therefore, finding suitable approaches to understand and investigate the mechanisms underpinning plant responses to salinity stress is essential to sustain agricultural production in saline soils. In this regard, the application of Se has been found to reduce the harmful effects of salinity and support the growth of maize (*Zea mays* L.), tomato (*Solanum lycopersicum* L.), and garlic (*Allium sativum* L.) through enhanced photosynthetic performance [29,30,122,129]. Moreover, enhanced growth and nutritional qualities of spinach (*Spinacia oleracea* L.), ryegrass (*Lolium perenne* L.), wheat (*Triticum aestivum* L.), and mung bean (*Vigna radiate* L.) have also been reported by exogenously applied Se under stressed and non-stressed conditions [22,106,142,146]. Further, a lower Na$^+$ concentration and higher K$^+$/Na$^+$ ratio was observed in selenite-treated plants as compared to untreated plants [27]. Se might have decreased the accretion of Na$^+$ ions which led to an increased K$^+$/Na$^+$ ratio in comparison to the untreated control plants of dill (*Anethum graveolens*) and garlic (*Allium sativum* L.) [129,147]. The addition of Se under salinity stress significantly improved the physico-biochemical properties such as the chlorophyll contents, carbohydrates, proteins, and carotenoids, of which adequate amounts are essential to regulate major metabolic processes such as photosynthesis in maize (*Zea mays* L.) [148]. The application of Se significantly improved the plant growth, photosynthetic activities such as the net photosynthetic rate, the actual photochemical efficiency of photosystem II (PSII), maximum quantum yield of PSII (*Fv/Fm*), photochemical quenching coefficient (*qP*), and non-photochemical quenching coefficient (*qN*) of tomato (*Solanum lycopersicum* L.) cultivars [30]. Similarly, Se application showed a positive effect on growth and improved the photosynthetic pigments and total amino acid contents in lemon balm (*Melissa officinalis* L.) and decreased Na$^+$, while increasing K$^+$ concentrations in the roots and shoots of dill (*Anethum graveolens*) plants [133,147]. Furthermore, many other researchers have shown that Se application to salt-stressed cucumber and tomato protected the cell membranes against lipid peroxidation, reduced oxidative stress by regulating the chloroplast, which is strongly linked with increasing the photosynthetic rates by improving the PSII, and thereby, enhanced plant stability [30,124]. Taken together, these findings suggest that Se played a significant role in improving the physiological and biochemical adaptation of plants, which eventually helped plants to survive better in stressed saline conditions.

It has been recognized previously that the amelioration of photosynthetic inhibition through Se supply might be a result of the cumulative impact on the antioxidative defense mechanisms, leading to the simultaneous alleviation of ROS effects, uptake and accumulation of important crop nutrients [149]. Recently, it was shown that a higher Se supply (10 μM) causes retardation in the growth and photosynthetic capacity of wheat (*Triticum aestivum* L.) seedlings [28], which might be attributed to decreased chlorophyll formation due to the inhibition of chlorophyll biosynthesizing enzymes and production of 5-aminolevulinic acid and protochlorophyllide [150]. An increment in Mn, Zn, and Fe contents in plant leaves under Se treatment [151] could also be the reason for the improved photosynthetic apparatus and avoidance of the degradation of chlorophyll [152]. Optimal supplementation of Se modulates photosynthetic functioning by enhancing CO_2 assimilation, photosynthetic rate, and chlorophyll fluorescence characteristics under normal and stressful conditions [149]. Moreover, a Se supply regulated proline accumulation by enhancing the activity of γ-glutamyl kinase (γ-GK) enzyme, leading to the enhanced synthesis of proline with subsequent declines in its degradation via the slowing down of the activity of proline oxidase [28,153]. In halophytic grasses, it has been demonstrated that increased accumulation of proline leads to enhanced photosynthetic efficiency and ATP production, resulting in greater water use efficiency [154]. The above discussion reveals that the application of a low concentration of Se could play an important role in the improvement of the physiological and defensive mechanisms of crop plants under salinity stress.

5.3. Se-Mediated Improvement in the Alleviation of ROS Effects

Plants produce an array of antioxidant enzymes once exposed to biotic and abiotic stresses and, interestingly, Se supplementation has been found to upscale these antioxidant enzyme activities to cope with experienced stresses [155]. Se has a significant role in numerous enzymatic processes—i.e., catalase (CAT), peroxidase (POD), superoxide dismutase (SOD), ascorbate peroxidase (APX), and glutathione peroxidase (GPX)—and non-enzymatic processes—i.e., phytochelatins and glutathione antioxidants—which help to combat the salt-induced overproduction of reactive oxygen species (ROS), which are responsible for agitating plant cell integrity (Figure 2). Molecular oxygen (O_2) works as an electron acceptor with a subsequent accretion of reactive oxygen species (ROS) such as singlet oxygen ($^1O_2$), hydroxyl radical (OH^-), superoxide radical (O^{-2}), and hydrogen peroxide (H_2O_2) under salt-stressed conditions. It has been well proven that lower concentrations of selenate (Na_2SeO_4) help to protect plants from ROS-stimulated oxidative damage, but a higher concentration of Se works as a pro-oxidant and stimulates the formation of ROS and induces oxidative stress [92]. Many researchers have described that Se is required to increase the scavenging activity of ROS, decreasing the concentration of MDA and membrane damage [156]. Moreover, decreased generation of H_2O_2 under Se supplementation has also been confirmed [157,158]. Under salinity stress, lowered H_2O_2 contents were observed in Se-treated canola (*Brassica napus* L.) plants [136]. Meanwhile, plants exposed to Se showed lower concentrations of MDA under NaCl stress, showing that Se was vital in bringing down the lipid peroxidation by amending the antioxidant enzymes and protecting the membranous structures of *Oryza sativa* L. [27], *Cucumis sativus* L. [124], *Brassica napus* L. [24], and *Anethum graveolens* [147]. In addition, it was noticed that lipid peroxidation (MDA) production was reduced by elevating Se concentration under salt stress [127]. A comprehensive impact of MDA on plant cells is lowering the fluidity of the membranes to elevate membrane leakiness and avoiding damage to membrane proteins, enzymes, and ion channels [159]. A suitable concentration of Se might be useful to limit the over-expression of lipidoxygenease for sustaining fatty acid formation in addition to the lessened ROS generation, which was led by the upregulation of antioxidant systems [28].

5.4. Se-Mediated Improvement in the Upregulation of Enzymatic and Non-Enzymatic Antioxidants

Under soil salinity stress, ROS can be detoxicated by antioxidant compounds (Figure 2; Table 2). It is believed that enzymatic and non-enzymatic antioxidants, such as SOD, POD, APX, CAT, GSH-Px, and GR, are positively interconnected in response to Se supplementation to induce salinity tolerance in crop plants [22,131]. Researchers postulate that an elevation in the Se-mediated antioxidant defense is one of the vital mechanisms that can save plants from salt-stimulated oxidative stress [58,134]. Antioxidant enzyme activities (SOD, APX, and CAT) significantly improved with exogenous Se treatment in rapeseed (*Brassica napus* L.) and dill (*Anethum graveolens*) seedlings under salinity stress [24,147]. In another study, the accumulation of lowered H_2O_2 contents in rice plants might have been due to Se-mediated higher levels of APX and CAT activities [27]. An increment in the activities of SOD, CAT, GST, APX, and GR has been noticed in different crops such as *Triticum aestivum* L., *Brassica juncea* L., *Avena sativa* L., and *Solanum lycopersicum* L. [13,144,160,161]. Recently, it was noticed that the translocation of minerals such as iron, zinc, and manganese was significantly increased in the shoots of rice (*Oryza sativa* L.) with Se application [151]. These minerals are essential components of antioxidant enzymes and responsible for increasing the activities of SOD, POD and CAT [162]. Under salinity stress, the exogenous supplementation of Se to maize (*Zea maize* L.) plants resulted in the upregulation of expression of mitogen activated protein kinase (*MAPK5* and *MAPK7*) and calcium-dependent protein kinase (*CPK11*) genes and stimulated the antioxidant defense system under salt stress [29,163]. It has been reported that *MAPK* flow is at the center of cell signal transduction and implicated in stress-related signal pathways [164]. Abscisic acid (ABA) accumulation could be stimulated under salinity stress [165], which in turn produces H_2O_2, causing the activation of *MAPK*, resulting in stimulated expression and activities of antioxidant enzymes [166]. Furthermore, NAD kinase-2 (NADK2) mutation impaired ABA-induced stomatal closure and ABA inhibition of light-promoted stomatal opening. NADK2 disruption also impaired the ABA-stimulated accumulation of H_2O_2 [167,168]. Elevation of SOD activity due to Se supplementation evolved in the quick transformation of the superoxide radicals into H_2O_2, which was produced at the chloroplast and mitochondrial electron transport chain. The evolving H_2O_2 was counteracted either by CAT in the cytoplasm or by APX in the ascorbate glutathione (AsA–GSH) pathway. Furthermore, increased SOD activity in Se-supplemented seedlings altered the chances of hydroxyl (OH^-) radical composition, following a better defense of chloroplast function [162].

Table 2. Selenium (Se) supplementation mitigates salinity-induced oxidative damage by changing different antioxidant enzymatic and non-enzymatic activities in the leaves of different salt-stressed plants (↑ indicates an increase, while ↓ indicates a decrease).

Salinity Stress	Plant Species	Se Dosages	Se Speciation	Experimental Details	↑↓ Antioxidant Activity	% Increase or Decrease	Reference
150 mM	*Oryza sativa* L.	2, 4, 6, 8, 10, 12 mg·L^{-1}	*Se (VI)*	Sand culture	↑SOD ↑APX ↑CAT ↑GR ↑GSH-Px	40.7% 92.7% 82.9% 77.2% 66.1%	[27]
0, 25, 50, 75 mM	*Fragaria × ananassa* Duch	0, 10, 20 mg·L^{-1}	Se-NPs	Reconstituted pot culture (perlite, peat, sand)	↑SOD ↑POD	35.9% 63.1%	[58]
100 mM	*Triticum aestivum* L.	5, 10 μM	*Se (VI)*	Reconstituted pot culture (Peat, compost, sand)	↑SOD ↑CAT ↑GST ↑APX ↑GR	16.2% 10.1% 16.2% 10.6% 22.1%	[28]
0, 30, 60, 90 mM	*Allium sativum* L.	0, 4, 8, 16 mg·L^{-1}	*Se (VI)*	Hydroponic culture	↑SOD ↑CAT ↓POX ↑PAL	81.0% minute minute ~15.0%	[127,129]
12 dS m^{-1}	*Zea mays* L.	0, 20, 40 mg·L^{-1}	*Se (VI)*	Sand culture	↑CAT ↑POD ↑SOD	~56.0% ~63.0% minute	[122]
0, 100 mM	*Phaseolus vulgaris* L.	0, 5, 10 μM	*Se (IV)*	Pot soil culture	↑SOD ↑POD ↑CAT	15.8% 313.3% 56.3%	[132]
8 dS m^{-1}	*Allium cepa* L.	0, 0.5, 1 kg·ha^{-1}	*Se (IV)*	Field experiment	↓CAT ↓POD	26.6% 10.0%	[131]
0, 25, 50 mM	*Lycopersicon esculentum*-Mill.	0, 5, 10 μM	*Se (IV)*	Hydroponic culture	↓POD ↑CAT	60.0% ~240.0%	[128]
0, 50 mM	*Vigna unguiculata* L.	5, 10 μM	*Se (VI)*	Sand-soil culture	↑SOD ↑POD ↑PAL	63.4% 238.1% 73.5%	[169]
0, 100 mM	*Vigna radiata* L. Wilczek	1, 2.5, 5 ppm	*Se (VI)*	Reconstituted pot culture(Soil, sand, farmyard manure)	↑SOD ↑CAT ↑APX ↑GR ↑GPX	14.2% 37.0% 34.8% 24.6% 41.0%	[170]
0, 10 dS m^{-1}	*Anethum graveolens* L.	0, 5 μM	*Se (VI)*	Hydroponic culture	↑CAT ↑SOD ↓APX	~40.0% ~19.0% minute	[147]
0, 100 mM	*Lycopersicon esculentum*-Mill. Shuangfeng 87-5	0.05 mM	*Se (IV)*	Hydroponic culture	↑GR ↓APX ↑DHAR ↑MDAR	~23.0% ~14.0% ~50.0% ~16.0%	[30]
0, 100 mM	*Glycine max* var. L17	0, 25, 50 mg·L^{-1}		Pot soil culture	↑CAT ↑POD ↑SOD	221.6% 85.0% 40.0%	[171]
0, 100 mM	*Cucumis melo* L.	0, 2, 4, 8, 16 μM	*Se (IV)*	Hydroponic culture	↑POD CAT ↑SOD	~29.0% unchanged ~106.0%	[172]
0, 100, 200 mM	*Brassica napus* L.	25 μM	*Se (VI)*	Semi-hydroponic culture	↑GSH ↑GSH/GSSG ↑DHAR ↑MDHAR ↑GST ↑GR	33.0% 86.0% 43.0% 45.0% 18.0% 40.0%	[24]

The values of % increase or decrease in antioxidant activities represent the NaCl and Se treatment dosages mentioned in bold characters. "~" indicates approximate values.

The scavenging of H_2O_2 and lipid peroxide (MDA) into water and lipid alcohol is done by two important enzymes: glutathione peroxidase (GSH-Px) and glutathione reductase (GR) [20]. GSH-Px is considered to be a vital enzyme, which is strongly activated by Se in different plants under various environmental stresses [173]. In the presence of Se, GSH-Px quenches H_2O_2 and then APX, CAT, and GR remove the leftover of H_2O_2. Under salinity stress, regardless of the mode of Se application, Se enhanced the GSH-Px and GR activity compared to controls [27,28]. Under the availability of Se, GSH-Px activity might be modulated due to higher selenocysteine formation at the catalytic site of GSH-Px [27,173]. The enhanced activity of GSH-Px and GR lowered the levels of H_2O_2 and MDA and improved the growth of rapeseed (*Brassica napus* L.) and rice (*Oryza sativa* L.) plants by overcoming ROS-stimulated oxidative damage under soil salinity stress [24]. APX lowers the level of H_2O_2, while GR impacts the preservation of GSH and AsA content resulting in reasonable cellular redox [72]. The supplementation of Se in wheat (*Triticum aestivum* L.) seedlings upregulated the AsA–GSH pathway by increasing the activities of APX and GR. Furthermore, elevating the AsA and GSH contents consistently evolved in the defense of the photosynthetic electron transport chain by sustaining better nicotinamide adenine dinucleotide phosphate ($NADP^+$) levels and limiting the composition of toxic radicals [28]. These results revealed that the wise use of Se could be beneficial to improving the plant antioxidative defense mechanism under soil salinity stress.

5.5. Se-Mediated Gene Expression Modifications under Salinity Stress

Very few studies have elucidated the role of Se in the alleviation of Na^+ accumulation and its hazardous impacts on plant growth and development at the gene level. In an experiment on maize (*Zea mays* L.), Jiang investigated the expression levels of associated genes such as *ZmMPK5*, *ZmMPK7*, and *ZmCPK11*, which are responsible for the antioxidant defense system in roots, while the expression of *ZmNHX1* gene clarified Se's involvement in Na^+ and K^+ homeostasis under salt-affected soils [29] (Figure 2). In previous studies, the contribution of genes to the removal of ROS has been well documented. It has been reported that H_2O_2 is the activator of *ZmMPK5*, and hence, the antioxidant defense system of maize leaves was enhanced [174]. Similarly, the expression of *ZmCPK11* increased the activities of APX and SOD in maize (*Zea mays* L.) leaves [175]. Moreover, under a stress salt environment, the *ZmMPK7* gene was found to be a good alleviator of ROS-induced damages in tobacco (Nicotonia tobaccum L.), resulting in low H_2O_2 accumulation [155]. Likewise, it was described that a small amount (1 µM) of Se (Na_2SeO_3) addition under osmotic stress enhanced the upregulation of *ZmMPK5*, *ZmMPK7*, and *ZmCPK11* genes in roots of maize (*Zea mays* L.) [29]. In many previous findings, *NHX* gene overexpression in transgenic plant species—i.e., rapeseed [176], tomato [81], and poplar [14]—is responsible for Na^+ compartmentalization and an enhancement of salt resistance. Recently, it was proven that *ZmNHX1* expression was significantly up-regulated in maize after 24 h of salinity stress exposure, which may contribute to Na^+ compartmentalization under osmotic stress [29].

Furthermore, it was reported that *OsNHX1* (vacuolar Na^+/H^+ antiporter gene) is responsible for maintaining plant osmotic balance by reducing the hindrance of Na^+ ions during water movement towards plant shoots [14], which might be due to the sequestration of sodium ions in vacuoles of roots and/or shoots [177]. Previously, this phenomenal mechanism was strengthened by research work on tomato (*Solanum lycopersicum* L.) and rapeseed (*Brassica napus* L.), respectively [81,176]. Recently, Se (Na_2SeO_4) was supplied to salinity-stressed rice (*Oryza sativa* L.) plants grown under a saline environment in a mixture of sand and polymer, and it was observed that plants receiving Se exhibited a higher transcription level of *OsNHX1* gene [27]. The researchers concluded that it could be imagined that a higher *OsNHX1* transcript level promoted Na^+ sequestration within the root vacuoles and therefore reduced the Na^+ accumulation in rice shoots, which ultimately improved plant growth and antioxidative defense mechanism. However, further research work is needed to explore how Se is involved in antioxidant defense genes and how these genes are up and downregulated to induce antioxidant defense systems in salt-stressed plants under Se supplementation.

6. Conclusions and Future Perspectives

Soil salinization has become an overwhelming environmental threat to world food production and agricultural sustainability. Selenium (Se) is recognized as an essential trace element for human beings and animals, although this is controversial for different plant species. However, based on published relevant literature, it is widely accepted that Se is capable of remediating various biotic and abiotic environmental stresses including soil salinity. The important mechanisms involved in Se-mediated salinity tolerance in crop plants include a reduction in Na^+ ion accumulation in plant parts through the overexpression of the Na^+/H^+ antiport, chelation and boosting of the antioxidative defense system in plants, Na^+ compartmentalization, improvement in various structural compositions, and the upregulation of Na^+ and Cl^- ions transporter genes. However, these salinity-tolerance mechanisms are still highly controversial and are influenced by growth conditions, growth medium (soil or water), stress duration, plant genotypes, plant species types, Se doses, speciation, and many more. Therefore, it is difficult to predict a general conclusion for the Se-mediated alleviation of salinity-induced phytotoxicity in crop plants. More precisely, at lower concentrations, Se can mediate plant growth and physiological characteristics (acts as a beneficial element), while at higher concentrations; it disturbs various plant metabolic processes, and thereby suppresses plant growth under salinity stress. Moreover, Se triggers the dismutation of ROS generated under salt stress and protects plants from oxidative damage. In conclusion, this review article has shed light on the hazardous impacts of soil-affected soils, various salinity tolerance mechanisms adopted by crops and the prospective mechanisms involved in Se-mediated salinity tolerance as well as improvements in the growth and productivity of various crop plants cultivated in salt-affected soils.

In this review paper, after critically reviewing the best available data to date, the authors anticipated that there would be an emergent interest in the scientific community to studying the mechanisms of Se-assisted salinity tolerance in plants in the near future; therefore, the following research gaps need to be explored in future.

On an instructive note, the suitable concentration of Se supplementation is still a matter of research. Complete interpretation of the role of Se as well as detailed protective mechanisms would be helpful for developing salinity tolerance in plants.

The Se transformations in the plants are still unclear. Therefore, future studies are required to explore the exact mechanisms involved in Se transformations inside plant species that enhance Se transfer to the plant shoots and its volatilization from aerial plant parts.

Previous researchers have focused on evaluating the role of Se in individual plant species grown in salt-affected soils; however, there is still a need to better understand its ameliorative roles in more plant species under various environmental factors for the confirmation of the Se-mediated amelioration of salinity-induced phytotoxicity on a larger scale.

According to the reviewed data, Se in most experiments was used under saline nutrient mediums (hydroponics). Such experimental results can overestimate the Na^+ uptake and translocation within the plant body. It is advised to conduct future experiments on natural saline soils (pots or field), as soil is a complex system, which will provide a better understanding of Se-mediated salinity-tolerance mechanisms. Moreover, such experiments will help the local farming community to learn about the use of Se in farming practices.

More importantly, to date, most soil-based experiments have been executed over the short term, which raises questions on Se's potential to remediate salt-affected soils in the long term. Therefore, well-planned, comprehensive, and long-term field experiments are needed to check the productivity and economic feasibility of Se-based ameliorations of saline soils.

Despite the widespread occurrence of Se deficiency globally, Se toxicity (selenosis) is a problem in some areas. Some soils and mineral deposits are naturally Se rich, and exploitation of these seleniferous soils can lead to toxic accumulation of Se in the environment. Therefore, effective enrichment of agricultural crops with Se via soil using Se-enriched fertilizers can be challenging due to varying soil

Se concentrations, soil types, soil redox potentials, soil pH, and microbiological activity. Furthermore, the high cost of Se fertilizer, in combination with the modest incorporation rate, should be considered.

Author Contributions: M.K., A.P., S.H., and J.-T.C. designed and wrote the manuscript. S.A. and M.K. compiled the tabulated data and carried out graphical work in the manuscript. Z.M., M.S.C., M.H.S., M.A., P.H., and J.-T.C. comprehensively revised and improved the quality of manuscript.

Funding: This research was funded by China Scholarship Council (CSC) under the grant number 2017GXZ024414.

Conflicts of Interest: The authors declare no conflict of interest, either financially or otherwise.

Abbreviations

Se-NPs	Selenium-nanoparticles
GSH	Reduced glutathione
GSSG	Oxidized glutathione
DHAR	Dehydroascorbate reductase
MDHAR	Monodehydroascorbate reductase
GST	Glutathione S-transferase
GR	Glutathione reductase
POX	peroxidase
PAL	Activity of phenylalanine ammonia-lyase
GSH-Px	Glutathione peroxidase
CAT	Catalase activity
APX	Ascorbate peroxidase activity
SOD	Superoxide dismutase activity
POD	Peroxidase activity
GPX	Glutathione peroxidase activity
MDAR	Monodehydroascorbate reductase activity
RWC	Relative water contents
TBARS	Thiobarbituric acid reactive substances
NPQ	Non-photochemical quenching
MDA	Malondialdehyde
RuBPCase	Ribulose-1,5-bisphosphate-carboxylase/oxygenase content
SPAD	Chlorophyll content in leaves
H_2O_2	Hydrogen peroxide
ATP	Adenosine triphosphate
$NADP^+$	Nicotinamide adenine dinucleotide phosphate
MAPK	Mitogen activated protein kinase gene
CPK	Calcium dependent protein kinase gen
NADK2	NAD kinase2 gene
ALMT	Aluminum-activated malate transporters
SULTRs	Sulfate transporters
γ-GK	γ-Glutamyl kinase
NHX	Sodium/hydrogen (Na^+/H^+) exchanger gene
PSII	Photosystem II
NPK	Nitrogen, phosphorous, and potassium

References

1. Gontia-Mishra, I.; Sasidharan, S.; Tiwari, S. Recent developments in use of 1-aminocyclopropane-1-carboxylate (ACC) deaminase for conferring tolerance to biotic and abiotic stress. *Biotechnol. Lett.* **2014**, *36*, 889–898. [CrossRef] [PubMed]
2. Abbasi, G.H.; Akhtar, J.; Ahmad, R.; Jamil, M.; Anwar-Ul-Haq, M.; Ali, S.; Ijaz, M. Potassium application mitigates salt stress differentially at different growth stages in tolerant and sensitive maize hybrids. *Plant Growth Regul.* **2015**, *76*, 111–125. [CrossRef]

3. Wallender, W.W.; Tanji, K.K. *Agricultural Salinity Assessment and Management*; American Society of Civil Engineers: Reston, VA, USA, 2011.

4. Munns, R.; Tester, M. Mechanisms of Salinity Tolerance. *Ann. Rev. Plant Biol.* **2008**, *59*, 651–681. [CrossRef] [PubMed]

5. Abdel Latef, A.A. Changes of antioxidative enzymes in salinity tolerance among different wheat cultivars. *Cereal Res. Commun.* **2010**, *38*, 43–55. [CrossRef]

6. Shabala, S.; Cuin, T.A. Potassium transport and plant salt tolerance. *Physiol. Plant.* **2008**, *133*, 651–669. [CrossRef]

7. Ruiz-Lozano, J.M.; Collados, C.; Barea, J.M.; Azcón, R. Arbuscular mycorrhizal symbiosis can alleviate drought-induced nodule senescence in soybean plants. *New Phytol.* **2001**, *151*, 493–502. [CrossRef]

8. Munns, R. Comparative physiology of salt and water stress. *Plant Cell Environ.* **2002**, *25*, 239–250. [CrossRef]

9. Wang, W.; Vinocur, B.; Altman, A. Plant responses to drought, salinity and extreme temperatures: Towards genetic engineering for stress tolerance. *Planta* **2003**, *218*, 1–14. [CrossRef]

10. Sheng, M.; Tang, M.; Chen, H.; Yang, B.; Zhang, F.; Huang, Y. Influence of arbuscular mycorrhizae on photosynthesis and water status of maize plants under salt stress. *Mycorrhiza* **2008**, *18*, 287–296. [CrossRef]

11. Flowers, T.J.; Galal, H.K.; Bromham, L. Evolution of halophytes: Multiple origins of salt tolerance in land plants. *Funct. Plant Biol.* **2010**, *37*, 604–612. [CrossRef]

12. Nabati, J.; Kafi, M.; Nezami, A.; Moghaddam, P.R.; Ali, M.; Mehrjerdi, M.Z. Effect of salinity on biomass production and activities of some key enzymatic antioxidants in Kochia (*Kochia scoparia*). *Pak. J. Bot.* **2011**, *43*, 539–548.

13. Ahanger, M.A.; Agarwal, R.M. Salinity stress induced alterations in antioxidant metabolism and nitrogen assimilation in wheat (*Triticum aestivum* L) as influenced by potassium supplementation. *Plant Physiol. Biochem.* **2017**, *115*, 449–460. [CrossRef] [PubMed]

14. Jiang, C.; Zheng, Q.; Liu, Z.; Xu, W.; Liu, L.; Zhao, G.; Long, X. Overexpression of Arabidopsis thaliana Na^+/H^+ antiporter gene enhanced salt resistance in transgenic poplar (*Populus* × *euramericana* "Neva"). *Trees* **2012**, *26*, 685–694. [CrossRef]

15. Mittal, S.; Kumari, N.; Sharma, V. Differential response of salt stress on *Brassica juncea*: Photosynthetic performance, pigment, proline, D1 and antioxidant enzymes. *Plant Physiol. Biochem.* **2012**, *54*, 17–26. [CrossRef] [PubMed]

16. Gengmao, Z.; Shihui, L.; Xing, S.; Yizhou, W.; Zipan, C. The role of silicon in physiology of the medicinal plant (*Lonicera japonica* L.) under salt stress. *Sci. Rep.* **2015**, *5*, 12696. [CrossRef] [PubMed]

17. Quintero, J.M.; Fournier, J.M.; Benlloch, M. Na^+ accumulation in shoot is related to water transport in K^+-starved sunflower plants but not in plants with a normal K^+ status. *J. Plant Physiol.* **2007**, *164*, 60–67. [CrossRef] [PubMed]

18. Noctor, G.; Foyer, C.H. Ascorbate and Glutathione: Keeping Active Oxygen Under Control. *Ann. Rev. Plant Physiol. Plant Mol. Biol.* **1998**, *49*, 249–279. [CrossRef]

19. Pitzschke, A.; Forzani, C.; Hirt, H. Reactive oxygen species signaling in plants. *Antioxid. Redox Signal.* **2006**, *8*, 1757–1764. [CrossRef]

20. Farooq, M.; Hussain, M.; Wakeel, A.; Siddique, K.H. Salt stress in maize: Effects, resistance mechanisms, and management. *A Rev. Agron. Sustain. Dev.* **2015**, *35*, 461–481. [CrossRef]

21. Aghighi Shahverdi, M.; Omidi, H.; Tabatabaei, S.J. Plant growth and steviol glycosides as affected by foliar application of selenium, boron, and iron under NaCl stress in Stevia rebaudiana Bertoni. *Ind. Crops Prod.* **2018**, *125*, 408–415. [CrossRef]

22. Iqbal, M.; Hussain, I.; Liaqat, H.; Ashraf, M.A.; Rasheed, R.; Rehman, A.U. Exogenously applied selenium reduces oxidative stress and induces heat tolerance in spring wheat. *Plant Physiol. Biochem.* **2015**, *94*, 95–103. [CrossRef] [PubMed]

23. Sieprawska, A.; Kornaś, A.; Filek, M. Involvement of selenium in protective mechanisms of plants under environmental stress conditions—Review. *Acta Biol. Crac. Ser. Bot.* **2015**, *57*, 9–20. [CrossRef]

24. Hasanuzzaman, M.; Hossain, M.A.; Fujita, M. Selenium-induced up-regulation of the antioxidant defense and methylglyoxal detoxification system reduces salinity-induced damage in rapeseed seedlings. *Biol. Trace Elem. Res.* **2011**, *143*, 1704–1721. [CrossRef] [PubMed]

25. Versini, A.; Di Tullo, P.; Aubry, E.; Bueno, M.; Thiry, Y.; Pannier, F.; Castrec-Rouelle, M. Influence of Se concentrations and species in hydroponic cultures on Se uptake, translocation and assimilation in non-accumulator ryegrass. *Plant Physiol. Biochem.* **2016**, *108*, 372–380. [CrossRef]
26. Babalar, M.; Mohebbi, S.; Zamani, Z.; Askari, M.A. Effect of foliar application with sodium selenate on selenium biofortification and fruit quality maintenance of Starking Delicious apple during storage. *J. Sci. Food Agric.* **2019**, *99*, 5149–5156. [CrossRef]
27. Subramanyam, K.; Du Laing, G.; Van Damme, E.J.M. Sodium selenate treatment using a combination of seed priming and foliar spray alleviates salinity stress in rice. *Front. Plant Sci.* **2019**, *10*, 116. [CrossRef]
28. Elkelish, A.A.; Soliman, M.H.; Alhaithloul, H.A.; El-Esawi, M.A. Selenium protects wheat seedlings against salt stress-mediated oxidative damage by up-regulating antioxidants and osmolytes metabolism. *Plant Physiol. Biochem.* **2019**, *137*, 144–153. [CrossRef]
29. Jiang, C.; Zu, C.; Lu, D.; Zheng, Q.; Shen, J.; Wang, H.; Li, D. Effect of exogenous selenium supply on photosynthesis, Na$^+$ accumulation and antioxidative capacity of maize (*Zea mays* L.) under salinity stress. *Sci. Rep.* **2017**, *7*, 42039. [CrossRef]
30. Diao, M.; Ma, L.; Wang, J.; Cui, J.; Fu, A.; Liu, H.Y. Selenium Promotes the Growth and Photosynthesis of Tomato Seedlings Under Salt Stress by Enhancing Chloroplast Antioxidant Defense System. *J. Plant Growth Regul.* **2014**, *33*, 671–682. [CrossRef]
31. Kusvuran, S.; Kiran, S.; Ellialtioglu, S.S. Antioxidant Enzyme Activities and Abiotic Stress Tolerance Relationship in Vegetable Crops. In *Abiotic and Biotic Stress in Plants–Recent Advances and Future Perspectives*; IntechOpen: London, UK, 2016.
32. Munns, R. Physiological processes limiting plant growth in saline soils: Some dogmas and hypotheses. *Plant Cell Environ.* **1993**, *16*, 15–24. [CrossRef]
33. Evelin, H.; Kapoor, R.; Giri, B. Arbuscular mycorrhizal fungi in alleviation of salt stress: A review. *Ann. Bot.* **2009**, *104*, 1263–1280. [CrossRef] [PubMed]
34. Hajiboland, R.; Aliasgharzad, N.; Laiegh, S.F.; Poschenrieder, C. Colonization with arbuscular mycorrhizal fungi improves salinity tolerance of tomato (*Solanum lycopersicum* L.) plants. *Plant Soil* **2010**, *331*, 313–327. [CrossRef]
35. Hussain, M.; Park, H.W.; Farooq, M.; Jabran, K.; Lee, D.J. Morphological and physiological basis of salt resistance in different rice genotypes. *Int. J. Agric. Biol.* **2013**, *15*, 113–118.
36. Hachicha, M.; Kahlaoui, B.; Khamassi, N.; Misle, E.; Jouzdan, O. Effect of electromagnetic treatment of saline water on soil and crops. *J. Saudi Soc. Agric. Sci.* **2018**, *17*, 154–162. [CrossRef]
37. Hamada, A.M. Alleviation of the adverse effects of NaCl on germination, seedling, growth and metabolic activities of maize plants by calcium salts. *Bull. Fac. Sci. Assiut Univ.* **1995**, *24*, 211–220.
38. Lee, D.G.; Park, K.W.; An, J.Y.; Sohn, Y.G.; Ha, J.K.; Kim, H.Y.; Bae, D.W.; Lee, K.H.; Kang, N.J.; Lee, B.H.; et al. Proteomics analysis of salt-induced leaf proteins in two rice germplasms with different salt sensitivity. *Can. J. Plant Sci.* **2011**, *91*, 337–349. [CrossRef]
39. Ibrahim, E.A. Seed priming to alleviate salinity stress in germinating seeds. *J. Plant Physiol.* **2016**, *192*, 38–46. [CrossRef]
40. Memon, S.A.; Hou, X.; Wang, L.J. Morphological analysis of salt stress response of pak choi. *Electron. J. Environ. Agric. Food Chem.* **2010**, *9*, 1.
41. Hanafy, M.S.; El-Banna, A.; Schumacher, H.M.; Jacobsen, H.J.; Hassan, F.S. Enhanced tolerance to drought and salt stresses in transgenic faba bean (*Vicia faba* L.) plants by heterologous expression of the PR10a gene from potato. *Plant Cell Rep.* **2013**, *32*, 663–674. [CrossRef]
42. Kapoor, K.; Srivastava, A. Assessment of Salinity Tolerance of Vigna mungo Var. Pu-19 Using ex vitro and in vitro Methods. *Asian J. Biotechnol.* **2010**, *2*, 73–85. [CrossRef]
43. Anwar-ul-Haq, M.; Akram, S.; Akhtar, J.; Saqib, M.; Saqib, Z.A.; Abbasi, G.H.; Jan, M. Morpho-physiological characterization of sunflower genotypes (*Helianthus annuus* L.) under saline condition. *Pak. J. Agric. Sci.* **2013**, *50*, 49–54.
44. Mallahi, T.; Saharkhiz, M.J.; Javanmardi, J. Salicylic acid changes morpho-physiological attributes of feverfew (*Tanacetum parthenium* L.) under salinity stress. *Acta Ecol. Sin.* **2018**, *38*, 351–355. [CrossRef]
45. Qados, A.M.S.A. Effect of salt stress on plant growth and metabolism of bean plant *Vicia faba* (L.). *J. Saudi Soc. Agric. Sci.* **2011**, *10*, 7–15.

46. Hajiboland, R.; Norouzi, F.; Poschenrieder, C. Growth, physiological, biochemical and ionic responses of pistachio seedlings to mild and high salinity. *Trees Struct. Funct.* **2014**, *28*, 1065–1078. [CrossRef]

47. Ahmadi, F.I.; Karimi, K.; Struik, P.C. Effect of exogenous application of methyl jasmonate on physiological and biochemical characteristics of *Brassica napus* L. cv. Talaye under salinity stress. *S. Afr. J. Bot.* **2018**, *115*, 5–11. [CrossRef]

48. Shi, Q.; Ding, F.; Wang, X.; Wei, M. Exogenous nitric oxide protects cucumber roots against oxidative stress induced by salt stress. *Plant Physiol. Biochem.* **2007**, *45*, 542–550. [CrossRef] [PubMed]

49. Netondo, G.W.; Onyango, J.C.; Beck, E. Sorghum and salinity: II. Gas exchange and chlorophyll fluorescence of sorghum under salt stress. *Crop Sci.* **2004**, *44*, 806. [CrossRef]

50. Sapre, S.; Gontia-Mishra, I.; Tiwari, S. Klebsiella sp. confers enhanced tolerance to salinity and plant growth promotion in oat seedlings (*Avena sativa*). *Microbiol. Res.* **2018**, *206*, 25–32. [CrossRef]

51. Rozeff, N. Sugarcane and salinity—A review paper. *Sugar Cane* **1995**.

52. Cantabella, D.; Piqueras, A.; Acosta-Motos, J.R.; Bernal-Vicente, A.; Hernández, J.A.; Díaz-Vivancos, P. Salt-tolerance mechanisms induced in Stevia rebaudiana Bertoni: Effects on mineral nutrition, antioxidative metabolism and steviol glycoside content. *Plant Physiol. Biochem.* **2017**, *115*, 484–496. [CrossRef]

53. Khan, M.I.R.; Asgher, M.; Khan, N.A. Alleviation of salt-induced photosynthesis and growth inhibition by salicylic acid involves glycinebetaine and ethylene in mungbean (*Vigna radiata* L.). *Plant Physiol. Biochem.* **2014**, *80*, 67–74. [CrossRef] [PubMed]

54. Munns, R. Genes and salt tolerance: Bringing them together. *New Phytol.* **2005**, *167*, 645–663. [CrossRef] [PubMed]

55. Hussain, S.; Zhang, J.H.; Zhong, C.; Zhu, L.F.; Cao, X.C.; YU, S.M.; Allen Bohr, J.; Hu, J.J.; Jin, Q.Y. Effects of salt stress on rice growth, development characteristics, and the regulating ways: A review. *J. Integr. Agric.* **2017**, *16*, 2357–2374. [CrossRef]

56. Hussain, S.; Zhong, C.; Bai, Z.; Cao, X.; Zhu, L.; Hussain, A.; Zhu, C.; Fahad, S.; James, A.B.; Zhang, J.; et al. Effects of 1-Methylcyclopropene on Rice Growth Characteristics and Superior and Inferior Spikelet Development Under Salt Stress. *J. Plant Growth Regul.* **2018**, *37*, 1368–1384. [CrossRef]

57. Lee, M.H.; Cho, E.J.; Wi, S.G.; Bae, H.; Kim, J.E.; Cho, J.Y.; Lee, S.; Kim, J.H.; Chung, B.Y. Divergences in morphological changes and antioxidant responses in salt-tolerant and salt-sensitive rice seedlings after salt stress. *Plant Physiol. Biochem.* **2013**, *70*, 325–335. [CrossRef] [PubMed]

58. Zahedi, S.M.; Abdelrahman, M.; Hosseini, M.S.; Hoveizeh, N.F.; Tran, L.P. Alleviation of the effect of salinity on growth and yield of strawberry by foliar spray of selenium-nanoparticles. In *Environmental Pollution*; Elsevier: Amsterdam, The Netherlands, 2019.

59. Gururani, M.A.; Venkatesh, J.; Tran, L.S.P. Regulation of photosynthesis during abiotic stress-induced photoinhibition. *Mol. Plant* **2015**, *8*, 1304–1320. [CrossRef]

60. James, R.A.; Blake, C.; Byrt, C.S.; Munns, R. Major genes for Na$^+$ exclusion, Nax1 and Nax2 (wheat HKT1;4 and HKT1;5), decrease Na$^+$ accumulation in bread wheat leaves under saline and waterlogged conditions. *J. Exp. Bot.* **2011**, *62*, 2939–2947. [CrossRef]

61. Singam, K.; Juntawong, N.; Cha-Um, S.; Kirdmanee, C. Salt stress induced ion accumulation, ion homeostasis, membrane injury and sugar contents in salt-sensitive rice (*Oryza sativa* L. spp. *indica*) roots under isoosmotic conditions. *Afr. J. Biotechnol.* **2011**, *10*, 1340–1346.

62. Hussain, S.; Bai, Z.; Huang, J.; Cao, X.; Zhu, L.; Zhu, C.; Khaskheli, M.A.; Zhong, C.; Jin, Q.; Zhang, J. 1-methylcyclopropene modulates physiological, biochemical, and antioxidant responses of rice to different salt stress levels. *Front. Plant Sci.* **2019**, *10*. [CrossRef]

63. Demiral, T.; Türkan, I. Comparative lipid peroxidation, antioxidant defense systems and proline content in roots of two rice cultivars differing in salt tolerance. *Environ. Exp. Bot.* **2005**, *53*, 247–257. [CrossRef]

64. Levitt, J. Responses of Plants to Environmental Stresses. *J. Range Manag.* **1985**, *38*, 480. [CrossRef]

65. Pattanayak, G.K.; Tripathy, B.C. Overexpression of protochlorophyllide oxidoreductase c regulates oxidative stress in arabidopsis. *PLoS ONE* **2011**, *6*, e26532. [CrossRef] [PubMed]

66. Santos, M.G.; Ribeiro, R.V.; Machado, E.C.; Pimentel, C. Photosynthetic parameters and leaf water potential of five common bean genotypes under mild water deficit. *Biol. Plant.* **2009**, *53*, 229–236. [CrossRef]

67. Foyer, C.H.; Noctor, G. Redox sensing and signalling associated with reactive oxygen in chloroplasts, peroxisomes and mitochondria. *Physiol. Plant.* **2003**, *119*, 355–364. [CrossRef]

68. Kamran, M.; Malik, Z.; Parveen, A.; Huang, L.; Riaz, M.; Bashir, S.; Mustafa, A.; Abbasi, G.H.; Xue, B.; Ali, U. Ameliorative Effects of Biochar on Rapeseed (*Brassica napus* L.) Growth and Heavy Metal Immobilization in Soil Irrigated with Untreated Wastewater. In *Journal of Plant Growth Regulation*; Springer: Berlin, Germany, 2019; pp. 1–16.

69. Adly, A.A.M. Oxidative stress and disease: An updated review. *Res. J. Immunol.* **2010**, *3*, 129–145.

70. Houot, V.; Etienne, P.; Petitot, A.S.; Barbier, S.; Blein, J.P.; Suty, L. Hydrogen peroxide induces programmed cell death features in cultured tobacco BY-2 cells, in a dose-dependent manner. *J. Exp. Bot.* **2001**, *52*, 1721–1730.

71. Farmer, E.E.; Mueller, M.J. ROS-Mediated Lipid Peroxidation and RES-Activated Signaling. *Ann. Rev. Plant Biol.* **2013**, *64*, 429–450. [CrossRef]

72. Asada, K. Production and Action of Active Oxygen Species in Photosynthetic Tissues. In *Causes of Photooxidative Stress and Amelioration of Defense Systems in Plants*; CRC press: Boca Raton, FL, USA, 1994; pp. 77–104. ISBN 0-8493-5443-9.

73. Drobot, L.B.; Samoylenko, A.A.; Vorotnikov, A.V.; Tyurin -Kuzmin, P.A.; Bazalii, A.V.; Kietzmann, T.; Tkachuk, V.A.; Komisarenko, S.V. Reactive oxygen species in signal transduction. *Ukrain'skyi Biokhimichnyi Zhurnal* **2013**, *85*, 209–217. [CrossRef]

74. Ramachandra Reddy, A.; Chaitanya, K.V.; Jutur, P.P.; Sumithra, K. Differential antioxidative responses to water stress among five mulberry (*Morus alba* L.) cultivars. *Environ. Exp. Bot.* **2004**, *52*, 33–42. [CrossRef]

75. Li, M.; Guo, S.; Xu, Y.; Meng, Q.; Li, G.; Yang, X. Glycine betaine-mediated potentiation of HSP gene expression involves calcium signaling pathways in tobacco exposed to NaCl stress. *Physiol. Plant.* **2014**, *150*, 63–75. [CrossRef]

76. Farhangi-Abriz, S.; Ghassemi-Golezani, K. How can salicylic acid and jasmonic acid mitigate salt toxicity in soybean plants? *Ecotoxicol. Environ. Saf.* **2018**, *154*, 1010–1016. [CrossRef] [PubMed]

77. Kamran, M.; Malik, Z.; Parveen, A.; Zong, Y.; Abbasi, G.H.; Rafiq, M.T.; Shaaban, M.; Mustafa, A.; Bashir, S.; Rafay, M.; et al. Biochar alleviates Cd phytotoxicity by minimizing bioavailability and oxidative stress in pak choi (*Brassica chinensis* L.) cultivated in Cd-polluted soil. *J. Environ. Manag.* **2019**, *250*, 109500. [CrossRef] [PubMed]

78. Krishnaiah, D.; Sarbatly, R.; Nithyanandam, R. A review of the antioxidant potential of medicinal plant species. *Food Bioprod. Process.* **2011**, *89*, 217–233. [CrossRef]

79. Cheeseman, J.M. The evolution of halophytes, glycophytes and crops, and its implications for food security under saline conditions. *New Phytol.* **2015**, *206*, 557–570. [CrossRef] [PubMed]

80. Mbarki, S.; Sytar, O.; Cerda, A.; Zivcak, M.; Rastogi, A.; He, X.; Zoghlami, A.; Abdelly, C.; Brestic, M. Strategies to mitigate the salt stress effects on photosynthetic apparatus and productivity of crop plants. In *Salinity Responses and Tolerance in Plants, Volume 1: Targeting Sensory, Transport and Signaling Mechanisms*; Springer: Berlin, Germany, 2018; ISBN 9-7833-1975-6714.

81. Zhang, H.X.; Blumwald, E. Transgenic salt-tolerant tomato plants accumulate salt in foliage but not in fruit. *Nat. Biotechnol.* **2001**, *19*, 765. [CrossRef]

82. Takahashi, R.; Liu, S.; Takano, T. Isolation and characterization of plasma membrane Na^+/H^+ antiporter genes from salt-sensitive and salt-tolerant reed plants. *J. Plant Physiol.* **2009**, *166*, 301–309. [CrossRef]

83. Zhang, J.; Wang, F.; Zhang, C.; Zhang, J.; Chen, Y.; Liu, G.; Zhao, Y.; Hao, F.; Zhang, J. A novel VIGS method by agroinoculation of cotton seeds and application for elucidating functions of GhBI-1 in salt-stress response. *Plant Cell Rep.* **2018**, *37*, 1091–1100. [CrossRef]

84. Demidchik, V.; Cuin, T.A.; Svistunenko, D.; Smith, S.J.; Miller, A.J.; Shabala, S.; Sokolik, A.; Yurin, V. Arabidopsis root K+-efflux conductance activated by hydroxyl radicals: Single-channel properties, genetic basis and involvement in stress-induced cell death. *J. Cell Sci.* **2010**, *123*, 1468–1479. [CrossRef]

85. Isla, R.; Aragüés, R. Yield and plant ion concentrations in maize (*Zea mays* L.) subject to diurnal and nocturnal saline sprinkler irrigations. *Field Crops Res.* **2010**, *123*, 1468–1479. [CrossRef]

86. Wakeel, A.; Farooq, M.; Qadir, M.; Schubert, S. Potassium substitution by sodium in plants. *Crit. Rev. Plant Sci.* **2011**, *30*, 401–413. [CrossRef]

87. Bhosale, R.; Boudolf, V.; Cuevas, F.; Lu, R.; Eekhout, T.; Hu, Z.; Van Isterdael, G.; Lambert, G.M.; Xu, F.; Nowack, M.K.; et al. A spatiotemporal dna endoploidy map of the arabidopsis root reveals roles for the endocycle in root development and stress adaptation. *Plant Cell* **2018**, *30*, 2330–2351. [CrossRef] [PubMed]

88. Lv, S.; Yu, D.; Sun, Q.; Jiang, J. Activation of gibberellin 20-oxidase 2 undermines auxin-dependent root and root hair growth in NaCl-stressed Arabidopsis seedlings. *Plant Growth Regul.* **2018**, *84*, 225–236. [CrossRef]

89. Younis, M.E.; El-Shahaby, O.A.; Nemat Alla, M.M.; El-Bastawisy, Z.M. Kinetin alleviates the influence of waterlogging and salinity on growth and affects the production of plant growth regulators in *Vigna sinensis* and *Zea mays*. *Agronomie* **2003**, *23*, 277–285. [CrossRef]

90. Javid, M.G.; Sorooshzadeh, A.; Moradi, F.; Sanavy, S.A.M.M.; Allahdadi, I. The role of phytohormones in alleviating salt stress in crop plants. *Aust. J. Crop Sci.* **2011**, *5*, 726.

91. Nawaz, K.; Ashraf, M. Improvement in salt tolerance of maize by exogenous application of glycinebetaine: Growth and water relations. *Pak. J. Bot.* **2007**, *39*, 1647–1653.

92. Sobahan, M.A.; Akter, N.; Ohno, M.; Okuma, E.; Hirai, Y.; Mori, I.C.; Nakamura, Y.; Murata, Y. Effects of exogenous proline and glycinebetaine on the salt tolerance of rice cultivars. *Biosci. Biotechnol. Biochem.* **2012**, *76*, 1568–1570. [CrossRef] [PubMed]

93. Athar, H.U.R.; Zafar, Z.U.; Ashraf, M. Glycinebetaine Improved Photosynthesis in Canola under Salt Stress: Evaluation of Chlorophyll Fluorescence Parameters as Potential Indicators. *J. Agron. Crop Sci.* **2015**, *201*, 428–442. [CrossRef]

94. Hasanuzzaman, M.; Nahar, K.; Fujita, M. Plant response to salt stress and role of exogenous protectants to mitigate salt-induced damages. In *Ecophysiology and Responses of Plants under Salt Stress*; Springer: Berlin, Germany, 2013; ISBN 9-78146-144-7474.

95. Xiang, L.; Liu, C.; Luo, J.; He, L.; Deng, Y.; Yuan, J.; Wu, C.; Cai, Y. A tuber mustard AP2/ERF transcription factor gene, BjABR1, functioning in abscisic acid and abiotic stress responses, and evolutionary trajectory of the ABR1 homologous genes in Brassica species. *PeerJ* **2018**, *6*, e6071. [CrossRef]

96. Long, L.; Yang, W.W.; Liao, P.; Guo, Y.W.; Kumar, A.; Gao, W. Transcriptome analysis reveals differentially expressed ERF transcription factors associated with salt response in cotton. *Plant Sci.* **2019**, *281*, 72–81. [CrossRef]

97. Uddin, M.N.; Hanstein, S.; Leubner, R.; Schubert, S. Leaf Cell-Wall Components as Influenced in the First Phase of Salt Stress in Three Maize (*Zea mays* L.) Hybrids Differing in Salt Resistance. *J. Agron. Crop Sci.* **2013**, *199*, 405–415. [CrossRef]

98. Menezes-Benavente, L.; Kernodle, S.P.; Margis-Pinheiro, M.; Scandalios, J.G. Salt-induced antioxidant metabolism defenses in maize (*Zea mays* L.) seedlings. *Redox Rep.* **2004**, *9*, 29–36. [CrossRef] [PubMed]

99. Rodríguez-Kessler, M.; Alpuche-Solís, A.G.; Ruiz, O.A.; Jiménez-Bremont, J.F. Effect of salt stress on the regulation of maize (*Zea mays* L.) genes involved in polyamine biosynthesis. *Plant Growth Regul.* **2006**, *48*, 175–185. [CrossRef]

100. Hawrylak-Nowak, B. Selenite is more efficient than selenate in alleviation of salt stress in lettuce plants. *Acta Biol. Crac. Ser. Bot.* **2015**, *57*, 49–54. [CrossRef]

101. Mehdi, Y.; Hornick, J.L.; Istasse, L.; Dufrasne, I. Selenium in the environment, metabolism and involvement in body functions. *Molecules* **2013**, *18*, 3292–3311. [CrossRef]

102. Shahid, M.; Niazi, N.K.; Khalid, S.; Murtaza, B.; Bibi, I.; Rashid, M.I. A critical review of selenium biogeochemical behavior in soil-plant system with an inference to human health. *Environ. Pollut.* **2018**, *234*, 915–934.

103. Galeas, M.L.; Zhang, L.H.; Freeman, J.L.; Wegner, M.; Pilon-Smits, E.A.H. Seasonal fluctuations of selenium and sulfur accumulation in selenium hyperaccumulators and related nonaccumulators. *New Phytol.* **2007**, *173*, 517–525. [CrossRef]

104. Hartikainen, H. Biogeochemistry of selenium and its impact on food chain quality and human health. *J. Trace Elem. Med. Biol.* **2005**, *18*, 309–318. [CrossRef]

105. White, P.J. Selenium accumulation by plants. *Ann. Bot.* **2015**, *117*, 217–235. [CrossRef]

106. Cartes, P.; Gianfreda, L.; Paredes, C.; Mora, M.L. Selenium uptake and its antioxidant role in ryegrass cultivars as affected by selenite seed pelletization. *J. Soil Sci. Plant Nutr.* **2011**, *11*, 1–14. [CrossRef]

107. Nakamaru, Y.M.; Altansuvd, J. Speciation and bioavailability of selenium and antimony in non-flooded and wetland soils: A review. *Chemosphere* **2014**, *111*, 366–371. [CrossRef]

108. Kikkert, J.; Berkelaar, E. Plant uptake and translocation of inorganic and organic forms of selenium. *Arch. Environ. Contam. Toxicol.* **2013**, *65*, 458–465. [CrossRef] [PubMed]

109. Bruggeman, C.; Maes, A.; Vancluysen, J. The interaction of dissolved Boom Clay and Gorleben humic substances with selenium oxyanions (selenite and selenate). *Appl. Geochem.* **2007**, *22*, 1371–1379. [CrossRef]

110. Shand, C.A.; Eriksson, J.; Dahlin, A.S.; Lumsdon, D.G. Selenium concentrations in national inventory soils from Scotland and Sweden and their relationship with geochemical factors. *J. Geochem. Exp.* **2012**, *121*, 4–14. [CrossRef]

111. Beckett, G.J.; Arthur, J.R. Selenium and endocrine systems. *J. Endocrinol.* **2005**, *184*, 455–465. [CrossRef] [PubMed]

112. Tan, J.; Zhu, W.; Wang, W.; Li, R.; Hou, S.; Wang, D.; Yang, L. Selenium in soil and endemic diseases in China. *Sci. Total Environ.* **2002**, *284*, 227–235. [CrossRef]

113. Song, Z.; Shao, H.; Huang, H.; Shen, Y.; Wang, L.; Wu, F.; Han, D.; Song, J.; Jia, H. Overexpression of the phosphate transporter gene OsPT8 improves the Pi and selenium contents in *Nicotiana tabacum*. *Environ. Exp. Bot.* **2017**, *137*, 158–165. [CrossRef]

114. Wang, P.; Menzies, N.W.; Lombi, E.; McKenna, B.A.; James, S.; Tang, C.; Kopittke, P.M. Synchrotron-based X-ray absorption near-edge spectroscopy imaging for laterally resolved speciation of selenium in fresh roots and leaves of wheat and rice. *J. Exp. Bot.* **2015**, *66*, 4795–4806. [CrossRef]

115. Yang, X.; Lu, C. Photosynthesis is improved by exogenous glycinebetaine in salt-stressed maize plants. *Physiol. Plant.* **2005**, *124*, 343–352. [CrossRef]

116. Huang, Q.Q.; Wang, Q.; Wan, Y.N.; Yu, Y.; Jiang, R.F.; Li, H.F. Application of X-ray absorption near edge spectroscopy to the study of the effect of sulphur on selenium uptake and assimilation in wheat seedlings. *Biol. Plant.* **2017**, *61*, 726–732. [CrossRef]

117. Malcheska, F.; Ahmad, A.; Batool, S.; Müller, H.M.; Ludwig-Müller, J.; Kreuzwieser, J.; Randewig, D.; Hänsch, R.; Mendel, R.R.; Hell, R.; et al. Drought-enhanced xylem sap sulfate closes stomata by affecting ALMT12 and guard cell ABA synthesis. *Plant Physiol.* **2017**, *174*, 798–814. [CrossRef]

118. Gigolashvili, T.; Kopriva, S. Transporters in plant sulfur metabolism. *Front. Plant Sci.* **2014**, *5*, 442. [CrossRef] [PubMed]

119. Winkel, L.H.E.; Vriens, B.; Jones, G.D.; Schneider, L.S.; Pilon-Smits, E.; Bañuelos, G.S. Selenium cycling across soil-plant-atmosphere interfaces: A critical review. *Nutrients* **2015**, *7*, 4199–4239. [CrossRef] [PubMed]

120. Wang, J.; Cappa, J.J.; Harris, J.P.; Edger, P.P.; Zhou, W.; Pires, J.C.; Adair, M.; Unruh, S.A.; Simmons, M.P.; Schiavon, M.; et al. Transcriptome-wide comparison of selenium hyperaccumulator and nonaccumulator Stanleya species provides new insight into key processes mediating the hyperaccumulation syndrome. *Plant Biotechnol. J.* **2018**, *16*, 1582–1594. [CrossRef] [PubMed]

121. Carey, A.M.; Scheckel, K.G.; Lombi, E.; Newville, M.; Choi, Y.; Norton, G.J.; Price, A.H.; Meharg, A.A. Grain accumulation of selenium species in rice (*Oryza sativa* L.). *Environ. Sci. Technol.* **2012**, *46*, 5557–5564. [CrossRef] [PubMed]

122. Ashraf, M.A.; Akbar, A.; Parveen, A.; Rasheed, R.; Hussain, I.; Iqbal, M. Phenological application of selenium differentially improves growth, oxidative defense and ion homeostasis in maize under salinity stress. *Plant Physiol. Biochem.* **2018**, *123*, 268–280. [CrossRef]

123. Butt, M.; Ayyub, C.M.; Amjad, M.; Ahmad, R. Proline application enhances growth of chilli by improving physiological and biochemical attributes under salt stress. *Pak. J. Agric. Sci.* **2016**, *53*, 43–49.

124. Hawrylak-Nowak, B. Beneficial effects of exogenous selenium in cucumber seedlings subjected to salt stress. *Biol. Trace Elem. Res.* **2009**, *132*, 259–269. [CrossRef]

125. Habibi, G.; Sarvary, S. The Roles of Selenium in Protecting Lemon Balm against Salt Stress. *Iran. J. Plant Physiol.* **2015**, *5*, 1425–1433.

126. Sattar, A.; Cheema, M.A.; Abbas, T.; Sher, A.; Ijaz, M.; Hussain, M. Separate and combined effects of silicon and selenium on salt tolerance of wheat plants. *Russ. J. Plant Physiol.* **2017**, *64*, 341–348. [CrossRef]

127. Astaneh, R.K.; Bolandnazar, S.; Nahandi, F.Z.; Oustan, S. Effects of selenium on enzymatic changes and productivity of garlic under salinity stress. *S. Afr. J. Bot.* **2019**, *121*, 447–455. [CrossRef]

128. Mozafariyan, M.; Kamelmanesh, M.M.; Hawrylak-Nowak, B. Ameliorative effect of selenium on tomato plants grown under salinity stress. *Arch. Agron. Soil Sci.* **2016**, *62*, 1368–1380. [CrossRef]

129. Astaneh, R.K.; Bolandnazar, S.; Nahandi, F.Z.; Oustan, S. The effects of selenium on some physiological traits and K, Na concentration of garlic (*Allium sativum* L.) under NaCl stress. *Inf. Process. Agric.* **2018**, *5*, 156–161. [CrossRef]

130. Atarodi, B.; Fotovat, A.; Khorassani, R.; Keshavarz, P.; Hawrylak-Nowak, B. Selenium improves physiological responses and nutrient absorption in wheat (*Triticum aestivum* L.) grown under salinity. *Toxicol. Environ. Chem.* **2018**, *100*, 440–451. [CrossRef]

131. Bybordi, A.; Saadat, S.; Zargaripour, P. The effect of zeolite, selenium and silicon on qualitative and quantitative traits of onion grown under salinity conditions. *Arch. Agron. Soil Sci.* **2018**, *64*, 520–530. [CrossRef]

132. Moussa, H.R.; Hassen, A.M. Selenium Affects Physiological Responses of Phaseolus vulgaris in Response to Salt Level. *Int. J. Veg. Sci.* **2018**, *24*, 236–253. [CrossRef]

133. Habibi, G. Selenium ameliorates salinity stress in Petroselinum crispum by modulation of photosynthesis and by reducing shoot Na accumulation. *Russ. J. Plant Physiol.* **2017**, *64*, 368–374. [CrossRef]

134. Khalifa, G.S.; Abdelrassoul, M.; Hegazi, A.; Elsherif, M.H. Mitigation of Saline Stress Adverse Effects in Lettuce Plant Using Selenium and Silicon. *Middle East J. Agric. Res.* **2016**, 347–361.

135. Abul-Soud, M.A.; Abd-Elrahman, S.H. Foliar selenium application to improve the tolerance of eggplant grown under salt stress conditions. *Int. J. Plant Soil Sci.* **2016**, *9*, 1–10. [CrossRef]

136. Hashem, H.A.; Hassanein, R.A.; Bekheta, M.A.; El-Kady, F.A. Protective role of selenium in canola (*Brassica napus* L.) plant subjected to salt stress. *Egypt. J. Exp. Biol.* **2013**, *9*, 199–211.

137. Selenium Induces Antioxidant Defensive Enzymes and Promotes Tolerance Against Salinity Stress in Cucumber Seedlings (*Cucumis sativus*). *Arab. Univ. J. Agric. Sci.* **2010**, *18*, 65–76.

138. Kong, L.; Wang, M.; Bi, D. Selenium modulates the activities of antioxidant enzymes, osmotic homeostasis and promotes the growth of sorrel seedlings under salt stress. *Plant Growth Regul.* **2005**, *45*, 155–163. [CrossRef]

139. Rietz, D.N.; Haynes, R.J. Effects of irrigation-induced salinity and sodicity on soil microbial activity. *Soil Biol. Biochem.* **2003**, *35*, 845–854. [CrossRef]

140. Carillo, P.; Mastrolonardo, G.; Nacca, F.; Fuggi, A. Nitrate reductase in durum wheat seedlings as affected by nitrate nutrition and salinity. *Funct. Plant Biol.* **2005**, *32*, 209–219. [CrossRef]

141. Szabados, L.; Savouré, A. Proline: A multifunctional amino acid. *Trends Plant Sci.* **2010**, *15*, 89–97. [CrossRef] [PubMed]

142. Shahzadi, I.; Iqbal, M.; Rasheed, R.; Arslan Ashraf, M.; Perveen, S.; Hussain, M. Foliar application of selenium increases fertility and grain yield in bread wheat under contrasting water availability regimes. *Acta Physiol. Plant.* **2017**, *39*, 173. [CrossRef]

143. Zhang, Y.; Wang, L.; Liu, Y.; Zhang, Q.; Wei, Q.; Zhang, W. Nitric oxide enhances salt tolerance in maize seedlings through increasing activities of proton-pump and Na^+/H^+ antiport in the tonoplast. *Planta* **2006**, *224*, 545–555. [CrossRef]

144. Ahmad, P.; Sarwat, M.; Bhat, N.A.; Wani, M.R.; Kazi, A.G.; Tran, L.S.P. Alleviation of cadmium toxicity in *Brassica juncea* L. (Czern. & Coss.) by calcium application involves various physiological and biochemical strategies. *PLoS ONE* **2015**, *10*, e0114571.

145. Shabala, S.; Pottosin, I. Regulation of potassium transport in plants under hostile conditions: Implications for abiotic and biotic stress tolerance. *Physiol. Plant.* **2014**, *151*, 257–279. [CrossRef]

146. Çakir, Ö.; Turgut-Kara, N.; Ari, Ş. Selenium induced selenocysteine methyltransferase gene expression and antioxidant enzyme activities in Astragalus chrysochlorus. *Acta Bot. Croat.* **2016**, *75*, 11–16. [CrossRef]

147. Shekari, F.; Abbasi, A.; Mustafavi, S.H. Effect of silicon and selenium on enzymatic changes and productivity of dill in saline condition. *J. Saudi Soc. Agric. Sci.* **2017**, *16*, 367–374. [CrossRef]

148. Gul, H.; Kinza, S.; Shinwari, Z.K.; Hamayun, M. Effect of selenium on the biochemistry of *Zea mays* under salt stress. *Pak. J. Bot.* **2017**, *49*, 25–32.

149. Alyemeni, M.N.; Ahanger, M.A.; Wijaya, L.; Alam, P.; Bhardwaj, R.; Ahmad, P. Selenium mitigates cadmium-induced oxidative stress in tomato (*Solanum lycopersicum* L.) plants by modulating chlorophyll fluorescence, osmolyte accumulation, and antioxidant system. *Protoplasma* **2018**, *255*, 459–469. [CrossRef] [PubMed]

150. Padmaja, K.; Prasad, D.D.K.; Prasad, A.R.K. Effect of selenium on chlorophyll biosynthesis in mung bean seedlings. *Phytochemistry* **1989**, *28*, 3321–3324. [CrossRef]

151. Moulick, D.; Santra, S.C.; Ghosh, D. Seed priming with Se mitigates As-induced phytotoxicity in rice seedlings by enhancing essential micronutrient uptake and translocation and reducing As translocation. *Environ. Sci. Pollut. Res.* **2018**, *25*, 26978–26991. [CrossRef]

152. Carvalho, E.R.; Oliveira, J.A.; Von Pinho, É.V.D.R.; Costa Neto, J. Enzyme activity in soybean seeds produced under foliar application of manganese. *Ciência e Agrotecnologia* **2014**, *38*, 317–327. [CrossRef]

153. Asgher, M.; Per, T.S.; Verma, S.; Pandith, S.A.; Masood, A.; Khan, N.A. Ethylene Supplementation Increases PSII Efficiency and Alleviates Chromium-Inhibited Photosynthesis Through Increased Nitrogen and Sulfur Assimilation in Mustard. *J. Plant Growth Regul.* **2018**, *37*, 1300–1317. [CrossRef]

154. Guo, C.Y.; Wang, X.Z.; Chen, L.; Ma, L.N.; Wang, R.Z. Physiological and biochemical responses to saline-alkaline stress in two halophytic grass species with different photosynthetic pathways. *Photosynthetica* **2015**, *53*, 128–135. [CrossRef]

155. Zong, X.J.; Li, D.P.; Gu, L.K.; Li, D.Q.; Liu, L.X.; Hu, X.L. Abscisic acid and hydrogen peroxide induce a novel maize group C MAP kinase gene, ZmMPK7, which is responsible for the removal of reactive oxygen species. *Planta* **2009**, *229*, 485–495. [CrossRef]

156. Shalaby, T.; Bayoumi, Y.; Alshaal, T.; Elhawat, N.; Sztrik, A.; El-Ramady, H. Selenium fortification induces growth, antioxidant activity, yield and nutritional quality of lettuce in salt-affected soil using foliar and soil applications. *Plant Soil* **2017**, *421*, 245–258. [CrossRef]

157. Filek, M.; Keskinen, R.; Hartikainen, H.; Szarejko, I.; Janiak, A.; Miszalski, Z.; Golda, A. The protective role of selenium in rape seedlings subjected to cadmium stress. *J. Plant Physiol.* **2008**, *165*, 833–844. [CrossRef]

158. Hawrylak-Nowak, B. Comparative effects of selenite and selenate on growth and selenium accumulation in lettuce plants under hydroponic conditions. *Plant Growth Regul.* **2013**, *70*, 149–157. [CrossRef]

159. Garg, N.; Manchanda, G. ROS generation in plants: Boon or bane? *Plant Biosyst.* **2009**, *143*, 81–96. [CrossRef]

160. Ahanger, M.A.; Agarwal, R.M.; Tomar, N.S.; Shrivastava, M. Potassium induces positive changes in nitrogen metabolism and antioxidant system of oat (*Avena sativa* L. cultivar Kent). *J. Plant Interact.* **2015**, *10*, 211–223. [CrossRef]

161. Saleem, M.H.; Kamran, M.; Zhou, Y.; Parveen, A.; Rehman, M.; Ahmar, S.; Malik, Z.; Mustafa, A.; Anjum, R.M.A.; Wang, B.; et al. Appraising growth, oxidative stress and copper phytoextraction potential of flax (*Linum usitatissimum* L.) grown in soil differentially spiked with copper. *J. Environ. Manag.* **2020**, *257*, 109994. [CrossRef] [PubMed]

162. Nouet, C.; Motte, P.; Hanikenne, M. Chloroplastic and mitochondrial metal homeostasis. *Trends Plant Sci.* **2011**, *16*, 395–404. [CrossRef] [PubMed]

163. Gao, W.; Xu, F.C.; Guo, D.D.; Zhao, J.R.; Liu, J.; Guo, Y.W.; Singh, P.K.; Ma, X.N.; Long, L.; Botella, J.R.; et al. Calcium-dependent protein kinases in cotton: Insights into early plant responses to salt stress. *BMC Plant Biol.* **2018**, *18*, 15. [CrossRef]

164. Liu, J.; Wang, J.; Lee, S.; Wen, R. Copper-caused oxidative stress triggers the activation of antioxidant enzymes via ZmMPK3 in maize leaves. *PLoS ONE* **2018**, *13*, e0203612. [CrossRef]

165. Sripinyowanich, S.; Klomsakul, P.; Boonburapong, B.; Bangyeekhun, T.; Asami, T.; Gu, H.; Buaboocha, T.; Chadchawan, S. Exogenous ABA induces salt tolerance in indica rice (*Oryza sativa* L.): The role of OsP5CS1 and OsP5CR gene expression during salt stress. *Environ. Exp. Bot.* **2013**, *86*, 94–105. [CrossRef]

166. Zhang, A.; Jiang, M.; Zhang, J.; Tan, M.; Hu, X. Mitogen-activated protein kinase is involved in abscisic acid-induced antioxidant defense and acts downstream of reactive oxygen species production in leaves of maize plants. *Plant Physiol.* **2006**, *141*, 475–487. [CrossRef]

167. Sun, L.; Li, Y.; Miao, W.; Piao, T.; Hao, Y.; Hao, F.S. NADK2 positively modulates abscisic acid-induced stomatal closure by affecting accumulation of H_2O_2, Ca^{2+} and nitric oxide in Arabidopsis guard cells. *Plant Sci.* **2017**, *262*, 81–90. [CrossRef]

168. Sun, L.R.; Wang, Y.B.; He, S.B.; Hao, F.S. Mechanisms for Abscisic Acid Inhibition of Primary Root Growth. *Plant Signal. Behav.* **2018**, *13*, e1500069. [CrossRef] [PubMed]

169. Manaf, H.H. Beneficial effects of exogenous selenium, glycine betaine and seaweed extract on salt stressed cowpea plant. *Ann. Agric. Sci.* **2016**, *61*, 41–48. [CrossRef]

170. Kaur, S.; Nayyar, H. Selenium fertilization to salt-stressed mungbean (*Vigna radiata* L. Wilczek) plants reduces sodium uptake, improves reproductive function, pod set and seed yield. *Sci. Hortic.* **2015**, *197*, 304–317. [CrossRef]

171. Ardebili, N.O.; Saadatmand, S.; Niknam, V.; Khavari-Nejad, R.A. The alleviating effects of selenium and salicylic acid in salinity exposed soybean. *Acta Physiol. Plant.* **2014**, *36*, 3199–3205. [CrossRef]

172. Hasanuzzaman, M.; Fujita, M. Selenium pretreatment upregulates the antioxidant defense and methylglyoxal detoxification system and confers enhanced tolerance to drought stress in rapeseed seedlings. *Biol. Trace Elem. Res.* **2011**, *143*, 1758–1776. [CrossRef] [PubMed]

173. Feng, R.; Wei, C.; Tu, S. The roles of selenium in protecting plants against abiotic stresses. *Environ. Exp. Bot.* **2013**, *87*, 58–68. [CrossRef]

174. Zhang, A.; Zhang, J.; Ye, N.; Cao, J.; Tan, M.; Zhang, J.; Jiang, M. *ZmMPK5* is required for the NADPH oxidase-mediated self-propagation of apoplastic H_2O_2 in brassinosteroid-induced antioxidant defence in leaves of maize. *J. Exp. Bot.* **2010**, *61*, 4399–4411. [CrossRef] [PubMed]

175. Ding, Y.; Cao, J.; Ni, L.; Zhu, Y.; Zhang, A.; Tan, M.; Jiang, M. *ZmCPK11* is involved in abscisic acid-induced antioxidant defence and functions upstream of *ZmMPK5* in abscisic acid signalling in maize. *J. Exp. Bot.* **2013**, *64*, 871–884. [CrossRef] [PubMed]

176. Zhang, H.X.; Hodson, J.N.; Williams, J.P.; Blumwald, E. Engineering salt-tolerant Brassica plants: Characterization of yield and seed oil quality in transgenic plants with increased vacuolar sodium accumulation. *Proc. Natl. Acad. Sci. USA* **2001**, *98*, 12832–12836. [CrossRef]

177. Almeida, D.M.; Gregorio, G.B.; Oliveira, M.M.; Saibo, N.J.M. Five novel transcription factors as potential regulators of *OsNHX1* gene expression in a salt tolerant rice genotype. *Plant Mol. Biol.* **2017**, *93*, 61–77. [CrossRef]

MDPI
St. Alban-Anlage 66
4052 Basel
Switzerland
Tel. +41 61 683 77 34
Fax +41 61 302 89 18
www.mdpi.com

International Journal of Molecular Sciences Editorial Office
E-mail: ijms@mdpi.com
www.mdpi.com/journal/ijms

9 783036 519548